Lecture Notes in Geoinformation and Cartography

About the Series

The Lecture Notes in Geoinformation and Cartography series provides a contemporary view of current research and development in Geoinformation and Cartography, including GIS and Geographic Information Science. Publications with associated electronic media examine areas of development and current technology. Editors from multiple continents, in association with national and international organizations and societies bring together the most comprehensive forum for Geoinformation and Cartography.

The scope of Lecture Notes in Geoinformation and Cartography spans the range of interdisciplinary topics in a variety of research and application fields. The type of material published traditionally includes:

- proceedings that are peer-reviewed and published in association with a conference;
- post-proceedings consisting of thoroughly revised final papers; and
- research monographs that may be based on individual research projects.

The Lecture Notes in Geoinformation and Cartography series also includes various other publications, including:

- tutorials or collections of lectures for advanced courses;
- contemporary surveys that offer an objective summary of a current topic of interest; and
- emerging areas of research directed at a broad community of practitioners.

More information about this series at http://www.springer.com/series/7418

Stan Geertman · Joseph Ferreira, Jr.
Robert Goodspeed · John Stillwell
Editors

Planning Support Systems and Smart Cities

Springer

Editors
Stan Geertman
Department of Human Geography and Planning
Utrecht University
Utrecht
The Netherlands

Joseph Ferreira, Jr.
Department of Urban Studies and Planning
Massachusetts Institute of Technology
Cambridge, MA
USA

Robert Goodspeed
Taubman College of Architecture and Urban Planning
University of Michigan
Ann Arbor, MI
USA

John Stillwell
School of Geography
University of Leeds
Leeds
UK

ISSN 1863-2246 ISSN 1863-2351 (electronic)
Lecture Notes in Geoinformation and Cartography
ISBN 978-3-319-18367-1 ISBN 978-3-319-18368-8 (eBook)
DOI 10.1007/978-3-319-18368-8

Library of Congress Control Number: 2015938314

Springer Cham Heidelberg New York Dordrecht London

Printed on acid-free paper

Springer International Publishing AG Switzerland is part of Springer Science+Business Media (www.springer.com)

Preface

This edited volume has been produced in conjunction with the 2015 Computers in Urban Planning and Urban Management (CUPUM) Conference, held in Boston, Massachusetts, on 7–10 July 2015. For more than 20 years, CUPUM has been one of the premier international conferences for the exchange of ideas on the use of computer technologies to address a range of economic, social and environmental problems. As interest and adoption of these technologies have been spread worldwide, so has the conference, and CUPUM 2015 in Boston marks the largest gathering yet.

We have chosen the title 'Planning Support Systems and Smart Cities' for this book with the intention of covering research and practice in two fields. Planning Support Systems is a relatively well-established sub-discipline, whilst Smart Cities is a new area of research where interest has burgeoned in recent years. Although many of the chapters involve research areas of long-standing interest to the CUPUM scholarly community, such as spatial analysis, urban modelling, simulation, public participation GIS, visualisation, and the quantitative analysis of urban phenomena such as demographic migration and energy consumption, a few contributions take up the challenge of linking the two fields.

The call for submissions for this book received an enthusiastic response, producing 84 abstracts of potential chapters for inclusion. After an extensive screening process, we invited 46 out of these 84 potential contributions to submit a full chapter for the book. The full chapters that we received were submitted to a further double-blind review, in which we, as editors, gratefully received help from the CUPUM Board of Directors and Advisors in the process of review and assessment. This process resulted in an invitation to the authors of 25 chapters for inclusion in the book.

We would like to thank the CUPUM Board members and advisors for their efforts in helping us to review the initial set of abstracts and the draft chapters that were submitted, and Lisanne de Wijs who provided able administrative support.

The book would not be possible without the contributors who have chosen to share their research here, and through their efforts helped us meet the tight deadlines required to bring this volume to fruition in advance of the conference.

Utrecht, The Netherlands — Stan Geertman
Boston, USA — Joseph Ferreira, Jr.
Ann Arbor, USA — Robert Goodspeed
Leeds, UK — John Stillwell
March 2015

Past CUPUM Conferences

#	Year	Place	Country
I	1989	Hong Kong	HKG
II	1991	Oxford	GBR
III	1993	Atlanta	USA
IV	1995	Melbourne	AUS
V	1997	Mumbai	IND
VI	1999	Venice	ITA
VII	2001	Honolulu	USA
VIII	2003	Sendai	JPN
IX	2005	London	GBR
X	2007	Iguazu Falls	BRA
XI	2009	Hong Kong	HKG
XII	2011	Lake Louise (Calgary/Banff)	CAN
XIII	2013	Utrecht	NLD

Board of Directors CUPUM

Name	Institute	Country	E-mail
Andrew Allan	University of South Australia	AUS	andrew.allan@unisa.edu.au
Michael Batty	University College London	GBR	m.batty@ucl.ac.uk
Joseph Ferreira, Jr.	Massachusetts Institute of Technology	USA	jf@mit.edu
Stan Geertman (Chair)	Utrecht University	NLD	s.c.m.geertman@uu.nl

(continued)

(continued)

Name	Institute	Country	E-mail
John Douglas Hunt	University of Calgary	CAN	jdhunt@ucalgary.ca
Karl Kim	University of Hawaii	USA	karlk@hawaii.edu
Kazuaki Miyamoto	Tokyo City University	JPN	miyamoto@tcu.ac.jp
Paola Rizzi	Università degli Studi di Sassari	ITA	paola.rizzi@uniss.it
Antonio N. Rodrigues da Silva	University of Sao Paulo	BRA	anelson@sc.usp.br
Anthony G.O. Yeh	University of Hong Kong	HKG	hdxugoy@hkucc.hku.hk

Advisors to the Board of Directors

Name	Institute	Country	E-mail
Ray Wyatt	University of Melbourne	AUS	raywyatt3@gmail.com
Michael Wegener	University of Dortmund	GER	mw@spiekermann-wegener.de
Dick Klosterman	University of Akron	USA	klosterman@whatifinc.biz

Contents

Chapter 1
Introduction to 'Planning Support Systems and Smart Cities'

Stan Geertman, Joseph Ferreira, Jr., Robert Goodspeed and John Stillwell

1 Introduction

Since their emergence, digital information and communication technologies (ICTs) have been applied in many urban planning and management contexts. Not only do they have the capability for collecting, managing, analysing, and storing information about cities more efficiently than ever before, new technologies also present planners and managers with opportunities to draw on this information to improve city life. These applications have developed alongside the expanding use of ICT by all sectors in cities. Published in conjunction with an international conference of researchers and practitioners interested in planning support systems, big data, data analytics and smart technologies, this book contains a variety of chapters documenting the many ways in which ICTs are used to produce new knowledge about cities, improve decision-making, and change the very fabric of city life.

S. Geertman (✉)
Department of Human Geography and Planning, Utrecht University, Utrecht, The Netherlands
e-mail: s.c.m.geertman@uu.nl

J. Ferreira, Jr.
Department of Urban Studies and Planning, Massachusetts Institute of Technology, 77 Massachusetts Avenue, Cambridge, MA 02139, USA
e-mail: jf@mit.edu

R. Goodspeed
Taubman College of Architecture and Urban Planning, University of Michigan, 2000 Bonisteel Blvd., Ann Arbor, MI 48109, USA
e-mail: rgoodspe@umich.edu

J. Stillwell
University of Leeds, Leeds, UK
e-mail: j.c.h.stillwell@leeds.ac.uk

S. Geertman et al. (eds.), *Planning Support Systems and Smart Cities*,
Lecture Notes in Geoinformation and Cartography,
DOI 10.1007/978-3-319-18368-8_1

The remaining sections of this introduction are organized as follows. First, we will elaborate on the two central concepts in the book—'smart cities' and 'planning support systems'—and explain our initial motivation in choosing these themes. Second, we provide an overview of the book's structure, providing summaries of each chapter and commenting on their contents. We close with observations about the nature of the research presented here, and some comments about the future of information infrastructure in cities.

2 Smart Cities and Planning Support Systems

In recent years, the concept of a 'smart city' has been taken up by many city leaders, IT companies and scholars worldwide, resulting in a flurry of professional (e.g., ARUP 2010; Washburn and Sindhu 2010), popular (Townsend 2013; Greenfield 2013) and scholarly (Deakin 2011; Glasmeier and Christopherson 2015) publications on the topic. As Hollands (2008) observed, the term 'smart city' and related terms like 'digital', 'wired', and 'intelligent' cities have been used in a variety of ways. One origin lies in the concept, developed by US-based scholars, of 'smart growth' (Harrison and Donnelly 2011; Daniels 2001; Burchell et al. 2000). Created as a reaction to excessive automobile-oriented urban sprawl which developed around many US cities, this sustainability-related concept promotes the adoption of urban growth management policies such as urban growth boundaries, rural land preservation and financial incentives to discourage growth at the urban periphery. On the other hand, the concept of smart cities also finds its origin in debates about how ICTs can contribute to the planning and management of cities (Goodspeed 2015). For some, a smart city refers to urban environments where 'pervasive' or 'ubiquitous' computing has introduced a range of digital devices for sensing, monitoring and managing the city (Kitchin 2013). This strand of research also emphasizes the digital infrastructure needed to collect and manage new sources of data, conduct analysis and ultimately use connected devices to manage cities in new ways (Batty et al. 2012).

Taking these two sources together, some authors have proposed a definition of a smart city that unites both ICT-based support and sustainability goals into one overarching concept. In line with this broader perspective, Caragliu et al. (2011, p. 70) defined the concept of smart city as: "*when investments in human and social capital and traditional (transport) and modern (ICT) communication infrastructure fuel sustainable economic growth and a high quality of life, with a wise management of natural resources, through participatory governance*." These theorists see the key to a smart city lying not only in ICT, but also in human, social and natural capital (Neirotti et al. 2014). From this perspective, examples of smart city projects include not only the use of digital devices to monitor and manage cities, but also broader issues such as governance and sustainability. As described further below, the collection of chapters in the second part of this volume span the full range of the emerging field of smart city practice and scholarship, drawing on empirical investigations to explore promising new ideas in this fast-emerging field.

In contrast, the concept of planning support systems (PSS) refers to a well-defined scholarly and professional field which dates to the late 1980s (Harris and Batty 1993; Klosterman 1997; Brail and Klosterman 2001; Brail 2008). This volume's predecessors contain a diverse array of PSS research (Geertman and Stillwell 2003, 2009; Geertman et al. 2013). PSS can be defined as geo-information technology-based instruments that are dedicated to supporting those involved in planning in the performance of their specific planning tasks (Geertman 2006). Although scholars and practitioners had worked for many years on developing computer-based planning instruments, Harris and Batty (1993) proposed these efforts to be categorised under the concept of PSS, promoting the integration of geographic information systems (GIS) with various types of models to produce tools uniquely useful for planning. Thus, the PSS concept emerged as a related but distinct field from GIS. While GIS are general-purpose tools that are applicable for many different spatial problems, PSS are distinctive in their specific focus on supporting specific professional planning tasks. PSS are also related to so-called spatial decision support systems (SDSS), which are also designed to aid particular decision tasks. These two types of systems differ in that PSS generally pay particular attention to long-range problems and strategic issues whereas SDSS are generally designed to support shorter-term policy making by independent individuals or business organizations (Clarke 1990).

Given this definition, Klosterman (1997) argued that PSS should serve as a single information framework integrating planning-related theory, data, information, knowledge, methods and instruments. Examples of PSS include geographically oriented websites that show interactive land-use maps that inform users about formal regulations and building restrictions, communication-oriented map-based touch tables that provide groups of professionals a mutual workbench to discuss and assess sketches of future layouts (Arciniegas et al. 2012), and more analytical oriented PSS like What If, CommunityViz and UrbanSim, which offer planners the ability to consider alternative possible future spatial scenarios (e.g., see Geertman and Stillwell 2003, 2009). In recent years, we have observed a growing theme in PSS research. Even while overall adoption of PSS has lagged behind researcher's expectations (Vonk et al. 2005), several PSS have reached maturity and become used more widely in planning practice. These trends together have sparked the development of a growing body of research examining the application of PSS itself. Examples of scholars working in this tradition include studies which have sought to define PSS performance metrics (Te Brommelstroet 2013), analysis of the role of PSS in facilitating professional communication (Pelzer and Geertman 2014), and studies examining PSS and group learning outcomes (Goodspeed 2013). As elaborated below, the contributions in this volume illustrate this exciting and diverse field of PSS research and development, and include accounts of new PSS instruments as well as studies of their development and application.

To summarise, at the moment there is a tremendous diversity of smart city definitions, goals and approaches, characteristic of an emerging field of study, whilst PSS, on the other hand, are the subject of an evolving scholarly tradition that has been in existence for two decades at least. Our initial motivation for this volume and for the conference as a whole was to shed light on how these two themes might be linked, for example a specific focus on PSS that are being developed or applied for use in smart cities around the world. In reading and compiling summaries of the contributions, it is apparent that whilst there is a tremendous amount of new research and development activity taking place in both of the theme areas, only a few researchers have tried to tackle the particular challenge of understanding their inter-relationship.

3 Structure of the Book

The chapters in this book are organized into four parts, the contents of which are summarised in the remainder of the chapter. We are of course well aware of the danger of imposing structure on such a diverse array of contributions. Nevertheless, we think that the structure provides some degree of organization on the basis of key themes. Part I is a collection of contributions that show the role of spatial data and ways in which these data can be transformed into useful knowledge. Part II contains chapters that investigate the concept of smart cities through theory, case studies and descriptions of new applications. Part III contains chapters in which PSS are used to support wider participation and/or assessment processes in planning practice. Part IV contains a series of chapters in which new methods and tools of PSS and their applications are described.

3.1 Part I: Spatial Data Analytics

Opportunities for urban planning-relevant spatial analyses have increased substantially with the advent of urban sensing, ubiquitous computing and the gradual standardization of embedded location information within administrative datasets about urban activities. The six chapters in Part I show how the new data can be captured, analysed and integrated into various types of PSS. In this way, the new data and related analytics can add to a city's capacity to be 'smart' by improving the data snapshots and behavioural models which feed PSS that enable everyday planning processes to be more intelligent. The first two chapters in Part I illustrate ways of tapping social media data streams to enhance the content and timeliness of PSS. In 'Development and Operation of Social Media GIS for Disaster Risk Management in Japan', *Kayoko Yamamoto* explores ways of integrating social media data streams into useful risk management tools for assisting with disaster awareness, relief and recovery. She presents several systems that her team has prototyped and tested, positions this work within the broader literature, evaluates the effectiveness of various system design and user interface features, and suggests development paths and user engagement strategies that can help integrate social media data into a practical PSS for disaster risk management. In the second chapter of Part I, 'The Role of Social Media Geographic Information (SMGI) in Spatial

Planning', *Michele Campagna, Roberta Floris* and *Pierangelo Massa* build measures of tourist interest in lodging and visitor attractions using volunteered geographic information (via Tweeter feeds, Instagram tags, TripAdvisor comments and the like). An Italian tourism planning example is used to demonstrate how data from these sources can be used to construct indicators for use in a PSS.

In the third chapter of Part I, *Michael Johnson, Justin Hollander* and *Eliza Devenport Whiteman* discuss 'Data and Analytics for Neighborhood Development: Smart Shrinkage Decision Modeling in Baltimore, Maryland'. The rapid spread of GIS technologies and related geospatial data have greatly enhanced the availability of spatially-detailed information about land use, urban infrastructure and the built environment. Johnson et al. take advantage of this new world of improved urban data by applying management science techniques to everyday community development questions. In their illustrative case, they help the US city of Baltimore, Maryland, to evaluate and prioritize alternative re-development options, showing how these methods might help the many communities which have experienced disinvestment and economic decline.

The fourth and fifth chapters of Part I integrate data about urban activities and preferences into dynamic models of urban change and resilience. *Lara-Britt Zomer, Winnie Daamen, Sebastiaan Meijer* and *Serge Paul Hoogendoorn* examine short-term questions of crowd control in 'Managing Crowds: The Possibilities and Limitations of Crowd Information During Mass Urban Events'. Through visitor surveys and behavioural models of events in the Netherlands, they calibrate and study alternative strategies for providing congestion and crowd control information to public participants at popular venues. *A. Yair Grinberger, Michal Lichter* and *Daniel Felsenstein* examine urban dynamics at very different temporal and physical scales. In 'Simulating Urban Resilience: Disasters, Dynamics and (Synthetic) Data', they examine the aftermath of natural disasters in terms of the changes in infrastructure and urban activity following natural disasters such as earthquakes. They use large-scale models of transportation and land-use interaction to explore disaster scenarios in Israel, simulating the long-term residential and work relocations that are likely to occur as a metropolitan area adjusts to the type of significant destruction and disruption that might result from a natural disaster.

In the last chapter in Part I, *Yi Zhu* and *Joseph Ferreira, Jr.* focus on the analytics and techniques that are required to translate the new, spatially-detailed data about urban activities into synthetic populations that are suitable for urban modeling and PSS. Their paper, entitled 'Data Integration to Create Large-Scale Spatially Detailed Synthetic Populations', uses the Singapore built environment as an example. They construct and utilize an ontology of building location, use and physical characteristics in order to make semi-automatic the integration of census and government statistics with real estate transaction data and other online sources. The approach is needed to construct a synthetic population of people and places with sufficient accuracy, spatial detail and reproducibility to support the next generation of urban dynamic models that are increasingly utilized in PSS.

3.2 Part II: Smart Cities

The next part of the book contains examples of the research on smart cities now underway. It begins with two chapters that consider the diverse theories that can be used to analyze smart cities. The next three chapters explore new infrastructures and new data sources often associated with smart cities, and the sixth chapter discusses a novel data infrastructure that links new data to analysis tools. Finally, the closing chapter in this part of the book considers how smart cities and PSS are related.

In the chapter 'Smart Cities: Concepts, Perceptions and Lessons for Planners', *Tuan-Yee Ching* and *Joseph Ferreira, Jr.* report the results of an empirical study of smart cities. To guide their study, they propose four alternative theories of a 'smart' city: smart machines; partnership and collaboration; learning and adapting; and investing in the future. Their detailed investigation of smart city initiatives in Boston, San Francisco, Amsterdam, Stockholm, Singapore and Rio de Janeiro reveals these cities are developing smart city practices appropriate to local capabilities and needs. For example, Boston has created new citizen-facing apps, San Francisco has created a new position in city hall for innovation, and Stockholm has invested in a broadband network. The chapter reminds us that while the idea of the smart city is a powerful inspiration for change, the ways cities use technology remain highly varied and are shaped by local capacity, needs and cultures.

The contemporary smart city debate often lacks connections with previous scholars who have considered how ICT might impact cities. *Kian Goh*'s chapter, 'Who's Smart? Whose City? The Sociopolitics of Urban Intelligence', addresses this gap by considering how smart city developments in Singapore and London can be analyzed from the perspective of two well-known urban scholars, Manuel Castells and William Mitchell. In some ways, she finds their theories are prescient, but concludes that neither adequately describes the heterogeneous developments in each city. In conclusions that echo those of Ching and Ferreira, the evidence suggests that cities are pursuing diverse goals with smart city technologies.

Shifting from theory to applications, the next chapters report on the technical details of several smart city projects. In the third chapter, 'Knowledge-Mining the Australian Smart Grid Smart City Data: A Statistical-Neural Approach to Demand-Response Analysis', *Omid Motlagh, Greg Foliente and George Grozev* present the Australian Smart Grid Smart City program and describe its use of big data. They present a novel statistical-neural approach to maximize knowledge extraction from large datasets, and demonstrate its use in evaluating the effectiveness of two different cost-reflective product offerings, named PeakRebate and PriceSmart. Their results show that users' energy consumption behaviour will change due to these offerings. In particular, participants changed their time of energy use behaviours after subscribing to these products.

In their contribution on 'Urban Emotions: Benefits and Risks in Using Human Sensory Assessment for the Extraction of Contextual Emotion Information in Urban Planning', *Peter Zeile, Bernd Resch, Jan-Philipp Exner* and *Günther Sagl* introduce their so-called 'Urban Emotions' approach. This method focuses on integrating humans' emotional responses to the urban environment in both time and geographical space to be able to incorporate these into planning processes. To detect and analyze these emotions and perceptions, they extract contextual emotion information from technical and human sensors, as well as georeferenced social media posts. This results in novel information for urban planners. In addition to technical and methodological aspects, data privacy issues and the potential of wearable technologies are discussed in this chapter. With the help of two case studies, the authors demonstrate how this approach can be translated into planning processes.

Of course, not all information needed for smart city applications is available. As an example, one of the most noticeable developments in cities of the global south has been the proliferation of mobile phones. Although this technology has improved urban life in many ways, two of their most common uses in cities—wayfinding and trip routing—have often lagged in the developing world where transit systems are semi-formal and good data are lacking. The chapter on 'Leveraging Cellphones for Wayfinding and Journey Planning in Semi-formal Bus Systems: Lessons from Digital Matatus in Nairobi' by *Jacqueline Klopp, Sarah Williams, Peter Waiganjo, Dan Orwa* and *Adam White* shows that these hurdles can be overcome with savvy use of existing technologies. The authors report on a project in which new digital data were created using smartphones to describe Nairobi's system of *matatus*, or semi-formal mini-buses. They then publish the data in the form of standardized GTFS data, as well as a new paper map, both of which are eagerly adopted. The chapter suggests an exciting new era of public participation GIS (PPGIS), where the results of data creation projects are linked through standards and data sharing into a growing shared information infrastructure for cities.

The proliferation of novel data sources and data analysis tools associated with smart cities has created new problems for urban analysts. One innovative response to these problems is described by *Chris Pettit, John Barton, Xavier Goldie, Richard Sinnott, Robert Stimson* and *Tom Kvan* in the chapter entitled 'The Australian Urban Intelligence Network Supporting Smart Cities'. The network they describe includes researchers, planners, and policymakers who have developed the Australian Urban Research Infrastructure Network (AURIN), a shared online workbench of databases and PSS tools. With support from the Australian Government and leading Australian universities, this group has established an impressive network of data hubs making available to users over 1000 datasets. Much more than simply a system for data storage and retrieval, AURIN also contains more than 100 statistical tools, the Online What If? PSS, and a novel walkability tool.

Concluding this section is a thoughtful chapter by *Yanliu Lin* and *Stan Geertman*, 'Smart Governance, Collaborative Planning and Planning Support Systems: A Fruitful Triangle?' that directly addresses the connection between smart cities and PSS by drawing on examples from China, Finland and the USA. Scholars, companies and city officials each promote their preferred terms and definitions for smart cities. Too often, municipal officials and ICT companies alike are eager to avoid the links between these new concepts and the realms of governance and planning, two fields which are shaped by institutions and, inevitably, by politics. Yin and Geertman observe that often PSS forms a link between smart governance and collaborative planning, and describe novel projects that seem to link the two, and therefore defy conventional categorization. In all three countries, innovative practitioners and activists are launching projects that integrate social media, participatory mapping and PSS to shape urban places. They make the important observation that technology alone is not enough—smart governance and collaborative planning each require well-designed institutions.

3.3 Part III: Planning Support Systems and Public Engagement

Continuing Yin and Geertman's focus on PSS projects, this section contains chapters on the development, application and assessment of PSS. This section reflects the new development in the PSS literature described above: namely, the growing body of research seeking to better understand social aspects of the development, application and use of PSS.

As discussions about smart cities focus on how ICT might be embedded within the urban environment, *Brian Deal, Varkki Pallathuchuril, Yong Wook Kim* and *Haozhi Pan* remind us in their chapter 'Sentient PSS for Smart Cities', that PSS also should be the focus for ongoing innovation. Inspired by theoretical and practical developments described by the term 'sentient computing', they argue that this concept should guide future PSS development. In the same way that smartphones dim their screens or zoom in or out of maps automatically, the authors of this chapter call for PSS that search the web for relevant data and adjust visualizations in response to user feedback, backgrounds and preferences. This concept provides a powerful conceptual framework for ongoing work to design PSS which match the high level of usability we increasingly take for granted in the consumer marketplace.

In their contribution 'Gaming, Urban Planning and Transportation Design Process', *Jayanth Raghothama* and *Sebastiaan Meijer* identify the discrepancies between two different approaches of using methods and tools for the analysis and design of urban systems. The first group is a technical-rational approach using computational and mathematical methods rooted in systems engineering. The second group is a participatory approach which uses qualitative methods to develop multiple narratives, values and perspectives. The authors propose integrating these approaches through gaming. They develop a framework to do this which they illustrate through two case studies in the cities of Paris and Stockholm. The authors conclude that games can create open spaces for dialogue, participation, experimentation.

The next contribution analyzes how ICT is changing the nature of planning itself. In the chapter, 'The Everyone City: How ICT-based Participation Shapes Urban Form', *Sara Levy, Karel Martens*, and *Rob van der Heijden* present a simulation model to explore how social media might change the planning process. Based on their modeling results, they conclude that ICT-facilitated participation will tend to bring building heights down and increase the number of buildings constructed.

The last two chapters of this part examine the issues of usability and facilitation, two key aspects of the participatory setting in which PSS are used. In their contribution 'Usability of Planning Support Systems: An Evaluation Framework' the authors *Patrizia Russo, Maria Francesca Costabile, Rosa Lanzilotti,* and *Christopher Pettit* recognize that previous research on PSS has shown that low usability of these tools is one of the reasons why they are not widely used by planning professionals. However, they identify a gap in the PSS literature for methods for evaluating PSS usability and performing usability tests. The authors develop a framework that aims at guiding usability evaluation of PSS. They then apply this evaluation framework to evaluate the usability of the user interface of three PSS, to determine how well the functionality can be used for specific circumstances and environments. The chosen PSS consist of CommunityViz, Envision and Online What If? Results of this user test are discussed, providing recommendations for the design of PSS.

In the final contribution in this part, 'Facilitating PSS Workshops: A Conceptual Framework and Findings from Interviews with Facilitators', *Peter Pelzer, Robert Goodspeed and Marco te Brömmelstroet* stress the importance of the facilitation of workshops where PSS are used in planning processes. They indicate that previous empirical studies have largely overlooked the important role of facilitation of workshops. By drawing on existing facilitation research, they identify four main categories of facilitation interventions: substantive, procedural, relational and tool related. Based on these categories, they propose a conceptual framework for the facilitation at PSS workshops. With the help of semi-structured interviews, they validate and develop this framework, concluding that successful facilitation of PSS workshops requires a careful balance of encouraging and limiting PSS use by participants.

3.4 Part IV: Planning Support Systems: New Methods and Tools

The evolution of PSS over the last three decades has taken place in tandem with the emergence of new methods and technologies. This final section of the book contains a collection of chapters that illustrate the application of new methods and tools in a range of contexts. The first three chapters are a continuation of the theme of Part III since they involve methods used to improve and assess the role and utility of PSS in public engagement but, collectively, they also demonstrate the value of particular tools designed for 3D visualisation, virtual reality and scenario planning in localities in different parts of the world. The fourth and fifth chapters contain descriptions of a system featuring graphical representation of spatial interaction data describing population changes in the UK, and a new tool for calculating and mapping the energy use of buildings in New York City. The final two chapters provide different perspectives on the use of new tools and systems in the context of transport planning in Boston, Massachusetts and in Perth, Western Australia.

The first chapter in this section describes a project featuring a PSS used to visualize climate change impacts. *Scott Lieske, Kari Martin, Ben Grant* and *Claudia Baldwin*, in ‘Visualization Methods for Linking Scientific and Local Knowledge of Climate Change Impacts’, consider the challenge of how PSS can be used to help communities in coastal areas of southeast Queensland (known as the Sunshine Coast) confront the combined effects of climate change, i.e. flooding, rising sea level, storm surges and other severe weather events. The chapter assesses the effectiveness of geographic visualization tools in aligning scientific knowledge with local knowledge by engaging with members of the coastal community in areas where residential development and infrastructure are vulnerable to the effects of climatic change under specific scenarios. The assessment of participants suggested that a combination of consultation techniques was important, including interactive 3D scenes, flood hazard maps, Photovoice, and interactive participatory mapping, but that 3D visualization was the most effective method for knowledge exchange about local climate change impacts. The chapter demonstrates that 3D geographic visualization is an important methodology for facilitating participation and the results confirm that despite certain limitations, there are clear benefits from this technology in promoting group understanding of environmental situations and coordinating actions and responses.

The value of using visualisation techniques is echoed in the following chapter on ‘Virtual Worlds as Support Tools for Public Engagement in Urban Design’ by *Anja Jutraz* and *Tadeja Zupancic* which explores the use of virtual worlds for public participation in urban design. In this case, the authors’ research is based on the use of the Terf virtual world visualisation tool which provides an immersive environment enabling the participants to experience 3D models of neighbourhoods as though they were pedestrians. This tool is used to explore interdisciplinary collaboration in urban design. The context in this case is a suburban neighbourhood in Slovenia’s capital city, Ljubljana, and semi-structured interviews were conducted with members of the public and with professional planners about their experience using Terf to assess alternative rearrangements for the urban area. The overall conclusion is that, when it comes to imagining the future urban landscape, people are served much better by tools that enable them to see the alternative environments in 3D and to experience those virtual worlds by walking through them. Moreover, the use of PSS featuring virtual reality helps improve the communication between the public and the urban design professionals.

The third chapter in this group is by *Jennifer Minner* and is entitled ‘Recoding Embedded Assumptions: Adaptation of an Open Source Tool to Support Sustainability, Transparency and Participatory Governance’. The chapter tells the story of the development and application of an open source planning support tool, Envision Tomorrow (ET), to support the assessment of planning scenarios at district, community and regional scales in the metropolitan region of Austin, Texas. Envision Tomorrow is an open source PSS tool that is an extension to ArcMap (a component of the ArcGIS suite of geospatial processing programs) and uses a set of linked Excel spreadsheets to construct alternative futures based on different sustainability indicators. Minner reports on the use of a number of methods of public engagement—visioning workshops, planning charrettes and open houses—in five different communities. Whilst these methods are not determined by Envision Tomorrow, the PSS role in this context is to provide planners with a method for performing planning from the modeling of individual buildings of different types through to constructing aggregate developments using Envision Tomorrow to develop and visualize scenarios in ArcMap. The use of spreadsheets means that the calculation of indicators (e.g. energy use, carbon emissions) is immediately transparent and the software has been recoded and extended to create new indicators to address equity and green infrastructure issues. The chapter provides a critical appraisal of transparency and adaptability and reveals that whilst there is much potential in the openness and flexibility of the PSS which is valuable for enhancing analytical capabilities, it also reveals some of the concerns associated with the application of Envision Tomorrow within planning processes. Balancing simplicity, transparency and ease-of-use versus the complexity, uncertainty, and sensitivity of 'real world' interactions is a continuing challenge for PSS designers, who risk hiding key assumptions and overstating the robustness of PSS-derived indicators.

The fourth chapter in Part IV shifts focus away from the public participation process and towards PSS tools themselves. 'Monitoring and Visualising Sub-national Migration Trends in the United Kingdom' by *John Stillwell, Nik Lomax* and *Nikola Sander* reports on work in progress on the development of a system for monitoring change in the components of demographic growth in local authorities throughout the United Kingdom. Whereas other chapters in this part of the book deal with systems and processes within regions, cities and particular localities, this chapter introduces a national system through which data can be extracted for the analysis of trends in population development over time and for the comparison of how the components of change (births, deaths, internal migration and international migration) are impacting on different urban or rural areas. The chapter describes a spreadsheet application which provides access to the data for analysis and visualization. In response to the problem of visualising flow data using conventional methods, the authors make use of the concept of circular plots and illustrate their application with examples of migration flows between city regions.

The next chapter describes a PSS to examine another pressing topic, urban energy use. Measuring energy consumption is a complex process, and in their chapter entitled 'Urban Data and Building Energy Modeling: A GIS-based Urban Building Energy Modeling System Using the Urban-EPC Engine', *Steven Jige Quan, Qi Li, Godfried Augenbroe, Jason Brown* and *Perry Pei-Ju Yang* explain the development and application of a GIS-based urban building energy modeling system, using the Urban-Energy Performance Calculator (EPC) simulation engine. Urban-EPC is a modeling system that is compatible with other planning tools and uses urban data related to the buildings, mutual shading, microclimate and occupant behaviour to generate the amount of energy used by every building in a city, which can then be mapped and analysed. The simulation method is applied to the Manhattan district of New York City to show its potential as an important PSS tool that can be valuable in assisting planners and policy makers in optimizing the urban energy system and achieving environmental objectives.

The final two chapters of the book are concerned with applications of PSS methods and tools in transportation management, a sector that is of critical importance in contemporary planning of sustainable urban areas and one with both a longstanding tradition of model-based PSS and is at the forefront in the adoption of smart city initiatives. One of the most important recent transportation developments has been the introduction of intelligent transportation systems that provide real-time information to transport operators, managers and travelers. These systems, used in combination with support tools such as traffic simulation models and decision support systems now provide traffic managers with the means to optimize the flow of traffic in real-time as well as prepare for various scenarios. Several metropolitan areas in the USA have joined up with federal agencies to adopt active traffic management (ATM) or integrated corridor management (ICM) approaches, including Seattle, Dallas, Minneapolis, Las Vegas and San Diego. In Boston, the Massachusetts Department of Transportation (MassDOT) is pursuing an innovative business model that provides real time travel time information to the public using dedicated highway signs covering over 700 miles of state highway. In the chapter entitled 'MassDOT Real Time Traffic Management System', *Russell Bond* and *Ammar Kanaan*, two transport practitioners, explain how MassDOT has invested in infrastructure to capture and archive real travel time data and freely provides these data to third party developers. In other states, these data are usually collected by and purchased from private companies. The authors explain how the development and operation of the MassDOT Real Time Traffic Management (RTTM) system has resulted in a shift towards new measures of system performance. There is also a commitment on behalf of MassDOT to expand how it interacts with its transportation network users through mobile technologies and social media, indicative of the extent to which the city is getting 'smarter'.

Land-use transport modeling (LUTM) dates from the 1950s when the first efforts were made in the USA to systematically study the relationships between transport and the spatial development of cities. Lowry's (1964) pioneering Model of Metropolis was the first attempt to model the interactions between land use and transportation systems. The number of applied LUT models has increased steadily and Wegener (2004) provides a valuable overview of the range of models developed whilst Hunt et al. (2005) offer a detailed review of six of the various alternative frameworks available for urban LUT interaction models in operation. LUT modeling in Western Australia is the focus of the final chapter by *Sharon Biermann, Doina Olaru, John Taplin* and *Michael Taylor*, in which the authors explain how three models were proposed and evaluated before PLATINUM (Perth LAnd and Transport INtegrated Urban Model) was selected as the preferred option. The chapter, entitled 'Pragmatic Incremental or Courageous Leapfrog [Re]Development of a Land-use and Transport Modelling System for Perth, Australia', considers the application of current theoretical insights into the critical success factors for PSS (Vonk 2006) to the design of PLATINUM as well as lessons learnt from other applications and from the development of the PSS in the particular planning and policy context of Western Australia.

4 Conclusions

The 25 remaining chapters included in this collection contain research with diverse theoretical and analytical approaches. Featuring contributions from scholars around the world, the chapters cover projects as varied as the cities they describe. In this way, this book continues a CUPUM tradition of drawing together a global scholarly community. However CUPUM has thrived not only because of a shared scholarly fascination with cities and their endlessly diversity, but also because of common connections among researchers. It is to these commonalities we will now turn. The conclusions first discuss the types of scholarly contributions contained in the volume before turning to some comments about trends in information infrastructure in cities.

4.1 Knowledge About Cities, Technology, and Planning

The scholarly contributions of the research included here fall into three distinct categories, and the chapters often contribute to more than one. The first category contains contributions that feature original knowledge about cities. Whether it is analyzing social media data (Chaps. 1, 2, 3, and 11), mapping their informal transportation infrastructures (Chap. 12) or visualizing population flows (Chap. 23), digital technologies have been indispensable for researchers to develop new insights into urban processes. In addition, with the growth of 'smart city' and other competing visions for how ICT should be used in cities, the book contains qualitative chapters analyzing practices in cities and their associated theoretical debates (Chaps. 8, 9, and 14).

The second category includes contributions to the evolving set of constructs, models, methods and instantiations which are the outputs of ICT research (March and Smith 1995). This category encompasses chapters featuring application of analytical techniques to urban problems (Chap. 4), new methods for data integration (Chap. 7), and urban modeling (Chaps. 6 and 26). The common purpose of these research outputs is to contribute to the evolving toolkit for urban research.

The CUPUM community has not been content to develop an in-depth understanding of cities through empirical studies and modeling as an end in itself. The third category includes contributions from research about the process of applying ICT to improve the planning and management of cities. As described above, this growing area of scholarship seeks to develop knowledge at the intersection of technology and social processes at the heart of planning and decision-making. Examples from this category in this volume include explorations of the potential for gaming and virtual worlds to improve communication among planning participants (Chaps. 16, 20, and 21), the role of facilitators in enabling stakeholder engagement with PSS (Chap. 19), and an analysis of the usability of PSS tools (Chap. 18). The results from this research is a growing body of knowledge practitioners can draw on when deciding where and how to apply ICT to a given context.

Surveying the contributions of the chapters, we next turn to an assessment of the two themes of this volume. So far as we have been able to tell, the concept of 'smart cities' remains an evocative slogan and is not a well-defined theoretical or empirical phenomenon. We hope this volume contributes to a robust dialogue among practitioners and scholars concerned with ICT and cities. We think the research traditions represented here can enrich the smart city debate, even as newcomers reinvigorate the field with fresh perspectives. PSS, on the other hand, has remained a remarkably durable concept for well over 20 years. The term has persisted despite shifting intellectual trends and rapidly evolving cities. We believe it has done so because the term's classic definitions contain a balance between specifics and abstraction. PSS is not a category of technology, but a perspective on what technology is asked to do. PSS should not only subject proposals to rigorous analysis (Harris and Batty 1993), but also facilitate the collective design and learning required by communities seeking to tackle long-range problems (Klosterman 1997). Most importantly, the term has served as a common reference point for researchers working in diverse planning traditions and has facilitated the development of the growing base of knowledge about how planning and decision-making can be improved with ICT in different ways.

4.2 Smart Cities and PSS: Towards New Information Infrastructures

In addition to illustrating the contours of research in this area, the chapters illuminate some broader trends as well. In particular, the chapters highlight the extent to which investment in urban information infrastructure has generated data streams that facilitate 'smart city' initiatives and PSS development in mutually synergistic ways. In Chap. 12, the digital cellphone infrastructure of Nairobi made it possible for researchers to generate a rich database from which the researchers could construct transit maps of the semi-formal matatus buses. More broadly, the evolution of digital standards for encoding road networks, street addresses, transit routes and the like has had a much broader impact beyond the census taking, navigation aids and traffic management applications that motivated and financed the original data collection and standardization. These new datasets have greatly facilitated the sharing and cross-referencing of spatially-detailed data about the urban facilities and the spatial-temporal activity patterns that constitute urban life. Current efforts to standardize additional records, such as land use information, are also impacting the quality, detail and accessibility of the types of data that make PSS much more affordable, timely and realistic.

We are only at the beginning of this virtuous cycle whereby improved data makes new uses practical in ways that generate yet more data, which enables yet more ways to understand how urban planning and management impacts urban activities and our quality of life. For example, Chaps. 2 and 3 illustrate how one can take advantage of volunteered geographic information on social media sites to improve disaster response systems and improve tourism planning. Likewise, many of the 'smart city' initiatives discussed are generating new data that can be used for research and analysis. In this way, 'smart city' efforts, while quite different in motivation and structure from most PSS applications, are helping to provide the data, infrastructure and institutional capacity that are needed for PSS applications to be more affordable, timely, realistic and routine.

Of course, these synergies are not guaranteed and many difficult issues have emerged concerning boundaries between public and private data, confidentiality, infrastructure financing, and intellectual property. Addressing these issues will be increasingly important for the beneficial evolution of PSS. In general, PSS applications have tended to focus on making everyday urban and metropolitan planning processes more practical, routine and information-based. As a result, most of these applications have been within-agency efforts (or bottom-up inter-agency collaborations) without the big budget and top-down mandate that might be needed to acquire new data or build elaborate models (however useful) of land use, transportation and environmental interactions. Promoting urban information infrastructure as a public good is helpful to accumulate useable 'city knowledge' (Carrera and Ferreira 2007). Doing this and tapping the new data streams from 'smart city' efforts, open data initiatives and social media portals are already impacting the quality and usefulness of PSS efforts.

Even as our cities—and our technologies—seem to change rapidly, this volume reminds us of the incremental development of the bodies of knowledge about cities, the urban research toolkit, and PSS applications. In light of the trends we observe, the challenge before us is not only to keep up with the pace of change, but also to develop the knowledge necessary to empower cities to better understand themselves, and in doing so, better direct their collective futures.

References

Arciniegas, G., Janssen, R., Rietveld, P. (2012). Effectiveness of collaborative map-based decision support tools: Results of an experiment. *Environmental Modelling & Software*(0). doi:10.1016/j.envsoft.2012.02.021.

ARUP. (2010). Smart cities: Transforming the 21st century city via the creative use of technology.

Batty, M., Axhausen, K., Fosca, G., Pozdnoukhov, A., Bazzani, A., Wachowicz, M., et al. (2012). Smart cities of the future. *UCL Centre for Advanced Spatial Analysis, 188*.

Brail, R. K. (2008). *Planning support systems for cities and regions*. Cambridge: Lincoln Institute of Land Policy.

Brail, R. K., & Klosterman, R. E. (2001). *Planning support systems: Integrating geographic information systems, models, and visualization tools*. Redlands, Calif.: ESRI Press.

Burchell, R., Listokin, D., & Galley, C. (2000). Smart growth: More than a ghost of urban policy past, less than a bold new horizon. *Housing Policy Debate, 11*(4), 821–879.

Caragliu, A., Bo, C. Del, & Nijkamp, P. (2011). Smart cities in Europe. *Journal of Urban Technology, 18*(2), 65–82.

Carrera, F., & Ferreira, J. (2007). The future of spatial data infrastructures: Capacity-building for the emergence of municipal SDIs. *International Journal of Spatial Data Infrastructures Research, 2*, 54–73.

Clarke, M. (1990). Geographical information systems and model-based analysis: Towards effective decision support systems. In H. Scholten & J. Stillwell (Eds.), *Geographical information systems for urban and regional planning* (pp. 165–175). Dordrecht: Kluwer.

Daniels, T. (2001). Smart growth: A new American approach to regional planning. *Planning Practice & Research, 16*(3–4), 271–279.

Deakin, M. (Ed.). (2011). Creating smart-er cities [Special issue]. *Journal of Urban Technology, 18*(2).

Geertman, S. (2006). Potentials for planning support: A planning-conceptual approach. *Environment and Planning B: Planning and Design, 33*(6), 863–881.

Geertman, S., & Stillwell, J. (2003). *Planning support systems in practice. Advances in spatial science*. New York: Springer.

Geertman, S., & Stillwell, J. (2009). *Planning support systems; Best practice and new methods*: Vol. 95. *The GeoJournal Library*. New York: Springer.

Geertman, S., Toppen, F., & Stillwell, J. (Eds.). (2013). *Planning support systems for sustainable urban development*. Heidelberg: Springer.

Giffinger, R., Fertner, C., Kramar, H., Kalasek, R., Pichler-Milanović, N., & Meijers, E. (2007). *Smart cities—Ranking of European medium-sized cities*. Vienna: Centre of Regional Science, Vienna Universitat Technology.

Glasmeier, A., & Christopherson, S. (2015). The smart city [Special issue]. *Cambridge Journal of Regions, Economy and Society, 8*(1).

Goodspeed, R. (2013). Planning support systems for spatial planning through social learning. PhD Thesis. Cambridge, MA: Massachusetts Institute of Technology.

Goodspeed, R. (2015). Smart cities: moving beyond urban cybernetics to tackle wicked problems. *Cambridge Journal of Regions, Economy and Society, 8*(1), 79–92.

Greenfield, A. (2013). Against the smart city (The city is here for you to use) (pp. 153).

Harris, B., & Batty, M. (1993). Locational models, geographic information and planning support systems. *Journal of Planning Education and Research, 12*(3), 184.

Harrison, C., & Donnelly, I. A. (2011). A theory of smart cities. In Proceedings of the 55th Annual Meeting of the ISSS 2011, Hull, UK.

Hollands, R. G. (2008). Will the real smart city please stand up? *City, 12*(3), 303–320.

Hunt, J. D., Kriger, D. S., & Miller, E. J. (2005). Current operational urban land-use–transport modelling frameworks: A review. *Transport Reviews, 25*(3), 329–376.

Johnston, E. W., & Hansen, D. L. (2011). Design lessons for smart governance infrastructure. In: A. T. Balutis, F. Buss, Ink, D. (Eds.), *Transforming American governance: Rebooting the public square*. National Academy of Public Administration.

Kitchin, R. (2013). The real-time city? Big data and smart urbanism. *GeoJournal, 79*, 1–14.

Klosterman, R. E. (1997). Planning support systems: a new perspective on computer-aided planning. *Journal of Planning Education Research, 17*(1), 45–54.

Lowry, I. S. (1964). A model of metropolis. RM-4035-RC. Santa Monica, CA: Rand Corporation.

March, S. T., & Smith G. F. (1995). Design and natural-science research on information technology. *Decision Support Systems, 15* (4), 251–266.

Neirotti, P., de Marco, A., Corinna, Cagliano, A., Mangano, G., & Scorrano, F. (2014). Current trends in smart city initiatives: Some stylised facts. *Cities, 38*, 25–36.

Pelzer, P., & Geertman, S. (2014). Planning support systems and interdisciplinary learning. *Planning Theory & Practice, 15*(4), 1–16.

Te Brommelstroet, M. (2013, September). Performance of planning support systems: What is it, and how do we report on it? *Computers, Environment and Urban Systems, 41*, 299–308.

Townsend, A. M. (2013). *Smart cities: Big data, civic hackers, and the quest for a new Utopia*. New York: W.W. Norton & Company.

Vonk, G. A. (2006). Improving planning support. The use of planning support systems for spatial planning. Ph.D. thesis, Netherlands Geographical Studies 340. Utrecht: Koninklijk Nederlands Aardrijkskundig Genootschap, Faculteit Geowetenschappen Universiteit Utrecht.

Vonk, G., Geertman, S., & Schot, P. (2005). Bottlenecks blocking widespread usage of planning support systems. *Environment and Planning A, 37*(5), 909–924.

Washburn, D., & Sindhu, U. (2010). Helping CIOs understand "smart city" Initiatives: Forrester Research.

Wegener, M. (2004). Overview of land-use transportation models. In D. A. Hensher & K. Button (Eds.), *Transport geography and spatial systems. Handbook 5 of the handbook in transport* (pp. 127–146). Kidlington: Pergamon/Elsevier Science.

Part I
Data Analytics

Chapter 2
Development and Operation of Social Media GIS for Disaster Risk Management in Japan

Kayoko Yamamoto

Abstract Since natural disasters frequently happen all over the world, we must make effective preparations for such disasters. As the implementation of sophisticated computerization expands, the society now benefits from ubiquitous network and cloud computing. Consequently, we can utilize a variety of information systems effectively for disaster reduction measures. Based on the experiences of natural disasters, among a variety of information systems, the roles of GIS (Geographic Information Systems) and social media are considered important for collection and transmission of disaster information. Against the above-mentioned backdrop, the present study aims to classify disaster risk management for natural disasters into three stages—normal times, disaster outbreak times, and times of recovery and reconstruction—to introduce the results of development and operation of social media GIS during each of these three stages. The social media GIS targeted residents who were more than 18 years old in the Tama region of Tokyo metropolis and the neighboring area in Japan for two months. Subsequently, the systems were evaluated based on the results of an online questionnaire survey to users, access surveys using log data during operation of the systems, and an analysis of the submitted information. Based on the results of the evaluation, measures for improvement of the development and operation of social media GIS can be summarized into three areas regarding (a) participation of various users and partnership with local communities, (b) usability, and (c) long-term actual operation.

K. Yamamoto (✉)
Graduate School of Information Systems, University of Electro-Communications,
1-5-1 Chofugaoka, Chofu, Tokyo 182-8585, Japan
e-mail: k-yamamoto@is.uec.ac.jp

S. Geertman et al. (eds.), *Planning Support Systems and Smart Cities*,
Lecture Notes in Geoinformation and Cartography,
DOI 10.1007/978-3-319-18368-8_2

1 Introduction

Since natural disasters frequently happen all over the world, we must make effective preparations for such disasters. According to the White Paper on Disaster Management (2012), measures for disaster prevention and reduction of the effects of natural disasters in Japan include "self-help", "mutual help (cooperation)", and "public help (rescue and assistance by public bodies)". "Self-help" refers to local residents, businesses, and other entities protecting themselves from disaster; "mutual help (cooperation)" refers to local communities helping each other; and "public help" refers to measures by government bodies such as national and local governments. Further, the most fundamental form of help was said to be self-help, which involves measures taken by individuals. Nowadays, anybody, anywhere, anytime can use an information system to easily send, receive, and share information, and through the effective use of information systems, disaster information possessed by local residents can be accumulated and shared.

Additionally, the Science Council of Japan (2008) divided "local knowledge" into "expert knowledge" based on scientific knowledge, and "experience-based knowledge" produced by the experiences of local residents, and indicated its importance. Concerning "local knowledge" that is the "experience-based knowledge" of local residents, and exists as "tacit knowledge" that is not visualized if it is not communicated to others, as a measure for disaster prevention and reduction, it is essential for the "experience-based knowledge" to be transformed into "explicit knowledge" which is of a form that can be accumulated, organized, utilized, and made publicly available through the use of information systems, and to have local related entities accumulate the knowledge together. Moreover, Committee for Policy Planning on Disaster Management—Final Report by the Central Disaster Management Council (2012) particularly specified the importance of the roles of GIS (Geographic Information Systems) and social media in the collection and transmission of disaster information, and the role of "public information commons", which provide free information services during disasters, is becoming more important.

Regarding the examples of disaster management in other parts of the world, Greene (2002) proposed five stages of disaster—identification and planning, mitigation, preparedness, response, and recovery—, and shows how GIS processes can be incorporated into each. Vivacqua and Borges (2012) considered the four phases of emergency management which are related in a cyclic fashion: mitigation, preparedness, response, and recovery. They discussed possibilities for the introduction of collective knowledge in emergency response systems. Mansouriana et al. (2006) addressed the role of Spatial Data Infrastructure (SDI) as a framework for the development of a web-based system to facilitate disaster management with emphasis on the response phase in Iran. Neuvel et al. (2012) proposed the network-centric concept of spatial decision support for risk and emergency management in Netherland.

In the present study, firstly, with reference to the White Paper on Disaster Management (2012), Committee for Policy Planning on Disaster Management—Final Report (2012), and Vivacqua and Borges (2012), disaster risk management is divided into three stages—normal times, disaster outbreak times, and times of recovery and reconstruction—. Further, the present study aims to introduce the results of development and operation of social media GIS for risk management in each of the above-mentioned stages, citing results concerning systems developed and operated by the present author and her co-researchers (Okuma and Yamamoto 2013; Murakoshi and Yamamoto 2014; Yamada and Yamamoto 2013).

2 Related Work

Focusing on research methods, existing research related to the present study can be broadly divided into four types: (1) Research involving the staging of workshops; (2) Research related to information system development and proposals; (3) Research related to Web-GIS design and development; and (4) Research related to social media development and use. In (1), studies involving the staging of workshops have been carried out by Matsuda et al. (2005), Miao et al. (2005), and Nagasaka et al. (2009). In these studies, through workshops in which disaster prevention radio dramas were created, and the use of regional disaster prevention capacity diagnostic sheets, local residents' awareness of issues and knowledge were shared. In (2), research related to developing and proposing information systems, Murakami et al. (2009) and Okano et al. (2009) developed disaster prevention activity support systems, and Kato et al. (2010) proposed a disaster management information mashup system. In (3), research related to the design and development of Web-GIS, Sato et al. (2004), Kajiki (2006), and Fujita et al. (2008) developed and published regional safety maps. Takatani et al. (2008) developed hazard maps. Kawasaki and Meguro (2010), and Inoguchi et al. (2011) showed the effectiveness of disaster information web mapping. Additionally, Yanagisawa and Yamamoto (2012) integrated Web-GIS, a Social Networking Service (SNS) and a wiki into a single system which enabled the accumulation of local safety information such as disaster information. In (4), research on social media development and use, Yamamori (2010) developed a bulletin board for use during disasters which utilized a regional SNS, and Korida et al. (2011) developed a community disaster prevention SNS. Wasaki (2012) demonstrated the usefulness of regional SNS in times of disaster. In addition, Yamamoto et al. (2012), and Yoshimura and Inoue (2012) analyzed the situation regarding information transmission and acquisition via Twitter during the Great East Japan Earthquake (2011), and demonstrated the possibility of real-time disaster information transmission via Twitter.

As outlined above, a large amount of existing research utilizes Web-GIS and social media, and the number of services that utilize some kind of information system is increasing. However, while various information systems related to disaster exist, they mainly focus on measures to reduce the effects of natural disasters, and deal separately with each stage of disaster risk management—normal times, disaster outbreak times, and times of recovery and reconstruction. Further, they only collect information and one-sidedly provide information. Excluding the results of research on system developed and operated by the present author and her co-researchers (Okuma and Yamamoto 2013; Murakoshi and Yamamoto 2014; Yamada and Yamamoto 2013), up till now, a series of systems designed for continuous disaster risk management which covers all stages—from normal times to times of recovery and reconstruction—has not been developed. Thus, the main objectives of systems in the series differ according to the stage of disaster risk management. The main objective from normal times to disaster outbreak times is support for accumulation and utilization of disaster information, and the main objective in times of recovery and reconstruction is information exchange; however, the series of systems is set such that by changing the mode for each stage of disaster risk management, the same system can be used continuously in all three stages. Thus, this system is an aggregate of multiple systems, and can be customized to suit the actual situation in a region where it is to be operated. Sections 4–6 describe the development and operation of the three types of social media GIS one by one.

3 Outline of Social Media GIS

3.1 Progress of Transformation to Information-Intensity in Japan

In Japan, the Basic Act on the Formation of an Advanced Information and Telecommunications Network Society (Basic IT Law) came into force in 2000. In "e-Japan" of the year 2000, a plan, strategy, and policies which aimed to realize a Japanese-style IT society were proposed. Further, "u-Japan" of the year 2006 aimed at realizing a society in which "anybody, anywhere, anytime, using any device" could easily connect to a network by the year 2010. In addition, in 2010, "i-Japan 2015", which declared the goal of realizing a "reassuring and vibrant digital society", was proposed. Now, Japan is shifting from being the ubiquitous network society which was aimed for in "u-Japan" to being a cloud computing society in which various information tools can be used to connect to the internet. Therefore, regardless of time and place, as long as there is an information environment which allows internet connection, people can use the internet, using devices such as mobile information terminals (e.g., smartphones and tablet-type terminals), as well as PCs, so anyone can easily send and receive information. Moreover, various social media such as blogs, Twitter, YouTube, Facebook, and Line can be used to transmit information in compound forms combining images, videos, and sounds, not just words.

3.2 Development of Social Media GIS

Information on the situation regarding damage in the Great Hanshin Earthquake that occurred in Japan in 1995 was put into a database utilizing GIS, and information for reconstruction planning was effectively provided. Therefore, the usefulness of GIS came to be acknowledged. Further, as legislation related to GIS, in addition to the Basic IT Law mentioned in Sect. 3.1, the Basic Act on the Advancement of Utilizing Geospatial Information came into force, and presently its application in various areas of our daily lives is highly anticipated.

As shown in Fig. 1, the GIS is based on a digital map, and broadly speaking has four basic functions—a database development function, an information analysis function, an information providing/sharing function, and a decision-making support function. The GIS uses these functions to connect the real world with the virtual world, and it can be said that the GIS is an information system that is intimately connected with people and society. Formerly, research which utilized database development functions and information analysis functions dominated, but together with the development of an information-intensive society, information providing/sharing function have come to be utilized often in research. Therefore, development of GIS which citizens participate in has proceeded, and GIS has come to be used as a tool for providing information to and sharing information with residents. According to Siebel (2006), the term "Public Participation Geographic Information Systems" (PPGIS) was first used in a workshop of the U.S. National Center for Geographic Information and Analysis (NCGIA) in 1996. Moreover, Godchild (2007) termed GIS which enable information provided voluntarily by ordinary people to be accumulated on a digital map, and provided and shared "Volunteered Geographic Information (VGI)", and demonstrated the importance of the role of ordinary people as social sensors.

The present author and her co-researchers took these civic-participation-style GIS a step further, and by integrating social media and Web-GIS, gave GIS a function for interactive communication between users, as shown in Fig. 1. Additionally, aiming for actual operation in local communities, in the series of studies mentioned in the previous section (Okuma and Yamamoto 2013; Murakoshi and Yamamoto 2014; Yamada and Yamamoto 2013), the present author and her co-researchers integrated social media with Web-GIS, and developed a social media GIS which was a disaster information system mainly for earthquake disaster prevention and reduction measures. Because a Web-GIS alone is limited to one-way information transmission utilizing a digital map, an SNS and Twitter were integrated with the Web-GIS to allow interactive information transmission and reception, and that is the greatest distinguishing feature of the system. If long-term operation using the communication feature of the social media GIS developed as outlined above can be achieved, it can be anticipated that information concerning the awareness and behavior of ordinary people as social sensors will be collected and accumulated on the GIS digital map, and using the information analysis

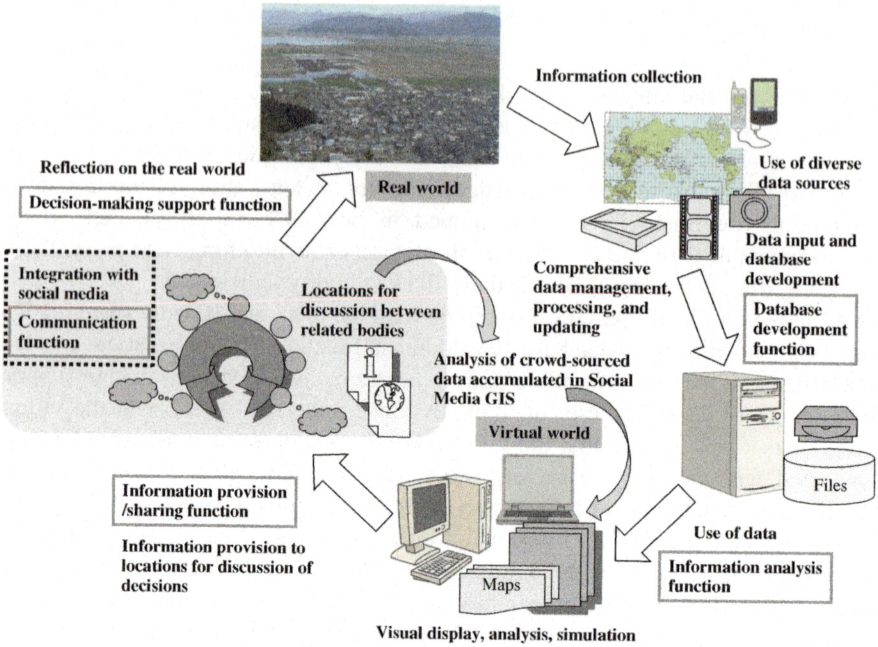

Fig. 1 Various functions of GIS to integrate relevant data by location

function, the information will be analyzed as crowd-sourced data tagged with spatial information.

As social media for integrating with the Web-GIS, firstly, an SNS was selected, and was designed and developed in a unique way in accordance with the primary objectives of systems able to handle each stage of risk management. This is because unlike with other forms of social media, with an SNS a system can be uniquely designed and developed in line with objectives of use, and detailed system configuration can be carried out to suit the actual situation in a region where the system is to be operated. However, considering the penetration rate of mobile information terminals (especially smartphones), for the two types of system developed and operated in 2012 for use in normal times and times of recovery and reconstruction, in addition to an SNS, Twitter was also added to the systems as a form of social media (Okuma and Yamamoto 2013; Yamada and Yamamoto 2013). This is because use from mainly mobile information terminals was anticipated. However, in the system developed and operated in 2013 that focused on normal times through to disaster outbreak times, not only an interface for PC use but also an interface optimized for mobile information terminals was provided (Murakoshi and Yamamoto 2014). Further, as social media for integration with the Web-GIS, only an SNS was used. This is because the results of the systems developed and operated in 2012 revealed that it was necessary to take into account users who did not have Twitter accounts.

3.3 Operation of Social Media GIS

Tables 1 and 2 show the operation of Social Media GIS Series in the present study, and Table 3 describes the outline of users and online questionnaire survey respondents. In accordance with the operation process in Table 1, each of the three types of social media GIS was actually operated after an operation test and an evaluation of the operation test had been conducted. As mentioned in Sect. 3.2,

Table 1 Operation process of Social Media GIS Series (2012, 2013)

Process	Aim	Period	Specific details
1. Survey of present conditions	To understand efforts related to disaster prevention and reduction in the region of operation	January–March	• Survey of government measures and internet services
			• Interview targeting municipal employees and officials of residents' councils
2. System configuration	Configure the system in detail to suit the region of operation	April–June	• Define system requirements
			• System configuration
			• Create operation system
3. Operation test	Conduct the system operation test	July	• Create and distribute pamphlets and operating instructions
			• System operation test
4. Evaluation of operation test	Reconfigure the system based on results of interviews with operation test participants	August–September	• Evaluation using interviews
			• System reconfiguration
			• Amendment of pamphlets and operating instructions
5. Operation	Carry out actual operation of the system	October–November	• Appeal for use of the system
			• Distribution of pamphlets and operating instructions to local residents
			• System operation management
6. Evaluation	Evaluate the system based on the results of online questionnaire survey to users, the results of access analysis which used log data during the period of actual operation, and the results of analysis of submitted information	December	• Evaluation using online questionnaire survey to users, access surveys using log data during operation of the systems, and an analysis of the submitted information
			• Identification of measures for improvement of the development and operation of the system even more effectively

Table 2 Outline of operation of social media GIS series

Relevant section in the present study	Period of stage	Region of operation	Number of users	User characteristics
Section 4	Normal times	Chofu City, Tokyo Metropolis	75	72 % of users were in their 20s, and 10 % were in their 30s and 40s
Section 5	From normal times to disaster outbreak times	Chofu City, Tokyo Metropolis	56	52 % of users were in their 20s, and 20 % were in their 40s. Further, 16 % were in their 50s or above
Section 6	Times of recovery and reconstruction	Tama region of Tokyo Metropolis and eastern Yamanashi Prefecture	45	93 % of users were in their 20s. There were 25 users in the Tama region of Tokyo Metropolis, and 20 users in eastern Yamanashi Prefecture

Table 3 Outline of users and online questionnaire survey respondents

	10–19	20–29	30–39	40–49	50–59	60–	Total
Section 4							
Number of users	4	54	8	7	2	0	75
Number of questionnaire respondents	4	43	2	2	0	0	51
Valid response rate (%)	100.0	79.6	25.0	28.6	0.0	–	68.0
Section 5							
Number of users	1	29	6	11	5	4	56
Number of questionnaire respondents	1	27	4	4	2	3	40
Valid response rate (%)	100.0	93.1	66.7	36.4	40.0	50.0	71.4
Section 6							
Number of users	1	42	2	0	0	0	45
Number of questionnaire respondents	1	30	2	7	2	3	33
Valid response rate (%)	100.0	71.4	100.0	0.0	0.0	0.0	73.3

though the system for use from normal times to disaster outbreak times was operated in 2013, two other types of systems were operated in 2012. As shown in Table 2, each of social media GIS was operated for a period of two months, targeting local residents who were more than 18 years old. The systems for use in normal times and disaster outbreak times were operated in Chofu City, Tokyo Metropolis, while the system for use in times of recovery and reconstruction was operated in two regions—the Tama region of Tokyo Metropolis, and eastern Yamanashi Prefecture. Use of the systems was appealed for the local residents in the regions for operation, using such means as the website of the present authors' laboratory. Further, in the regions of operation, the local governments such as Tokyo Metropolis, Yamanashi Prefecture, and Chofu City helped the present author to distribute system pamphlets and operating instructions to the local residents. Each system was respectively operated according to the main objective mentioned in Sect. 2 in normal time. After the operation of each system for two months, online questionnaire surveys were conducted to users.

It can be seen from Tables 2 and 3 that the majority of users of all three types of system were in their twenties, meaning many users were of a generation proficient in using a variety of information systems in their daily lives. After operation, the systems were evaluated based on the results of an online questionnaire survey to users, access surveys using log data during operation of the systems and an analysis of the submitted information. The questionnaire items were related to the operability especially regarding the functions and the effects of use to evaluate the usability of the three types of social media GIS.

4 System Designed for Accumulating Disaster Information in Normal Times

When a disaster occurs, the helping hand of fire brigades and other rescue groups will not extend to all disaster victims. In order for disaster damage to be kept to a minimum by disaster prevention and reduction measures taken in normal times, it is necessary for each person to have a high level of awareness of disaster information in their daily life. Accordingly, it is important that people always have an accurate understanding of what places are dangerous to pass through during a disaster, where the evacuation sites are in their region of residence, and that this information be collected as geotagged disaster information. Further, in order to link this to "mutual help" and "public help", it is necessary for local governments and residents to accumulate and share disaster information to a sufficient extent during normal times.

In Japan, representative examples of disaster prevention and reduction measures are government efforts such as the "Hazard Map Portal Site" developed and operated by the Ministry of Land, Infrastructure, Transport and Tourism, and the disaster prevention maps and hazard maps of local governments. However, these resources have little detailed information which local residents actually need during a disaster, so they are not very user-friendly, and in the case of resources published in PDF format, the information cannot be viewed all at once. In order to solve the above-mentioned problems, Okuma and Yamamoto (2013) developed a social media GIS specially tailored to accumulate disaster information on digital maps for the purpose of disaster prevention and reduction measures in normal times. To achieve this, two types of social media, an SNS and Twitter, were added to a Web-GIS, and those three applications were integrated into a single system. Due to the integration, this system fundamentally has the information submission and viewing

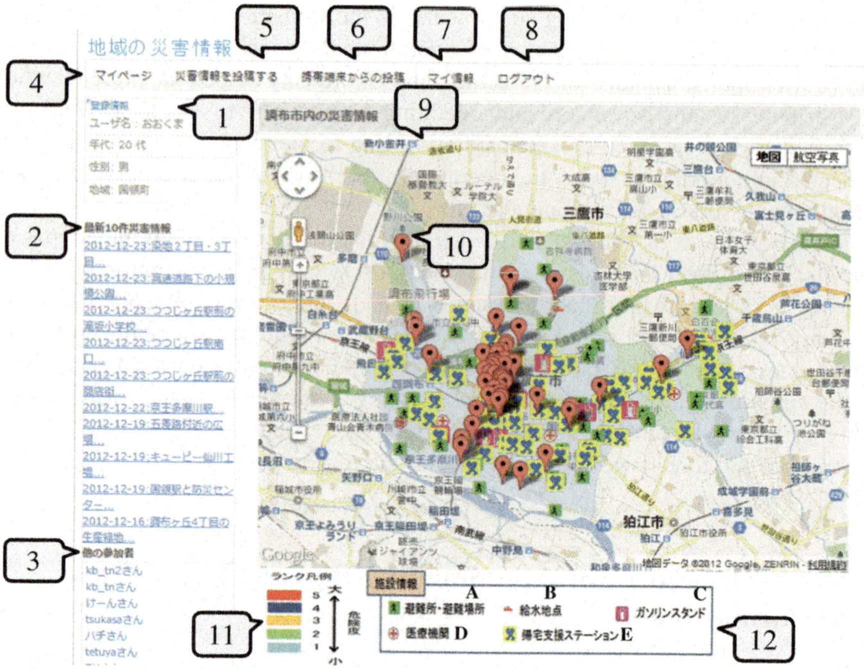

No.	Description
1	User profile publication
2	The ten most recent items of information submitted from computers and mobile information terminals using Twitter
3	Other users
4	Go to my page
5	Go to the page where information can be submitted from computers
6	Go to the page where information can be submitted from mobile information terminals using Twitter
7	Go to the page where change and registration of personal data can be made
8	Logout
9	Disaster information is displayed on Web-GIS digital map in the region of operation(ChofuCity)
10	Submitted information by residents
11	General degree of risk
12	Explanation of disaster information provided by local governments A: Evacuation assembly area, B: Water station, C: Petrol station, D: Medical institution, E: Homecoming support station

Fig. 2 PC interface of system developed by Okuma and Yamamoto (2013)

functions. Moreover, with reference to disaster prevention maps produced by local governments, information about the support facilities (evacuation sites, stations which provide support for people returning home, water supply bases, etc.) in addition to the general degree of risk was accumulated in the database in advance. Therefore, disaster information provided by local governments and residents was mashed up on a GIS base map. Figure 2 shows the PC interface of the system.

In this system, because the system consists of not just Web-GIS, but rather integrates a Web-GIS with social media, the various functions of social media can be used to collect and transmit disaster information which is the experience-based knowledge of local residents, and this disaster information can be accumulated and shared as explicit knowledge on the Web-GIS digital map. Through this, it can also be anticipated that local residents will be able to appreciate the weaknesses of the area they live in, and their awareness of disaster prevention and reduction will be heightened through the experience of accumulating and sharing disaster information. Accordingly, "self-help" may lead to "mutual help" and "public help". Further, based on the concept of resilience, utilizing the accumulated disaster information, the possibility that this system can be applied to pre-disaster outbreak advance reconstruction efforts involving measures and preparations for recovery and reconstruction can be anticipated.

According to Okuma and Yamamoto (2013), the online questionnaire survey showed the usability of the system in terms of its operability and effects of use. It also showed that the system was effective to heighten local residents' awareness of disaster information, and it will be continuously used as disaster prevention and reduction measures in normal times. Additionally, it was clear that most of users used PCs rather than mobile information terminals (especially smartphones) as a means of information submission, and approximately 30 % of users only viewed the disaster information provided by local governments and residents. The access survey showed that users in their 30s and 40s submitted information only from PCs, while those in their 20s submitted information to approximately the same extent from both PCs and mobile information terminals. Additionally, the 92 pieces of submitted information were classified into five types, among which 73 % of information was provided about evacuation sites and routes. Further, 24 % of information was provided about food and water supply bases in times of disaster, and 3 % of information was related to the Great East Japan Earthquake.

5 System Designed for Supporting Utilization of Disaster Information from Normal Times to Disaster Outbreak Times

In Japan, as a disaster countermeasure, local governments provide information to local residents in the form of disaster prevention maps and hazard maps which show local hazardous places, evacuation sites and so on. However, this information is mainly published as maps that are in paper form or in PDF format on the website. Therefore, it is difficult to update the information on disaster prevention maps and hazard maps in real time, and these forms of information are not very suited to being shared during a disaster outbreak. Further, so that information can be efficiently accumulated and shared during a disaster outbreak, it is desirable that information systems which people are accustomed to using in normal times can be used as is during disasters. However, when a disaster occurs, a situation where the amount of submitted information increases, and there is an excessive amount of information can be expected; therefore, it is necessary for systems to automatically classify submitted information.

Based on awareness of the above-mentioned issues, Murakoshi and Yamamoto (2014) developed a social media GIS which integrated Web-GIS with an SNS, and was specially tailored to mashup the information that local governments and residents provide to support information utilization from normal times to disaster outbreak times. This social media GIS employed the system developed by Okuma and Yamamoto (2013) as a base, and extended the period of use of the system from normal times, during which the system would be used for disaster prevention and reduction measures, to include disaster outbreak times, during which the system would be used for evacuation activity support and support for people facing difficulty in returning home due to disaster. The strongest reason for extending the period of use of the system was that it could be anticipated that through people using the system in normal times and becoming familiar with it, they would also continue to use the system at the stage when a disaster occurred and the situation was very urgent. As mentioned in Sect. 3.2, Twitter was not included in this system. In addition to a PC interface (Fig. 3), an interface optimized for mobile information terminals (Fig. 4) was provided.

In this system, the fundamental functions—the information submission and viewing functions, and the mashup of disaster information provided by local governments and residents—are almost the same as those of the system developed by Okuma and Yamamoto (2013). However, this system also has a function for classifying submitted information, and a function for checking support facilities in times of disaster. Using the former function, based on text information, the system automatically determines whether submitted information is related to either danger or safety. Further, on the Web-GIS digital map, the system indicates danger-related information using semitransparent red, and safety-related information using semitransparent green. Therefore, evacuees can determine with a single glance which areas are dangerous and which areas are safe, even when looking at a small screen on a mobile information terminal when they are evacuating during a disaster. Concerning the latter function, based on the information provided by local governments, users can search for a facility which provides support during a disaster by freely specifying a category of facility and a distance from their present location, and display the search results on the Web-GIS digital map.

According to Murakoshi and Yamamoto (2014), the online questionnaire survey showed the usefulness of the system in terms of its operability, especially regarding the above-mentioned two specific functions, and the possibility to provide disaster information mainly for mobile information terminals in disaster outbreak times. Since it also showed the large extent of the effects of use particularly related to local residents' awareness of disaster information, we expected that this system will be continuously used according to the main objective from normal times to disaster outbreak times. The access survey showed that users continuously accessed the system and 181 pieces of disaster information were distributed throughout the whole region of operation. Additionally, it was clear that 83 % of users accessed the system mainly using PCs, and mobile information terminals were used to assist the use of

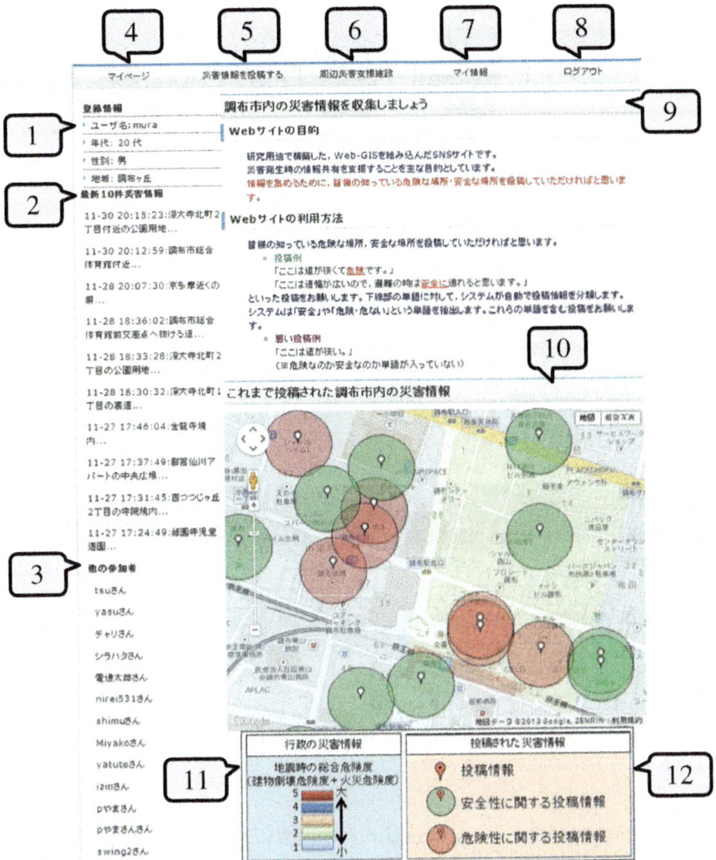

No.	Description
1	User profile publication
2	The ten most recent items of information submitted from computers and mobile information terminals
3	Other users
4	Go to my page
5	Go to the page where information can be submitted from computers
6	Go to the page where support facilities in times of disaster can be checked
7	Go to the page where change and registration of personal data can be made
8	Logout
9	Explanation to use this system
10	Disaster information is displayed on Web-GIS digital map using markers
11	General degree of risk
12	Explanation of disaster information submitted by users

Fig. 3 PC interface for system developed by Murakoshi and Yamamoto (2014)

PCs. Among the disaster information, danger-related information occupied 28 %, safety-related information occupied 67 %, and other information occupied 5 %.

What is thought to be a problem at this stage is the internet communication environment (internet connection, electricity, use of information terminals, etc.). However, technological development in related fields is proceeding rapidly, so it is highly likely that this problem will be solved in the near future. For example,

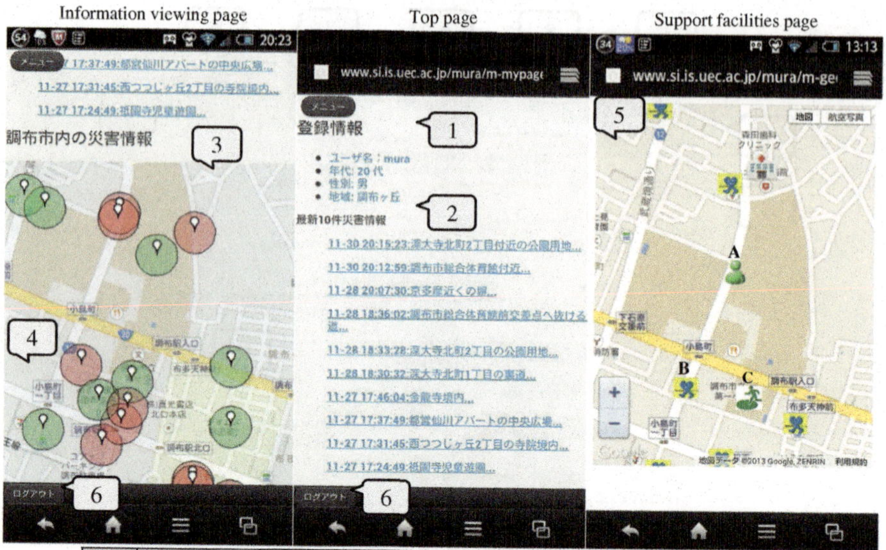

No.	Description
1	User profile publication
2	The ten most recent items of information submitted from computers and mobile information terminals
3	Submitted information is displayed on Web-GIS digital map using markers
4	General degree of risk
5	Support facilities in times of disaster A: User's present location, B: Homecoming support station, C: Evacuation assembly area
6	Logout

Fig. 4 Mobile information terminal interface for system developed by Murakoshi and Yamamoto (2014)

technological development is proceeding rapidly in the area of securement of internet access through such means as mobile internet base stations fitted into cars, wireless communication via solar portable base stations, and satellite communication; in the area of electricity supply which utilizes hybrid cars and the like; and in the area of extension of the length of time which mobile information terminal batteries last.

6 System Designed for Information Exchange Between Regions in Times of Recovery and Reconstruction

Yamada and Yamamoto (2013) developed a social media GIS that enabled information exchange between multiple regions. Figure 5 shows the PC interface of the system. The information submission and viewing functions, and the basic configuration of the system are almost the same as those of the system developed by Okuma and Yamamoto (2013); however, the system also includes a comment function, a button function, and a ranking function. By using the comment function, users in multiple regions or within a particular region can communicate with each

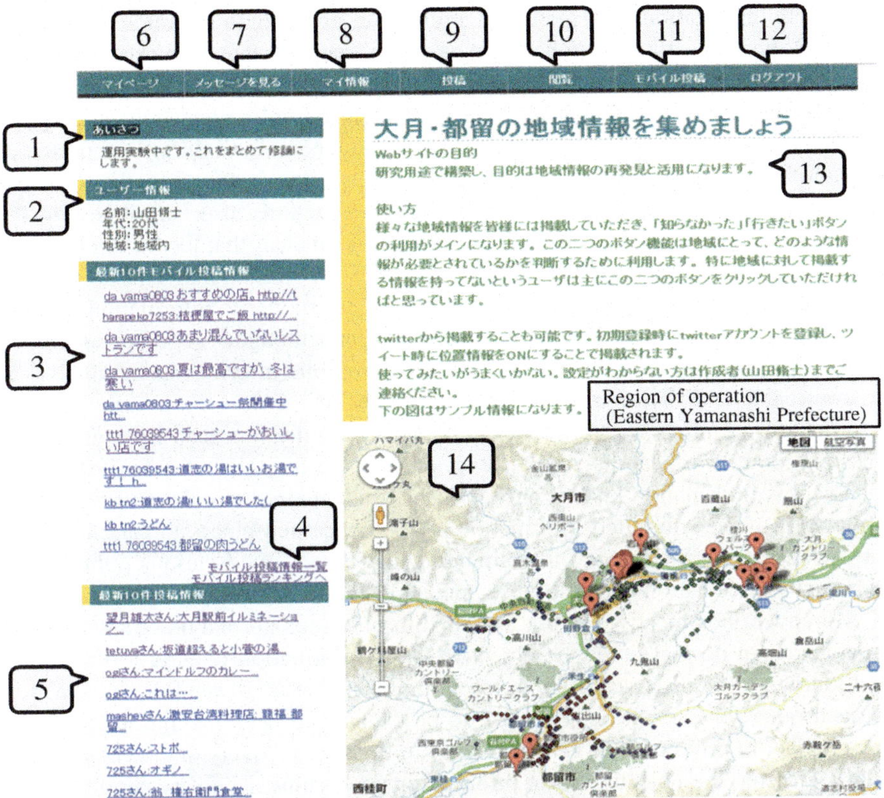

No.	Description
1	User greeting
2	User profile publication
3	The ten most recent items of information submitted from mobile information terminals using Twitter
4	Go to a list of information submitted from mobile information terminals using Twitter and the ranking page
5	The ten most recent items of information submitted from computers
6	Go to my page
7	Go to the page which contains messages from administrators
8	Go to the page where change and registration of personal data can be made
9	Go to the page where information can be submitted from computers
10	Go to the page where information submitted from computers can be viewed
11	Go to the page where information submitted from mobile information terminals using Twitter can be viewed
12	Logout
13	Explanation to use this system
14	Submitted information is displayed on Web-GIS digital map using markers

Fig. 5 PC interface for system developed by Yamada and Yamamoto (2013)

other. By using the button function, users can conduct simple communication with each other, and also evaluate the importance of submitted information. Further, regarding the submitted information list, the ranking function can be used to display submitted information in descending order starting with information which has attracted the highest button function usage frequency, so that important submitted information does not get lost amongst other information.

According to Yamada and Yamamoto (2013), the online questionnaire survey showed the high level of operability, the high frequency of use during the period of operation, and the large extent of the effects of use, and it was clear that the high evaluation of the above-mentioned three unique functions also contributed to the high level of the operability. Additionally, it was also clear that the button function was most highly evaluated by users among the three functions. This is because users were able to easily indicate their intentions in response to information viewed, by using the button to reflect their experiences and thoughts. The access survey results showed that while the access count from mobile information terminals was no more than 12 % of 383 access count, the submission count from mobile information terminals was 41 %, accounting for slightly less than half of 85 submission count. Further, the results revealed that the majority of submitted information was information known only to local residents in the regions of operation. Moreover, of the submitted information, 41 % concerned places to eat and drink, 38 % concerned scenery, and the remaining 11 % concerned various public facilities such as recreation centers, museums, and science centers.

Moreover, the objectives of this system differ between normal times and times of emergency. In normal times the objective is for information relating to the region and tourism to be discovered within each region, for users to send the information to each other, and for the information to be exchanged between multiple regions. Meanwhile, in times of emergency, such as in times of recovery and reconstruction after a disaster, the objective is for information concerning necessary people, things and so on to be exchanged by those inside and outside the disaster-stricken region, and for what is necessary to be rapidly sent from outside the disaster-stricken region to the places where it is needed inside the disaster-stricken region. Therefore, by having local governments which conduct regional exchange such as sister city arrangements with each other use this system to exchange information in normal times, the possibility of achieving effective support for disaster-stricken regions in times of recovery and reconstruction after a disaster can be anticipated.

7 Conclusion

Thus, the present study classified disaster risk management into three stages—normal times, disaster outbreak times, and times of recovery and reconstruction; cited results concerning systems developed and operated by the present author and her co-researchers designed for purposes of mainly earthquake disaster prevention and reduction measures; and presented results concerning the development and operation of social media GIS that can handle each one of the above-mentioned stages. In developing systems for disaster risk management, a particularly essential condition is that the system must actually be usable in the event of a disaster.

Therefore, based on the principles and techniques of inclusive design, it is important to appropriately reflect the opinions of the people who actually use the system, such as local governments and residents, and civic groups, starting from the design stage of the system. It is also important to take into account social change and progression in information communication technology, and continuously improve the system.

An example of an issue that came up based on the results of the evaluations which was common to all three systems was firstly to involve not just local governments and residents, and civic groups, but also to cooperate with fire brigades, police and so on, and operate the system with the participation of a wider user base. In order to achieve this, as well as further strengthening ties with local communities and holding courses on the use of the systems and information tools, it is also necessary to consider having people of a wide range of ages imagine disaster outbreaks and actually use the systems at disaster drills and evacuation drills, and to consider creating systems in which people such as members of the younger generation who are proficient in using information tools support other people, particularly the aged, in using the systems.

A second issue was to further improve the usability of the systems by customizing the interfaces to suit the needs and preferences of each user. A third issue was to achieve long-term actual operation of the systems, as mentioned in Sect. 3.2, and thereby collect submitted information in the form of crowd-sourced data tagged with spatial information; and by analyzing the data using methods such as text mining combined with GIS information analysis functions, to conduct more effective information provision for disaster prevention and reduction measures in local communities.

Acknowledgments In the operation of the social media GIS and the online questionnaire survey of the present study, enormous cooperation was received from those in the Tama region of Tokyo metropolis and the neighboring area in Japan. I would like to take this opportunity to gratefully acknowledge them.

References

Cabinet Office, Government of Japan. (2012). *White paper on disaster management 2012*. Tokyo.

Central Disaster Management Council. (2012). *Committee for policy planning on disaster management—Final report: Toward reconstruction for Japan that is unshakable*. Tokyo.

Fujita, H., Kato, R., Shinohara, S., Seki, M., Yoshida, S., & Sadohara, S. (2008). Construction of the safety map by GIS in large-scale apartment complex: Intended for Wakabadai apartment complex, Yokohama. *Proceedings of the Annual Conference of the Institute of Social Safety Science, 23*, 79–82.

Godchild, M. F. (2007). Citizens as sensors: The world of volunteered geography. *GeoJournal, 69* (4), 211–221.

Greene, R. W. (2002). *Confronting catastrophe: A GIS handbook*. Redlands: ESRI Press.

Inoguchi, M., Tamura, K., Furuya, T., Kimura, R., & Hayashi, H. (2011). Proposal of effective on-demand mashup among spatial information from the activity of emergency mapping team:

A case study of the 2011 off the pacific coast of Tohoku Earthquake. *Journal of Social Safety Science, 15*, 219–229.

Kajiki, N. (2006). Examination of making regional safety map by partnership between elementary school and community: Case study of advanced efforts made by West-Suma elementary school in Kobe. *Summaries of Technical Papers of Annual Meeting Architectural Institute of Japan*, F-1, pp. 539–540.

Kato, T., Kobayashi, M., Sato, N., & Yotsuyanagi, T. (2010). Prototype development of "geospatial disaster management mashup system". *Monthly Journal of the Institute of Industrial Science, University of Tokyo, 62*(4), 377–380.

Kawasaki, A., & Meguro, K. (2010). New movement on the geospatial utilization for disaster response in the Haiti earthquake. *Monthly Journal of the Institute of Industrial Science, University of Tokyo, 62*(4), 409–416.

Korida, K., Kikuchi, T., Yoshiyama, N., Shibata, Y., Takahashi, M., Takenaka, M., et al. (2011). Development of community disaster prevention SNS based on the relationship of mutual trust of local inhabitants. *The Special Interest Group Technical Reports of Information Processing Society of Japan, CSEC, "Computer Security"*, 2011-CSEC-52(16), 1-6.

Local Research Committee, Science Council of Japan. (2008). *Recommendations: Towards the accumulation and utilization of "local knowledge"*. Tokyo.

Mansouriana, A., Rajabifardb, A., Valadan Zoej, M. J., & Williamsonb, I. (2006). Using SDI and web-based system to facilitate disaster management. *Computers and Geosciences, 32*(3), 303–315.

Matsuda, Y., Itotani, T., & Okada, N. (2005). Study on method for improving capacity of regions to cope with disasters by using regional diagnostic questionnaires. *Proceedings of the Annual Conference of the Japan Society of Civil Engineers, 60*(4), 3–4.

Miao, Q., Arima, M., & Kawamukai, H. (2005). How to use JMP in revealing issue structure from a questionnaire survey in a local community. *Proceedings of the SAS Forum Users Group Academic Session, 2005*, 425–436.

Murakami, M., Shibayama, A., Hisada, Y., Ichii, T., Zama, S., Endo, M., et al. (2009). Development of web GIS-based support system for collecting and sharing information by collaboration between local government and residents. *Journal of Japan Association for Earth Engineering, 9*(2), 200–220.

Murakoshi, T., & Yamamoto, K. (2014). Study on a social media GIS to support the utilization of disaster information: For disaster reduction measures from normal times to disaster outbreak times. *Socio-Informatics, 3*(2), 17–31.

Nagasaka, T., Tsubokawa, H., Sunaga, Y., Usuda, Y., Taguchi, H., Okada, S., et al. (2009). Risk communication in a process of making plans for disaster response training: In case of Yamakoshi area of Nagaoka City. *Proceedings of the Annual Conference of the Society for Risk Analysis Japan, 22*, 28–29.

Neuvel, J. M. M., Scholten, H. J., & van den Brink, A. (2012). From spatial data to synchronized actions: The network-centric organization of spatial decision support for risk and emergency management. *Applied Spatial Analysis and Policy, 5*, 51–72.

Okano, S., Yasui, S., Okazawa, K., Ebara, H., & Okada, H. (2009). A hybrid distribution method of local disaster information for community networks. *IEICE Technical Report, IN, Information Network, 108*(458), 499–504.

Okuma, T., & Yamamoto, K. (2013). Study on a social media GIS to accumulate urban disaster information: Accumulation of disaster information during normal times for disaster reduction measures. *Socio-Informatics, 2*(2), 49–65.

Sato, T., Motosaka, M., & Nakamura, K. (2004). An educational trial of earthquake disaster prevention for school children by making regional map. *Proceedings of AIJ Tohoku Chapter Architectural Research Meeting, 67*, 183–186.

Siebel, R. (2006). Public participation geographic information systems: A literature review and framework. *Annals of the Association of American Geographers, 96*(3), 491–507.

Takatani, T., Iwaki, T., & Yoshinori, K. (2008). City planning based on a regional disaster prevention supporting system. *Bulletin of Maizuru National College of Technology, 43*, 34–50.

Vivacqua, A. S., & Borges, M. R. S. (2012). Taking advantage of collective knowledge in emergency response systems. *Journal of Network and Computer Applications, 35*(1), 189–198.

Wasaki, H. (2012). Disaster and regional SNS: The connection effect for rescue/recovery that came to light through the Sayou Town torrential rain disaster. *Bulletin of the National Museum of Ethnology, 106*, 127–146.

Yamada, S., & Yamamoto, K. (2013). Development of social media GIS for information exchange between regions. *International Journal of Advanced Computer Science and Applications, 4*(8), 62–73.

Yamamori, K. (2010). An implementation of the bulletin board for disaster. *Bulletin of Faculty of Education Mie University, 61*, 13–19.

Yamamoto, M., Ogasawara, H., Suzuki, I., & Furukawa, M. (2012). Information propagation network for 2012 Tohoku Earthquake and tsunami on Twitter. *Information Processing Society of Japan Magazine, 53*(11), 1184–1191.

Yanagisawa, T., & Yamamoto, K. (2012). Study on information sharing GIS to accumulate local knowledge in local communities. *Theory and Applications of GIS, 20*(1), 61–70.

Yoshimura, N., & Inoue, T. (2012). The use of Twitter around governmental accounts in and after the Great East Japan Earthquake. *The Special Interest Group Technical Reports of Information Processing Society of Japan, GN, "Groupware and Network Service"*, 2012-GN-83(5), pp. 1–8.

Chapter 3
The Role of Social Media Geographic Information (SMGI) in Spatial Planning

Michele Campagna, Roberta Floris, Pierangelo Massa, Anastacia Girsheva and Konstantin Ivanov

Abstract This contribution reports on ongoing research carried on by the authors on the role of Social Media Geographic Information in spatial planning, design, and decision-making. Explicit and Implicit Volunteered Geographic Information (VGI) from social media platforms, namely Social Media Geographic Information (SMGI) resources, were used to explore novel methods and tools for analysis and knowledge construction. The results concern three main research streams carried on with the common feature of integrating social media and other volunteered and authoritative sources of information from Spatial Data Infrastructures (SDI). These findings demonstrated that the integration of SMGI with more traditional Authoritative Geographic Information (A-GI) may offer a high potential for eliciting pluralist knowledge for spatial planning.

M. Campagna (✉) · R. Floris · P. Massa
DICAAR, Università Di Cagliari, Via Marengo 2, 09123 Cagliari, Italy
e-mail: campagna@unica.it

R. Floris
e-mail: roberta.floris@unica.it

P. Massa
e-mail: pmassa@unica.it

A. Girsheva · K. Ivanov
Università Di Cagliari, Via Marengo 2, 09123 Cagliari, Italy
e-mail: grisheva@unica.it

K. Ivanov
e-mail: konstantin.ivn@gmail.com

K. Ivanov
Tomsk Polytechnic University, 30 Lenin Avenue, Tomsk, Russia 634050

S. Geertman et al. (eds.), *Planning Support Systems and Smart Cities*,
Lecture Notes in Geoinformation and Cartography,
DOI 10.1007/978-3-319-18368-8_3

1 Introduction

Since last decade, advances in the Information and Communication Technologies (ICT), the Internet, and more recently, in the Web 2.0 technologies are increasingly channeling digital Geographic Information (GI) into daily life of a growing number of users. This wealth of GI may foster notable innovations in spatial planning methodologies and practices, for the majority of information required to support analysis, design, and decision is inherently spatial in nature. However, this hypothesis should be carefully tested and much work is still needed to develop methods and tools capable to offer planners reliable and user-friendly methods and tools.

Since the late 1990s, developments in Spatial Data Infrastructures (SDI) granted the access to digital data, produced and maintained by public or private organizations for institutional or business purposes. In Europe, the Directive 2007/02/CE establishing a shared INfrastructure for SPatial InfoRmation (INSPIRE), is leading to the development of National and Regional SDIs in the Member States according to common data, technology, and shared standards, so enabling public access and reuse of available official spatial data, or Authoritative Geographic Information (A-GI). The term 'authoritative' refers to spatial data produced by experts, professionals, organizations, and mapping agencies for a mission under institutional or legal frameworks (Ball 2010; Goodchild and Glennon 2010). The production of A-GI by highly trained experts complies with specific requirements and quality assurance procedures, in order to guarantee accuracy and quality standards (Goodchild and Glennon, ibidem; Elwood et al. 2012). Furthermore, the authoritativeness of A-GI is assured by metadata, which describe content, quality, accuracy, authorship, conditions of use and other characteristics of this information (Nogueras-Iso et al. 2004).

At the same time, many platforms continue to flourish online thanks to continuous advances in Web 2.0 technologies, which enabled the production, collection, and diffusion of user-generated contents (Krumm et al. 2008), wherein the community plays a more fundamental role in data production (Bruns 2006). Hence the popularity of the term Volunteered Geographic Information (VGI) (Goodchild 2007), which refers to the user-generated contents with a geospatial component created by citizens acting as volunteer sensors. The concept of VGI encompasses a wide range of activities and practices, which may provide pluralist sources of both experiential knowledge from local communities and expert knowledge from professionals, generating unprecedented opportunities for enhancing democratic decision-making in spatial planning processes. In several countries worldwide, the use of VGI has been proven useful in many application domains such as emergency management (Zook et al. 2010), environmental monitoring, spatial planning (Poser and Dransch 2010), crisis management (Roche et al. 2013), as well as participatory processes within Citizen Science initiatives (Haklay 2013).

In addition, widespread popularity of social media platforms is fostering the diffusion of geo-referenced multimedia (Sui et al. 2013), or Social Media

Geographic Information (SMGI). The latter information sources may be easily accessed and shared by users, which seamlessly become producers and consumers of personal geo-referenced contents on location-aware social networks. SMGI represents a deviation from an early vision of VGI, inasmuch dissemination of geographic information is not the final purpose of production (Stefanidis et al. 2013). As a matter of fact, SMGI for its nature may be classified as implicit VGI, in contrast with explicit VGI, whose main purpose is the diffusion of geographical contents (Craglia et al. 2012).

Despite this distinction, SMGI could lead toward innovative scenarios for gathering and disseminating geographic information among million of users worldwide, eventually providing valuable insights about user perceptions or needs, opinions on places, daily-routine events, so helping to get better insights on local identities (Campagna 2014), flows of information, and social networking within societies (Stefanidis et al., ibidem).

However SMGI, unlike A-GI, due to its peculiar user-generated mode of production features Big Data characteristics (Caverlee 2010) and, consequently, traditional spatial analysis methodologies and techniques may be not fully adequate to take advantage of the enclosed knowledge potential. A possible way to address this challenge is given by computational social science, a new emerging discipline aiming at finding new methods and tools to tackle the complexity of Big Data management issues (Lazer et al. 2009). The management of geographic Big Data, their integration with A-GI, and the use of advanced analytics may enable the extraction of relevant knowledge to support decision-making in diverse fields including spatial planning and design. This approach might also inform smart city initiatives by supplying real-time dynamic pluralist knowledge on people's perception of places.

In the light of these premises, the authors present a critical review of their research findings on the integrated use of A-GI, VGI, and SMGI in the domain of spatial planning. The remainder of the chapter is organized as follows. In the next section, a brief discussion about the main components of smart city initiatives and the way such strategies could be affected by SMGI is given. In Sect. 3 the authors introduce a novel approach to SMGI analytics, proposing its application in three different case studies. Finally, Sect. 4 draws conclusions discussing the results and the relevance of SMGI for spatial planning.

2 Digital Geographic Information for Smart City Strategies

The wealth of digital geographic information about facts, opinions and concerns of users, made available by the Internet and Web 2.0 technologies, could affect current practices in spatial planning and smart city strategies offering the possibility for real-time monitoring of the needs and aspirations of local communities. Nowadays, the label 'smart city' identifies several strategies for dealing with problems generated by rapid urbanization and population growth in cities. In literature, several definitions of smart city can be found, providing a wide set of components that should be considered for the success of such strategies.

The Internet and Web 2.0 technologies play a central role to deal with several societal challenges, such as urban welfare, societal participation, environmental sustainability, and quality of life (Schaffers et al. 2010). Likewise, Information and Communication Technology (ICT) should be considered fundamental to integrate, connect, and make efficient the global system of infrastructures and services (Washburn and Sindhu 2010), or to improve livability and sustainability in the urban systems (Toppeta 2010). Technology is also fundamental to make 'smart cities' a source of spatial enablement for citizens, in order to improve access, sharing, and integration of spatial data with services (Roche et al. 2012). At the same time, technologies should allow innovative forms of communication, governance, and organization for the community engagement in evaluating and solving urban key problems (Batty et al. 2012). Therefore, several factors, such as governance, policies, and the community, enclosed in the political dimension, may substantially contribute to 'smart' development.

More specifically, local communities may play a critical role in the development of smart cities, due to the fact that these strategies directly affect the quality of citizens' life; hence, their needs and opinions play a major role for the transparency of such strategies. Therefore, SMGI may represent a valuable source of information regarding opinions, needs and perceptions of local communities, which could be used to inform 'smart city' strategies. However, the current lack of shared and reliable user-friendly methods for analysis and knowledge mining from SMGI could prevent to exploit the full potential from these sources. In order to address this issue, in the next section a novel approach for SMGI analytics is introduced and then discussed in the light of the results of three different case studies.

3 A Novel Approach to SMGI Analytics

The increasing SMGI production and availability over the web is paving the way to innovative analysis scenarios in spatial planning and geodesign, which in turn could be used to increase smart city performances. As introduced earlier, the integration of SMGI with A-GI may offer potentially boundless and affordable sources of information regarding not only geographic facts, but also perceptions, opinions, and feelings of local communities in space and time, or in other words it may help to depict the local identity or the genius-loci.

However, there is a lack of a shared analytical framework to collect, manage, and process this information for different purposes. In the next paragraphs, the authors present the findings of three projects which explore and formalize a novel approach to SMGI analytics, based upon the spatial, temporal, and textual analysis of SMGI, or STTx analytics. The first case study deals with the use of the geographic social media platform Place, I care! which enables the users to publish and interact with SMGI in a geo-browser, wherein the working space is an interactive map. Secondly, an original user-friendly tool for the collection and analysis of SMGI from social media is presented. Lastly, a wider case study demonstrates the role of SMGI in tourism planning.

3.1 Place, I Care! Crowdsourcing the Sense of Place and Supporting Community Dialogue

The first experience of the authors concerning VGI was developed back in 2011 in order to investigate how the opportunities generated by the innovation in social processes granted by new technologies could bring innovation in spatial planning and decision-making. After few attempts, trial and errors pilots using existing free tools, a brand new web application, called "Place, I care!" (PIC!), was built by the authors in order to fulfill the emerging requirements in a novel and integrated way. The results of the first pilots using PIC! were successful for framing a new SMGI analytics and developing the methods and tools presented in the following sections.

PIC! was originally designed as a geographic social networking platform to be used in urban and regional planning processes. The idea was to create a VGI planning support tool for collecting information from concerned citizens about the physical, environmental, and socio-cultural space, and for supporting the dialogues about urban and environmental issues, in a collaborative and participatory manner.

PIC! 1.0 is a web application which enables each user to easily create its own VGI projects in few steps, offering a flexibility that allows implementing a number of use-cases to fit a variety of working settings. While other similar platforms and applications were available on the web at the time of PIC design, no other was found which would allow creating as easily and flexibly many different use-cases not only to express and describe individual issues of concern or appreciation, but also to improve the possibility for collaborative discussion. This result was achieved integrating geo-web tools within a social network paradigm. Unlike in other geobrowsers and web mapping tools, in PIC! each post can be "Liked/Disliked" and commented by other users supporting the discussion.

Data collected with PIC! in several pilots enabled to understand "what and where" the main interests and concerns of the participants were. The more a post got comments or Like/Dislike, the more the issue was considered hot by the interacting community. If this is a common feature for a social network, the novelty here is the discussion is georeferenced and visualized in the map. As the number of post grows to tenths or hundreds it would be always possible to visualize what issues were important to the community, when, and where they are located, and to investigate patterns in the developing discourse through spatial analysis and statistics functions.

In summary, PIC! 1.0 was designed aiming at combining two main requirements: ease of use and availability to all, and robustness, which other geobrowsers could not guarantee to the final user. Hence, the main features of PIC! are the followings:

- project creation in few clicks
- flexible profiles and permissions management
- user-friendly multimedia posting
- like/dislike-ing and commenting
- advanced post query
- customizable layer management
- data export for the final user (to be implemented).

The implementation of the first PIC! pilots involving small groups between 60 and 100 participants produced only a limited amount of data (i.e. a few thousand georeferenced multimedia posts including point, line and polygon placemarks), but it immediately showed a great potential for discovering a new analytics which may be applied to much bigger data volumes. In Fig. 1, an example of a multimedia post with comments is shown in the PIC! interface.

From an analytical perspective, the integration of GIS and Social network data models allows one to discover more useful hints than would either A-GI or SMGI alone because (1) it integrates opinions and perceptions about facts and (2) it integrates the spatial and the thematic dimensions with time, multimedia (i.e. text, images, audio, and video), and the user, whose behavior eventually becomes a new dimension of analysis.

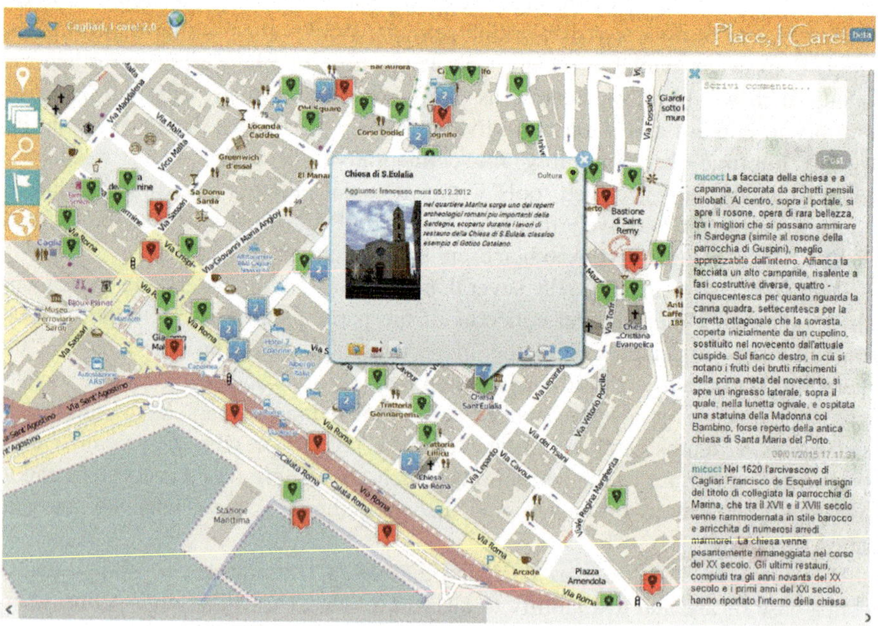

Fig. 1 Example of multimedia post in PIC!

The first exploratory analyses soon led to framing integrated analytics which, thanks to the specificity of the SMGI data model, includes:

- spatial analysis of user interests
- temporal analysis of users interests
- spatial statistics on user preferences
- multimedia content analysis on texts, images, audios, or videos
- user behavioral analysis
- combination of two or more of the previous such as the Spatio-Temporal-Textual Analysis (STTx) which enables to elicit what people discuss in space and time.

The analytic framework above shaped the design of the tools presented in Sect. 3.2, as well as, the study presented in Sect. 3.3.

3.2 SMGI: Spatext Tool

The early experiences with SMGI analytics informed the development of an original user-friendly tool, called Spatext, which enables one to extract information from multiple social media platforms and to seamlessly integrate it in a GIS environment for analysis. The tool provides several SMGI analysis methods including Spatial, Temporal and Textual (STTx) analysis. This SMGI Analytics suite is implemented as Python 2.7 add-in for ESRI ArcGIS©, offering a number of tools, which can be used mainly to (1) retrieve social media data from social media (including so far Twitter, YouTube, Wikimapia, and Instagram); (2) to geocode or georeference data; to carry on integrated (3) spatial, (4) temporal and (5) textual analyses. The number of analytical methods available in the tool is steadily increasing to include spatial-temporal clustering, in order to achieve user profiling, user movement analysis, and land use detection. Beside the desktop tool, Spatext analytical methods can also be used as Web Processing Services (WPS), in order to enable SMGI analytics via web interfaces. In Table 1 the analytical tools supplied by Spatext are classified by analytical functions.

The SMGI analytics application by Spatext is discussed through a number of examples conducted at the regional and local scale, which investigate both local communities' perceptions on relevant topics for spatial planning and the geography of places.

The first example concerns the analysis of georeferenced YouTube video metadata related to the term 'landscape' in Sardinia (Italy). The analysis of metadata spatial patterns shows that the term landscape in some Provinces (i.e. Cagliari and Olbia Tempio) is related to the coastal areas while in some other (i.e. Nuoro) concerns inner mountain areas, so expressing different local vocations. Similarly, in a second example, the spatial and textual analysis of video metadata contents in Cagliari (Italy) enabled the investigation of differences in perception regarding a number of neighborhoods by YouTube users. The results of the STTx define a clear

Table 1 Set of available tools in the Spatext suite

Spatext suite tools			
Harvesting SMGI from social media			
Tool name	Function	Textual query	Spatial query
Twitter extractor	Twitter SMGI extraction to table	Yes	No
YouTube extractor	YouTube SMGI extraction to feature class	Yes	Yes
Instagram extractor	Instagram SMGI extraction to feature class	No	Yes
Wikimapia extractor	Wikimapia SMGI extraction to feature class	Yes	Yes
Geocoding SMGI			
Tool name	Function	Manual	Automatic
Geocode address	Geocoding place/address from string	Yes	No
Geocode table	Batch-geocoding place/address from table	No	Yes
Georeferencing	Georeferencing SMGI coordinates by extractor (YouTube, Instagram, WikiMapia)	No	Yes
Textual analysis of SMGI			
Tool name	Function		
Attribute to string	Creation of a text file from an attribute field in a SMGI feature class for textual analysis		
Attribute to table	Creation of a table from an attribute field in a SMGI feature class for textual analysis		
Attribute to tag-cloud	Tag-clouding analysis from an attribute field in a SMGI feature class		
Selection to tag-cloud	Tag-clouding analysis from a spatial/attribute selection in a SMGI feature class		
Text to tag-cloud	Tag-clouding analysis from a text file		
Temporal analysis of SMGI			
Tool name	Function		
Identify month/weekday/day/hour	Add the month/weekday/day/hour of creation in a new field of SMGI feature class		
Trend day/hour	Creation of a 24 h/60 min time graph and statistic report from SMGI feature class		
Spatial analysis of SMGI			
Tool name	Function		
Decompose feature	Creation of a new feature class for each group in SMGI feature class (e.g. User)		
DBScan (density-based scan)	Run DBScan algorithm Ester et al. (1996) on SMGI feature class to detect density clusters and add the cluster group in a new field of SMGI feature class		
Feature-based DBScan	Run DDBscan algorithm Ester et al. (1996) on SMGI feature class to detect density clusters for each group in SMGI feature class (e.g. User)		

(continued)

Table 1 (continued)

Spatial analysis of SMGI	
Tool name	Function
Advanced SQL maximum	SQL selection on a SMGI feature class to detect maximum values for a specific group (e.g. User cluster exposing max number of points)
Google™ static maps	Add the Google static map URL in a new field of SMGI feature class

picture of the overall perception of the users for each neighborhood at the local scale: the Castello neighborhood was depicted as the historical one, while the Marina as the retail and leisure one. As a matter of facts, both are historical neighborhoods, but the functional characterization clearly emerged in the analysis of the users perceptions. The results of the spatial analysis and the textual analysis are displayed in Fig. 2 and Table 2, respectively.

More interestingly, while it may be argued that this study on perception would be representative of the YouTube user community only, the same results were confirmed by another pilot study named Cagliari, I Care! 2.0, carried on using PIC! with a completely different pool of participants. In spite of different datasets, SMGI origin, involved users, and time period, the analyses gave the same results for the examined neighborhoods, raising interesting questions about SMGI reliability and community representativeness for analysis, which should be further investigated with new ad hoc studies.

In the last example of SMGI analytics with Spatext, the geography of Iglesias municipality (Italy) was investigated through the spatial and temporal distribution of Instagram photos. The purpose of the analysis was the identification and classification of SMGI clusters, relying on spatial and temporal component of users'

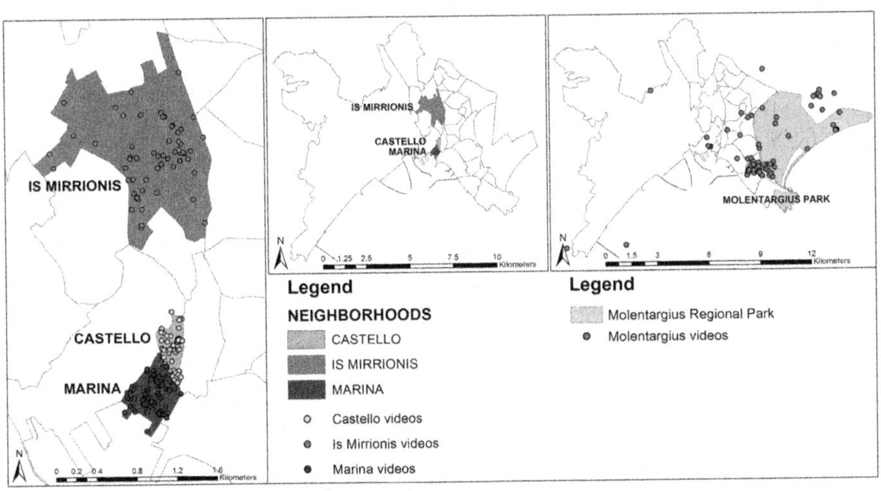

Fig. 2 Spatial distribution of videos related to Cagliari's neighborhoods

Table 2 Interpretation of major interests and concerns related to Cagliari's neighborhoods

Textual analysis of video title and description	
Neighborhoods	Keywords (frequency) translated
Castello	*Cagliari* (48), Urban (10), Safe (10), Concert (6), *Marmora* (6), Palace (5), Royal (5), Historical (4), Cathedral(2), Bastion (2)
Is Mirrionis	*Cagliari* (40), Tournament (10), *Monteclaro* (7), *CUS* (6), Soccer (4), Football Club (4), Sound (4), Final (4), Music (3), Park (2)
Marina	*Cagliari* (61), *Santa Lucia* (13), Concert (9), New Year's Day (6), Celebration (6), *San Sepolcro* (5), Music (5), Festival (4), Harbor (4), Church (4)
Molentargius natural park	*Cagliari* (62), *Quartu Sant'Elena* (15), Salt mine (10), Flamingo (8), Park (8), *Poetto* (5), Conference (5), Service (5), Pond (4), *Monte Urpinu* (3)

contributions, in order to detect buildings not appearing in official datasets. A one-year sample of approximately 14,000 photos from 1243 anonymized users was collected and georeferenced with Spatext, and analyzed in space and time. Spatext integrates the DBScan (Ester et al. 1996) and a Feature-based DBScan which were used to detect 290 clusters and 368 clusters, respectively. The difference between the results may be explained considering how the first tool relies exclusively on points' spatial density, while the second tool adds the grouping factor to the analysis, thus performing a separate DBScan analysis for points of each user in the SMGI dataset. Then, the Instagram dataset was integrated with the latest official buildings dataset available in the Regional Spatial DI. For each user's cluster the overlaps of the cluster centroid, of the photos barycenter, and of the shape with the buildings' footprint were evaluated. The analysis resulted in 113 clusters without any overlap, which were then visually assessed through the satellite photo viewer integrated in Spatext, in order to find un-mapped buildings in official information. An example of the results of the analysis is provided in Fig. 3, showing five different clusters (i.e. A, B, C, D and E), their centroids, the existing buildings footprints from the official dataset, and the Instagram point dataset. In the example, the manual investigation through the Google Map's satellite image allowed to detect two buildings which are not mapped in official dataset, identified by the cluster A and clusters C and D respectively. At the same time the analysis in the example confirmed the building in cluster B and allowed to detect a public space (cluster E), which is used by local residents for leisure. This example demonstrated how Instagram data may be used to elicit information related to topography of places and it may potentially integrate A-GI.

Altogether, these examples contribute to demonstrate how SMGI can be used to elicit information, not only about the physical geography of places to integrate existing A-GI, but also to express the perceptions of places and issues in time and spaces by the involved community, which may add a pluralist perspective of great relevance for spatial planning and decision-making.

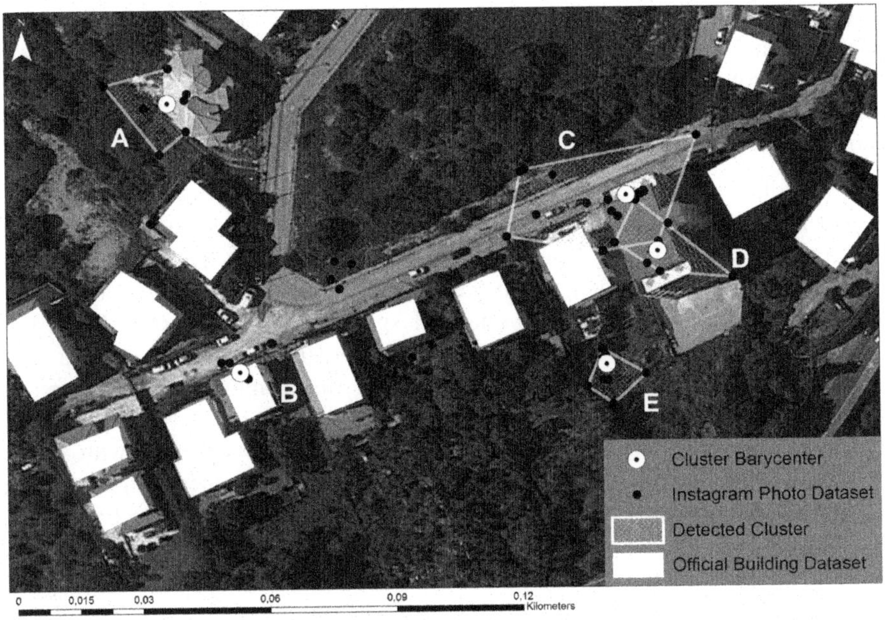

Fig. 3 Results of manual investigation on SMGI. Identification of buildings not mapped in the official dataset

3.3 The Use of SMGI in Tourism Planning

A wider case study was also developed in order to investigate the potential of the use of SMGI in tourism planning. The aim was to spatially analyze tourist preferences on (1) destinations and (2) tourism industry services, using tourists' ratings and comments collected by two major tourism social networks (i.e. Tripadvisor.com and Booking.com). Overall the study explores the following questions related to Sardinia's costumers' preferences:

- Which are the most popular destinations and why?
- What does attract tourists' attention and what do tourists appreciate/disregard?
- Why tourists choose those destinations?
- How this information can be used as tourism planning support?

From an operational perspective, the challenge was answering these questions relying on both data from the local SDI and from tourism SMGI, in order to discover novel and useful knowledge.

First of all, at the regional level, spatial analysis and statistics techniques for investigating the spatial distribution of tourists' preferences were carried on, identifying clusters of positive or negative preferences or hotspots of interest by tourist profiles. Then at the local level, large scale statistical and STTx analyses were carried on aiming at understanding, qualitatively, the reasons beneath patterns and singularities. Finally, a properly spatially calibrated model, analyzing spatial non-stationarity (Fotheringham et al. 2003), was implemented in order to express the impact of spatial variation in the relationships among dependent variable (tourists' preferences) and explanatory variables. Operationally, the study was carried on according to the following workflow:

- data collection and geocoding: data were extracted by TripAdvisor.com and Booking.com, geocoded, and integrated in a geodatabase including a one-year full set of data about 992 Tourism Lodging Services (TLS). The dataset includes TLS name, location, category, as well as quantitative and textual evaluations
- regional preferences dynamics analysis: spatial analysis and spatial statistics techniques, as well as STTx, were applied in order to detect and analyze preference clusters and hot/cold spots in Sardinia, using municipalities as unit of analysis
- local preferences dynamics analysis: data were further combined with spatial data themes from the regional SDI in order to earn deeper insights on the relationships among tourist preferences, local territorial features, and quality of industry services in selected destinations
- geographically weighted regression analysis: a model was developed in order to investigate how the detected patterns varied across different census tracts.

The last three steps are carried on iteratively on the relevant clusters and spots.

Spatial patterns of the TLS typology, together with their reviews analysis, offered interesting clues to characterize different destinations for tourism planning purposes. As an example, Cagliari TLS supply is characterized by the success of B&B, while for Alghero and Olbia, other two coastal city destinations, hotels and resorts are more popular among travelers. Analyzing tourists' nationality demonstrates also that tourists from Spain, Norway, and Ireland privileged the macro area of Alghero, while tourists from France and Germany preferred Cagliari, not surprisingly due to the majority of fly connections to and from these countries. In addition, Russian and Eastern European tourist showed a substantial preference for the Cagliari area.

Analyzing tourists' comments and evaluations enables investigation of the satisfaction level with destination and services. According to the results of the comments analysis, most words in the posts refer to spatial features and tourism structures, such as locations, hotel, or beaches. Frequent words also refer to the level of satisfaction with both destination leisure services and local traditions. Overall from the analysis, the main reason for tourists to visit Sardinia seems to be related to both its natural attractions and the presence of a unique cultural heritage.

The tourism preferences analyses in space were applied to investigate the preferences' patterns on both the territorial resources and the tourism industry features at the local level, across the whole region. The analysis started by mapping the Tourist Positive Preferences Incidence (TPPI) as the ratio between the positive scores and the number of TLS by municipality. The result shows an overall high and diffused concentration of preferences in the North-East coast. The Costa Smeralda area appears as the only area in Sardinia where tourist preferences overall fulfill visitors expectations. However, the analysis shows that the Alghero and the Cagliari areas also expose high TPPI rates.

After the spatial patterns analysis at the regional scale, the analyses shift to the local scale relying on spatial analysis and spatial statistics techniques on an integrated SMGI/A-GI geodatabase with the aim of investigating the local success factors within the single destinations. Alghero has been recognized as a best-selling destination from different tourists' typologies. Thus, the following question to answer was "What exactly does attract the tourists attention in Alghero and why?" Focusing on the local scale, the historic city center of Alghero clearly represents the major hotspot of tourists' attraction, while the modern residential districts in the outskirts represent a cold spot (Fig. 4).

The STTx analysis on tourists textual comments enables to understand both what tourists think and where. More than 880.000 long textual reviews were extracted (of which 1050 in English) related to Alghero. Table 3 indicates how the majority of the words in the posts refer to spatial or physical features, such as "location", "beaches" and "city center". Other frequent words are related to tourism structures, such as "hotel" and "staff". The textual analysis results also indicate a high satisfaction level with the destination leisure services. According to the results, the main reason for tourists to visit Alghero seems to be related to both its natural attractions, which include landscape features, such as beaches, and the presence of a unique cultural heritage. These facts generate a positive tourism destination image, which is the most influential psychological factor when tourists decide where to travel (Van Raaij 1986; Buhalis 2000).

Lastly, the study was also supported by the integration of SMGI with other A-GI describing topography, transport infrastructures, cultural heritage sites, and socio-economic features. The spatial relationships and the explanatory factors behind observed spatial patterns were modeled using the Geographic Weighted Regression analysis (GWR) (Fotheringham et al. ibidem , p. 9). The GWR is a technique which extends ordinary linear regression models by taking into account local variations in rates of changes.

In this case, the objective of the GWR analysis is twofold: the analysis is performed to investigate quantitatively why visitors' preferences are mostly located in the Alghero city center rather than in other locations in the municipality, and to discover which factors contribute to the Alghero high TPPI rate in those areas. The model was applied to a sample of 131 TLS distributed over 89 of the 471 census tracts in Alghero.

In this model, the dependent variable (local TPPI) was calculated as the sum of the positive score of the TLS normalized by the total number of comments per census tract. For each city census tract, the measure of each independent variable was calculated and normalized by the total area of the census tract. Preliminary results of the statistical tests suggested to exclude some of the explanatory variables originally chosen for the model, because they were not statistically significant. Eventually, the following candidate variables normalized by total area of census tract were included:

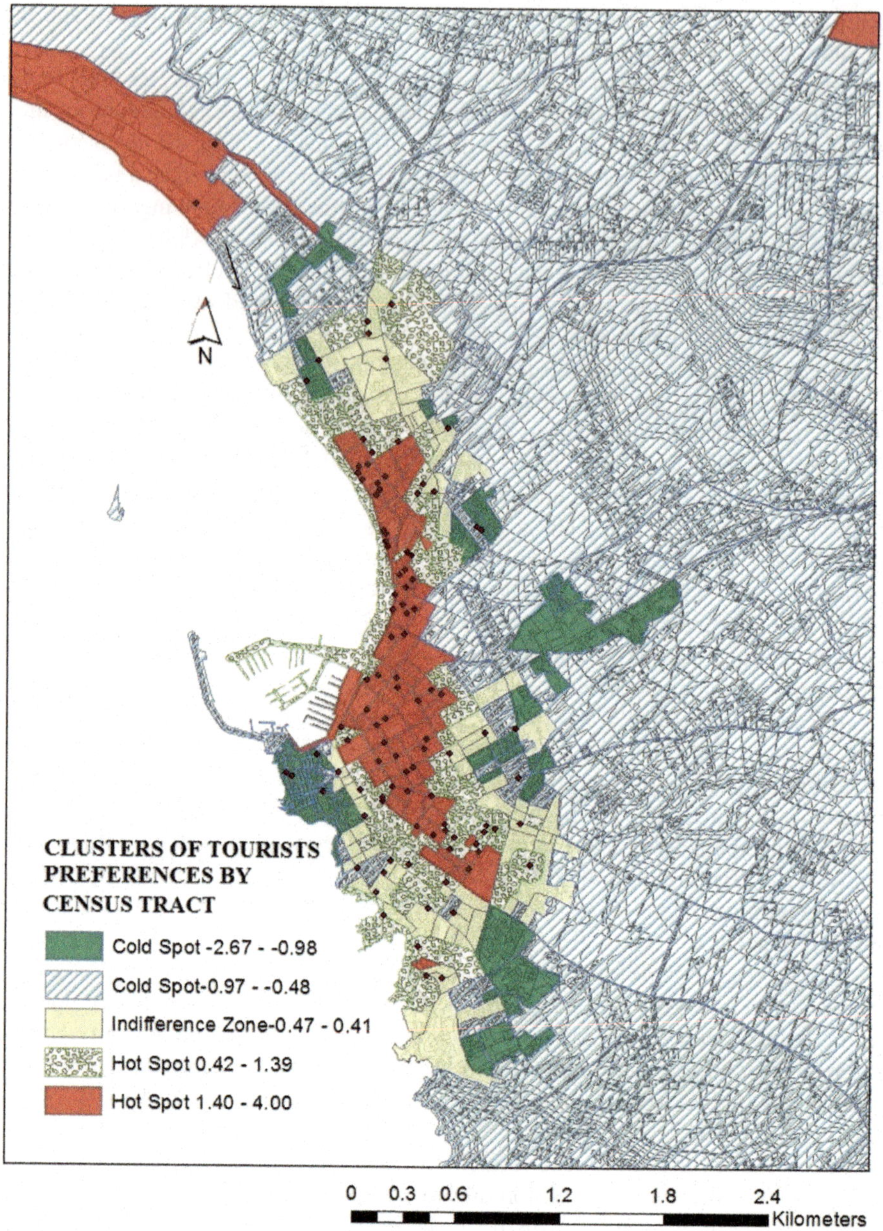

Fig. 4 Results of SMGI analysis: significant patterns of tourist positive preferences incidence in Alghero municipality by census tracts

Table 3 Results of the textual analysis: top 15 most used words in the Alghero cluster divided by category

Category	Words (frequency)
Geographic location	Location (1010), town (476)
Services	Staff (890), restaurant (643), room (459), hotel (469), pool (230), food (180)
Accessibility	Mapped (250), harbor (237), proximity (164), walking (146)
Natural and no natural components	City center (426), beach (378), old city (132)

1. number of historical buildings;
2. number of restaurants and facilities;
3. hectares of natural protected areas;
4. distance from the main transport nodes;
5. proximity to the historical city center;
6. distance from the most popular beach.

The assumption is that if the value of the TPPI at the local level is similar to the values that it takes in the closest spatial units, the variable is characterized by spatial autocorrelation. This issue can be addressed by adding a spatially-lagged dependent variable to the set of covariates (Anselin 1988, Anselin et al. 1996).

The presence of autocorrelation of the dependent variable (i.e. the normalized local TPPI) of a model is detected through the Moran's test. The local Moran's Index for the second order of queen contiguity was 0.06636, which is quite meaningful; the p-value was 0.0000146 ($0.01 \leq p < 0.05$) hence statistically significant; the z-score featured positive sign, meaning that local spatial autocorrelation in the dependent variable is higher than it would occur randomly (Table 4). Thus, the regression is estimated using the six dependent variables and the tourists' preferences weight matrix obtained by the Moran's test.

In order to better understand the local variation of the explanatory variables, the coefficient raster surfaces created by GWR in ArcGIS were taken into account. The analysis allows to investigate how spatially consistent relationships between the dependent variable and each explanatory variable are across the study area. In addition, the analysis of the coefficient distribution as a surface shows where and how much variation exists. Figure 5 shows the residual values of the model, obtained by the differences between the fitted and the observed values of the dependent variable.

The map of the residual values is classified with the standard deviation method. As it is possible to note, for the majority of tracts, the values of the standardized

Table 4 Summary output of the Moran's index

R squared	Constant a	Std err a	t-stat a	p-value b	slope b	Std err b	t-stat b	p-value b
0.0393	0.0106	0.0151	0.703	0.482	0.0664	0.0151	4.38	0.0000146

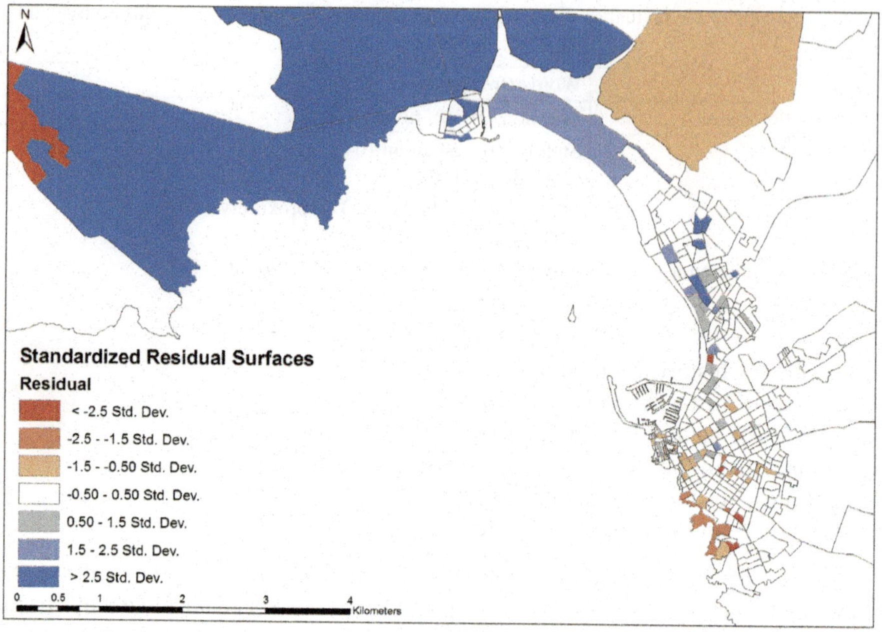

Fig. 5 GWR results: standardized residual surfaces

residuals are in a range between −1.5 and 1.5; few census parcels show standardized residual values higher than 1.5 or smaller than −1.5. Not surprisingly, many of the census tracts are placed in the city center, where the presence of historical buildings coupled with typical restaurants and leisure is more significant.

The results concerning the goodness of fit of the regression are significant: R-squared is very high, 0.943992, which indicates that the GWR model explains more than a 94 % of the variance of the positive tourists' preferences at the local level. The GWR model coefficients of the variables show the relationships between the dependent variable and each explanatory variable. The coefficients of the variables related to location are almost always significant (with p values less than 5 %) and show positive sign. The variables "hectares of natural protected areas" and "number of natural sites" are not significant, for the p value is higher than 10 %, while the variable "number of restaurants", related to service quality, shows a significant coefficient (with p value = 0.0167321) and positive sign. Table 5 shows the coefficient value for each considered variable.

Overall, these findings suggest that the spatial interest of the participant is quantitatively influenced by the chosen explanatory variables. The selected variables give a more or less significant contribution to the explanation of the tourists' preferences through their coefficient. In the model, the inclusion of the only tracts where the TLS are present, allows saying that the values of the coefficients in the tracts reflect the positive effects of the geographic position and the presence of facilities.

Table 5 Results of the GWR model: influence of each explanatory variable on dependent variable (normalized tourist's preference)

Variable	Coefficient	Std: Error	z-value	p-value
W_N_TPPI	0.058552	0.035614	1.644075	0.100161
CONSTANT	−0.012852	0.008821	−1.457048	0.145103
N_H_BUILD	0.753521	0.076918	9.796376	0.000000
DIST_BEACH	1.015170	0.100169	10.134590	0.000000
N_RESTAUR	0.089016	0.037206	2.392542	0.016732
H_NATURAL	0.005126	0.032138	0.159486	0.873286
D_AIRPORT	0.448991	0.126307	3.554754	0.000378
PROX_C_CENTRE	−0.005772	0.010428	−0.005536	0.005799

The results provide insights on the tourism preferences dynamics in Alghero, which would have been not possible to obtain through more traditional data sources for tourism planning. In addition, these findings confirm that the success of tourist destination is closely dependent not only on the quality of the tourist industry offer, but also on the territorial setting of the destinations, including the natural, cultural, and the physical character of the places, as well as the infrastructure and services. So far, the literature on TLS distribution dealt with several relevant sustainability issues, however often the spatial dimension of tourists' subjective perception was neglected. Including the latter dimension may open new opportunities for planners and offer new research challenges for a pluralist customer-oriented view on strategic tourism development.

4 Conclusion

The earlier results of this study offer an overview of possible uses of active and passive SMGI platforms to investigate what people observe, evaluate, and how they behave in space and time. The underlying endeavor of the study was to develop novel SMGI analytics methods and tools, which may help to access data sources and extract meaningful knowledge for spatial planning.

A number of case studies are discussed which demonstrate how SMGI, from active and passive sources, may be integrated with A-GI and used to understand people perceptions, contributing to define a pluralist model of local identity. In addition, several examples demonstrated how SMGI extracted from popular social media platforms may be used to detect changes in topography, as in the case of Instagram images, as well as, social and economic processes in the case of tourism planning. In the Sardinia tourism case study the underlying assumption was that the same methods and tools can be used successfully in urban and regional planning as much as in tourist planning, for in both cases they contribute to take into account a pluralist view on strategic development issues.

Overall the case studies show how qualitative and quantitative analysis can be applied to SMGI using spatial analysis and statistics combined with STTx techniques, which can be used to verify hypothesis unleashing the knowledge enclosed in the huge amount of qualitative descriptive social media comments.

Moreover, active and passive SMGI platforms may enable such scenarios where a city planner would aim to "listen" to what the local community feels about community issues or to interact with them to ask what alternative projects, or development options, would be welcome to the community. This would be a possible common use-case in the tradition of Public Participation/Participatory GIS (PPGIS) domain. However, until recently, PPGIS initiative required substantial endeavor in order to set-up technology and management. The availability of geographic social network platforms may ease the process both in technology and social terms. As a matter of facts, while on the one hand almost no technology set-up is needed, on the other hand the social network functions require less commitment by the potential participants, which increasingly use the media in their daily lives. The participation to the initiatives would blur with the everyday social networking activity. Accordingly, we can use such platforms as PIC! to create dynamic dialogues and monitor people interests about places on a routine base, or launch time-limited initiatives to ask the assessment of development alternatives.

A first conclusion may be that, as for the subject of observation (i.e. the features for places), the tools required to collect and extract knowledge are very local in nature. Globalization may occur in the future to this respect, but for the time being both A-GI and VGI sources are differently available in diverse places, contributing to define the local identity even in the knowledge communication media.

What appeared clear from the development, as well as from the usage of Spatext, is that any SMGI source is characterized by a specific data model and by different rates in geographic diffusion. Therefore different combinations of analytical approaches are required in order for the results to interpret the local context appropriately.

Several issues are still unresolved and need further investigation such as those related to data representativeness. Nevertheless, the presented case studies show how, in certain cases, similar images for a given place were depicted by different groups of users with different profiles, exposing invariance in the perception of places. While this issue should be further investigated, the potential of these new data sources already far exceeds those of more traditional community inquiring methods. To what extent SMGI may contribute to enhance the quality of knowledge, and eventually bring innovation to spatial planning, is still to be verified, but this question represents a very stimulating issue for further research.

Acknowledgments The authors wish to thank the anonymous reviewers and the editors for the valuable comments on the early version of this chapter. The work presented in this chapter was developed by the authors within the research project "Efficacia ed efficienza della governance paesaggistica e territoriale in Sardegna: il ruolo della VAS e delle IDT" [Efficacy and efficiency of landscape and environmental management in Sardinia: the role of SEA and of SDI] CUP: J81J11001420007 funded by the Autonomous Region of Sardinia under the Regional Law no. 7/2007 "Promozione della ricerca scientifica e dell'innovazione tecnologica in Sardegna".

References

Anselin, L. (1988). *Spatial econometrics: Methods and models* (Vol. 4). Berlin: Springer.

Anselin, L., Bera, A. K., Florax, R., & Yoon, M. J. (1996). Simple diagnostic tests for spatial dependence. *Regional Science and Urban Economics, 26*(1), 77–104.

Ball, M. (2010). What's the distinction between crowdsourcing, volunteered geographic information and authoritative data? http://www.sensysmag.com/spatialsustain/whats-the-distinction-between-crowdsourcing-volunteered-geographic-information-and-authoritative-data.html. Accessed Jan 15, 2015.

Batty, M., Axhausen, K. W., Giannotti, F., Pozdnoukhov, A., Bazzani, A., Wachowicz, M., et al. (2012). Smart cities of the future. *European Physical Journal-Special Topics, 214*(1), 481–518.

Bruns, A. (2006). Towards produsage: Futures for user-led content production. In F. Sudweeks, H. Hrachovec & C. Ess, (Eds.), *Cultural attitudes towards communication and technology 2006* (pp. 275–284). June 28–July 1, Tartu, Estonia.

Buhalis, D. (2000). Marketing the competitive destination of the future. *Tourism Management, 21* (1), 97–116.

Campagna, M. (2014). The geographic turn in social media: opportunities for spatial planning and geodesign. In *Computational science and its applications—ICCSA 2014* (pp. 598–610). Berlin: Springer.

Caverlee, J. (2010). A few thoughts on the computational perspective. In *Presented during the specialist meeting on spatio-temporal constraints on social networks, December 2010*. Santa Barbara: CA, USA.

Craglia, M., Ostermann, F., & Spinsanti, L. (2012). Digital earth from vision to practice: Making sense of citizen-generated content. *International Journal of Digital Earth, 5*(5), 398–416.

Elwood, S., Goodchild, M. F., & Sui, D. Z. (2012). Researching volunteered geographic information: Spatial data, geographic research, and new social practice. *Annals of the Association of American Geographers, 102*(3), 571–590. doi:10.1080/00045608.2011.595657.

Ester, M., Kriegel, H. P., Sander, J., & Xu, X. (1996). A density-based algorithm for discovering clusters in large spatial databases with noise. In *Kdd* (Vol. 96, pp. 226–231).

Fotheringham, S. A., Brunsdon, C., & Charlton, M. (2003). *Geographically weighted regression: The analysis of spatially varying relationships*. New York: Wiley.

Goodchild, M. F. (2007). Citizens as voluntary sensors: Spatial data infrastructure in the world of web 2.0. *International Journal of Spatial Data Infrastructures Research, 2*, 24–32.

Goodchild, M. F., & Glennon, J. A. (2010). Crowdsourcing geographic information for disaster response: A research frontier. *International Journal of Digital Earth, 3*(3), 231–241.

Haklay, M. (2013). Citizen science and volunteered geographic information: Overview and typology of participation. In *Crowdsourcing geographic knowledge* (pp. 105–122). Netherlands: Springer.

Krumm, J., Davies, N., & Narayanaswami, C. (2008). User-generated content. *IEEE Pervasive Computing, 7*(4), 10–11.

Lazer, D., Pentland, A., Adamic, L., Aral, S., Barabasi, A. L., Brewer, D., & Van Alstyne, M. (2009). Life in the network: The coming age of computational social science. *Science, 323* (5915), 721.

Nogueras-Iso, J., Zarazaga-Soria, F. J., Lacasta, J., Béjar, R., & Muro-Medrano, P. R. (2004). Metadata standard interoperability: Application in the geographic information domain. *Computers Environment and Urban Systems, 28*(6), 611–634.

Poser, K., & Dransch, D. (2010). Volunteered geographic information for disaster management with application to rapid flood damage estimation. *Geomatica, 64*(1), 89–98.

Roche, S., Nabian, N., Kloeckl, K., & Ratti, C. (2012). Are 'smart cities' smart enough? In *Global geospatial conference 2012*. German: Global Spatial Data Infrastructure Association.

Roche, S., Propeck-Zimmermann, E., & Mericskay, B. (2013). GeoWeb and crisis management: Issues and perspectives of volunteered geographic information. *GeoJournal, 78*(1), 21–40.

Schaffers, H., Guzmán, J. G., Navarro, M., & Merz, C. (2010). *Living labs for rural development: Results from the C@ R integrated project*. Madrid, Spain: TRAGSA.

Stefanidis, A., Crooks, A., & Radzikowski, J. (2013). Harvesting ambient geospatial information from social media feeds. *GeoJournal, 78*(2), 319–338.

Sui, D., Elwood, S., & Goodchild, M. (2013). Crowdsourcing geographic knowledge. Springer.

Toppeta, D. (2010). *The smart city vision: How innovation and ICT can build smart, "Liveable", sustainable cities, the innovation knowledge foundation*. http://www.inta-aivn.org/images/cc/Urbanism/background%20documents/Toppeta_Report_005_2010.pdf. Accessed Jan 21, 2015.

Van Raaij, W. (1986). Consumer research on tourism mental and behavioral constructs. *Annals of Tourism Research, 13*(1), 1–9.

Washburn, D., Sindhu, U. (2010). Helping CIOs understand "smart city" initiatives. Forrester Research, Inc. http://www.uwforum.org/upload/board/forrester_help_cios_smart_city.pdf. Accessed Apr 20,2015.

Zook, M., Graham, M., Shelton, T., & Gorman, S. (2010). Volunteered geographic information and crowdsourcing disaster relief: A case study of the Haitian earthquake. *World Medical and Health Policy, 2*(2), 7–33.

Chapter 4
Data and Analytics for Neighborhood Development: Smart Shrinkage Decision Modeling in Baltimore, Maryland

Michael P. Johnson, Justin B. Hollander and Eliza Davenport Whiteman

Abstract Many older cities in the United States confront the problem of long-term decline in population and economic activity resulting in blighted conditions that make conventional revitalization initiatives unlikely to succeed. Smart shrinkage, a planning approach that emphasizes alternative land uses while preserving quality of life, offers a way for cities to remain desirable places to live and work. However, there is little research on empirical methods to support planning decisions consistent with smart shrinkage. We present results from two studies with planners from the City of Baltimore that provide novel insights regarding ways in which planners can perform vacant property redevelopment using methods from data analytics and decision science. This study provides a foundation for practitioners to make better use of large volumes of data describing blighted communities, accommodate diverse attitudes about policy and planning responses to blight, and judiciously apply advanced methods in data analysis and decision models.

M.P. Johnson (✉)
Department of Public Policy and Public Affairs, University of Massachusetts Boston, 100 Morrissey Blvd, Boston, MA 02125, USA
e-mail: michael.johnson@umb.edu

J.B. Hollander
Department of Urban and Environmental Policy and Planning, Tufts University, 97 Talbot Avenue, Medford, MA 02155, USA
e-mail: Justin.Hollander@tufts.edu

E.D. Whiteman
Doctoral Program in City and Regional Planning, School of Design, University of Pennsylvania, 210 S 34th St, Philadelphia, PA 19104, USA
e-mail: elizadw@design.upenn.edu

S. Geertman et al. (eds.), *Planning Support Systems and Smart Cities*, Lecture Notes in Geoinformation and Cartography,
DOI 10.1007/978-3-319-18368-8_4

1 Introduction

1.1 Shrinking Cities and Smart Shrinkage

Many communities in the U.S. face sustained economic and social decline as a result of declining populations, increasing housing vacancies, and unsustainable costs for maintaining poor-quality infrastructure, which is used by fewer and fewer residents, businesses, and visitors. While innovative urban planning and community development initiatives may enable some of these distressed communities to enjoy economic stabilization and provide increased opportunities for residents to enjoy affordable, decent quality housing and improved safety, portions of other communities face exceedingly long odds for traditional revitalization.

Many American cities face shrinkage; most prominent of these are the post-industrial Rust Belt cities of Akron, OH, Buffalo, NY, Cleveland, OH, and Detroit, MI (Dewar and Thomas 2013). According to the U.S. Census, for example, the city of Detroit lost 25 % of its population from 2000 to 2010. However, some of the largest and most striking recent population losses can be found in Sun Belt locales that exploded in population in the 1990s and early 2000s, including cities such as Las Vegas, NV, Atlanta, GA, Modesto, CA, and Phoenix, AZ (Hollander et al. 2011).

No single rationale or explanation exists as to why a place depopulates. Depopulation has been explained by everything from natural disasters (Vale and Campanella 2005) to deindustrialization (Bluestone and Harrison 1982), suburbanization (Jackson 1985), globalization (Hall 1997) and, of course, the natural economic cycle of boom and bust (Rust 1975).

Local leaders and community activists face three choices when presented with population decline: public redevelopment, no action, or smart shrinkage. With public redevelopment, these agencies attempt to manipulate exogenous factors to encourage private investment to create new jobs and generate new demand for real estate. Communities may also simply do nothing. The third alternative, smart shrinkage, is defined by Popper and Popper (2002) as "planning for less—fewer people, fewer buildings, fewer land uses" (23). It is a way to accommodate population loss in a way that focuses on quality of life improvements in a neighborhood.

For communities facing long-term declines in population and economic activity, decreased levels of traditional investments, paired with increased investments in nontraditional land uses such as urban farming, recreation, public art and environmental remediation, may be the best use of public funds to ensure improved quality of life for a region's citizens. This new development paradigm has been the subject of scholarly inquiry (e.g. Popper and Popper 2002; Wiechmann 2008; Hollander et al. 2009; Hollander and Németh 2011) as well as popular and practitioner-focused reporting (Stohr 2004; Swope 2006; Tortorello 2011; Gallagher 2013).

1.2 Research Questions

This chapter is intended to apply the emerging theory and practice of smart shrinkage to a contemporary case, linking a pilot study on decision models for vacant property acquisition and re-use in Baltimore, MD (Johnson and Hollander 2014) to field research documenting decisions made by planning practitioners regarding specific collections of vacant parcels (Davenport Whiteman 2014). An extended examination of smart shrinkage with a focus on contemporary mixed-methods analytic approaches that includes the Baltimore case is contained in (Hollander et al. 2015). In this chapter we will address the following research questions: How do planning practitioners conceptualize the problem of smart shrinkage? In what ways do the decision processes they apply in practice diverge from the theory of decision modeling? How can planners make most appropriate use of information and decision sciences to develop sustainable and politically-feasible strategies for smart decline? Answers to these questions will enable practitioners to make better use of large volumes of data, to accommodate diverse attitudes of multiple stakeholders, and to apply appropriate methods in data analysis and decision models. As such, this chapter is intended to contribute to research and practice in urban modeling, urban analytics and urban planning and decision support that are foci of the conference for which this volume documents particular presentations.

1.3 Data Analytics and Decision Modeling

'Analytics' is a term generally understood to encompass a wide range of methods and techniques that enable professionals and researchers to design solutions to challenging problems in industry and policy that are typically represented through large volumes of data, often quantitative in nature (Liberatore and Luo 2010). In contrast to corporate firms, which may place primary interest on a fairly small set of metrics related to profitability that are straightforward to quantify and collect, government and nonprofit organizations are often interested in a larger set of metrics related to individual and community health and capacity, and economic and social progress. Some of these metrics may be difficult to operationalize. Moreover, limited technical resources may make it impossible for a not-for-profit organization to develop applications that may allow the analysis and sharing of data they desire (Wallace 2014; Stone and Cutcher-Gershenfeld 2002).

Decision modeling, also known as operations research and management science, is one component of data analytics. Decision modeling is the process of representing certain problems or social/physical phenomena in order to generate recommendations, which reflect the preferences of decisionmakers. The aim of the process is to achieve, to the greatest extent possible ('optimize'), one or more goals ('objectives') which embody principles or values that are important to the decisionmakers and/or the organization for which they work (Winston and Albright 2012). Decision modeling has a long history of applications to problems in housing, community development and urban and regional planning (Johnson 2011). Recent research has proposed an application of decision science, under the rubric of community-based operations research, to public-sector applications for distressed and underserved communities in such a way as to support local development and social change (Johnson 2012). Decision modeling has been previously applied to smart shrinkage to develop neighborhood-level investment strategies in order to balance concerns of social benefit and equity (Johnson et al. 2014).

1.4 Chapter Description

In the remainder of this chapter, we will first profile the city of Baltimore, emphasizing the characteristics that make it a promising candidate for the application of decision modeling for smart shrinkage. Next, we will describe the decision modeling strategy we used during our engagement with the City of Baltimore Department of Planning. Decision modeling results for a sample of distressed communities within a portion of Baltimore follow. We then describe a field study, concurrent with our engagement with the Department of Planning, to better understand the rationales provided by planners to justify selection decisions regarding specific vacant parcels. The results of this study, contrasted with the results of our decision modeling experiment, support a presentation of principles of decision modeling for smart decline.

2 Profile of Baltimore

The population of the city of Baltimore has declined by 34.6 % between 1950 and 2010, while the population of the Baltimore Metropolitan Statistical Area has increased by 115 % over the same period (U.S. Census Bureau 2015a). . According to the City, there are 30,000 vacant properties in the city today—16,000 of which are vacant buildings and 14,000 of which are vacant lots (Baltimore Department of Planning 2013). Of these vacant properties, 40 % are city owned. A large part of this decline and abandonment has been attributed to the loss of manufacturing jobs in the city, particularly from the closing of the Bethlehem Steel plant. Between 1950 and 1990 two thirds of the city's manufacturing jobs were lost (Friedman 2003). The city's population continued to decline even after the manufacturing industry stabilized, losing more than 115,000 people from 1990–2010.

There has been a long history of housing segregation in Baltimore, as in many other U.S. cities (Davenport Whiteman 2014). Today the city remains highly residentially segregated by race. Baltimore also has a higher concentration of vacant housing in both neighborhoods that are majority black and neighborhoods below the median income (Ransome 2007). Neighborhoods with a black population of greater than 50 % accounted for more than 80 % of vacancy notices in the city, while neighborhoods with higher than 50 % white populations accounted for less than 16 % of vacancy notices (Open Baltimore 2015; U.S. Census Bureau 2015b).

The City of Baltimore recognizes, through the Housing Department's Vacants to Value initiative (http://www.vacantstovalue.org/), that innovative methods, such as the Neighborhood Housing Market Typology, can be used to identify portions of the city that may qualify for different housing and community development interventions. In particular, the Vacants to Value strategy titled "Maintain, Clear, Hold and Identify Non-Housing Uses" acknowledges that levels of blight in weaker and distressed housing markets are not conducive to conventional residential development. The city's Growing Greener initiative (http://www.baltimoresustainability.org/growinggreen) has identified a variety of planning initiatives, currently pursued by the city, that can support creative re-use of vacant and/or blighted parcels to support long-term physical and social sustainability.

The City of Baltimore Department of Planning partnered with Johnson and Hollander (2014) to apply principles of smart shrinkage and public-sector decision modeling in select neighborhoods in order to assist the city in identifying specific parcels for acquisition and redevelopment for uses other than residential and commercial development. This work is intended to support the efforts of professionals within the Department of Planning who have identified certain collections of parcels ('clusters') that are candidates for alternative development efforts. Specifically, these professionals would like to know which 'clusters' to acquire earlier rather than later, and the kinds of specific potential uses that might be appropriate for these clusters, consistent with city development priorities. The City also collaborated with Davenport Whiteman (2014) to learn more about the nature and justification for specific redevelopment decisions, and to clarify distinctions between stated principles for vacant parcel redevelopment, and principles for public-sector decision modeling that might enable the city to better meet its Vacants to Value and Growing Greener goals.

3 Decision Modeling Strategy

Our decision modeling process started with defining the unit of analysis: the *cluster*. A cluster is a collection of parcels that are adjacent and have similar physical characteristics. The clusters were chosen through an inter-agency review team consisting of representatives from Police, Planning, Housing and Community Development, and the Department of General Services. Clusters were selected based on:

- Blight elimination
- Public costs for maintenance and demolition

- Minimizing resident relocation
- Opportunities for re-using the sites for environmental or economic development benefits.

The clusters used for this analysis were located in the East Baltimore neighborhoods of Better Waverly, Broadway East, Coldstream Homestead Montebello, East Baltimore Midway and Oliver. We started our analysis with 26 clusters within these neighborhoods for a pilot study (reported in Davenport Whiteman 2014), and later expanded the analysis to 118 clusters (reported in Johnson and Hollander 2014).

To ensure that clusters defined as blighted would be selected by our analysis to the greatest extent possible, we defined a set of alternative uses for acquisition candidates, consistent with the Vacants to Value initiative, consisting of urban agriculture, stormwater management, and future development. *Urban agriculture* refers to the use of city-owned land for the purposes of commercial farming, typically through a long-term lease to a farmer. *Stormwater management* refers to the use of city-owned land by the Department of Public Works to increase the amount of non-impervious surfaces to aid in water infiltration. *Future development* refers to vacant land that will be held and managed by Baltimore Housing under the expectation that new residential or commercial structures will be developed on them in the near future. *Blight elimination* is a classification of clusters suggested by the City of Baltimore that indicates a particularly severe set of physical decay and social disorder problems, and for which a variety of alternative land uses may be considered in the future.

Working with the Department of Planning, we defined criteria by which given clusters might qualify for any of the three land uses, as well as the blight elimination category. For example, to qualify for urban agriculture, a cluster would have to be ½ acre or larger, with a ground slope of less than 5 %, and have less than 30 % of its area covered by trees. Johnson and Hollander (2014) contains details on criteria for all land uses and planning classifications, and site suitability analysis results.

We used geographic information systems (GIS) to determine the land uses and planning classification for which a given cluster might qualify. We implemented this site suitability analysis in Microsoft Excel, using the logical 'and' and 'or' functions to apply the rules for cluster selection. Of the 118 clusters used for the full study within the five neighborhoods of East Baltimore, we determined that 10 qualified for urban agriculture, 38 qualified for stormwater mitigation, 35 qualified for future development, and 76 qualified for blight elimination. Moreover, 71 clusters qualified for a single use/classification; only two qualified for all four uses/classifications, and nine qualified for no uses/classifications. We found that clusters that qualified for various uses showed quite distinct spatial configurations, and that estimated redevelopment cost was strongly associated with cluster size.

Finally, we designed a decision model to determine which of the 118 candidate clusters should be acquired and re-purposed to achieve a range of policy and planning goals consistent with the site suitability analysis. We chose between two competing decision modeling approaches. The first approach is *multi-objective optimization*. Here, one develops a mathematical optimization model to choose values for decision variables (select a cluster and assign it to a use and/or classification) that jointly optimize multiple objectives (e.g. maximize social value, minimize perceived inequity), subject to constraints on cluster selection (i.e. do not exceed the budget available). This approach, presented for example by Cohon (1978), makes relatively few assumptions about the structure of decision-maker utility functions. The second approach is *multi-attribute decision analysis*. In this case, one ranks decision alternatives (particular clusters, or defined collections of clusters) according to their aggregate value, or utility. Value or utility functions are defined in terms of metrics associated with multiple characteristics, or attributes, and the relative importance placed upon the various attributes. This can be done using many competing methods (see e.g. Eiselt and Sandblom 2004), all of which generally require detailed assumptions about decision-maker preferences.

Because of the importance of a constraint on total funds available, the uncertainty of the impact of this budget constraint upon the configuration of potential collections of clusters, the lack of clarity regarding decision-maker preferences, and the large number of potential strategies, we chose to pursue a multi-objective optimization modeling approach. To formulate this model, we first defined objective functions. The Department of Planning determined that the primary metric for success in cluster acquisition and re-purposing would be the total area, in acres, of clusters chosen for a particular use/classification, and the primary metric to determine the feasibility of any potential acquisition strategy would be the total cost of acquisition and re-purposing.

The area of each cluster was defined by the City of Baltimore using GIS; the cost of acquisition and re-purposing for each cluster was defined as comprising only those estimated dollar values associated with acquisition, demolition, and relocation of occupants of a cluster (if any) to owner-occupied homes (if the cluster contained owner-occupied homes) or to rental properties (if the cluster contained rentals). The City of Baltimore further decided that these costs would be independent of the uses to which clusters might be assigned. This logic derived from the fact that clusters that would be re-purposed would have the future costs associated with particular uses borne by entities other than the city.

The vacant planning decision model whose solutions we describe below has the following structure: for clusters indexed $i = 1, \ldots, 118$ and categories indexed $j = \{$urban agriculture, stormwater mitigation, future development, blight elimination$\}$ choose clusters and assign them to categories to jointly optimize the following functions: area of land devoted to urban agriculture, stormwater mitigation, future development and blight elimination, respectively. Any assignment strategy must satisfy the following constraints: do not exceed the total budget; any selected cluster must be assigned to a single land use, and a selected cluster cannot be assigned to the blight elimination category unless it has already been selected for a land use. In this application, the Department of Planning set the budget for our model at \$3.5 million. The rationale for this number was that in 2013, the city of Baltimore received \$10 million in a mortgage settlement towards the Vacants to Value program (Sharrow 2013). Since our pilot study addressed 118 clusters in five target East Baltimore neighborhoods out of approximately 640 clusters identified by the

Department of Planning citywide, the Department would expect to use only a fraction of the $10 million if our pilot study were used for actual decision-making.

4 Decision Modeling Results

The vacant planning decision model, including data on clusters, is implemented in Microsoft Excel. The multiobjective program is coded using the 'add-in' Premium Solver product of Frontline Systems, Inc. (Frontline Systems 2011). We use the weighting method of Cohon (1978), in which each of the four objectives is multiplied by a numerical weight corresponding to the relative importance placed on the objective by a decision-maker, and the weighted objectives are added together to create a single composite objective. The resulting single-objective integer program is solved using the 'branch-and-bound' method by which successive linear 'relaxations' of the original problem are solved until the values of all decision variables are integral (Frontline Systems 2015; Winston and Albright 2012). By varying the values of weights associated with each objective, we are able to identify alternative acquisition and re-use strategies that are Pareto-optimal ('non-dominated'). That is, in the absence of detailed information about decision-maker utility functions, each of these strategies is equally well-preferred. Collette and Siarry (2003) describe a variety of advanced methods to solve multi-objective optimization models in addition to the weighting method used here. Our discussion of modeling results below follows the tradition of an analyst who provides decision-makers with promising opportunities from which the decision-makers identify a most-preferred strategy that best reflects their own needs ('a posteriori methods', Collette and Siarry 2003; 'generating techniques', Cohon 1978).

We report our results in two ways: in 'objective space', in which we show how alternative solutions vary according to the values of the four objectives, and in 'decision space', in which we display on maps the locations and category assignments of various clusters. In Fig. 1, we represent objective-space solutions using an

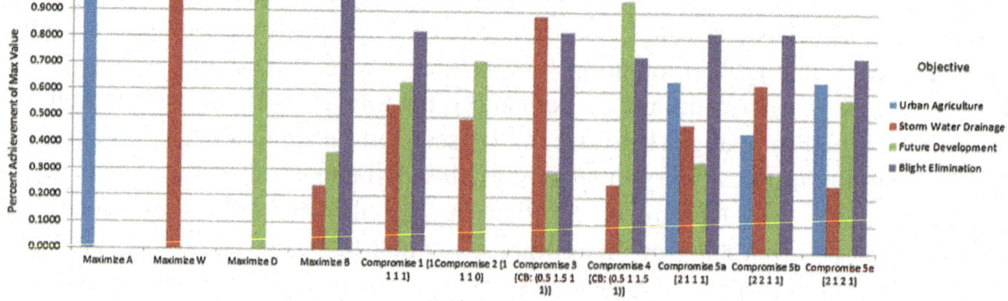

Fig. 1 Value chart, compromise solutions to vacant planning decision model

adaptation of Cohon's (1978) 'value path' visualization, in which we are able to compare the relative performance of the four objectives across all non-dominated solutions.

We first generated a set of 'corner solutions', i.e. solutions to the vacant planning decision model in which weights were set to one for one objective and zero for all others. For example, the solution "Maximize W" (area devoted to storm water management) corresponds to the weightset (0, 1, 0, 0), as stormwater mitigation is defined to be the second of four uses/classifications. By identifying corner solutions, we were able to verify the internal consistency of our decision model and to get a sense of the range of possible values for the various objectives.

Of greater interest to the City of Baltimore, however, were model results associated with weighting schemes more reflective of actual practice. We defined a range of objective function weights, starting with equal values for all objectives, corresponding to equal preferences for each objective. The City of Baltimore then suggested two additional weight-sets. In the first, labeled 'Compromise 3', greater emphasis was put on stormwater mitigation and less emphasis was put on urban agriculture, as compared to the equal-weights case. The weightset for Compromise 3 is (0.5, 1.5, 1, 1). In the second, labeled 'Compromise 4', greater emphasis was put on future development and less emphasis was put on urban agriculture. The weightset for Compromise 4 is (0.5, 1, 1.5, 1). Observing that no clusters were acquired for use as urban agriculture in these solutions, as well as the compromise solution with equal objective function weights, we defined additional compromise solutions (Compromises 5a, 5b and 5e in Fig. 1) intended to generate nonzero levels of all uses/classifications, including urban agriculture.

As expected, compromise solutions generate nonzero values for two or more objectives, as compared to three of the first four corner solutions. (The corner solution that maximizes blight elimination has nonzero values for two use objectives because our model requires that any cluster selected by the model for blight elimination must also be selected for an actual 'use' category.) Note that the compromise solutions that generate nonzero values for urban agriculture use weight-sets that imply priorities for objectives that are different from those the city itself said it favored. We will return to this point when we discuss sensitivity analyses.

We now turn to decision-space results. Of the many maps corresponding to each non-dominated solution, we show two, one corresponding to a compromise solution associated with Baltimore Planning-defined weights, and another corresponding to a weight-set chosen to generate all four use and planning categories.

Each of the panels in Fig. 2 shows the location of various clusters and their uses, as proposed by our decision model. As the weights assigned to the four objectives varies across these compromise solutions, the branch-and-bound solution algorithm chooses different clusters to recommend for acquisition, and different purposes to which acquired clusters are to be put, to optimize the aggregate single objective, subject to the same budget constraint of \$3.5 million. Uses for various clusters are color-coded: green for urban agriculture, blue for stormwater management, orange for future development. Since blight elimination is defined to occur contemporaneously with the three uses, clusters that qualify for blight elimination as well as a

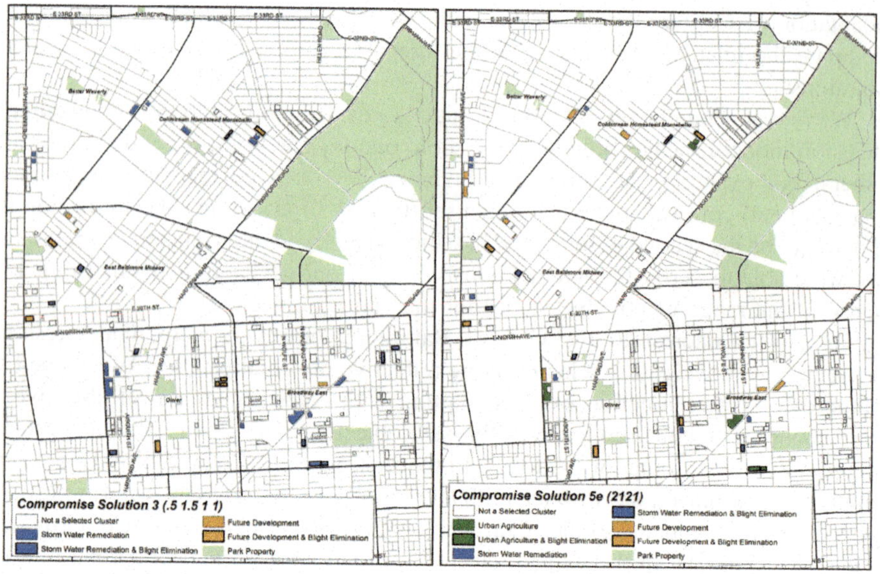

Fig. 2 Results maps, compromise solutions

particular use are shown in heavy black outline. We see that the identity of selected clusters and especially the assigned uses varies significantly across these two compromise solutions.

Our final area of analysis is changes to assumptions about our decision model. The Department of Planning first expressed a desire that a specific cluster, intended for urban agriculture, be included in all non-dominated solutions. Developing variants such as this one as compared to a 'base case' decision model is common in practice. (A variant of this sensitivity analysis would require a non-zero number of acres assigned to a particular land use or cluster category.) We found, consistent with theory, that imposing this constraint degraded the quality of our solutions: there were fewer total acres acquired, and, in most cases, somewhat less funds expended for acquisition and re-purposing, as compared with the base case (unconstrained) solutions. Another model variation addressed the notion of treating blight elimination as a 'use', even if there may be no known planning activities or interventions for any cluster so assigned. We addressed this concern by dropping the constraint that required any cluster assigned to blight elimination also be assigned to a defined land use. Again consistent with theory, we generally found that for all compromise solutions, the total area of clusters chosen is greater, and the amount of funds expended for cluster acquisition and re-purposing greater for the modified model than for the original model.

5 Practitioner Perspectives on Vacant Property Selection

We now describe an experiment intended to generate insights regarding decision-maker preferences regarding vacant cluster selection. The current process by which the Department of Planning selects clusters for acquisition and re-purposing is as follows. A selection team consisting of professionals from the Department of Planning's Office of Sustainability and the Housing Department's offices of Land Resources and Permits and Code Enforcement uses maps of vacant clusters, highlighted by values from a city-developed Housing Market Typology that classifies neighborhoods according to market conditions (City of Baltimore 2010), to identify high-priority clusters on the basis of acquisition cost, relocation cost and the likelihood of neighborhood stabilization. Using this process, the Department of Planning selected 24 vacant clusters during the first round of analysis in 2012. We wanted to learn what choices professionals from the Department of Planning would make regarding cluster selection and potential re-purposing if provided with a list of vacant clusters, designated by our vacant parcel decision model as eligible for selection and re-purposing. By doing so, we could uncover 'implicit' decision criteria used by planners and thus identify opportunities for decision models to improve the vacant parcel selection process in practice.

The choice set for this experiment consisted of ten vacant clusters in the same five neighborhoods of East Baltimore used in our vacant parcel decision model. Nine of these were eligible for selection by the decision model. An earlier 'pilot version' of our model applied to 26 clusters generated acquisition and re-purposing outcomes for the ten vacant clusters used here. For each cluster, we asked planners if they would select it for acquisition and re-purposing and why. We found that planners selected only the first four clusters. Stated reasons for these choices included community requests, pre-existing city priorities, blight elimination and supporting local development. Planners chose not to select the last six clusters because they assessed excessive acquisition and/or relocation costs, saw the clusters as second-tier priorities with respect to the goals of the Department of Planning, felt a need to spend funds equitably across East Baltimore neighborhoods, or felt that the clusters represented rehabilitation potential or historic value. Additional probing yielded that planners were generally more knowledgeable about clusters 1–4 than 5–10 due to stakeholder advocacy.

We observed during this experiment that planner decisions did not appear to reflect principles for selection that many of the same planners with whom we had engaged during development of the vacant cluster planning model had previously described to us. For example, no planners in this experiment expressed acquisition and cost concerns as formal criteria, e.g. value thresholds. There was little expressed concern for alternative land uses. There was no formal expression of 'strategic' value with respect to redevelopment choices. There was no weighting of priorities or evaluation of tradeoffs. Without such information about the planners' implicit decision processes, how would planners know if a particular strategy they have designed is preferred, on objective grounds, to an alternative strategy?

6 Formal Model of Decision Modeling-Driven Smart Shrinkage

Given results of our field study with professionals from the Department of Planning, and our work to develop the vacant parcel planning model, we propose a framework by which planning professionals might adapt the problem-structuring approaches of decision modeling, without necessarily formulating and solving decision problems, skills that are not common among professional planners (Table 1).

We start by proposing that planners identify *core themes* that drive their efforts for vacancy planning. Based on our discussions with planners and our knowledge of the relevant research literature, we propose two such themes: “strategic demolition and rehabilitation” and “budgetary constraints and challenges”. There are differing justifications for the each of these themes; they represent our understanding, from planners at the City of Baltimore, our first-hand observations of the East Baltimore neighborhoods, and the smart shrinkage literature, as to social and planning goals that planners are likely to find compelling. Next, we propose *criteria* by which one might determine whether a particular parcel, if selected, would be likely to support any of the core themes. For each criterion, we identify one or more *metrics* that can be quantified using data we believe can be collected from the field, derived from administrative datasets, or generated using policy analytic techniques.

We believe that our framework for decision modeling for smart shrinkage, as well as the vacant cluster planning model itself, should be viewed as suggestive, not prescriptive. It can be best understood as a way to encourage explicit consideration of policy and planning goals by quantifying social impact objectives that are expressed by planners themselves as important. By using clear criteria to assess cluster eligibility for selection, and ultimately, choice for actual acquisition, planners can recognize the role that advocacy can play in cluster selection and identify conflicts between objectives and policy tradeoffs between alternative strategies. In addition, we feel that a planning process rooted in decision modeling principles can enable participation by community stakeholders at all phases, and enable planners to identify shortcomings in formal model design on the basis of an explicit understanding of planning goals.

However, our preferred process for vacant parcel acquisition and re-purposing is not limited to a collective process of interactive discussions informed by decision modeling principles. We believe that parcel acquisition for smart shrinkage is in fact a difficult enough problem that explicit mathematical modeling tools (multi-objective optimization; multi-attribute decision analysis) are necessary to generate alternative strategies that can be shown to be superior to those produced using conventional methods. Initially, having discussed the values that form a basis for property acquisition and re-purposing in blighted neighborhoods, planners may find it useful to develop fairly simple and stylized representations of this problem applied to relatively small datasets, as we have done here. Such representations

Table 1 A decision modeling framework for vacant cluster selection, informed by planning practice

Fundamental objective	Why important	Criteria	Metrics
Strategic demolition and rehabilitation	Support existing homeowners and bolster redevelopment efforts	Blight elimination	Crime hotspots
		Support existing redevelopment	Code enforcement problem areas
	Improve quality of life for Baltimore residents	Whole block outcomes	Proximity to areas with current redevelopment projects
		Neighborhood stability	Blocks with high vacancy concentration
			Proximity to areas with high owner-occupancy
	Stabilize neighborhoods	Targeted investment	Proximity to areas with current redevelopment projects
			Proximity to areas with high owner occupancy
			Areas with high vacancy concentration
			Housing market typology (distressed categories)
Apply limited funding in an equitable and maximally effective way	Budgetary constraints and challenges	Cost	Minimal acquisition
			Minimal relocation
			Minimal need for structural supports
	Social equity	Equitable distribution of funds	
	Political pressure	Targeted investment	
			Request from the community/ community
			organizers/city councilmen, etc.
			Historic value/preservation

would defer until a later point such modeling complications as contingencies, varying time horizons, policy-analytic methods to estimate social impacts associated with particular criteria, and uncertainties in the values of key structural parameters. Solutions to the resulting basic decision model, such as those we have shown above, will enable planners to better understand tradeoffs associated with alternative problem instances (e.g. objective function weights) and model variants (different formulations of objective functions, or constraints).

At this point, having gained confidence in the policy and planning insights to be gained from a structured, mixed-method analytic approach, planners may choose to develop more complex yet realistic models, applied to larger datasets. Additional expertise from multiple disciplines may be required to formulate and solve these problems, and to communicate the solutions clearly. However, appropriate methods have been well-studied in the literatures of policy analysis (Weimer and Vining 2011), OR/MS (Collette and Siarry 2003) and planning (Timmermans 1997). Solutions to these more advanced models have the potential to provide specific, robust and tangible guidance for neighborhood-level interventions that can generate beneficial social outcomes desired by stakeholders.

7 Conclusion

In this chapter, we have shown that decision modeling can make contributions to challenging problems in land-use planning for distressed communities. We have done so through a formal modeling exercise, using a multi-objective math optimization model developed in collaboration with professional planners with the City of Baltimore, as well as discussions with planners about the processes they use to select among candidate clusters for acquisition and re-purposing. Based on these results, we propose a decision modeling framework that is flexible enough to accommodate math programming models as well as improved deliberations on the basis of professional planning principles. These methods are not intended to replace community practitioners or planners, but rather to enhance their effectiveness in addressing the problems they already solve in the real world. However, where the number of alternatives is large, where there are competing objectives, where the attributes of concern may be difficult to compute and visualize, and where resources are limited, analytics and decision science can be useful. We hope that use of these methods may fulfill the promise of community development that is “integrated, broadly collaborative, data-driven, and focused on what works, and entrepreneurial” (Seidman 2012, 367). There are many extensions to the work described in this chapter we wish to pursue. One of these would be to develop more sophisticated decision models that explicitly address issues of nonlinearity in objective functions and constraints and uncertainty in structural parameters. Another would be to explore objective function formulations that incorporate policy-analytic methods by which one might associate planning decisions with specific neighborhood or population outcomes. A third might be to ground our model formulation process more closely in the experiences and perspectives of neighborhood stakeholders.

References

Baltimore Department of Planning (2013). Homegrown Baltimore: Grow local. Office of Sustainability, Baltimore, MD. http://www.baltimoresustainability.org/homegrown-baltimore-grow-local. Accessed November 30, 2014.

Bluestone, B., & Harrison, B. (1982). *The deindustrialization of America*. New York: Basic Books.

City of Baltimore (2010). Housing market typology: Purpose and methodology. Department of Planning, Baltimore, MD. http://archive.baltimorecity.gov/government/agenciesdepartments/planning/masterplansmapspublications/housingmarkettypology.aspx. Accessed December 1, 2014.

Cohon, J. L. (1978). *Multiobjective programming and planning*. New York: Academic Press.

Collette, Y., & Siarry, P. (2003). *Multiobjective optimization: Principles and case studies*. Berlin: Springer.

Davenport Whiteman, E. (2014). *Decision-making strategies for vacant lot reuse in Baltimore, Maryland*. M.A. Thesis. Medford, MA: Tufts University.

Dewar, M., & Thomas, J. M. (Eds.). (2013). *The city after abandonment*. Philadelphia: University of Pennsylvania Press.

Eiselt, H. A., & Sandblom, C.-L. (2004). *Decision analysis, location models, and scheduling problems*. Berlin: Springer.

Friedman, E. (2003). *Vacant properties in Baltimore: Strategies for reuse*. Submission to the Abell Foundation.

Frontline Systems (2011). *Premium Solver pro version 11.5.2.0*. Incline Village, NV.

Frontline Systems (2015). Basic solver—integer programming. http://www.solver.com/content/basic-solver-integer-programming. Accessed January 7, 2015.

Gallagher, J. (2013, January 8) *Plan to save Detroit unveiled: Vision for a smaller, stronger Motor City*. Detroit Free Press.

Hall, P. (1997). Modeling the post-industrial city. *Futures, 29*(4/5), 311–322.

Hollander, J. B., Johnson, M. P., & Davenport Whiteman, E. (2015). Algorithms for abandonment: The application of math modeling to shrinking cities. Cambridge, MA: MIT Press (in press).

Hollander, J. B., & Németh, J. (2011). The bounds of smart decline: A foundational theory for planning shrinking cities. *Housing Policy Debate, 21*(3), 349–367.

Hollander, J. B., Pallagst, K., Schwarz, T., & Popper, F. (2009). Planning shrinking cities. *Progress in Planning, 72*(4), 223–232.

Hollander, J. B., Polsky, C., Zinder, D., & Runfulo, D. (2011). The new American ghost town: Foreclosure, abandonment and the prospects for city planning. Working paper. Lincoln Institute of Land Policy. http://www.lincolninst.edu/pubs/download.asp?doc_id=1211&pub_id=1897. Accessed January 7, 2015.

Jackson, K. T. (1985). *Crabgrass frontier: The suburbanization of the United States*. New York: Oxford University Press.

Johnson, M. P. (2011). Housing and community development. In J. Cochran (Ed.), *Wiley encyclopedia of operations research and management science*.

Johnson, M. P. (Ed.). (2012). *Community-based operations research: Decision modeling for local impact and diverse populations*. New York: Springer.

Johnson, M. P., & Hollander, J. B. (2014). Baltimore vacant lot redevelopment project: Final report. (With H. A. Rohman, S. Bahirwani, P. Ciurzak, A. Engel, K. Keeny, E. Lambergs, A. Levine, J. Tu, E. Whiteman, M. Hughes, H.-J. Lee, & O. Usidame). Baltimore: City of Baltimore Department of Planning.

Johnson, M. P., Hollander, J. B., & Hallulli, A. (2014). Maintain, demolish, re-purpose: Policy design for vacant land management using decision models. *Cities: Special Issue: Vacant Land: The New Urban Green, 40B*, 151–162.

Liberatore, M. J., & Luo, W. (2010). The analytics movement: Implications for operations research. *Interfaces, 40*(4), 313–324.

Open Baltimore (2015). https://data.baltimorecity.gov/. Accessed January 30, 2015.
peakbagger.com (2014). Historical metropolitan populations of the United States. http://www.peakbagger.com/pbgeog/histmetropop.aspx#tables. Accessed January 7, 2015.
Popper, D. E., & Popper, F. J. (2002). Small can be beautiful: Coming to terms with decline. *Planning, 68*(7), 20–23.
Ransome, K. T. (2007). *Housing abandonment and vacancy in Baltimore city. Master's Thesis.* Baltimore: Morgan State University.
Rust, E. (1975). *No growth: Impacts on metropolitan areas*. Lexington, MA: Lexington Books.
Seidman, E. (2012). Integration and innovation in a time of stress: Doing the best for people and place. In N. O. Andrews & D. J. Erickson (Eds.), *Investing in what works for America's communities: Essays on people, place and action* (pp. 354–376). San Francisco: Federal Reserve Bank of San Francisco & Low Income Investment Fund.
Sharrow, R. (2013, September 18). Baltimore puts $10 M towards Vacants to Value program. *Baltimore Business Journal*. http://www.bizjournals.com/baltimore/blog/realestate/2013/09/baltimorevacantstovalueprogram.html. Accessed January 30, 2015.
Stohr K (2004, February 5) Shrinking city syndrome. *The New York Times*.
Stone, M. M., & Cutcher-Gershenfeld, S. (2002). Challenges of measuring performance in nonprofit organizations. In P. Flynn, & V. A. Hodgkinson (Eds.), Measuring the impact of the nonprofit sector (pp. 33–57). New York: Kluwer Academic/Plenum Publishers.
Swope, C. (2006, November). Smart decline. *Governing Magazine*. www.governing.com/topics/economic-dev/Smart-Decline.html. Accessed November 30, 2014.
Timmermans, H. (Ed.). (1997). *Decision support systems in urban planning*. London: E & FN SPON.
Tortorello, M. (2011, August 3) Finding the potential in vacant lots. *New York Times*.
U.S. Census Bureau (2015a). *Census of population and housinp*. https://www.census.gov/prod/www/decennial.html. Accessed April 26, 2015.
U.S. Census Bureau (2015b). *American fact finder*. http://factfinder.census.gov/faces/nav/jsf/pages/index.xhtml. Accessed January 30, 2015.
Vale, L. J., & Campanella, T. J. (2005). *The resilient city: How modern cities recover from disaster*. New York: Oxford University Press.
Wallace, N. (2014, February 24). Nonprofits are taking a wide-eyed look at what data could do. *The Chronicle of Philanthropy*.
Weimer, D. L., & Vining, A. R. (2011). *Policy analysis* (5th ed.). Boston: Longman.
Wiechmann, T. (2008). Errors expected—aligning urban strategy with demographic uncertainty in shrinking cities. *International Planning Studies, 13*(4), 431–446.
Winston, W., & Albright, S. (2012). *Practical management science* (4th ed.). Mason, OH: South-Western/CENGAGE Learning.

Chapter 5
Managing Crowds: The Possibilities and Limitations of Crowd Information During Urban Mass Events

Lara Britt Zomer, Winnie Daamen, Sebastiaan Meijer and Serge Paul Hoogendoorn

Abstract This chapter, based on a mixed method research approach, offers insights into possibilities and limitations of using ICT measures for crowd management and distribution during urban mass events (UMEs). Based on literature, practical applications and analyses of research results, we propose crowd management should consider characteristics of both crowds and UMEs to increase information effectiveness. In relation to urban planning, results show that possibilities to influence a crowd's behavior depend on available (and known) choice sets offered in various locations, while distances towards locations across city centers appear less important. Limitations appear to be related to scarce knowledge on what drives crowd members to adapt or adhere to their activity choice behavior. Such insights are essential for smart cities striving for an optimal use of infrastructural capacity, as both the ambiguous effects of ICT measures, as well as a crowd's self-organizing capacity should be taken into account for delaying, solving and preventing disruptions of pedestrian flows in city centers.

L.B. Zomer (✉)
Transport Science, KTH Royal Institute of Technology, Stockholm, Sweden
e-mail: zomer@kth.se

W. Daamen · S.P. Hoogendoorn
Department of Transport and Planning, Delft University of Technology Delft, Delft
The Netherlands
e-mail: W.Daamen@tudelft.nl

S.P. Hoogendoorn
e-mail: S.P.Hoogendoorn@tudelft.nl

S. Meijer
School of Technology and Health, KTH Royal Institute of Technology, Stockholm, Sweden
e-mail: sebastiaan.meijer@sth.kth.se

S. Geertman et al. (eds.), *Planning Support Systems and Smart Cities*,
Lecture Notes in Geoinformation and Cartography,
DOI 10.1007/978-3-319-18368-8_5

1 Introduction

Due to urbanization, increased pedestrian and traffic flows stretch city centers' physical infrastructure capacity to the max. Infrastructural capacity is especially challenged whenever mass events take place in city centers. Increasingly, such events gain in popularity as a means for city branding. Crowd management in urban environmentsis believed necessary whenever self-organization within pedestrian flows reduces because of increasing densities in the streets. Due to inefficient pedestrian choice behavior, infrastructural capacity drops and uneven distributions over the city center's network or even blockades and gridlocks occur (Still 2000; Hoogendoorn and Daamen 2004; Helbing et al. 2005; Helbing and Mukerji 2012).

This chapter applies the term **urban mass events** (UMEs) to indicate (non-) recurring (un-)organized events attended by sufficient numbers of visitors within city boundaries to strain the planning and response resources of a community, state or nation (WHO 2008). Typically, from a demand side, such events are characterized by high numbers of visitors (pedestrians) gathering within a communal urban area during a specific timespan. From a supply perspective, there will be an urban area, constrained in infrastructural capacity containing various (scheduled) attraction locations (Mamoli et al. 2012). Besides the planning of the physical environment, to manage crowds during UMEs and to use the city's capacity and infrastructure efficiently, it is essential to both understand pedestrian flow behavior and the extent to which information measures can influence such behavior (Fruin 1993; Still 2000; Helbing et al. 2005).

In line with smart city concepts, ICT technologies enable municipal authorities to influence movement patterns during UMEs by applying novel styles of demand management, e.g. online communication platforms. However, little is known about the actual impact of these mobile and real-time information and communication measures on crowd behavior in cities. The same applies to group behavior within crowds, or, the impact of a crowds' collective intelligence and resilience in delaying, preventing and solving disruptions (Tonkin et al. 2012; Treurniet 2014).

This chapter discusses the findings of explorative research on the effectiveness of information measures regarding crowd management during UMEs, in relation to crowd distribution over city centers. To this end, using the real-life context of Nijmegen Vierdaagsefeesten 2013,[1] a mixed method study has been conducted on decision-making behavior of social groups within a crowd to gain insight into the relations between crowd characteristics and effectiveness of information regarding activity choice behavior (ACB). Additionally, and to generalize obtained results for

[1]An annually organized 7-day free-access UME in Nijmegen city center, daily attracting up to 300.000 visitors to various attraction locations across the city center. Manifold information measures (both high and low-tech) are deployed to inform, advise, guide, steer and even enforce the crowd.

similar UMEs, a stated choice (SC) experiment is used to investigate ACB and the possibilities and limitations of information measures in a controlled environment.

The remainder of this chapter consists of five parts. Based on literature, the next section discusses crowds' decision-making, ACB and the ways in which information measures may influence these processes. The methodology section discusses how the impact of information on ACB has been researched in a meaningful and reliable way. Third, the results section presents analyses of qualitative and quantitative data regarding the Vierdaagsefeesten and the SC experiment. Based on this section, next, findings on the effectiveness of information measures for crowd management during UMEs are discussed as well as the contribution to the body of knowledge on crowd management. The chapter concludes by offering recommendations both for research and for practical applications of ICT measures during UMEs.

2 Literature Overview

This section discusses how information measures may enhance crowd management, paying special attention to optimizing city centers' infrastructural capacity. Next, attributes for decision-making and the expected impact of information on ACB will be put forward.

2.1 The Role of Information in Crowd Management

From a demand perspective, crowd management aims for behavioral change by applying information measures, especially to optimize the use of available infrastructural capacity. Based on literature on behavior adaptation, human factors and crowd- and traffic management, (Still 2000; Tertoolen et al. 2012; Hoogendoorn 2013), crowd management involves measures and actions instigated by authorities to mitigate visitors' demands by providing them with information to confirm or adapt their behavior during UMEs. Tertoolen et al. (2012) introduce five steps: (1) inform, (2) advise, (3) guide, (4) steer and (5) enforce to argue that the effectiveness of both management and control measures will improve if aligned to the willingness of the users. Guidance constitutes the transition between management (inform and advise) and control (steering and enforcement).

Crowd management (and -control) during UMEs is mainly studied in the wake of disasters. In the case of the 2009 Duisburg Love Parade, ensuing research concluded failing information measures contributed to deteriorating critical densities culminating in crowd turbulence causing the disaster (Helbing and Mukerji 2012). This chapter provides a different perspective, focusing on the effectiveness of information measures for equally distributing visitors across the city's infrastructural network to prevent blockades. To study ACB in crowds it is necessary to understand crowd members' behavior in social groups at a microscopic level (Moussaïd et al. 2010), while insights into macroscopic crowd behavior can be enriched by analyzing percentages and premises of the crowd typologies (Berlonghi 1995; Challenger et al. 2010).

2.2 The Role of Information in Decision-Making Processes

ACB concerns decision-making at individual level. Within crowd management, the impact of ACB has been scarcely researched. Therefore, knowledge from adjoining scientific domains, mostly pedestrian dynamics, transportation, traffic demand management and land-use modelling are applied to study ACB in crowds (Wegener 2004). A conceptual framework has been developed to relate ACB during UMEs to the influence of information. Figure 1 represents expected relationships between behavioral characteristics, ACB, information and distribution across the city. The complexity is caused by interactions between all four elements, continuously influencing one another. Crowd information is based on the interpretation and knowledge of both distribution across the city as well as crowd's behavior and is used to inform, advise and guide visitors. Similarly visitors' ACB will be influenced by their knowledge and interpretation of the distribution (densities) across the city and the crowd information provided.

From research on pedestrian dynamics, it turns out visitors seek their optimum journey (location and route choice) and, to optimize in-group communication, walking formations of social groups depend on group size and street densities (Helbing and Molnar 1998; Hoogendoorn and Bovy 2004). According stated choice research, individuals will choose what they perceive to render maximum utility based upon their experiences and current knowledge about available options (Chorus et al. 2013). To seek the optimum journey during UMEs, the sum of the utilities of the sequence of locations is maximized.

At group level, decision-making varies between compromise, competitive and cooperative styles. Within group decision-making processes, it is assumed that the larger the group and the more decisions to be made, chances increase of not everybody getting their own way. To get a group into action, a shared vision is needed as strong as the binding within the group. A shared vision enhances group cohesion and the number of group members willing to change/adapt their personal behavior in favor of the group (Pan et al. 2007).

Though research on the effect of information on (pedestrian) location choice behavior remains scarce, in traffic management, research is conducted on the effects of information regarding route choice behavior. Important attributes include travel time and -delay, congestion, distance, habit, experience, reliability of information, attractiveness of destination, purpose of travel, personal preferences and socio-economic status (Ben-Akiva and Lerman 1985; Kitamura 1988; Abdel-Aty et al. 1997; Kemperman 2000; Kurose et al. 2001; Hensher 1994; Muizelaar 2011).

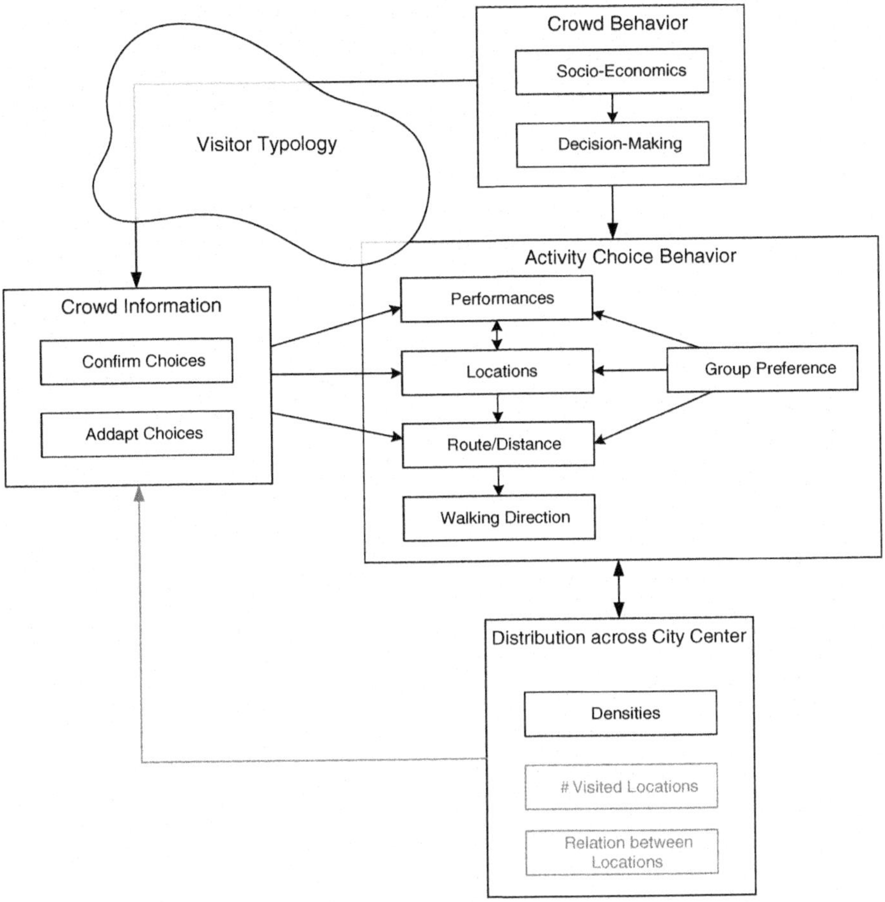

Fig. 1 Conceptual framework on crowd management during UMEs

Traffic management insights are used to investigate the effects of information on ACB, leading to seven attributes of interest to investigate perceived utility of amusement locations: density at locations, performance and location types and distance (travel time). Based on pedestrian dynamics and crowd behavior, the influence of a crowd's walking direction and group preference can be studied (Pan et al. 2007). Based on theories on behavioral change, the effect of information is expected to differ for confirming or adapting location choice behavior. To adapt location choice behavior, information measures have to be able to impact perceived utilities attributed to specific amusement locations. To this end, both static measures, such as city maps and information leaflets as well as dynamic (real-time) measures, such as location bound LED screens, mobile information on social media and heatmaps, will be investigated as to their impact on the crowd members' decision-making behavior during UMEs (see Fig. 2). The attributes of the

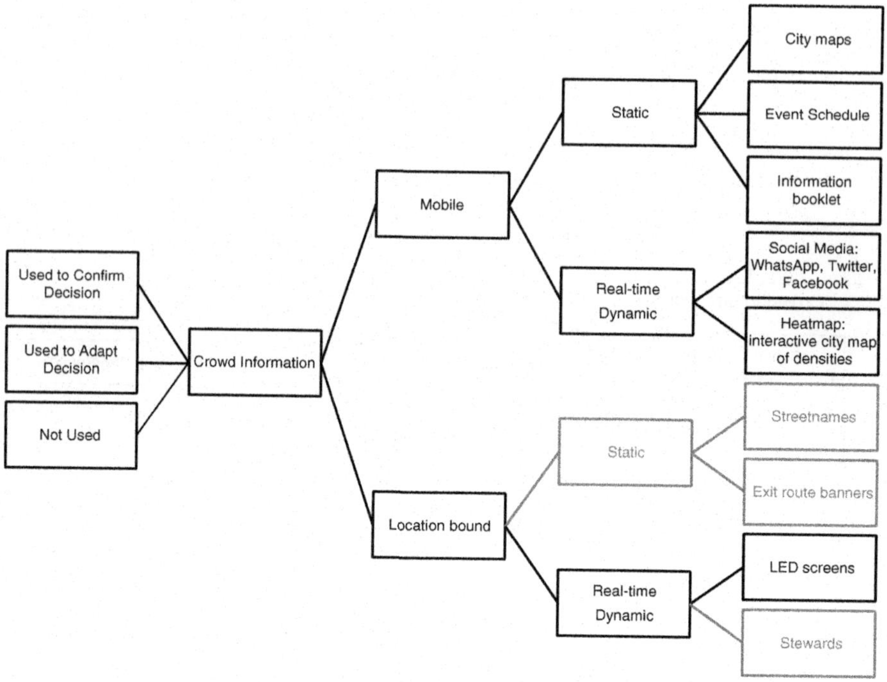

Fig. 2 Diagram for crowd information during UMEs

developed conceptual framework have been operationalized in Sect. 3 to investigate the expected relationships.

3 Methodology

Based on research into ACB and effectiveness of traffic information (Abdel-Aty et al. 1997; Kemperman 2000; Muizelaar 2011), a state preference method has been selected for data gathering. An online survey minimizes interference with visitors' behavior while gaining background information on personal data without dealing with privacy restrictions or camera motion patterns, as these would increase planning, costs and data calibration. Field observations during the event showed information to inform or advise visitors to avoid crowded locations is easily misinterpreted, as (some) visitors specifically aim for crowded locations. Based on literature and occurrences of high densities during the Vierdaagsefeesten, a guidance (warning) message is hypothesized to be more appropriate to raise visitors' awareness about potential negative consequences of incompliant behavior. To investigate the effect of an improved information measure to manage crowds, a tradeoff has been made between a controlled experimentand capturing the "state-of-mind" during UMEs. The results aim to capture the questions: What attributes influence activity (attraction location) choice behavior of crowds during UMEs, and, can this behavior be controlled or influenced by providing crowd information?

The operationalization of attributes, used as reference of the online survey, is presented in Fig. 3. The survey consists of two sections; Qualitative data providing insights into decision-making within crowds during UMEs, visitor typologies and the influence of information amongst visitors of the Vierdaagsefeesten 2013. By means of quantitative data from a Stated Choice (SC) experiment concerning a fictive UME, more generic insights to influence ACB are obtained, specifically regarding the impact of the warning message: "Be careful, this location is too crowded".

Respondents' consent to take part in the research, as well as their e-mail addresses have been gathered by the first author in July, 2013, during fieldwork at the Vierdaagsefeesten. Targeting those who visited the first six evenings of the Vierdaagsefeesten, a survey was developed, piloted, revised and published online in September 2013. Out of 1450 acquired e-mail addresses from visitors of the Vierdaagsefeesten, 386 respondents completed the qualitative part of the survey and 316 respondents finished both the qualitative and quantitative parts of the survey.

3.1 Qualitative Data

Data from the first part of the survey has been used to analyze dependencies between group size/-composition and crowd behavior, familiarity with the Vierdaagsefeesten as well as how visitors used (what kind of) information and the effects of using this information. These results are derived by descriptive, statistical and correspondence analyses (CA) using SPSS, as discussed in Sects. 4.1 and 4.2.

CA is used to define visitor typologies, analyzing (the strength of) similarities and differences between attributes of crowd information usage, crowd behavior and ACB. CA calculates row (personal characteristics) and column points (information types and ACB) on a two-dimensional map. The stronger the positive association between rows and columns, the closer to each other and further away from the origin they are placed on the map. On the other hand, the stronger the negative association between rows and columns, the further apart from each other they are placed on the map. CA holds a strong advantage over factor analysis, because it encompasses nominal variables and is able to describe relationships between both categories of each variable, as well as the relationship between variables, without a need for interval data or large numbers of observations (Meulman and Heiser 1989). The effectiveness of ten information measures are discussed in Sect. 4.3.

3.2 Quantitative Data: SC Experiment

The second part of the online survey consists of an SC experiment, used to determine the independent influence and relations between different attributes on the observed (theoretical) choices between two or more alternatives by sampled respondents undertaking the experiment. Based on the theory of utility maximization, each individual is likely to select the alternative (amusement location) perceived to yield the highest utility (Ben-Akiva and Lerman 1985; Chorus et al. 2013).

The expected relations as presented in the conceptual framework can be tested in a controlled environment. The designed SC experiment allows respondents to choose between three alternatives (two different amusement locations and opt-out). The two amusement locations are characterized by seven attributes, at two levels, an example is presented in Table 1. The aim is to find out whether choice behavior differs per location typology and if information measures can impact visitors' choice behavior raising awareness about negative consequences of incompliant behavior. Activity choice refers to three generic attributes (performance and location type and distance to location), leading to eight (2^3) typologies of amusements locations; the remaining four attributes consist of alternative specific attributes and refer to the respondents' expectations.

The SC experiment has been analyzed using discrete choice modeling and estimation with multinomial models using BIOGEME. Section 4.4 presents the results of a multinominal logit (MNL) model; its utility function, to calculate the goodness-of-fit, is shown below (Ben-Akiva and Lerman 1985).

Table 1 Example of a choice situation during the SC experiment "The Summer Experience"

	Performance A	Performance B	Opt-out
Performance type	Spectacular performance	Spectacular performance	
Location type	Small location	Small location	
Distance to location	Around the corner	Around the corner	
Density	Very crowded	Crowded	
Information	Informative: "Be careful, it is too crowded"		
Group preference	Towards performance A	NOT towards performance B	
Walking direction of the crowd	Towards performance A	NOT towards performance B	

Tonight, the Summer Experience takes place, a free-access mass event in Nijmegen. You are visiting this event together with the same people as during the Vierdaagsefeesten. What performance do you want to go to?

$$P_q(i) = \frac{e^{V_{iq}}}{\sum_{j \in C_q} (e^{V_{jq}})} \quad (1)$$

In which:

$P_q\ (i)$ probability of P of choosing alternative i by individual q
V_{iq} systematic utility of alternative i and individual q
C_q choiceset of individual q

Results are interpreted based on model performance, -correctness and -significance. The model estimation is executed in three steps:

- MNL1, model comprising all attributes of interest.
- MNL2, model comprising only attributes leading to an increased adjusted ρ^2.
- MNL3, model comprising only attributes leading to an increased adjusted ρ^2 and significant dummy variables for certain attributes. One attribute (crowd information) has been found considered a dummy variable.

4 Results

Sections 4.1–4.3 indicate the complexity of changing visitors' location choice behavior, differences in preferences and usage of information during the Vierdaagsefeesten 2013. Next, results on discrete choice behavior and the effect of a warning message in a controlled environment are presented in Sect. 4.4.

4.1 Socio-Economics

Personal and group characteristics and attitudes towards information and decision-making behavior during the Vierdaagsefeesten are presented in Table 2. Most respondents are between 20–29 years old, remember having visited five amusement locations during this UME in a group ranging from 3–6 persons. Most respondents state to have decided collectively (democratically and "flat hierarchy") within the group what amusement locations to visit. For democratic social groups performance type is more important compared to more hierarchical groups, they also appear to have more people familiar with the event and/or city. Just over 25 % of the respondents indicate they occasionally use information provided by the authorities during the event. Most respondents prove to be satisfied with the provided information, whether they used this information or not.

Table 2 Characteristics of respondents

Characteristic	Levels	Number of respondents
Gender	Male	201
	Female	185
Age	<20	46
	20–29	234
	30–39	40
	40–49	28
	>50	38
Education	Secondary school	56
	MBO	59
	HBO	125
	University	146
# Visits to Nijmegen	Every day	161
	>Once per month	91
	>Once per year	95
	<Once per year	39
# Visits to UMEs	>Once per month	73
	>Once per year	295
	<Once per year	18
Group Size	<3 People	91
	3–6 People	204
	>6 People	91
Group familiarity with Nijmegen	Me and other group members	309
	Some are not familiar	65
Leadership	Democratic	338
	Flat hierarchy	22
	Strong leadership	22
Information usage during UME	Never	235
	Sometimes	64
	Often	72
	Very often	15
Satisfaction with existing information	Very dissatisfied	15
	Dissatisfied	13
	Not satisfied/not dissatisfied	34
	Satisfied	174
	Very satisfied	48
	No opinion	102

4.2 Group Behavior and Information

Based on differences and similarities between personal characteristics, preferences for information usage and ACB four visitor typologies have been identified using a CA. Figure 4 presents the visitor typologies and their differences in need and usage of information measures. The effectiveness of information measures in getting the message across is expected to vary over the typologies; some visitor typologies may be better guided by other means. Decision-making of ***Goal-Oriented*** visitors is mainly done by one person, associated with strong leadership. ***Elder/Unfamiliar*** visitors are associated with hierarchical decision-making, allowing other people in their social group to be in charge of decision-making. The ***Average*** visitors

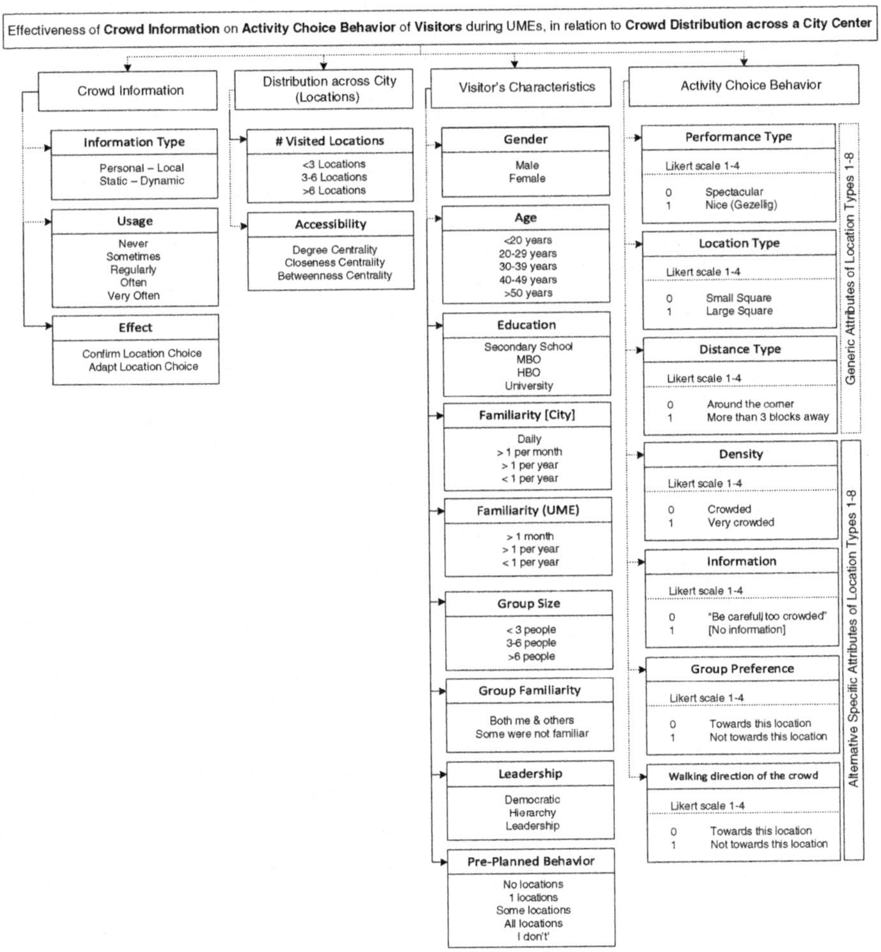

Fig. 3 Operationalization of attributes

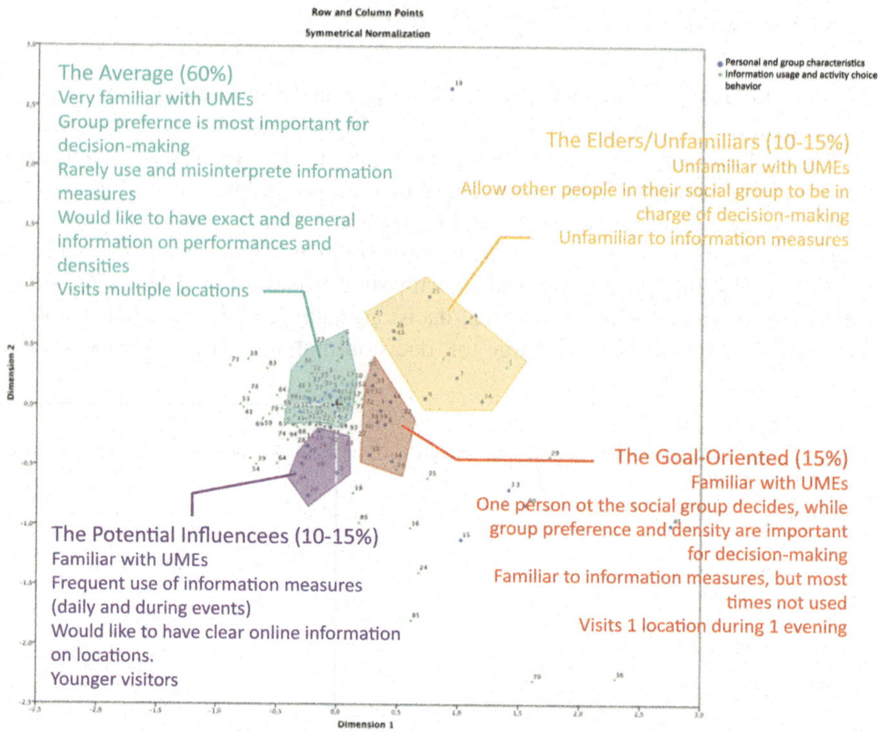

Fig. 4 CA of personal characteristics (*row points*) versus information and activity choice (*column points*)

associate in social groups consisting of more democratic group structures, with a stronger focus on group preference. Differences and similarities, however, are not "strong". This may result from the fact that during the Vierdaagsefeesten visitors and social groups, simultaneously, belonged to multiple typologies (Fig. 4).

4.3 *Effectiveness of Information*

As presented in Table 3, for ten information services, respondents have been asked to indicate their usage and the effect (confirm or adapt behavior) regarding their location choice. Effectiveness is calculated by multiplying usage percentages with percentages of expressed adapting or confirming effects regarding location choice.

Generally, information measures have been used more to confirm location choice than to adapt location choice behavior. WhatsApp and LED screens turn out to be the most effective information measures for both adapting and confirming location choice. To improve effectiveness, assuming the content of information

Table 3 Effectiveness of 10 information measures

Impact on location choice	Adapt	⟷								Confirm
Information measures	Negative tweets of friends	Heatmap (app)	Tweets by Police	Tweets by organization	City map (app)	LED screens	Event schedule (app)	Info booklet	Positive tweets of friends	WhatsApp messages
Adapt location choice (%)	9.0	6.1	2.1	3.3	1.9	13.8	4.6	4.7	4.7	14.5
Confirm location choice (%)	6.0	6.1	3.1	4.9	7.8	20.7	18.4	18.7	18.9	33.8
Δ (%)	3.0	0.0	−1.0	−1.6	−5.9	−6.9	−13.8	−14.0	−14.2	−19.3

plays an important role, the question on what constitutes effective information in case of adaptation or confirmation needs further investigation.

Most effective for adapting location choice are negative tweets from friends and the heatmap. It should be mentioned the heatmap, an interactive city map emphasizing 'hotspots'—locations with (too) high densities of people in the city intended the make visitors aware of these densities in order to adapt their location choice—turns out to be equally effective in confirming location choice. This is expected to relate to little familiarity with the heatmap and ambiguity in understanding the information. For instance, several visitors reported to perceive the hotspots on heatmap as "nicely" crowded locations.

4.4 Stated Choice Experiment

The SC experiment is analyzed using discrete choice modeling and estimation with multinomial models using BIOGEME. The experiment found various combinations of significant relationships for ACB regarding eight location typologies. The results for all location typologies are presented in Table 4. For three location typologies (1, 3 and 4), the crowd's direction has not turned out significantly important. As to spectacular performances at large squares (5 and 6), relations between group preference and density are not significant. The negative warning message, however, seems to be effective for most spectacular performance locations.

Based on the estimated models, a goodness-of-fit test has been conducted to calculate the alternative with the highest probability for each respondent of the SC experiment. Hence, it can be determined whether this alternative would actually have been chosen during crowded UMEs. The MNL3 models of Location Typology 1, 2 and 6 (respectively, a city center main stage, side stage and a main stage out of city center) have been used to investigate the effect of a warning message to decrease choosing that attraction location during very crowded situations. For simplification, all densities are fixed at "very crowded" and it is assumed that distances to amusement locations do not change, e.g. the main stage remains "nearby". The estimation of the effects of deploying a warning message is shown in Table 5

Starting at the base situation, without initial preferences or warning messages, it appears the highest utility is gained for choosing main stage, resulting in a preference and flow towards main stage. The situation changes as the expected utility of choosing main stage increases. 75.1 % of the visitors perceive choosing main stage as their highest utility. Because the initial state is already very crowded, this will lead to a jammed density at main stage. If, at this moment, a negative warning message is deployed, this will effectively decrease the utility for choosing main stage, and, simultaneously, increase both utilities for the side stage and extra main stage out of city center. As a result, because of a preference and flow towards side stage and extra main stage, the warning message at main stage can be abolished.

Table 4 Results BIOGEME estimation location typology choice MNL3

Variable	Location typology 1		Location typology 2		Location typology 3		Location typology 4	
	Spectacular, small, nearby		Nice, small, nearby		Nice, large, nearby		Nice, large, far away	
	Estimate	t statistic	Estimate	t statistic	Estimate	t statistic	Estimate	t statistic
ASC1 (Loc A)	0		0		0		0	
ASC2 (Loc B)	−0.835	−5.75	−1.75	−9.76	–	–	0.451	3.46
ASC3 (opt-out)	–	–	–	–	–	–	–	–
$\beta_{density}$, β_4	−0.558	−3.91	−0.351	−2.16	−0.429	−3.04	−0.286	−2.09
$\beta_{information}$, β_{5_0}	1.85	9.53	*1.31*	5.76	*0.894*	5.30	–	–
$\beta_{information}$, β_{5_1}	1.37	7.06	*1.49*	6.39	*0.964*	5.83	0.197[a]	1.53
β_{group}, β_6	−0.584	−4.09	0.256[a]	1.60	0.779	-5.45	−0.368	−2.71
β_{crowd}, β_7	–	–	0.497	3.01	–	–	–	–
# observations		323		316		312		316
Null LL		−354.852		−347.161		−342.767		−347.161
Final LL		−299.296		−243.886		−315.867		−335.564
ρ^2		0.157		0.297		0.078		0.033
Adjusted ρ^2		0.142		0.280		0.067		0.016

Variable	Location typology 5		Location typology 6		Location typology 7		Location typology 8	
	Spectacular, large, nearby		Spectacular, large, far away		Spectacular, small, far away		Nice, small, far away	
	Estimate	t statistic	Estimate	t statistic	Estimate	t statistic	Estimate	t statistic
ASC1 (Loc A)	0		0		0		0	
ASC2 (Loc B)	−1.14	−7.34	−0.992	−6.54	–	–	−1.26	−7.06
ASC3 (opt-out)	–	–	–	–	–	–	–	–
$\beta_{density}$, β_4	–	–	−0.275[a]	−1.90	−0.298	−2.15	−1.07	−6.17
$\beta_{information}$, β_{5_0}	1.02	5.83	*1.30*	7.08	*0.736*	5.70	2.20	9.24
$\beta_{information}$, β_{5_1}	1.04	6.09	*1.18*	6.59	–	–	2.15	9.35

(continued)

Table 4 (continued)

Variable	Location typology 5		Location typology 6		Location typology 7		Location typology 8	
	Spectacular, large, nearby		Spectacular, large, far away		Spectacular, small, far away		Nice, small, far away	
	Estimate	t statistic	Estimate	t statistic	Estimate	t statistic	Estimate	t statistic
β_{group}, β_6	–	–	–	–	0.531	-3.76	-1.12	-6.41
β_{crowd}, β_7	−0.363	−2.31	−0.703	−4.74	0.281	2.12	0.381	2.33
# observations		305		325		321		310
Null LL		−335.077		−357.049		−352.655		−340.570
Final LL		−295.041		−314.190		−328.060		−242.353
ρ^2		0.119		0.120		0.070		0.288
Adjusted ρ^2		0.108		0.106		0.058		0.271

[a]Numbers in italics indicate non-significant estimates at 95 % level

Table 5 Results of deployment of a warning message

Visualization	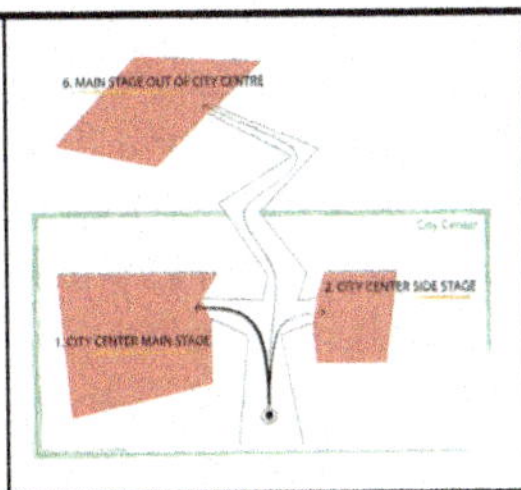			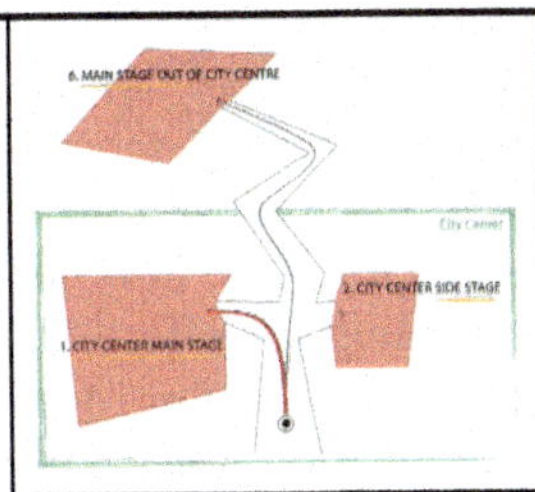			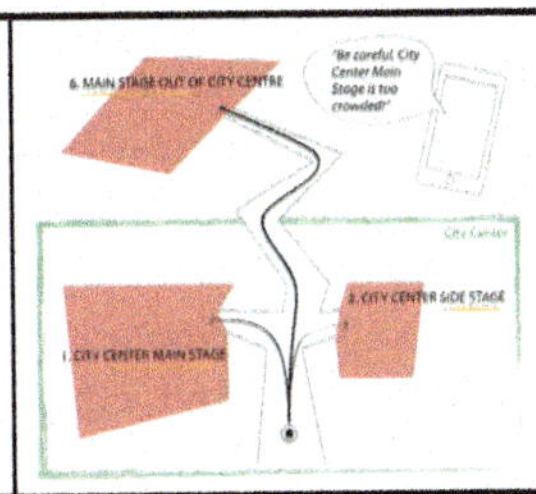		
Actions	[Initial state]			Preference and flow towards 1			"Be careful, main stage is too crowded"		
Location	1.	2.	6.	1.	2.	6.	1.	2.	6.
Utility	1.015	−0.44	0.308	1.801	−0.44	0.308	−0.023	−0.44	0.308
Exp.	2.76	0.64	1.36	6.06	0.64	1.36	0.98	0.64	1.36
%	57.9	13.5	28.7	75.1 ↑	7.99 ↓	16.9 ↓	32.8 ↓	21.6 ↑	45.6 ↑

Visualization	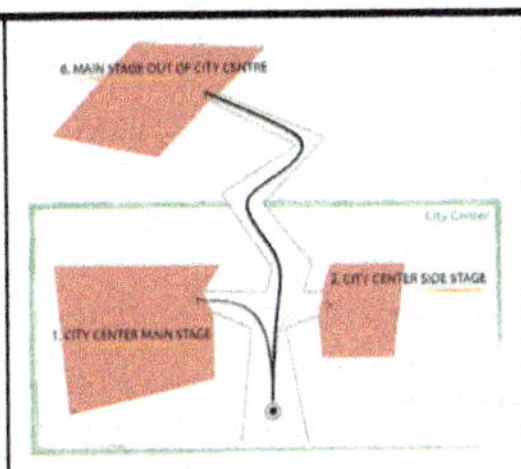			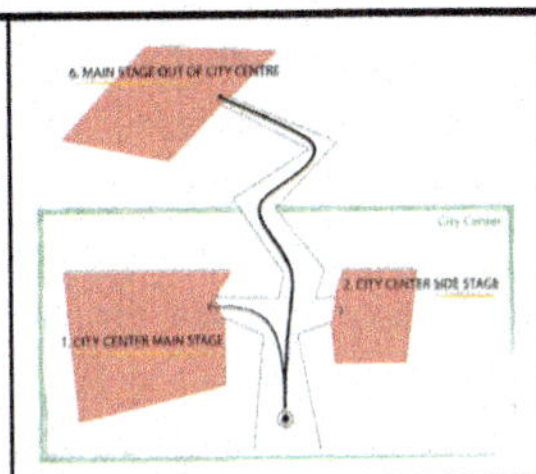		
Actions	Warning message of, preference and flow towards 2 and 6			Preference and flow towards 6 [stable]		
Location	1.	2.	6.	1.	2.	6.
Utility	1.015	-0.184	1.488	1.015	-0.44	1.488
Exp.	2.767	0.83	4.43	2.76	0.64	4.43
%	34.4	10.4 ↓	55.2 ↑	35.2	8.2	56.5

*↑ indicates an expected increase in number of visitors deciding to go to that location ** ↓ indicates an expected decrease in number of visitors deciding to go to that location.
*** Red cells indicate a jammed density at that location

Afterwards the utilities of the three choices remain rather stable, so no guidance information is needed.

Results of the experiment show a warning message to be effective when the available choice set consists of similar performance types, while the distances towards locations are less important. Providing a warning message at main stage is effective because the preference and flow are changed from main stage towards the extra main stage (further away). An "extra main stage further away" is expected to attract more visitors than a "side stage nearby". The latter will only become an alternative for main stage during the time a warning message is deployed, while "extra main stage further away" will also yield an increased preference after a warning message has been abolished.

5 Conclusions

This chapter demonstrates possibilities and limitations of ICT measures to inform, advise and guide crowds during UMEs, in relation to optimizing crowd distribution across city centers. Combining knowledge from different fields, a framework is proposed to help determine appropriate information and to coherently design integral sets of measures. Regarding the effectiveness of ICT measures on crowd management during UMEs, the body of knowledge has been added to and such insights are essential to create smarter cities. The results of a case study on the Nijmegen Vierdaagsefeesten 2013 indicate the complexity of changing visitors' location choice behavior. To a certain extent, visitors' reactions to information measures turn out to be ill-predictable, as they may easily misinterpret information measures whenever—to them—this seems more suited to achieve their goals. A SC experiment reveals that effective information relates mostly to an available choice set consisting of similar performance types, while distances towards locations across city centers are deemed less important. Using a mixed method approach, the results can be generalized for comparable UMEs characterized by high densities and familiar visitors, to generalize results for emergency situations, further research is recommended.

Although no significant relations between group size and information usage have been found, an indirect relation may exist, since significant relations, applying also to information usage, were noted between density, group preference and pre-planned behavior. Additionally, these findings suggest other methods may be considered and/or developed to investigate the effects of information measures on social groups (dynamics) and the moment or occasional circumstances of information intake. Secondly, differences regarding the potential of a negative warning message and the limitations to confirming location choice indicate more research into the relation between effectiveness and content of information measures is essential.

Although the effectiveness of the innovative heatmap has been calculated at 0 %, its actual influence may be larger. Especially, since the heatmap enables novel crowd management and -control measures, such as, sending push-messages to visitors and planning for police deployment based on measured densities across the city center. As most respondents complained about insufficient network capacity of the telecommunication, the effects of measures on the demand and supply should be investigated. Hence, a holistic approach appears necessary to assess the effectiveness of crowd management in general, in which the organizational structure and characteristics of both crowd and UME should be considered.

A general disadvantage to stated preference (SP) methods is reliability, as expressed preference may not be consistent with actual behavior. It is assumed respondents will try to justify their actual (operational) behavior or provide socially acceptable answers, on, e.g. the walking direction of the crowd or their actual information usage (Sanko 2001). Location typologies with nice performances at large locations (3 and 4) and spectacular performance at small locations (7) show lower model performances than other location typologies. This is assumed to be the result of using an orthogonal design for the SC experiment, instead of an efficient design. To respondents feasible choices may seem unrealistic, leading to ambiguous decision-making (Hoogendoon-Lanser 2005).

However, regarding their location choice behavior, SP experiments offer insights into strategic and tactical decision-making by respondents. To increase reliability, revealed preference (RP) and SP data may be combined (Ben-Akiva et al. 1994), to investigate the complete chain of decision-making.

Socio-economic data serve as additional input for the utility estimation in BIOGEME. Additional goodness-of-fit estimations, using these data, indicate providing inappropriate or incorrect information to be counter-effective regarding equal distributions across locations, and, with increasing heterogeneity within the crowd and/or across performance types, information measures will be less suited to mitigate such crowds.

5.1 Recommendations

The possibilities and limitations of the "right" information at the "right" time will be affected by contingency planning and provision of suitable alternatives. In the Netherlands, road authorities, as Rijkswaterstaat, ensure continuous R&D, evaluationsof deployed measures, and integrated network management and cooperation between (road) traffic management regions. Smart cities would benefit from a similar organizational structure providing for integrated crowd management during UMEs.

ACB in crowds and the impact of information on behavioral change constitutes a complex adaptive system in which unexpected things will certainly occur and small disturbances may lead to disproportional consequences. Because of the ill-predictability of crowd's behavior, resilient methods are required to investigate potential impacts of interventions prior to their deployment, especially in the wake of riots, disasters and emergency situations. In this respect, it is recommended to investigate and validate the possibilities of a participatory simulation platform (PSP), to design and evaluate resilient and integral crowd management, before putting new technologies into practice. In line with Bharosa et al. (2010) such research would be innovative because of the increased reliability of decision-support dashboards for ill-predictable and complex-adaptive processes of both technical and social aspects in integrated crowd management during UMEs and multi-actor decision-making during different situations. Hence, using a semi-controlled participatory experiment, aims to increase validity of results by capturing situational awareness and including social (group) dynamics.

Acknowledgments This research is mainly based on research conducted as part of a master thesis at Delft University of Technology, in collaboration with Royal Haskoning/DHV. We thank Jan

Anne Annema, Hans Marinus and Marian Weltevreden for their valuable comments during the process. In addition we thank the anonymous reviewers for their comments and suggestions.

References

Abdel-Aty, M., Kitamura, R., & Jovanis, P. (1997). Using stated preference data for studying the effect of advanced traffic information on drivers' route choice. *Transportation Research Part C: Emerging Technologies, 5*(1), 39–50.

Berlonghi, A. E. (1995). Understanding and planning for different spectator crowds. *Safety Science, 18*(4), 239–247.

Ben-Akiva, M. E., & Lerman, S. R. (1985). *Discrete choice analysis: Theory and application to travel demand.* Cambridge: MIT Press.

Ben-Akiva, M. E., Bradley, M., Morikawa, T., Benjamin, J., Novak, T., Oppewal, H., & Rao, V. (1994). Combining revealed and stated preferences data. *Marketing Letters, 5*(4), 335–349.

Bharosa, N., Janssen, M., Meijer, S., & Brave, F. (2010). Designing and evaluating dashboards for multi-agency crisis preparation: a living lab. In *Electronic Government* (pp. 180–191). Berlin, Heidelberg: Springer.

Challenger, R., Clegg, C., & Robinson, M. A. (2010). *Understanding crowd behaviours, Vol. 1 Practical guidance and lessons identified.* London: The Stationery Office (TSO).

Chorus, C., Walker, J., & Ben-Akiva, M. (2013). A joint model of travel information acquisition and response to received messages. *Transportation Research Part C: Emerging Technologies, 26*, 61–77.

Dijk, J., Boeschoten, T., Tije, S., & Wijngaert, L. (2013). *De weg naar Haren: de rol van jongeren, sociale media, massamedia en autoriteiten bij de mobilisatie voor Project X Haren: Deelrapport 2.*

Fruin, J. J. (1993). The causes and prevention of crowd disasters. In *Engineering for crowd safety*, (pp. 99–108). Amsterdam: Elsevier.

Helbing, D., & Molnar, P. (1998). Self-organization phenomena in pedestrian crowds. arXiv preprint cond-mat/9806152.

Helbing, D., Buzna, L., Johansson, A., & Werner, T. (2005). Self-organized pedestrian crowd dynamics: Experiments, simulations, and design solutions. *Transportation science, 39*(1), 1–24.

Helbing, D., & Mukerji, P. (2012). Crowd disasters as systemic failures: Analysis of the love parade disaster. *EPJ Data Science, 1*(1), 1–40.

Hensher, D. A. (1994). Stated preference analysis of travel choices: The state of practice. *Transportation, 21*(2), 107–133.

Hoogendoorn, S. P., & Bovy, P. H. (2004). Pedestrian route-choice and activity scheduling theory and models. *Transportation Research Part B Methodological, 38*(2), 169–190.

Hoogendoorn, S. P., & Daamen, W. (2004). Self-organization in walker experiments. In *Proceedings of the 5th Symposium on Traffic and Granular Flow*, (pp. 121–131).

Hoogendoorn, S. P. (2013, Novemeber 21–23). *Why traffic management works…and why coordinated traffic management will work even better.* Presentation at the 3rd EU-US NAE Frontiers of Engineering Symposium, Chantilly.

Hoogendoon-Lanser, S. (2005). Modeling travel behavior for multi-modal transport networks. *Dissertation*, TRAIL Research School, TU Delft.

Meulman, J. J., & Heiser, W. J. (1989). IBM SPSS Categories 20.

Kemperman, A. (2000). *Temporal aspects of theme park choice behavior.* Eindhoven, The Netherlands: Technische Universiteit Eindhoven.

Kitamura, R. (1988). An evaluation of activity-based travel analysis. *Transportation, 15*(1–2), 9–34.

Kurose, S., Borgers, A. W. J., & Timmermans, H. J. P. (2001). Classifying pedestrian shopping behaviour according to implied heuristic choice rules. *Environment and Planning B: Planning and Design, 28*(3), 405–418.

Mamoli, M., Michieletto, P., Bazzani, A., & Giorgini, B. (2012). Venice as pedestrian city and tourist magnet mass events and ordinary life. *Ara Journal of Tourism Research,* 93–102.

Moussaïd, M., Perozo, N., Garnier, S., Helbing, D., & Theraulaz, G. (2010). The walking behaviour of pedestrian social groups and its impact on crowd dynamics. *PloS one, 5*(4), e10047.

Muizelaar, T. J. (2011). Non-recurrent traffic situations and traffic information: determining preferences and effects on route choice. *Dissertation*, TRAIL Research School, Twente University.

Pan, X., Han, C. S., Dauber, K., & Law, K. H. (2007). A multi-agent based framework for the simulation of human and social behaviors during emergency evacuations. *AI & SOCIETY, 22* (2), 113–132.

Sanko, N. (2001). *Guidelines for stated preference experiment design*. Ecole Nationale des Ponts et Chaussées: Master of Business Adminstration diss.

Still, G. K. (2000). Crowd dynamics. *Dissertation*, University of Warwick.

Tertoolen, G., Grotenhuis, J., Lankhuijzen, R. (2012). Human factors en dynamisch verkeers-management: Waar psychologie en techniek samenkomen. In *Colloquium Vervoersplanologisch Speurwerk*, Amsterdam, 22–23 November 2012.

Timmermans, H. J. P. (2009). *Pedestrian behavior: Models, data collection and applications*. Gloucester: Emerald Group Publishing.

Tonkin, E., Pfeiffer, H. D., & Tourte, G. (2012). Twitter, information sharing and the London riots? *Bulletin of the American Society for Information Science and Technology, 38*(2), 49–57.

Treurniet, W. (2014). Shaping Comprehensive Emergency Response Networks. In *Network Topology in Command and Control: Organization, Operation, and Evolution: Organization, Operation, and Evolution*, 26.

Wegener, M. (2004). Overview of land-use transport models. *Handbook of Transport Geography and Spatial Systems, 5*, 127–146.

World Health Organization. (2008). Communicable disease alert and response for mass gatherings: Key considerations. Accessed 20 May, 2013 from http://www.who.int/csr/mass_gathering/en/print.html.

Chapter 6
Simulating Urban Resilience: Disasters, Dynamics and (Synthetic) Data

A. Yair Grinberger, Michal Lichter and Daniel Felsenstein

Abstract An agent based (AB) simulation model of urban dynamics following a disaster is presented. Data disaggregation is used to generate 'synthetic' data with accurate socio-economic profiling. Entire synthetic populations are extrapolated at the building scale from survey data. This data is coupled with the AB model. The disaggregated baseline population allows for the bottom-up formulation of the behavior of an entire urban system. Agent interactions with each other and with the environment lead to change in residence and workplace, land use and house prices. The case of a hypothetical earthquake in the Jerusalem CBD is presented as an illustrative example. Dynamics are simulated for a period up to 3 years, post-disaster. Outcomes are measured in terms of global resilience measures, effects on residential and non-residential capital stock and population dynamics. The visualization of the complex outputs is illustrated using dynamic web-mapping.

1 Introduction

Urban resilience is invariably conceptualized as a cities' ability to 'bounce back', post-disaster, to some pre-existing equilibrium (Campanella 2008; Godschalk 2003; Müller 2011). This pre-shock state embodies spatial and temporal relationships, direct and indirect effects and short and long term process. Disentangling these in order to isolate those factors promoting urban resilience is particularly challenging. In addition, urban resilience is more than just the sum of the parts of its' inhabitants' individual resilience. While cities are agglomerations of individuals, they are also

A.Y. Grinberger (✉) · M. Lichter · D. Felsenstein
The Hebrew University of Jerusalem, Jerusalem, Israel
e-mail: asherya.grinberger@mail.huji.ac.il

M. Lichter
e-mail: mlichter@gmail.com

D. Felsenstein
e-mail: daniel.felsenstein@mail.huji.ac.il

S. Geertman et al. (eds.), *Planning Support Systems and Smart Cities*,
Lecture Notes in Geoinformation and Cartography,
DOI 10.1007/978-3-319-18368-8_6

much more. For example, they represent the accumulation of rounds of fixed capital expressed in infrastructure and other hard investment. These are generally expressed in terms of stock variables (roads, buildings etc.). In the event of a shock, resilience of stock will be expressed in static terms, for example by maintaining function (transport flows, energy provision, providing shelter) in the face of adverse conditions.

However, cities are also much more than accumulations of capital stock. They comprise complex network and flow systems such as input-output relations between producers and consumers, origin-destination traffic patterns and so on. These represent a dynamic and long term view of resilience that involves not just maintaining the existing state of the city but also recovering in order to reach a desired state. Flows of information, labour and capital have the ability to not just maintain current conditions but to change the urban growth trajectory by increasing productivity over the medium to long term. Dynamic resilience therefore contributes variable inputs to urban development and the more inputs are variable, the greater the likelihood of inefficient allocation of urban resources (Rose 2009).

This chapter presents a disaggregated agent-based (AB) simulation model of urban resilience in the wake of a disaster. The activities of multiple agents create a computable system in which the actions of individual agents affect each other and the system as a whole. The result is a complex network of behavior patterns that could not have been predicted by simply aggregating individual agent behavior. The system can be simulated and subjected to various exogenous shocks. The motivation for the study is to show how a bottom-up simulation modeling approach combined with an initial population created from synthetic data at the building scale, can be used to aid urban rejuvenation in the aftermath of a disaster. We also illustrate how web-based technology can be used for communicating these findings to planners, policy makers and the public.

2 Agent-Based Modeling and Urban Disasters

Urban disasters occur randomly in time and space. They affect both individuals and the environments that they populate. Large scale disasters are generally not one-time disturbances but generate a series of sub-incidents such as aftershocks in the case of earthquakes or secondary contamination in the case of pandemics. These keep the disaster environment in state of flux. Individuals therefore operate in a randomly changing context. By 'agentizing' this environment (Axtell 2000) and reducing its elements to autonomous programmable entities, it becomes amenable for management.

Agent-based simulations have been applied in a variety of disaster contexts such as flooding, fires and earthquakes (Chen and Zhan 2008; Crooks and Wise 2013; Dawson et al. 2011). The AB framework lends itself to these situations. A high level of agent heterogeneity can be programmed and applied differentially to the various stages of an urban disaster from mitigation and preparation through response and on to recovery. This yields a rich array of human behaviors.

For example, AB models have been coupled with network models for simulating evacuation (Chen et al. 2012). GIS tools and crowdsourced data has been combined with agent based modeling to assist with post disaster recovery analysis (Crooks and Wise 2013). Kwan and Lee (2005) merged network analysis, GIS and 3D visualization tools to provide a realtime micro-scale simulation tools for emergency response at the individual building or city block level. In the field of traffic modeling, Chen and Zhan (2008) have used the AB approach to evaluate different evacuation strategies under different road network and population density regimes. The emergency response literature also uses AB modeling in urban contexts to provide a simulation capability for the public health and medical communities. This allows for the efficient management of medical and evacuation resources under conditions of severe uncertainty and stress. Invariably these systems use hybrid architecture that integrates a simulator with GIS, databases and rule based protocols for agents (Narzisi et al. 2006; Zimmerman et al. 2010).

3 Method

To account for the spatio-temporal dynamics of urban disasters we present an agent-based framework that is driven by synthetic spatial data (Fig. 1). In this framework, residential choice, workplace and activity location are determined

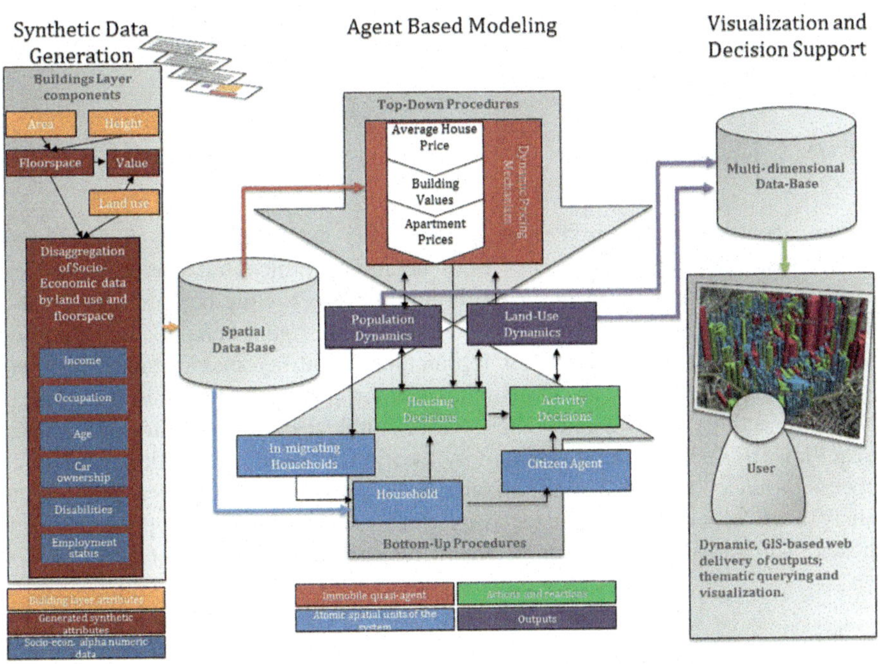

Fig. 1 Simulating resilience: a general framework

bottom-up while land use and house prices are fixed top-down. As agents are inherently mobile, their behavioral response is articulated in varying temporal and spatial dimensions. While much of the data for observing agents behavior is only available at coarse administrative units, we use a GIS-based method for the accurate socio-economic profiling of the population under such circumstances. This involves moving from a database describing only hundreds or thousands of spatial units to one containing records of millions of buildings and individuals over time. In the resultant spatial data, every individual in a city is synthetically represented by a single specific record. This database is input to an agent-based model of urban disasters as the initial physical properties of the urban environment and the distribution of the units (Fig. 1). Repast Simphony 2.0, an open source, Java-based programming platform, is used as the simulator (Crooks and Castle 2012). Model outputs are delivered and visualized using web-GIS.

The urban system is modeled as the outcome of interactions between agents and the environment (Fig. 1). In this section we describe the mechanics of the agent-based procedures that give the database a dynamic, multi-dimensional nature. These mechanics rely on (necessary) simplifying assumptions. While this limits realism, it is required for modeling an agent-rich environment. To increase the validity of results, the model's mechanics are 'structurally stochastic', as the random preference element allows behavior to vary in relation to the basic behavioral structure (Reichert and Mieleitner 2009). While the assumptions we make are not ungrounded, the stochastic element diminishes possible diversion from realistic behavior, given the large number of agents in our model. Common validation processes (such as backstacking) are inapplicable in the case of long-term effects. Hence the need for a solid base for the mechanics of the model.

3.1 Behavior of Agents

Citizen agents are organized at two levels—as individuals and clustered into households. Each agent embodies both socio-economic and spatial properties. Residential decisions are made at the household level while activities and workplace location decisions are executed by individuals. Agent behavior is not ad hoc but grounded in standard behavioral principles of utility maximization and risk evasiveness (Lancaster 1966), satisficing (Simon 1952), preferences for scale in economic activity (Fujita and Thisse 2002) and segregative residential choice (Schelling 1971). Residential and activity location decisions are guided by a search process grounded in 'satisficing' behavior. In this process the first location found to satisfy a set of constraints and a utility threshold that represents preferences, is chosen. This threshold is randomly drawn for each agent from the range [0,1].

3.1.1 Place of Residence

At any given moment, a household may decide to relocate or to move out of the city. In most cases this choice is probabilistic and dependent upon exogenous probabilities for out-migration/relocation. In exceptional cases a household is forced out of home due to land-use dynamics such as residential building use becoming commercial or due to the direct destruction resulting from the disaster. In this case the choice between relocation and out-migration is entirely random. The choice of new place of residence (Eq. 1) is guided by two elements: the affordability of a dwelling and its attractiveness (Chen et al. 2012). These are evaluated in relation to the household's willingness to allocate up to one third of monthly income to housing (a budget constraint) and preference for residential segregation represented by limited tolerance to change in living environment:

$$b_h = j \Rightarrow \left[\frac{I_h}{3} > HP_j\right] * \left[k_h > \frac{\Phi\left(\frac{\bar{I}_j - \bar{I}_h}{I_{\sigma_h}}\right) + \Phi\left(\frac{\bar{A}_j - \bar{A}_h}{A_{\sigma_h}}\right)}{2}\right] = 1 \qquad (1)$$

where

- b_h is the new residential location for household h randomly drawn from a choice set that includes all vacant buildings and partially occupied residential buildings,
- j is the building considered,
- $[\,]$ is a binary expression with value of 1 if true and 0 otherwise,
- I_h is household h's monthly income,
- HP_j is monthly housing cost of an average housing unit in building j,
- k_h is the tolerance level for household h,
- Φ is the standard normal cumulative probability function,
- $\bar{I}_j, \bar{A}_j$ are the average household income and average age of individuals in building j, respectively
- $\bar{I}_h, \bar{A}_h$ are average household income and average age of individuals in residential buildings within 100 m of current residential location of household h,
- $I_{\sigma_h}, A_{\sigma_h}$ are standard deviations of household income and of resident age in residential buildings within 100 m from current home location of household h, respectively.

A random-order search process is initiated whenever a household relocates or when a new household migrates into the city. The volume of in-migration is proportional to the number of vacant dwelling spaces and to an exogenous ratio of in-migration to out-migration. The search process is terminated if 100 iterations do not lead to relocation. In this case the household (whether in-migrant or native) leaves the city. If the conditions for relocation are fulfilled, the dwelling unit is removed from the set of vacant units.

3.1.2 Workplace Location

Location of workplace is related to land-use as each employment sector is associated with a particular use (e.g. commercial, industrial, governmental etc.). Apart from this constraint, locational choice (Eq. 2) is dependent on distance-minimizing and preference for scale (representing more opportunities):

$$WP_i = b_j \Rightarrow [LU_j = ELU_i] * \left[k_i > \frac{D_{ij}/\max D_i + 1 - FS_j/\max FS}{2}\right] = 1 \qquad (2)$$

where

WP_i is the workplace location of individual i,
ELU_i is the employment-sector-related land-use for individual i,
k_i is the preferences index,
D_{ij} is the distance between building j and individual i's place of residence,
$\max D_i$ is the distance of the building within the study area furthest away from individual i's place of residence.

Workplace is not part of the initial database and agents are assigned to locations within the model. These locations are assumed to be stable unless the building changes use. Only in such a case is the search for a new workplace initiated.

3.1.3 Location of Activities

Each day an individual agent participates in a varying number of activities. This number is dependent on individual attributes promoting or inhibiting mobility and accessibility (e.g. age, car ownership, disability) as well as employment status and personal preferences:

$$\begin{aligned} NAc_i = \Bigg\| a * \left(\frac{k_i}{0.5}\right) * (1 + car_h * 0.33) * (1 - dis_i * 0.33) \\ *(1 + [age_i = 2] * 0.33) * (1 - [age_i \neq 2] * 0.33) \Bigg\| + emp_i * loc_i \end{aligned} \qquad (3)$$

where

NAc_i is the number of activities for resident i,
k_i is a preference index,
car_h is a binary variable equal to 1 if the household h owns a car and 0 otherwise,
dis_i is a binary variable equal to 1 if individual i is disabled and 0 otherwise,
age_i is the age group of individual i,
emp_i is a binary variable equal to 1 if i is employed and 0 otherwise,

loc_i is a binary variable equal to 1 when i's workplace is located within the study area and 0 otherwise,

$\|x\|$ indicates the nearest integer number to x,

a is the average number of activities based on employment status; equals 2.5 for employed residents and 3 for non-employed.

The number of activities thus ranges between 0 and 12. The location of each activity is set by distance from previous location (starting from home) minimizing preferences, risk evasive behavior and preferences for scale. Risk evasiveness is embodied in the tendency to avoid areas in which a large proportion of the buildings are vacant and volume of floor-space represents preferences for scale.

$$a_{t+1,i} = b_j \Rightarrow [b_j \neq a_{t,i}] * [k_i \geq Att(b_j)] = 1 \tag{4}$$

where

$a_{t,i}$ is the current location of individual i,

$a_{t+1,i}$ is the next location of activity of individual i,

k_i is a randomly drawn number between [0,1] reflecting activity location preferences,

$Att(b_j)$ is the attractiveness score for building j, calculated as follows:

$$Att(b_j) = \frac{1 - {}^{\Sigma E_j}/_{\Sigma B_j} + 1 - {}^{D_{ij}}/_{\max D_i} * (1 + 0.33 * (-car_h + dis_i + [age_i = 3])) + [LU_j = non\text{Res}] * {}^{FS_j}/_{\max FS}}{2 + [LU_j = non\text{Res}]} \tag{5}$$

where

ΣE_j is the number of unoccupied buildings within a 100 m buffer of building j,

ΣB_j is the number of buildings within a 100 m buffer of building j,

D_{ij} is the distance of building j from the current location of individual i,

D_{ij} is the distance of the building within the study area furthest away from the current location of individual i,

LU_j is the land-use of building j,

$nonRes$ is non-residential use,

FS_j is the floor-space volume of building j,

max FS is the floor-space volume of the largest non-residential building within the study area.

3.2 Environmental Processes

In most agent-based models the environment is a passive backdrop which changes only as a direct consequence of agents' actions (for example, changes in resource levels due to consumption by agents). However many components of the urban environment while not pro-active are at least reactive. For the sake of simplicity these are not modeled individually but in aggregate, they are treated as components of environmental sensitivity. Individual spatial elements such as census tracts, buildings and dwelling units are characterized as quasi-agents. These are not autonomous or mobile but are sensitive to environmental changes. This mechanism operates top-down as the effects of aggregate trends trickle down to the level of individual quasi-agents. We employ this mechanism in the areas of land-use dynamics and house prices.

3.2.1 Commercial Land-Use Dynamics

In the context of these dynamics, we enlist three assumptions. First, revenue levels required by a commercial function in order to be profitable are proportional to its floor-space volume. Second, actual revenues at a location are proportional to local flows of customers and third, flows are proportional to the traffic loads[1] in the vicinity of the function. This set of assumptions allows for formalizing the (logistic) probability of land-use change (Eq. 6) as related to the congruence between floor-space volume and traffic loads at a location (Eq. 7). This congruence, formalized as the difference in the relative position within an exponential distribution (Eq. 8), thus represents demand or supply surplus:

$$P_{j,t}(\Delta x_{j,t}) = \frac{\mathrm{e}^{-\Delta x_{j,t}}}{1 + \mathrm{e}^{-\Delta x_{j,t}}} \tag{6}$$

$$\Delta x_{j,t} = \frac{z_{TR_{j,t}} - z_{FS_{t,t}}}{|z_{FS}|} \tag{7}$$

$$z_{y_{t,j}} = \hat{\lambda}_t * \left(\mathrm{e}^{-\hat{\lambda}_t * y_{j,t}} - \mathrm{e}^{-\hat{\lambda}_t * y_{med_t}}\right) \tag{8}$$

where of land-use change for building j at time t,

$P_{j,t}$ is the probability

$\Delta x_{j,t}$ is the relative difference in standardized values for traffic load and floor-space for building j at time t.

---

[1]We do not employ a shortest-path algorithm for movement routes but use a computationally less-demanding model where agents move at each step to the not-already-visited node closest to the destination (in aerial distance; loops are removed from the path). This also represents satisficing behavior.

$z_{y_{j,t}}$ is the standardized value y (in relation to the median value y_{med_t}) for building j at time t drawn from the exponential distribution $y \sim \mathrm{Exp}\left(\hat{\lambda}_t\right)$, $\hat{\lambda}_t = \frac{1}{\bar{y}}$.

Traffic loads are calculated as the average daily load (citizens per meter) on roads within 100 m radius for the building for the preceding 30 days. P values of 0.99 and above are set to represent a demand surplus. For such values, residential or vacant buildings become commercial functions. In such a case, any residents in the location relocate according to principles discussed in Sect. 3.1.1. Values within the range $[P(1) - 0.01, P(1)]$ are used to identify supply surplus, or unprofitability, which results in commercial buildings becoming vacant. Such formalization and critical values limit the sensitivity of large commercial functions and small residential uses, thus eliminating a possible bias for changing initial uses in these cases.

3.2.2 Dynamic House Pricing

Housing decisions (Sect. 3.1.1) are made in relation to prices. These prices represent demand, supply and the locational (dis)advantages of a specific dwelling unit (DU) and building. In order to capture the unique contribution of such market-level dynamics to the value of the individual unit, we formalize a three-stage mechanism. Within this mechanism the effects of global changes trickle down from the census-tract (CT) level to the building and DU level. Average housing values per meter in CTs (Eq. 9) change daily with changes to supply, demand and accessibility to services (supply of non-residential functions) within them. These prices set the value of individual buildings, along with local accessibility levels (Eq. 10). Assuming equal size for all DUs within a building, the monthly cost of housing is derived in relation to the average willingness to pay of the population (Eq. 11). Changes to commercial values are the result of a similar simpler process which is dependent on supply only and ends at the building level (see Eqs. 10 and 11):

$$AV_{CT,t+1} = AV_{CT,t} * \left(1 + \log\left(\frac{\left({pop_{CT,t+1}}/{pop_{CT,t}} + {res_{CT,t}}/{res_{CT,t+1}}\right) * [LU = \mathrm{Res}] + \left({n\mathrm{Res}_{CT,t+1}}/{n\mathrm{Res}_{CT,t}}\right)^{-1+2*[LU=\mathrm{Res}]}}{1 + 2 * [LU = \mathrm{Res}]}\right)\right) \tag{9}$$

where

$AV_{CT,t}$ is average value (commercial or house prices) per meter in CT at time t,

$pop_{CT,t}$ is population in CT at time t,

$res_{CT,t}$ is the number of residential buildings in CT at time t,

$n\mathrm{Res}_{CT,t}$ is the number of non-residential buildings in CT at time t,

$[LU = \text{Res}]$ is a binary expression which is true (equals 1) if AV relates to house prices and 0 otherwise

$$V_{j,t} = AV_{CT,t} * FS_j * \left({SL_{j,t}}/{SL_{CT,t}} \right)^{[LU=\text{Res}]} \tag{10}$$

where:

$V_{j,t}$ is the house price of a dwelling unit in building j at time t,

$SL_{s,t}$ is the service level within area s at time t—the ratio of non-residential buildings to residential buildings in the census tract CT if $s = CT$ or, if $s = j$, within a 100 m buffer of building j.

$$P_{DU,t} = \frac{{\bar{I}_t}/{3} * \left(1 + \frac{{V_{j,t}}/{\#Ap_j} - \sum_{l=1}^{L_t} V_{l,t} \Big/ \sum_{l=1}^{L_t} \#Ap_l}{P_{\sigma_t}} \right)}{c} \tag{11}$$

where

$P_{DU,t}$ is the monthly cost of living in dwelling unit DU at time t,

$\bar{I}_t$ is the average household income in the study area at time t,

$\#Ap$ is the number of DUs within a building. If the building is initially residential, this is equal to its initial population size. Otherwise it is the floor-space volume of the building divided by 90 (the average DU size in meters),

L_t is the number of residential buildings in the study area at time t,

P_{σ_t} is the standard deviation of DU prices within the study area at time t,

c is a constant.

4 Context and Data

The simulation involves a hypothetical earthquake in the CBD of Jerusalem. While the CBD lies in a relatively stable seismic area, the city itself is located only 30 km east of the active Dead Sea Fault Line. Moreover, the majority of the buildings in the city center were constructed prior to the institution of seismic-mitigation building regulations, hence they are prone to damage (Salamon et al. 2010). This study area covers 1.45 km^2 and includes two major commercial spaces: the Mahaneh Yehuda enclosed street market and the CBD (see Fig. 2). The heterogeneous mix of land uses is represented by residential buildings (243 Th m^2, 717

Table 1 Variables used in the model

Variable	Source	Spatial unit
Residential building value per m^2	National Tax authority 2008–2013	SA
Non-residential plant and machinery value	Local authorities financial data (CBS) (Beenstock et al. 2011)	Local authority
Number of households	CBS 2008	SA
Number of inhabitants	CBS 2008	SA
Average monthly household earnings	CBS 2008	SA
Labor force participation	CBS 2008	SA
Employment by sector	CBS 2008	SA
Percent disabled	CBS	SA
Age	CBS	SA
Workplace location	GPS survey 2014	Survey of individuals

structures and 22,243 inhabitants), commercial buildings (505 Th m^2, 119 structures) and government/public use buildings (420 Th m^2, 179 structures). Three major transportation arteries traverse the area and generate heavy traffic volumes: Agripas and Jaffa (light railway route) Streets run north-west to the south-east and King George Street runs north-south.

The available data on buildings and population in the study area is aggregate data provided at the level of the Statistical Areas (SA).[2] The data drives the model at three different spatial resolutions: buildings, households and individuals. The variables used to populate the buildings can be grouped into three categories and are defined in Table 1:

- Building level: land-use, floor-space, number of floors, building value, households.
- Household level: inhabitants, earnings, car ownership.
- Individual level: Household membership, disability, participation in the work force, employment sector, age, workplace location.

We use a GIS buildings layer to provide the distribution of all buildings nationally with their inherit attributes such as aerial footprint, height and primary land use. These attributes are utilized to calculate the floor-space of each building. We then calculate the value of buildings according to their land use and floor space. Inhabitants' socio-economic attributes are proportionally allocated to each building, using a methodology described in Lichter and Felsenstein (2012). The disaggregated building level data serves as the basis for the further disaggregation at the

[2]A statistical area (SA) is a uniform administrative spatial unit defined by the Israeli Central Bureau of Statistics (CBS). It corresponds to a census tract and has a relatively homogenous population of roughly 3000 persons.

level of the individual. This begins with assigning each individual in the database a unique id, so that it is represented as an entity tied to a building in the database. Next, each person is allocated a random point location (a lat, lon coordinate) in the building with which it is associated. Demographic attributes (labor force participation, employment sector, disabilities and age group), assigned to buildings in the previous stage, are allocated to each individual so that they comprise the entire distribution in the building. Individuals are then clustered into households according to the household size (number of persons) in each building. Households are also represented as unique entities in the database and are associated with buildings. The clustering introduces heterogeneity in terms of the age distribution to closely represent a "traditional family household" having both adults and children when these are present in the building. A household entity represents the sum or average of the attributes of its members and is further assigned attributes such as earnings and car ownership in the same way these were assigned to individuals. The distribution of work locations of inhabitants by employment sector is derived from a GPS-basedtransport survey carried by the Jerusalem Transport Master Plan Team (Oliveira et al. 2011). We use this data to create a distribution of inhabitants working within and outside the study area, by sector of employment.

4.1 Case Study Specifications

The impact of the earthquake is modeled here as a one-time shock diffusing from a focal point. The effects of this shock result in physical damage to buildings and the road network. The probability that a building will collapse is proportional to its height and distance from the epicenter (Carenno et al. 2012) and is compared to a randomly-generated physical resilience score:

$$b_j = damaged \Rightarrow \left[R_j < \frac{c * 10^{mag}}{D_j * \left|\log(D_j)\right| * F_j} \right] = 1 \qquad (12)$$

where

R_j is the resilience score for building j,
c is a constant,
mag is the earthquake magnitude (similar to the Richter scale),
D_j is distance of building j from the earthquake epicenter,
F_j is number of floors in building j.

Every building that collapses becomes vacant and unusable. All residents choose between migrating and relocating (see Sect. 3.1.1). A collapsed building blocks the closest road. Buildings are restored to pre-shock size and the duration required for this recovery (and the attendant re-opening of the blocked road) is proportional to floor-space (Carenno et al. 2012). Upon restoration, the building does not automatically retain pre-shock use.

The earthquake is simulated 25 times[3] with its epicenter randomly located in order to avoid location-based bias in the results. Each simulation comprises 1000 iterations (ticks) where each tick represents one activity day of resident-agents. In each simulation, the earthquake occurs after 50 iterations in order to let market dynamics kick-in and stabilize. The results below describe the averages of all simulations.

5 Results

The results presented below relate to both short and long term impacts. They differentiate between global impacts and their temporal and spatial distributions and between effects on population flows and housing stocks.

5.1 Aggregate Patterns and Equilibrium

Folke et al. (2002) conceptualize resilience as the ability of a system to reorganize itself following a change. This stresses the notion of moving beyond recovery to pre-shock state and attaining stability. The global indices of resilience presented below are constructed in this spirit. They quantitatively assess the tendency of a system to achieve a stable equilibrium. A system meets stability criteria if it registers consistent value levels over a consecutive period of days. Specifically, we relate to the difference between current value and the average value over the preceding 50 days. Attainment of equilibrium is defined as the first day (counting back from the end of the simulation) when the day-to-day change is not significant. Table 2 presents the frequencies of achieving equilibrium, along with average durations and final changes in value (final to pre-shock value ratio) for a variety of indicators.

The study area is resilient to the shock across most dimensions but not in the classical recovery sense of the term. Residential and non-residential capital stock both stabilize in terms of size (number of buildings) and value in the simulations but usually on values different to those existing under pre-shock conditions (the exception being average residential value). Population, the most mobile element in the model, does not show such stability. In the majority of cases (14/25, i.e. 56 %), income presents a continuing change, while in the other cases equilibrium is achieved quite late. This suggests that in spite of population size stabilizing, migration flows keep on affecting the composition of population.

---

[3]This arbitrary number was chosen in order to balance between computing loads and convergence of results.

Table 2 Global resilience measures

Parameter	Variable	Frequency of equilibrium (out of 25 simulations)	Average duration to achieve equilibrium (days)	Average final change (% of pre-shock value)
Population	Population	24	397	67.85
	Average income	11	950	50.53
Residential stock	Residential stock size (# buildings)	25	332	88.34
	Average residential value	22	677	96.12
Non-residential stock	Non-residential stock size (# buildings)	23	670	142.43
	Average non-residential value	25	385	78.61

The new situation of a larger, yet cheaper, non-residential capital stock is not fully explained by the decreasing value of non-residential stock, as stock size stabilizes long after values. Floor-space volume, which also affects values, may account for this trend as it decreases by 20 % on average. Most of this decrease happens within 300 days of the shock and a negligible difference (0.56 %) is registered between values at day 350 and day 1000, correlating with the time required for non-residential values to achieve equilibrium. Therefore, these values experience a large initial shock. The new non-residential functions that appear over time are smaller in terms of floor-space and do not further affect values.

Residential stock presents a mirror image with average values lagging behind size. Supply and demand (population size) dynamics do not temporally correlate with the recovery of average values. This suggests that these effects may be attributed to other elements in the house price mechanism such as service levels (number of non-residential buildings). These stabilize just before residential values and floor-space volume. The latter increases by 12.7 % on average. These two trends are sufficient to contain the effects of the sharp decrease in population thereby creating a recovery scenario in relation to housing values.

5.2 *Spatial and Temporal Distribution of Effects on Stocks*

As outlined in the previous section, the average post-shock picture is one of smaller and cheaper commercial functions along with slightly larger residential buildings.

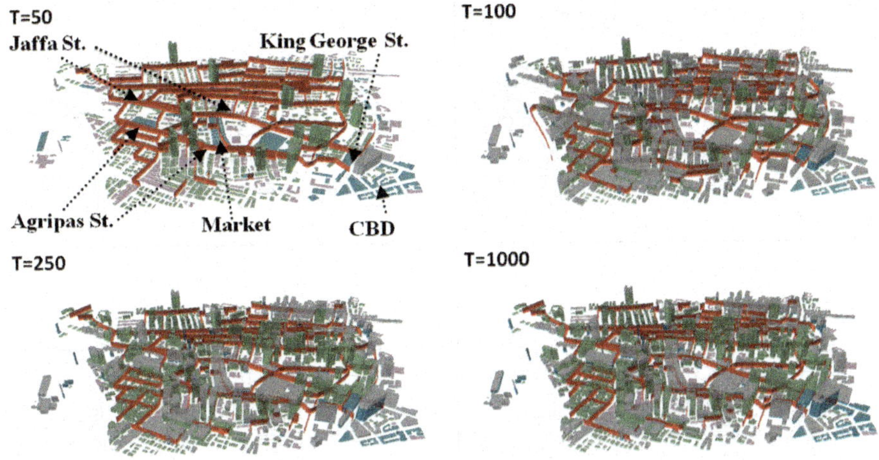

Fig. 2 Spatial distribution of land-use change. Height represents the number of simulations in which land-use change (*colored*) and vacancy (*in grey*) occurs. Building color indicates initial land-use: residential (*green*), commercial (*blue*), or public (*pink*). Road heights represent average traffic load. Roads in *red* denote values above the average

Yet inequalities in the distribution of these changes may exist, as some areas may enjoy/suffer their consequences more than others. Furthermore, these distributions may or may not be stable over time. Figure 2 represents the distribution of land-use over these two dimensions. The unequal spatial pattern is evident as the propensity for new commercial functions to emerge is greater in the areas south-west and north-east of the market. These new land uses tend to stabilize over time, as indicated by the diminishing vacancy rate. This phenomenon may be attributed to centripetal and centrifugal forces. The centrifugal force is set in motion by the physical damage of the earthquake blocking movement paths. In search of new routes, traffic patterns disperse from pre-shock state to the south-west and north-east[4] (Fig. 2, T = 100). As traffic loads are the locus behind the spatial pattern of commercial uses (see Sect. 3.2.1), their dispersal increases the potential profitability of new locations, attracting new activities away from previous clusters. Yet, the new emerging functions exert a self-enhancing centripetal force through a cyclical process of influence: new functions attract more traffic, which increases their profitability and attracts more uses, which further attract traffic loads. This centripetal force continues to work long after the effects of the initial shock have subsided thus perpetuating some of the new traffic patterns even after blocked roads start to open up (T = 250, T = 1000).

[4]This is also apparent in change to the average standard deviation of traffic loads (agents per meter). Over time, the average s.d. decreases by 73.58 % in relation to pre-shock state, suggesting a more even dispersal of traffic loads.

5.3 Spatial and Temporal Population Dynamics

In order to characterize population flows we calculate a normalized weighted composite 'Social Vulnerability Index' (SVI; see Lichter and Felsenstein 2012) at the level of the building. The relative weights of the elements comprising the index reflect their contribution to aggregate socio-economic vulnerability.

$$SVI_{b,t} = 0.5 * \bar{I}_{b,t} + 0.1 * Cr_{b,t} - 0.2 * \bar{A}_{b,t} - 0.2 * Ds_{b,t}$$

where

$\bar{I}_{b,t}$ is the average monthly income of households residing in building b at time t,
$Cr_{b,t}$ is the rate of car ownership of households residing in building b at time t,
$\bar{A}_{b,t}$ is the average age within households residing in building b at time t,
$Ds_{b,t}$ is the share of residents who suffer disability building b at time t.

Assessing the spatio-temporal distribution of population flows is achieved via a two-step procedure. First, we interpolate the individual SVI values of buildings onto a continuous surface[5] and then we calculate the Local Indicators of Spatial Autocorrelation[6] (LISA; Anselin 1995) for each cell. We do this at temporal intervals of 50 days. This procedure allows for identifying clustering and dispersal patterns over time. Figure 3 shows the significant clusters of similar values (high LISA values).

As in the case of land use, the results suggest the existence of a similar centripetal-centrifugal tension for population flows. The shock breaks down the initial divide of a less vulnerable western cluster and vulnerable eastern areas (T = 50) into a pattern of clusters surrounded by areas of mixed population (T = 150). The tendency of households to choose living environments that preserve previous conditions (see Sect. 3.1.1) acts as a centripetal force that makes pre-shock clusters more attractive. As the centrifugal effect of the shock is only temporary, clustering continues and the clusters grow and become more punctuated (T = 250 and T = 1000).

In spite of these similarities, there is a subtle yet important difference between effects on flows and on stocks. While the dissolution of previous clusters into new agglomerations may be interpreted as a sign of recovery and efficiency in an economic system, in the social context this is not necessarily the case. This process of re-grouping may in fact exacerbate vulnerability as existing communal support systems may stop functioning. This is especially true in the case of vulnerable populations. As they are less mobile due to greater budget constraints, they are exposed to the effects of more resilient, in-migrant households. If such households are characterized by high tolerance to change, they can easily relocate to more

---

[5]We use an Inverse Distance Weighting procedure. The parameters used are: pixels of 10 × 10 m, 100 m search radius and 2nd order power function.

[6]Neighborhood is defined using a queen contingency matrix.

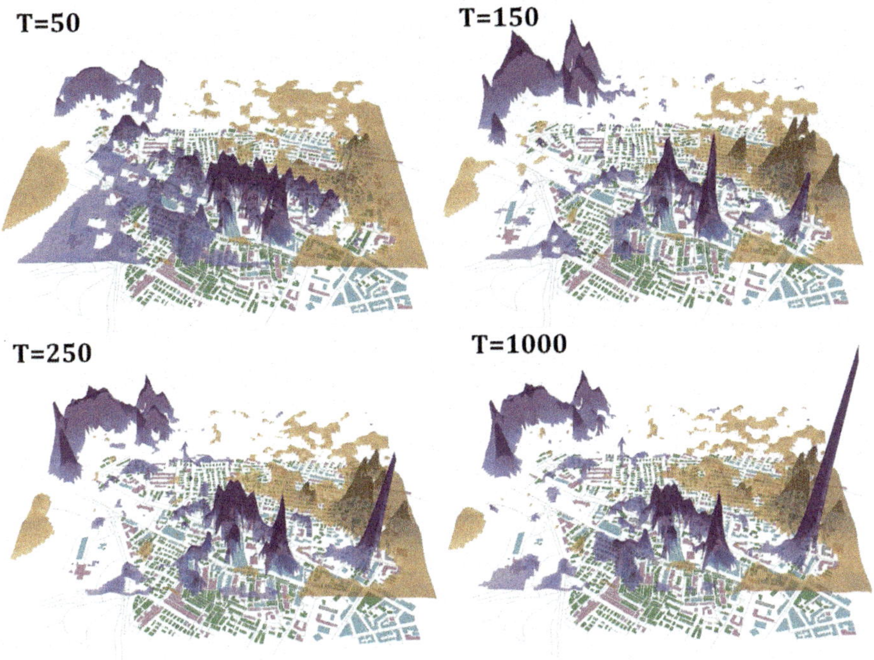

Fig. 3 Spatio-temporal distribution of SVI. *Purple* indicates higher absolute SVI values, *brown* indicates lower. Height represents LISA values. Building color indicates land-use at time T in the majority of the simulations: residential (*green*), commercial (*blue*), or public (*pink*)

vulnerable areas. By doing so, they act as agglomeration nuclei, changing the nature of their environment and attracting more population similar to them. As social mixing increases, social cohesiveness of the neighborhood decays and its institutions break down. Vulnerable populations are thus faced with the choice of remaining with no support or migrating to 'ghettos'.

5.4 Interactive Web-GIS Application for Visualization of Results

As the simulation outputs are multi-dimensional and include vast amounts of information on urban dynamics both spatial and temporal, we use web-GIS to communicate the results. The complexity of outputs is hard to internalize or visualize in their entirety using traditional graphic representations. We develop a web-based application to allow interactive visualization and querying of the multi-dimensional output in an intuitive and user-friendly fashion. (see http://ccg.huji.ac.il/AgentBasedUrbanDisaster/index.html.) The site serves solely as a visualization tool for pre generated results and not as a vehicle for distributing the model. Using a

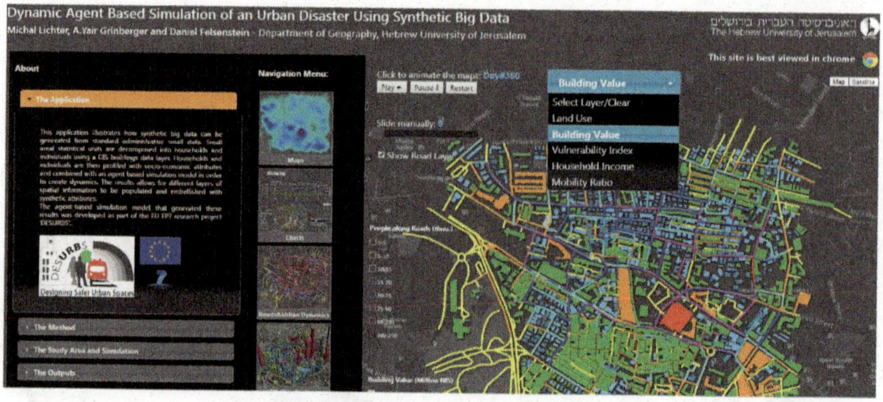

Fig. 4 Web-based querying and visualization application of selected variables on a dynamic web-map (see http://ccg.huji.ac.il/AgentBasedUrbanDisaster/index.html)

simple web browser, users can generate time lapse visualizations in the form of maps and charts without any previous knowledge in handling spatial data or using GIS. They can choose a variable of interest, visualize its change over space and time and create location-specific information. This interaction is facilitated by simply clicking on the map or the chart or using gauges such as buttons and sliders (Fig. 4). When initiating such an event, a single click can trigger complex querying of the database in the background. This necessitates database design and construction in a way that allows for fast and efficient data extraction. We create a dedicated database for the output results of time series from the model simulation. This is achieved by using DB design that does not always follow strict design guidelines but rather contains some flat tables to enable lateral data charting, displayed in pop-ups, graphs and charts. The visualization includes time lapse representation of human mobility (household level), changes in passengers along roads, changes in building land use and value, household socio-economic change and so on.

The web-mapping platform is Google Maps API. Middleware functionalities are added to the application based on JavaScript libraries and APIs. These functions interact with the web-mapping platform to provide ancillary capabilities (Batty et al. 2010) such as time laps animation, action buttons, sliders, interactive graphs etc.

6 Conclusions

The findings above have looked at urban resilience at both the metropolitan and local scales. With respect to the former, we observe that the resilience patterns of residential and non-residential (commercial) capital stock are very different. Post-disaster, non-residential stock attains equilibrium based on a pattern of smaller, less expensive units, while the opposite holds for residential stock. Population levels stabilize much faster than income, indicating demographic turnover and churning. At the local scale, we look at the difference between the resilience of stock variables, exemplified by land use and that of flow variables, represented by a composite measure of social vulnerability. Our main finding is that in the advent of a shock, both stock and flow variables disperse and re-aggregate over time. However more resilient socio-economic groups cope better with dispersing and then re-clustering. Less resilient populations are more in need of community support systems and cannot rejuvenate quickly.

To add further realism, future work will need to relax some of the strong behavioral assumptions underlying the model. For example, the demand for housing is currently determined by affordability and attractiveness of the units on offer. This is a slightly mechanistic representation of a process that generally involves bidding, expectations and perceptions of opportunities. On the supply side, it would be useful to explicitly include the behavior of building contractors. At present, housing supply is driven by land use change and in particular by commercial land use becoming residential. Furthermore, migration behavior in the model is currently motivated by steady-state probabilities of movement augmented by the destruction of buildings. This results in mass flight followed by stabilization at a lower level equilibrium. We do not capture the psychological over-reaction of population movement identified in the literature (Stein et al. 2010; Whitehead et al. 2000). In this respect, our results may be downwardly-biased.

If stronger populations have the resources to accommodate the negative impacts of a disaster, then urban resilience is thus as much about economic welfare as it is about engineering or morphology. From a socioeconomic perspective, it is not the magnitude of the disaster that is important but the ability to cope with its results. Vulnerable populations or communities can be disproportionately affected by unanticipated disasters which are more likely to push them into crisis relative to the general population. Much of this can only be detected at the micro level such as the household or building. This is often smoke-screened in studies dealing with aggregate city-wide impacts. The use of highly disaggregated and accurately profiled data is thus critical in understanding urban resilience.

Two policy implications arise from the above findings. First, as an exogenous shock has no predetermined outcomes, a disaster may elicit wildly diverging responses in different urban environments. This has policy implications calling into question much of the popular literature advocating a 'one size fits all' approach to urban resilience (Prasad et al. 2009; UNISDR 2012). While well intentioned, the standard check-list approach to promoting resilience may be misleading. Second, the dynamic simulation outcomes point to differential rates of recovery over time across the components of the urban system (for example residential and commercial capital stock). In an effort to 'get things done' in the aftermath of a disaster, public policy for urban recovery is often characterized by knee-jerk (over) reaction that involves time-compressing rebuilding and rejuvenation measures (Olshansky et al. 2012). The redevelopment opportunities for large scale urban change over a short period of time, afforded by disaster, fail to recognize the existence of multiple and unstable urban equilibria resulting from different activities recovering at different rates. Urban resilience is as much about judiciously synchronizing recovery across the urban system as it is about getting cities to 'bounce back'.

Acknowledgments Thanks to two referees for valuable comments on an early draft.

References

Anselin, L. (1995). Local indicators of spatial association—LISA. *Geographical Analysis, 27*(2), 93–115.

Axtell, R. (2000). *Why Agents? On the varied motivations for agent computing in the social sciences* (Working Paper 17). Washington, DC: Center on Social and Economic Dynamics. http://www2.econ.iastate.edu/tesfatsi/WhyAgents.RAxtell2000.pdf. Accessed March 1, 2015.

Batty, M., Hudson-Smith, A., Milton, R., & Crooks, A. (2010). Map mashups, web 2.0 and the GIS revolution. *Annals of GIS, 16*(1), 1–13.

Beenstock, M., Felsenstein, D., & Ben Zeev, N. (2011). Capital deepening and regional inequality: An empirical analysis. *Annals of Regional Science, 47*(3), 599–617.

Campanella, T. J. (2008). Urban resilience and the recovery of New Orleans. *Journal of the American Planning Association, 72*(2), 141–146.

Carenno, M. L., Cardona, O. D., & Barbat, A. H. (2012). New methodology for urban seismic risk assessment from a holistic perspective. *Bulletin of Earthquake Engineering, 10*, 547–565.

Chen, Y., Li, X., Wang, S., & Liu, X. (2012). Defining agents' behavior based on urban economic theory to simulate complex urban residential dynamics. *International Journal of Geographic Information Systems, 26*(7), 1155–1172.

Chen, X., & Zhan, F. B. (2008). Agent-based modeling and simulation of urban evacuation: Relative effectiveness of simultaneous and staged evacuation strategies. *Journal of the Operational Research Society, 59*(1), 25–33.

Crooks, A. T., & Castle, C. J. E. (2012). The Integration of agent-based modeling and geographical information for geospatial simulation. In A. J. Heppenstall, A. T. Crooks, L. M. See, & M. Batty (Eds.), *Agent-based models of geographical systems* (pp. 219–251). Dordrecht: Springer.

Crooks, A. T., & Wise, S. (2013). GIS and agent based models for humanitarian assistance. *Computers, Environment and Urban Systems, 41*, 100–111.

Dawson, R. J., Peppe, R., & Wang, M. (2011). An agent-based model for risk-based flood incident management. *Natural Hazards, 59*(1), 167–189.

Folke, C., Carpenter, S., Elmqvist, T., Gunderson, L., Holling. C.S., & Walker, B. (2002). Resilience and sustainable development: building adaptive capacity in a world of transformations. *AMBIO: A Journal of the Human Environment, 31*(5), 437–440.

Fujita, M., & Thisse, J. F. (2002). *Economics of agglomeration: Cities, industrial location and regional growth*. Cambridge, UK: Cambridge University Press.

Godschalk, D. R. (2003). Urban hazard mitigation: creating resilient cities. *Natural Hazards Review, 4*(3), 136–143.

Kwan, M. P., & Lee, J. (2005). Emergency response after 9/11: The potential of real-time 3D GIS for quick emergency response in micro-spatial environments. *Computers, Environment and Urban Systems, 29*(2), 93–113.

Lancaster, K. J. (1966). A new approach to consumer theory. *Journal of Political Economy, 74*, 132–157.

Lichter, M., & Felsenstein, D. (2012). Assessing the cost of sea-level rise and extreme flooding at the local level: A GIS-based approach. *Ocean and Coastal Management, 59*, 47–62.

Müller, B. (2011). Urban and regional resilience—a new catchword or a consistent concept for research and practice? In B. Müller (Ed.), *German annual of spatial research and policy 2010* (pp. 1–13). Berlin-Heidelberg: Springer.

Narzisi, G., Mysore, V., Rekow, D., Triola, M., Halcomb, L., Portelli, I., et al. (2006). Complexities, catastrophes and cities: unraveling emergency dynamics. In H. Schärfe, P. Hitzler, & P. Øhrstrøm (Eds.) *International conference on complex systems, Boston, MA, June 2006* (Vol. 4068). Lecture notes in computer science (Lecture notes in artificial intelligence). Berlin, Heidelberg: Springer.

Oliveira, M. G. S., Vovsha, P., Wolf, J., Birotker, Y., Givon, D., & Paasche, J. (2011). Global positioning system-assisted prompted recall household travel survey to support development of advanced travel model in Jerusalem, Israel. *Transportation Research Record: Journal of the Transportation Research Board, 2246*(1), 16–23.

Olshanky, R. B., Hopkins, L. D., & Johnson, L. (2012). Disaster and recovery: Processes compressed in time. *Natural Hazards Review, 13*(3), 173–178.

Prasad, N., Ranghieri, F., Shah, F., Trohanis, Z., Kessler, E., & Sinha, R. (2009). *Climate resilient cities: A primer on reducing vulnerabilities to disasters*. Washington, DC: World Bank publications.

Reichert, P., & Mieleitner, J. (2009). Analyzing input and structural uncertainty of nonlinear dynamic models with stochastic, time-dependent parameters. *Water Resources Research, 45* (10), W10402. doi:10.1029/2009WR007814.

Rose, A. (2009). *Economic resilience to disasters* (Community and Regional Resilience Research Report 8). Oak Ridge, TN: Oak Ridge National Laboratory. http://research.create.usc.edu/published_papers/75. Accessed March 1, 2015.

Salamon, A., Katz, O., & Crouvi, O. (2010). Zones of required investigation for earthquake-related hazards in Jerusalem. *Natural Hazards, 53*(2), 375–406.

Schelling, T. (1971). Dynamic models of segregation. *Journal of Mathematical Sociology, 1*(2), 143–186.

Simon, H. (1952). A behavioral model of rational choice. *Quarterly Journal of Economics, 69*(1), 99–118.

Stein, R. M., Duenas-Osorio, L., & Subramanian, D. (2010). Who evacuates when hurricanes approach? The role of risk, information and location. *Social Science Quarterly, 91*(3), 816–834.

UNISDR (United Nations International Strategy for Disaster Reduction). (2012). *How to make cities more resilient—a handbook for local government leaders*. Geneva, Switzerland: UNISDR.

Whitehead, J. C., Edwards, B., Van Willigen, M., Maiolo, J., Wilson, K., & Smith, K. T. (2000). Heading for higher ground: Factors affecting real and hypothetical hurricane evacuation behavior. *Environmental Hazards, 2*, 133–142.

Zimmerman, B., Nawn, D., Wang, Y., Kuhlman, B., Sochats, K., Luangesorn, L., et al. (2010). Dynamic model generation for agent-based emergency response simulation. ESRI International User Conference, Center for National Preparedness, University of Pittsburgh.

Chapter 7
Data Integration to Create Large-Scale Spatially Detailed Synthetic Populations

Yi Zhu and Joseph Ferreira, Jr.

Abstract Many planning support systems and, indeed, some 'smart city' initiatives begin with time consuming efforts to integrate cross-agency data describing current conditions in sufficient detail to support 'what-if' exploration of urban development options. Integrating data from different sources has become increasingly challenged as available datasets, and the relevant urban modeling efforts, become more disaggregated and spatial-temporally detailed. Open data initiatives, with unprecedented amounts of embedded georeferenced information, have made web services and crowdsourcing attractive. However, data from such sources are typically imperfect and their integration is complicated by syntactic and semantic differences. We develop an ontology-based data integration mechanism to fuse data from different sources in generic ways that can utilize semantic information to minimize the labor involved and facilitate updating as new data are acquired. As a test application, we evaluate, filter, adjust, and integrate building information from heterogeneous data sources for use in an agent-based microsimulation model of transportation and land-use dynamics in Singapore. Third-party data about building size, age and use added substantial value to the official datasets generally available from government agencies.

Y. Zhu (✉)
Singapore-MIT Alliance for Research and Technology, 1 CREATE Way, #09-02 CREATE Tower, Singapore 138602, Singapore
e-mail: zhuyi@mit.edu

J. Ferreira, Jr.
Department of Urban Studies and Planning, Massachusetts Institute of Technology, 77 Massachusetts Avenue, Cambridge, MA 02139, USA
e-mail: jf@mit.edu

S. Geertman et al. (eds.), *Planning Support Systems and Smart Cities*, Lecture Notes in Geoinformation and Cartography, DOI 10.1007/978-3-319-18368-8_7

1 Introduction

Many planningsupportsystems and, indeed, some 'smart city' initiatives begin with time consuming efforts to integrate cross-agency data describing current conditions in sufficient detail to support 'what-if' exploration of urban development options. The models used for these what-if explorations are data demanding and large-scale land-use and transportation models are no exception. Detailed information on the urban built environment, demographic and socio-economic profiles, human activities and trips, and real estate are required for model calibration, validation and prediction. As modeling approaches evolve from aggregated to disaggregated, and from static to dynamic, the need for data with high spatial and temporal resolution and detail is growing. More detailed spatial data would enable better characterization of urban built environment that are further distinguished by increasingly complicated urban policies and developments. Furthermore, to ensure the temporal consistency of different models, large-scale land-use and transportation modeling require data that are temporally more precise for calibration and validation.

In fact, a growing number of large-scale land-use and transportation models, including MATSIM (Balmer et al. 2009), DynaMIT (Ben-Akiva et al. 2010), UrbanSim (Waddell 2002) and ILUTE (Salvini and Miller 2005), have adopted agent-based microsimulation approaches to forecast the travel demand and land use patterns under different scenarios. These models generally require complete lists of agents like households, persons and firms to be initialized with realistic attributes, locations, relationships and behaviors at the beginning of a simulation. Although in most cases the detailed information of full populations can be found in census surveys, real estate tax records, building licenses, etc., privacy reasons and policy restrictions usually make such data inaccessible to researchers. Population synthesis approaches have been developed to combine microdata samples that lack spatial detail with spatially aggregated marginal data about population characteristics in order to expand the microdata sample into a fully attributed, albeit approximate, synthetic population. These synthetic populations are crucial to generate a realistic starting point for microsimulations that can support scenario analyses of urban development options. If done well, the synthetic populations have the added advantage of being more readily shareable since the attributes of any particular synthetic person or place cannot be linked to a real-world person or place with 100 % accuracy.

The interest in generating synthetic population microdata of small geographical areas from aggregated data can be dated back to 1980s (Birkin and Clarke 1988). However, as modeling efforts have become more and more disaggregated and spatial-temporally detailed, the data collection and integration burden needed to construct viable synthetic populations has not only increased but also expanded beyond traditional interpolation methods. Data from single sources are often imperfect and cannot fulfill the requirements of model development. Often, information about the same entity from multiple sources may be complementary with different attributes and different strengths and weaknesses. Open data initiatives that many government agencies have embraced have enabled unprecedented amounts of spatially-referenced data to become available from web services and crowdsourcing. In principle, data fusion efforts can produce an improved and more complete dataset more suited to modeling needs. However, the data typically involve heterogeneous formats, qualities and semantics. Better tools are needed to fuse the data in generic ways that take advantage of newly emergent data sources, utilize the semantic information in ways that minimize the labor involved, and can more readily be updated as new data are acquired. This need is especially true for the new generation of urban land-use and transportation models that may simulate urban dynamics with different starting points, which require generating different synthetic populations representing snapshots of the real world for various scenarios.

In this chapter, we describe an ontology-based data integration mechanism to evaluate, filter, adjust and integrate geospatial information from heterogeneous data sources. The ontology defines the vocabularies and concepts that are commonly understood in the real estate and transportation domains. The focus is not to build a comprehensive ontology for all urban planning support purposes. Instead, we develop a prototype that allows semantic-level matching of local data and a simplified ontology both to demonstrate the benefits and applicability of the proposed data integration approach and to test its usefulness in a particular 'real world' setting. We generate a synthetic population of households and real estate for use in an agent-based microsimulation model of transportation and land-use change dynamics in Singapore, a city-state occupying 725 km^2 with a population of over five million persons and approximately 100 thousand buildings.

The remainder of the chapter is organized as follows. In Sect. 2, we discuss different types of data heterogeneity. In Sect. 3, we review previous research efforts on data integration, especially the usage of ontology, in the context of urban planning. Section 4 presents the proposed ontology-based data integration approach. In Sect. 5, the proposed integration approach is applied to the population synthesis in the Singapore case. Finally, Sect. 6 contains the conclusions.

2 Heterogeneity in Multiple Datasets

Although the volume of available data from emerging sources like ubiquitous urban sensors, personal electronic devices and web-based services are growing rapidly, only small fractions of these data are readily discoverable and accessible, much less usable. Whilst the emergent data appear to be able to assist urban modeling, their values can be extracted fully only when being coupled with other datasets. One of the biggest challenges in data integration is how to reconcile various types of heterogeneities stemming from different data sources. A diverse range of heterogeneities and conflicts can be found among datasets from different sources, as listed in Table 1.

Table 1 Types of data heterogeneities and examples

Type of heterogeneity	Description	Examples
Syntax	A set of rules to encode data	• Different tuple (entity, identifier)
Structural	Schemes of organizing and storing data (hierarchical, network, relational etc.)	• Different cardinality between entities
		• Different entity-identifier relationship
Semantic	Rules symbolically defining the relationships between data and real world entities	• Polysemy, synonymy and homonym
		• Different spatial/temporal scale and representation
		• Different measurement unit
		• Different abbreviation
		• Different categorization
Data quality	Whether data properly represent the real-world construct and are fit to serve its purpose in a given context	• Accuracy
		• Completeness
		• Source trustworthiness
Data type	Types and formats of data	• Different data format and type (integer, float, Boolean etc.)
		• Types of raw information (structured, unstructured, etc.)

While variations in data models and schema lead to syntax and structural differences among datasets, semantic difference is more common due to different naming or categorization conventions of data providers. In addition, heterogeneity in data quality, in the timing of data snapshots, and in privacy and security requirements also hinder the integration of relevant information from different data sources.

3 Related Work

3.1 Existing Data Integration Approaches

To tackle the issue of data interoperability and integration, computer scientists and software engineers have developed several approaches. Software vendors and consortiums like W3C (World Wide Web Consortium) and OGC (Open Geospatial Consortium) set standards on data formats and representations. Examples are XML for web service based data sharing and CityGML for the storage and exchange of virtual 3D city models. But such standards only address the syntax heterogeneity of datasets of the same type. They can facilitate the structuring of online data warehouses but they focus more on the consistent formatting of integrated datasets rather than on the fusion of data from heterogeneous sources.

Federated database approaches to data integration use a global schema and mediators to reconcile semantic differences of local database management systems (DBMS) (Ziegler and Dittrich 2004). Mediator-wrapper architecture is used to translate user defined queries to queries that can be carried out by wrappers built for local databases. Most applications of this type are generally based on relational data models and global schemas tightly coupled with tasks or applications (Ziegler and Dittrich 2004). Schema matching is typically performed manually by database or domain experts. Traditional data manipulation built on such relational database schema have difficulty integrating the semi-structured or unstructured data that are increasingly common as online websites and 'open data' efforts vastly expand the availability of relevant data about spatially and temporally-detailed urban settings.

This deluge of new data can greatly enhance traditional data integration efforts if the various semantics, temporal and spatial differences, and quality issues can be systematically addressed. The data fusion steps needed in such cases require more complicated filtering and merging operations involving comparisons, cross-validation, corrections, realignment and aggregation.

End users need not know the nuts and bolts of every dataset, if they can understand and trust the merge and filter operations. It is also unrealistic to expect all end users to have the expert skills to tackle each dataset. As the size and the variety of relevant data accumulate over time, it becomes more important to have a single entry point to the integrated data based on transparent and trusted operations that can be updated and tuned with minimal amounts of human intervention. Successful data integration requires a system that recognizes the meaning, representation and structure of the data as part of the data fusion process (Nathalie 2009).

3.2 Ontology-Based Data Integration

As suggested in the last section, data integration approaches relying on mediator-wrapper architecture and middleware are usually task-specific and will become more complex and cumbersome when facing complicated urban modeling and analysis. Meanwhile, the data processing and schema matching procedures will have to be repeated for each planning support system or planning analysis that is undertaken. It will be desirable if the data integration approach can map local datasets from different sources to a purpose-generic but domain-specific global schema which provides a common representation and interpretation of the information needed for the research and application in the domain, and a single entry point for users. Ontology-based methods offer a promising approach that can provide urban planners and researchers with a single entry point to rich 'snapshots' of current or past conditions in an urban area.

As defined by Mars (1995, p. 9), ontology is "*a structured, limitative collection of unambiguously defined concepts*". In general, ontology explicitly defines the meanings, representations and the structures of the concepts in a domain, which forms the mutually understandable, sharable and reusable domain knowledge repositories. Many researchers have shown that ontologies can be a valid tool for reconciling conflicts in the semantics and syntax of datasets that come from multiple heterogeneous sources (Partridge 2002; Nathalie 2009).

An integrated ontology from one or several domain-specific ontologies can benefit the community by providing guidance on generic vocabularies and correspondences to represent data and modeling related knowledge. Ontology research has been active in the field of geographic information systems since the late 1990s when Frank (1997) discussed the potential of ontology to improve interoperabilityamong various geographical encoded datasets. Later, Fonseca et al. (2002) proposed to use a multiple-layer ontology to integrate geographical information and remote sensing information, based on semantic values even if the geometric representations involved different coordinate systems and granularity. They argued the classification process facilitated by the ontology make the geospatial data not just static polygons with pixel values but semantic features and corresponding values, which could enable a broad set of inquiries. More recently, Ramos et al. (2013) used ontology to identify sets of homologous features in the volunteered geographic information (VGI) like OpenStreetMap and the information from authoritative sources. This is a step toward the objective of drawing on VGI to improve existing geospatial data and make it more complete and up-to-date.

However, although ontology-based approaches have reached a good level of maturity in other fields, their applications in the domain of urban planning and modeling is limited (Benslimane et al. 2000). Most existing efforts focus on the construction of ontology as a shared vocabulary to improve communication or to facilitate knowledge management (Teller et al. 2007).

4 Proposed Approach for Data Integration

Large-scale land-use and transportation models deal with a large amount of objects and concepts in an urban system that can take on various representations and relationships. Different classification and naming conventions as well as different reference systems adopted by different agencies add to the complexity of the already intricate integration tasks. For example, buildings can be geometrically represented as points or footprint polygons, but in other cases as 3Dvolumes or nested polygons representing footprints at various floor levels. In transit network analysis, a node may not be the same as a vertex. Edges could refer to links, connections, or routes, which can be confusing for inexperienced data users. As the amount of data grows, it is not only desirable but also imperative to build a knowledge base to explicitly support interpretation and exchange of entity representations and correspondences in a field. The proposed ontology-based framework covers ontology building, ontology mapping, and custom tools to support integrating, sharing and exchanging the information needed for population synthesis and potentially other parts of large-scale land-use and transportation modeling.

4.1 General Framework

Figure 1 illustrates the proposed framework of the ontology-based data integration approach used in this study. The framework centers on a domain ontology, which is used as an unambiguous and persistent knowledge reservoir consisting of the concepts and relationships used in the activity-travel analysis. In the domain ontology, concepts and relationships are elaborated and annotated by the definition and terminology that are generally accepted in the field. As outlined in Fig. 1, the approach contains a series of sub-modules like data cleaning and standardization, schema mapping, data query and integration based on matched schemas, and the imputation of missing attribute. Each of these sub-modules is subsequently discussed in detail.

4.2 Creation of Ontology to Support Semantic Interoperability

Creation of a domain specificontology requires great effort and collaboration on the design of knowledge representations as well as annotations on concepts, relationships and axioms. The focus of this study is not on developing a comprehensive generic ontology for all urban planning and modeling, but on demonstrating the

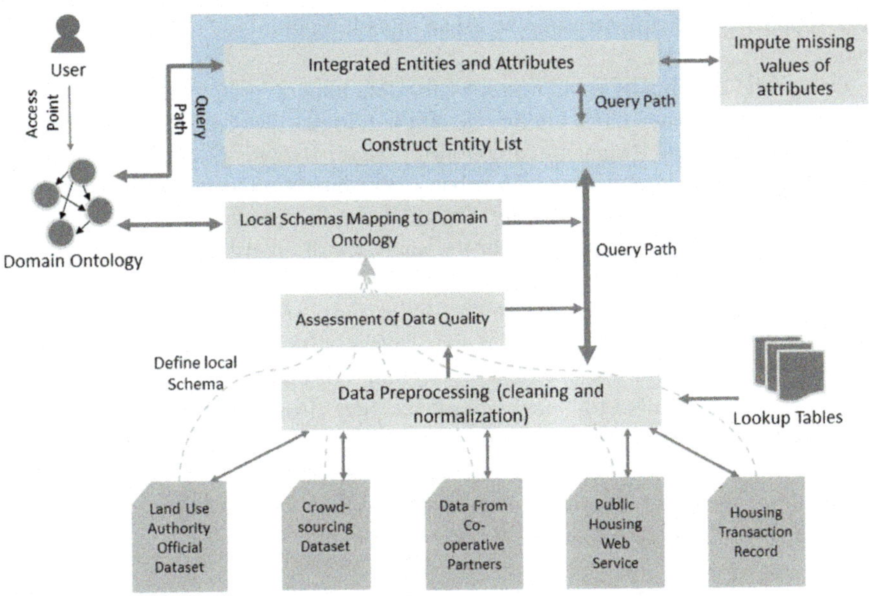

Fig. 1 Framework of the proposed ontology-based data integration approach

utility and applicability of the proposed approach by building and using a task-oriented ontology example.

The data model adopted in this ontology example is compliant with the Resource Description Model (RDF) triples <Subject, Predicate, Object>, a metadata model specified by the W3C. The subject and object are classes representing real-world entities, and the predicate denotes a relationship between the subject and the object. RDF is usually written in an XML-like language. Thus, it is machine-readable and can be easily shared and integrated with other applications and platforms, either online or offline.

Much of the power of ontology comes from its capability and flexibility in defining the correspondences between classes that capture complex relationships in the real world. In addition to entity-attribute association commonly seen in the traditional relational database, ontology often includes other types of correspondences such as subsumption, mereology, equivalence and constraint. For example, ontology enables one to explicitly define relationships like "A-class road is equivalent to expressway" or "block number is a part of address" in the context of Singapore. These relationships enable a richer and more precise matching with increasingly sparse, complicated and heterogeneous data. For example, in the traditional relational database schema, public housing buildings are often listed in the "buildings" table to avoid the redundancy. However, public buildings typically have specific properties and policy constraints that only apply to them. In this sense, a data schema that allows hierarchical definitions of the building entity is more appropriate, because it is able to elaborate not only the superclass-subclass inheritance relationship between generic buildings and public residential buildings, but also the attributes specific to public housing buildings. Table 2 lists a number of correspondences used in creating the urban modelling ontology example used in the study.

One of the benefits of using ontology for data integration is that at the ontology creation stage, we can start with the relevant ontologies designed by others because the modification and extension of ontology is relatively straightforward. A few ontologies have been developed for urban management and planning (Teller et al. 2007). However, none of them are designed for the modeling and analytics of activity-travel research, especially for the analytics of emerging big data. Figure 2 presents an urban building ontology example targeted at the fusion of building information for Singapore. One of the major purposes of using ontology is to reconcile semantic conflicts. Many pairs of classes have an "equivalent as" relationship between them. For example, "year built" of buildings from one dataset is assumed to be equivalent to "lease commence year" from another dataset. Also, as mentioned above, this ontology enables a more flexible representation of the hierarchical and intricate relationships among different types of buildings as well as the relations between buildings and other entities. For example, it is clearly exhibited in the diagram that "multiple-story car park (MSCP)" is a subclass of

Table 2 Types of correspondences specified in the urban modelling ontology example

Type	Relationship	Description	Example
-subclass_of -superclass_of	Subsumption	Every instance of subclass is also an instance of superclass	HDB[a] building is a *subclass of* residential building. Residential building is a *superclass of* condominium building
-has_ attribute	Entity-attribute	One class is a property of another class	Building has height
-has_key	Identifier	Every key uniquely identifies a class instance	Building has key address
-part_of -have	Mereology	One class is a part of the other class	Bridges are *part* of roads Buildings *have* units
-equivalent_to	Equivalence	Two classes are equivalent	HDB building is *equivalent to* public housing building
-constrained by	Constraint	One class is constrained by another class	The classification of children is constrained by age (under 20)
-from_time -to_time	Temporal correspondence	Temporal validity of a class	Activities start *from time1* Activities last *to time2*
-located_in -contain	Spatial relationship	Spatial containment relationship between two classes	Buildings are located in zones Parcels contain buildings

[a]HDB Building refers to public residential flats built and managed by the Housing and Development Board (HDB) of Singapore

"building" but also a subclass of "parking". Ontology allows classes at the same level to be overlapped with each other unless the disjointedness between classes is explicitly stated. This makes it easier to classify buildings like "shop-house", which has both residential and commercial uses.

The example ontology used in this study is generated by using Protégé, an interactive ontology editor developed by the Stanford Center for Biomedical Informatics Research at the Stanford University School of Medicine (Knublauch et al. 2004; Tudorache et al. 2008). Figure 3 shows the screen prints of four different components in Protégé that can be used to assist ontology creation and reasoning. Readers can refer to Horridge et al. (2004) for a comprehensive introduction of Protégé. Figure 3a is the class editing panel that allows metadata to be added and annotation of classes such as class name and property. The class-property matching panel (Fig. 3b) is used to add relationships among classes, whilst Fig. 3c shows the graphical visualization of the ontology and provides a way to visualize the classes and their relationships defined for an ontology. The SPARQL query panel (Fig. 3d) is a function to write a SPARQL query, a semantic query language for databases recommended by W3C, to retrieve and manipulate data stored in Resource Description Framework (RDF) format.

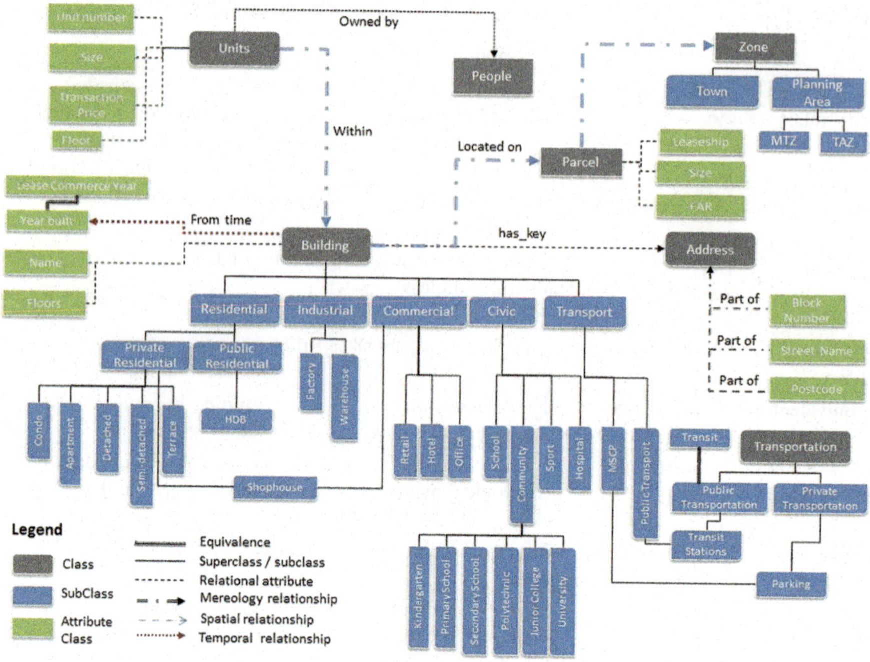

Fig. 2 An example of the urban building ontology

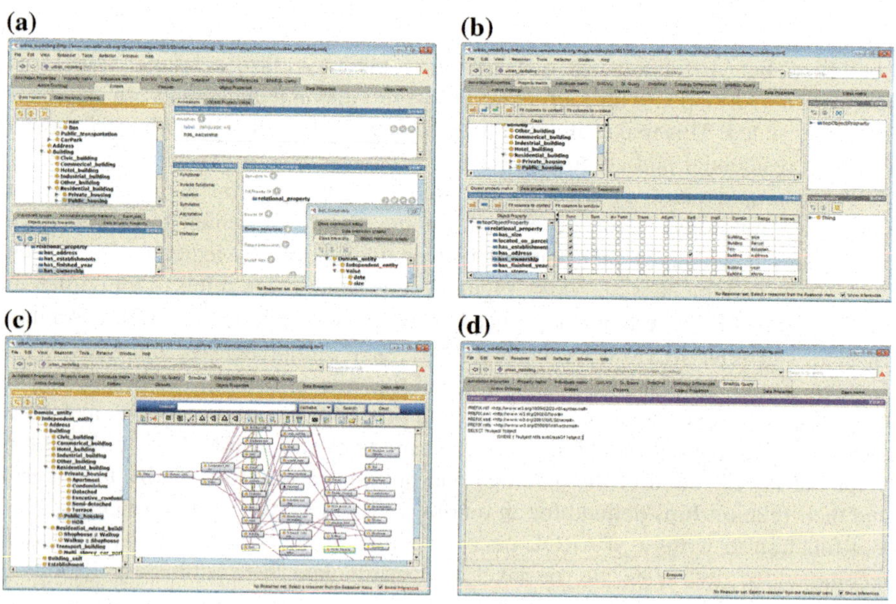

Fig. 3 Screenshots of the Protégé software. **a** Class editing panel. **b** Class-property matching panel. **c** Ontology graphical visualization. **d** SPARQL query panel

4.3 Schema-Mapping Between Data Sources and Ontologies

One of the most critical challenges for data integration is to fit the information from individual data sources to the global data schema, i.e. ontology designed for tasks. This challenge is addressed by schema mapping, which aims at finding correspondence between semantically related entities based on the mapping between local schemas and global ontologies. Many schema mapping approaches have been proposed (Kokla 2006; Nathalie 2009). These mapping approaches can be grouped into two types: lexical matching and structural matching. While lexical matching focuses on using linguistic and text techniques to detect the correspondence between ontology labels and local column names, structural matching relies on the structural relationship outlined in the ontology to sort out the relationships among information contained in various datasets. The proposed data integration approach adopts the structural matching between local datasets and the global ontology.

It is relatively convenient and straightforward to convert a static ontology schema to a graph data structure of metadata with the vertices of the graph representing classes, and the directed edges representing correspondences. The analysis of graph data is supported by a rich set of algorithms, like tree traverse and shortest-path search. These algorithms have been implemented in various programming and statistical analysis packages, which make the ontology mapping more plausible for automatic computer processing. For example, the data integration process can reason through relationships not explicitly defined in the ontology by measuring the connectivity or adjacency between relevant entities. This is especially helpful when the size of the domain ontology grows with more details, which will make the manual inference and consistency check difficult.

In order to facilitate the schema matching process, for each local dataset, a local-global schema mapping needs to be specified compliant to the format drafted in Table 3. The idea is to list the information provided by the dataset that matches the entity relationships specified in the ontology. Each row in Table 3 represents information from local dataset matches to an edge (correspondence) in the global ontology. For example, the first row of the table matches with the edge indicating "subclass HDB building has an attribute finished year". "Node1" and "Node2" correspond to two classes "HDB" and "Finish_Year" in the ontology. "Field1" and "Field2", instead, list the names of the fields corresponding to "Node1" and "Node2" in the local datasets. In this case, the "address" field in the "hdb_sale" dataset is used as the primary key of HDB building and the "lease_commence" field is assumed to have the information on the finished year of a building. The first row of Table 3 indicates that the information about the "Finish_Year" attribute of "HDB" buildings can be found in the "lease_commence" field of the "hdb_sale" dataset with the HDB buildings identified by "address" field. In the schema mapping look-up table, the temporal tags of the datasets are included in order to help users to retrieve the information that is temporally valid. The last column in Table 3 is a subjective index of dataset reliability that is used to establish a hierarchy for reconciling data conflicts and imputing missing values.

Table 3 A look-up table example for schema mapping between local datasets and global ontology

Node1	Node2	Dataset	Field1	Field2	From_time	To_time
HDB	Finish_Year	hdb_sale	Address	lease_commence	lease_commence	Null
HDB	Finish_Year	hdb_resale	Address	lease_commence	lease_commence	Null
Private_housing	Building_type	private_street_dir	Address	Type	Null	Null
Condominium	Building_type	condo_sax_dir	Address	prop_type	Year	Null
Condominium	Building_type	condo_guru_dir	Address	Type	Null	Null
Private_housing	Building_type	realis_private	Address	Type	Year	Null

4.4 Integration Operations

Because an ontology has a richer representation of correspondences among classes, it enables users to formulate more specific and tailored queries based on custom defined transitivity rules. In this study, three operations are designed and developed to facilitate the integration of multiple datasets.

- *Construct entity list*: constructs the most comprehensive list for the entity given the available datasets and the targeted time periods. When building the list, the algorithms are designed to traverse the graph-structure ontology to include all instances of targeted entities and instances of its subclass entities that can be found in the datasets. For example, when constructing the list of private residential buildings, the function automatically includes instances of condominiums, apartments and detached-houses etc., which are subclasses of the private residential buildings.
- *Retrieve attribute values*: selects and complements values of the targeted attribute for an entity. Based on the list of instances belonging to the entity, the algorithms are implemented to search and retrieve corresponding attribute values from all relevant entity classes (including superclasses and subclasses). For example, when querying the "built year" of private housing buildings, in addition to searching the datasets for the private housing buildings, the function would also look for the "built year" information in the building (superclass) datasets and as the apartment (subclass) datasets. The sequence of information retrieval is dependent on the assessed quality of each dataset.
- *Impute missing values*: for the attributes of some geospatial objects, missing values of attributes of the targeted instances are substituted by the existing attribute values of instances that are spatially similar.

5 Integrating Building Information for Population Synthesis

For the Singapore case study, the research is part of the SimMobility development project funded by the Singapore National Research Foundation (NRF) through the Future Urban Mobility (FM) research group at the Singapore-MIT Alliance for Research and Technology (SMART) Center. SimMobility is a large-scale agent-based model at individual household and building scale and requires an initial population of households, persons and firms to be initialized with realistic attributes, locations, relationships and behaviors at the beginning of simulation.

5.1 Data Sources

Official datasets were available for building footprints, census statistics, and real estate transactions, and we had the Government's 2008 travel survey for a 1 % sample of all households with known demographics and postcode locations of their residence and workplaces. Much of the built environment data had good spatial detail since the data were tagged with postcodes or addresses that distinguish over 110 thousand different places in Singapore. However, for privacy reasons, the socio-economic data from the census bureau were not as spatially detailed and, even for the real estate layers, building heights and floor space were not available—but could, in some instances, be obtained from other sources. For example, large buildings are often associated with a single postcode and real estate websites and transaction databases typically report the floor on which a purchased apartment is located as well as the building's address and postcode. Systematic integration of the various data sources helps to improve the estimates of each building's volume and floor space and those estimates, in turn, can generate more realistic assignments of households to residential and workplace locations when we use iterative proportional fitting and related statistical methods to infer the detailed characteristics of the full synthetic population from the 1 % sample of residents reported in the travel survey (Zhu and Ferreira 2014).

Table 4 lists a number of datasets pertinent to buildings that have been gathered from different sources, including the building directory from the Singapore Land Authority (SLA), the housing unit transaction information, and the semi-structured building lists collected from various websites. These datasets differ in building types, numbers, attributes and data qualities. Although the official building data provided by SLA are very reliable and complete in terms of the number and the

Table 4 A list of datasets available for building information integration

Data type	Coverage	Source	Records	Attributes	Year
Building list	All types of building	SLA	162,491	Postcode, address, type	2008
Building list	All types of buildings	Street directory	111,591	Postcode, address, type	2013
Building list	HDB, condos and apts	FourSqaure	14,256	Postcode, address, type, year	2013
Building list	HDB blocks	HDB web service	9276	Postcode, address, type, year, unit type and amount	2013
Housing unit transaction record	HDB and private housing units	HDB web service/ realis	54,375	Address, type, lease_commence_date, floor area, transaction price, min/max level	2013
Building list	All types of buildings	EMPORIS	6868	Address, postcode, year, floor, height	2013

spatial locations (represented by address and postcode) of buildings, the attributes of buildings of great interest to analysts like the year built and the floor information are absent. Besides, the classification of building types in the dataset from SLA is ambiguous. Therefore, information like building types, capacities, finished year, property value and unit types and areas need to be retrieved from other datasets, which are considered less authoritative and error-prone but more complete.

5.2 Integration of Building Data

The first task is usually to identify the reference attributes of the entity class. In the case of this study, because addresses or postcodes are available in most datasets and are semantically consistent, it becomes an ideal attribute for cross-referencing between datasets. Then, to improve the coherence of values for the same entities and attributes from different data sources, a series of data cleaning and pre-processing routines were applied to the attributes like address, postcode, building type and year.

The quality of data is evaluated at the attribute level as opposed to the dataset level because there are significant variations in the reliability of information for different attributes from the same data source. When assessing an attribute, the values of the attribute from a dataset that is most trustworthy are picked out as reference for the attribute values from other datasets to compare against. The most important attribute of building is probably the function type. Types of buildings are tricky to deal with because of its hierarchical classification structure as shown in the building ontology diagram in Fig. 2. For example, Emporis classifies buildings at a very detailed level (e.g. primary school, church, fire station and theme park etc.). Building lists from SLA have mixed levels of classification with some buildings tagged as "Residential" while others are tagged as "Walk-up". Based on our anticipated analytic needs, we determine that an accurate classification of buildings at medium aggregated level (HDB, Apartment, Factory, Warehouse, Retail and Office, etc.) is desired.

In this study, categories of building types from local data sources are mapped to building classes and subclasses pre-defined in the example ontology. The building classes are semantically enriched by the building types that appear in local data sources. A generic R function is written and applied to carry out the fusion of building type information. Because the graphic-based schema mapping allows the data user to easily traverse the graph, the returned building types not only include the values from the inquired datasets but also their superclasses. For example, if an inquired building record is tagged as "primary school", then the returned building record will not only be tagged as "primary school", but also be tagged as "community" and "civic", which are the parent and grandparent classification types of "primary school" in the ontology.

Table 5 shows the contributions of different datasets in providing useful building type information for the integrated data. As we can see from the table, the street-

Table 5 Number of building records that are classified in the process of integration

Dataset	Category 1	Category 2	Category 3
HDB Web service data	9276	9276	9276
+Housing unit transaction	33,687	33,687	18,327
+Streetdirectory	119,768	84,451	84,451
+SLA	144,604	135,329	95,409
+FourSquare	144,900	135,707	95,929
+Emporis	144,908	135,709	96,343

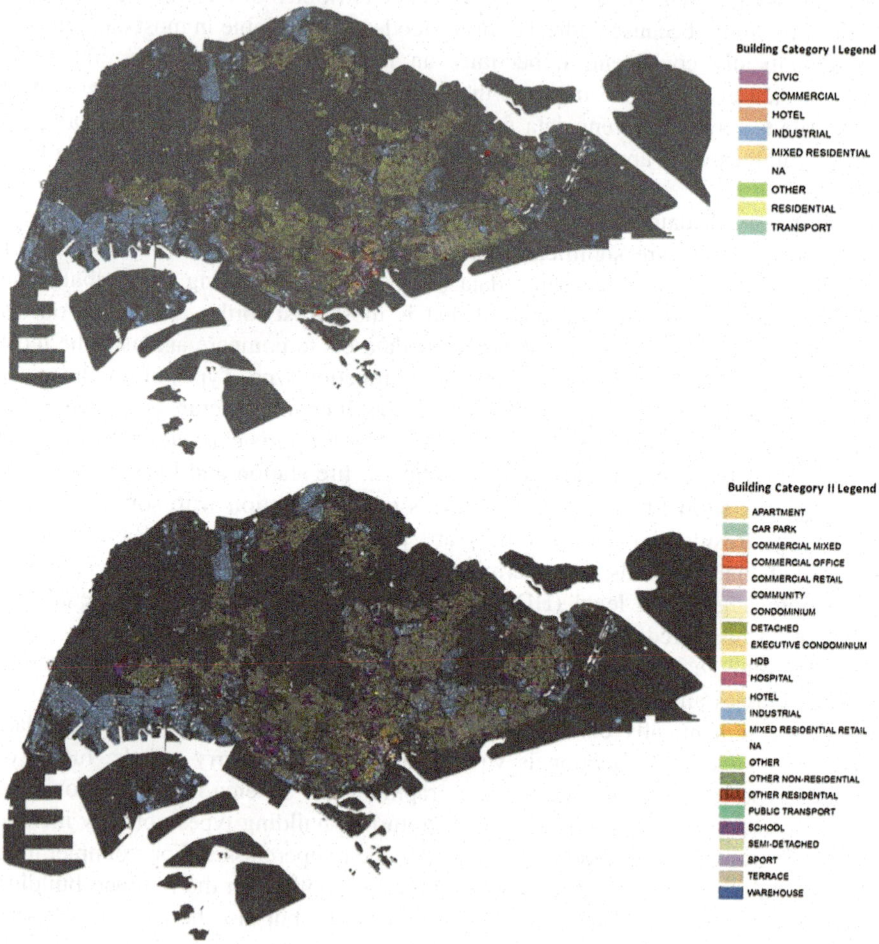

Fig. 4 Different levels of building type classification in the integrated dataset

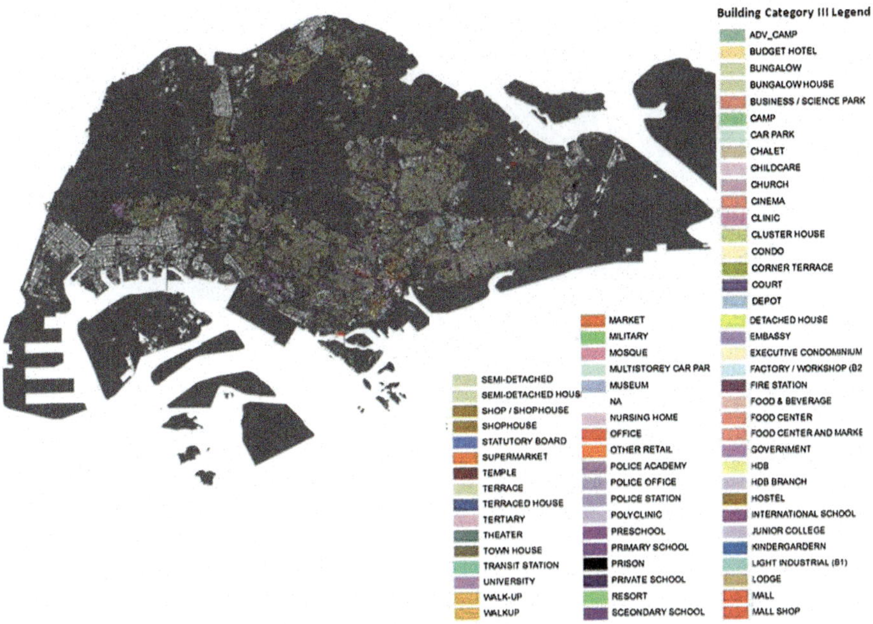

Fig. 4 (continued)

directory appears to have the most significant contribution because of its very detailed classification of building types. In comparison, the Emporis dataset provided very limited building type information. Figure 4 shows the thematic maps of integrated buildings by different levels of categorizations. Category 1 building types are very generic and only have eight types, which leads to around 12.1 % of buildings unclassified. In contrast, Category 3 building types are quite specific, which results in a large number of buildings with missing Category 3 information.

Likewise, other attributes like building finished year are also merged following the same procedures. However, because 76.7 % of buildings did not find finished year information from any of these datasets, a spatial similarity-based imputation is performed to infer the finished year of remaining buildings. In this case, a set of heuristic rules were designed to impute the missing finished year values by using the corresponding available information of nearby buildings. These rules dictate that the search will first be directed to the buildings of the same type in the same development project as the targeted building, followed by the buildings of the same type on the same street. If these two searches fail, the values for the nearest building of the same type will be used. This imputation approach helped to bring the percentage of buildings with missing completion year values down to 15.7 % from 76.7 % in the original fused building data. Most of these buildings are industrial buildings in recently developed areas, as shown in Fig. 5.

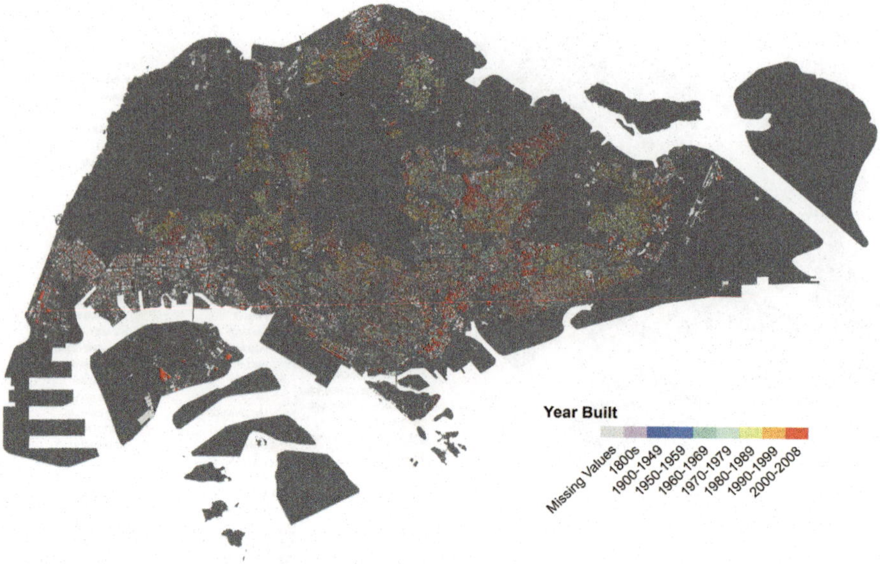

Fig. 5 Finished year of buildings

6 Discussion and Conclusion

The scarcity of spatiotemporal data restricts the depth and scope of research that can be conducted. Although rich data are becoming available from more and more sources, merging information from multiple datasets is held up by data interoperability and quality issues. In general, considerable time and effort are spent on data assembly and preparation when building planning support systems, and this is especially true when developing large-scale land use and transportation models. This step has become increasingly time-consuming due to the needs for spatially-detailed representations of current conditions and the growth in available data sources.

This chapter presented an ontology-based schema mapping approach to reconcile the existing semantics and data representation conflicts among datasets. The approach explicitly builds domain knowledge into a task-specific ontology to help integrate data about the built environment and activity-travel research. The ontology is regarded as a global schema using the generic representation of entities and relationships among them, which are independent of the available data. To address the heterogeneity and scarcity of information from various data sources, a local dataset is matched with global schema via a graphic structure-matching table consisted of nodes corresponding to entities and edges corresponding to relationships. Then, needed information encapsulated in local datasets can be identified, retrieved and integrated based on the interpretation from the agreed concepts and correspondences laid out in the ontology.

We demonstrated the approach by applying it to gather information about buildings for population synthesis for a large scale agent-based micro-simulation model. A significant improvement is observed in complementing the missing values of attributes for buildings, as evidenced by the examples of building type and completion year. But more importantly, the ontology and the associated data fusion steps result in a semi-automatic data processing pipeline and has yielded plausible synthetic populations that we have begun to use in our urban modeling. The process has been especially useful in saving time as we constantly discover new sources and can now integrate the new data more efficiently. Our example also suggests ways to improve the general utility of our emerging urban information infrastructure so that urban modeling and various planning support systems can be better supported and more widely used.

In fact, ontology can play an important role in facilitating the development and use of planning support systems. Once integrated into the urban information infrastructure, ontology tools can accelerate the task-specific customization of urban datasets that are needed at the start of each planning project. Towards this end, ontology development can be viewed as a 'smart city' investment in the tools needed to repurpose existing datasets for new uses in order to provide a richer, consistent knowledge base and hence stimulate more connections and collaborations among researchers and planners. In addition, Ontologies are a key component of the semantic web, which is defined by W3C (2013) as the "*common framework allowing data to be shared and reused across application, enterprise and community boundaries*". They are also likely to be a critical part of the web 3.0 initiative, which stresses the shareability, interoperability and connectivity of distributed datasets based on the semantic web and linked data technologies (Hendler 2009). Thus, it is proactive and timely to consider the application of ontology-based data integration approaches as part of 'smart city' efforts to improve urban information infrastructure so that planning support systems can more easily obtain the customized and spatially detailed data that they need about current conditions.

Acknowledgments This chapter draws on a portion of the PhD Dissertation of Yi Zhu, submitted to the Massachusetts Institute of Technology (Zhu 2014). We acknowledge the Singapore Land Authority and other public and private entities who have provided the datasets used in this research. We also acknowledge helpful comments and suggestions from the editors and reviewers and partial support of the Singapore National Research Foundation through the Future Urban Mobility program of the Singapore-MIT Alliance for Research and Technology (SMART) Center.

References

Balmer, M., Rieser, M., Meiste, K., Charypar, D., Lefebvre, N., Nagel, K., & Axhausen, K. (2009). MATSim-T: Architecture and simulation times. In A. L. C. Bazzan (Ed.), *Multi-agent systems for traffic and transportation engineering* (pp. 57–78). Hershey, PA: Information Science Reference, IGI Publishing.

Ben-Akiva, M., Koutsopoulos, H. N., Antoniou, C., & Balakrishna, R. (2010). Traffic simulation with DynaMIT. In J. Barceló (Ed.), *Fundamentals of traffic simulation* (pp. 363–398). New York: Springer.

Benslimane, D., Leclercq, E., Savonnet, M., Terrasse, M. N., & Yétongnon, K. (2000). On the definition of generic multi-layered ontologies for urban applications. *Computers, Environment and Urban Systems, 24*(3), 191–214.

Birkin, M., & Clarke, M. (1988). SYNTHESIS—A synthetic spatial information system for urban and regional analysis: methods and examples. *Environment and Planning A, 20*(12), 1645–1671.

Fonseca, F. T., Egenhofer, M. J., Agouris, P., & Câmara, G. (2002). Using ontologies for integrated geographic information systems. *Transactions in GIS, 6*(3), 231–257.

Frank, A. U. (1997). Spatial ontology: A geographical information point of view. In O. Stock (Ed.), *Spatial and temporal reasoning* (pp. 135–153). Netherlands: Springer.

Hendler, J. (2009). Web 3.0 Emerging. *Computer, 42*(1), 111–113.

Horridge, M., Knublauch, H., Rector, A., Stevens, R., & Wroe, C. (2004). A practical guide to building OWL Ontologies using the Protegé-OWL plugin and CO-ODE tools, edition 1.3, University of Manchester. Accessed February 28, 2015. https://mariaiulianadascalu.files.wordpress.com/2014/02/owl-cs-manchester-ac-uk_-eowltutorialp4_v1_3.pdf.

Knublauch, H., Fergerson, R. W., Noy, N. F., & Musen, M. A. (2004). The Protégé OWL plugin: An open development environment for semantic web applications. In S. A. McIlraith, D. Plexousakis, & F. V. Harmele (Eds.), *The semantic Web–ISWC 2004* (pp. 229–243). Berlin, Heidelberg: Springer.

Kokla, M. (2006). Guidelines on geographic ontology integration. In *Proceedings of the ISPRS Technical Commission II Symposium*, Vienna, Austria.

Mars, N. J. I. (1995). What is an ontology? In A. Goodall (Ed.), *The impact of ontologies on reuse, interoperability, and distributed processing* (pp. 9–19). Uxbridge, Middlesex: Unicom Seminars Limited.

Nathalie, A. (2009). Schema matching based on attribute values and background ontology. In *Proceedings of 12th AGILE International Conference on Geographic Information Science* (Vol. 1, No. 1, pp. 1–9).

Partridge, C. (2002). The role of ontology in integrating semantically heterogeneous databases. Rapport technique, 5(02). Accessed February 28, 2015. http://www.borosolutions.co.uk/research/content/files/O-02-SE.pdf.

Ramos, J. M., Vandecasteele, A., & Devillers, R. (2013). Semantic integration of authoritative and volunteered geographic information (VGI) using ontologies. In *Proceeding of Association of Geographic Information Laboratories for Europe (AGILE) Conference 2013*, Leuven.

Salvini, P. A., & Miller, E. J. (2005). ILUTE: An operational prototype of a comprehensive microsimulation model of urban systems. *Networks and Spatial Economics, 5*, 217–234.

Teller, J., Keita, A. K., Roussey, C., & Laurini, R. (2007). Urban ontologies for an improved communication in urban civil engineering projects. Presentation of the COST Urban Civil Engineering Action C21"TOWNTOLOGY". *Cybergeo: European Journal of Geography.* doi:10.4000/cybergeo.8322.

Tudorache, T., Noy, N. F., Tu, S., & Musen, M. A. (2008). Supporting collaborative ontology development in Protégé. In A. P. Sheth, S. Staab, M. Paolucci, D. Maynard, T. Finin, & K. Thirunarayan (Eds.), *The semantic Web—ISWC 2008* (pp. 17–32). Berlin Heidelberg: Springer.

W3C Semantic Web Activity (2013). Accessed February 28, 2015. http://www.w3.org/2001/sw/.

Waddell, P. (2002). UrbanSim: Modeling urban development for land use, transportation, and environmental planning. *Journal of the American Planning Association, 68*(3), 297–314.

Zhu, Y. (2014). Spatiotemporal learning and geo-visualization methods for constructing activity-travel patterns from transit card transaction data. PhD Dissertation, Massachusetts Institute of Technology, DSpace: http://hdl.handle.net/1721.1/93807.

Zhu, Y., & Ferreira, Jr J. (2014). Synthetic population generation at disaggregated spatial scales for land use and transportation microsimulation. In *Transportation Research Record*, (Vol. 2429, pp. 168–177).

Ziegler, P., & Dittrich, K. R. (2004). Three decades of data integration-all problems solved? In *18th IFIP World Computer Congress, Volume 12, Building the Information Society* (pp. 3–12). Dordrecht: Kluwer Academic Publishers.

Part II
Smart Cities

Chapter 8
Smart Cities: Concepts, Perceptions and Lessons for Planners

Tuan-Yee Ching and Joseph Ferreira, Jr.

Abstract What is a "smart" city? This paper examines concepts and perceptions of city officials from six "smart" cities, Boston, San Francisco, Amsterdam, Stockholm, Singapore and Rio de Janeiro. Their "smart" efforts, gathered through interviews and secondary sources, are analyzed against four theories of "smart" cities; (a) "smart machines" and informated organizations, (b) partnerships and collaboration, (c) learning and adaptation, and (d) investing for the future. The findings show that instead of converging toward a single definition of being "smart", the cities have taken different approaches in planning and implementation, and adopt different combinations of elements from the theories. The cities' experiences and elements of being "smart" are distilled and presented as learning points and pathways for other cities.

1 Introduction

The "smart city" buzzword is captivating city leaders and planners worldwide. It is commonly associated with the application of information and communications technologies (ICTs) to reap efficiencies and benefits. Yet, the definition of "smart" cities remains diverse; according to Caragliu et al. (2009), "a fuzzy concept". Will inadequate understanding result in cities making poor investments in technology and infrastructure? Vanolo (2013) argues that the lack of definition allows cities to use the buzzword to support their own agenda, and hence, any examination will need to be "contextualized and related to specific cases". In this vein, this chapter

T.-Y. Ching (✉)
Urban Redevelopment Authority, Singapore 069118, Singapore
e-mail: ching_tuan_yee@ura.gov.sg

J. Ferreira, Jr.
Department of Urban Studies and Planning, Massachusetts Institute of Technology, 77 Massachusetts Avenue, Cambridge, MA 02139, USA
e-mail: jf@mit.edu

S. Geertman et al. (eds.), *Planning Support Systems and Smart Cities*, Lecture Notes in Geoinformation and Cartography,
DOI 10.1007/978-3-319-18368-8_8

explores how city planners are conceptualizing "smart" cities and whether these concepts consistent with theories of being a "smart" city.

2 Methodology

The concepts and perceptions of officials from six "smart" cities—Boston, San Francisco, Amsterdam, Stockholm, Singapore and Rio de Janeiro—were examined. Data was collected through phone and email interviews, and secondary sources (e.g. news articles, blogs, city reports, etc.). Cities' efforts were analyzed against four theories of "smart" cities; (a) "smart machines" and informated organizations, (b) engaging communities, technology providers and research institutions, (c) learning and adaptation, and (d) investing for the future. As the cities differ in size, mode of governance, planning and management, their approaches were examined in light of their own contexts. Their experiences and elements of being "smart" were synthesized and distilled to draw several learning points and conclusions.

3 A Brief Survey of "Smart" Cities and Technology Providers

Surveys of "smart" cities show that notions of being "smart" vary. Neirotti et al. (2014) analyzed from seventy cities that "smart" concepts are applied in diverse domains—"hard" (e.g. transportation, energy and resource management), "soft" (e.g. education, innovation, social inclusion) and "in between" (e.g. healthcare, public safety)—and concluded that there is no unique definition of "smart" cities. Similarly, Ching (2013) observed from over fifty cities that initiatives are applied in different fields, with some cities having implemented them while others remain in a conceptual or development phase. Batty et al. (2012) categorize cities' efforts into those which are "badging" or "regenerating" themselves as "smart", the development of technopoles, and the application of ICTs in urban services, urban intelligent functions, and online and mobile forms of participation. For their "smart" initiatives, many cities partner with firms providing expertise and technology, including IBM, CISCO and Siemens. The global market for investments in "smart" technology and infrastructure is projected to grow, exceeding $39 billion in 2016 (ABI Research 2012), and totaling $108 billion between 2010 and 2020 (Pike Research 2011). Multidisciplinary firms such as Arup provide consultancy services for cities on investments in "smart" initiatives.

4 Four Theories of "Smart" Cities

As an adjective, "smart" is associated with being clever and intelligent, possessing acumen, learning and being adept. In the context of modern technology, "smart" is associated with intelligent autonomy achieved through computer programming or guidance.[1] For cities, "smart" concepts can be grouped under four theories.

4.1 "Smart Machines" and Informated Organizations

Assumptions: The "smart" city uses ICTs for automation and intelligent functions, and structures processes, organization and governance to take advantage of the technologies.

In her theory of "smart machines", Zuboff (1988) highlighted *automation* as one of two dimensions in the application and impact of intelligent or information technology (IT) in workplaces. Automation breaks down human tasks, translating human actions into software instructions, i.e. information, that guides machines to perform tasks repeatedly and reliably. When city functions employ "smart machines", made intelligent through the use of data sensors and computing algorithms, they are envisaged to perform more efficiently, accurately and reliably than what could have been done by humans, if humans could perform such functions at all. For example, the Integrated Operations Center in Rio de Janeiro is made to predict the amount of rainfall more accurately than standard weather forecast systems and more efficiently alert city departments for flood mitigation operations (Singer 2012).

Zuboff's second dimension, *informating*, takes advantage of "smart machines" generating new digital information about underlying processes, creating potential for organizations to exploit and innovate their organizational structures and processes. Brynjolfsson and McAfee (2011) believe that humans and "smart machines" combined, through the re-engineering of processes to exploit ICTs, can improve business organizational models and reap benefits. Good organization, governance and management are also underscored as essential foundations for "smart" cities (Morier 2012; Belissent 2011). Most 'smart city' efforts involve some level of ICT-enabled automation, and many also claim some level of informating although significant achievement along this second dimension is much more elusive.

[1]Sources: Collins Dictionary, 2012, "Smart"; Merriam-Webster Dictionary, 2012, "Smart"; Merriam-Webster Learner's Dictionary, 2012, "Smart"; Oxford English Dictionary, 2012, "Smart, adj.".

4.2 Beyond "Smart Machines": Partnerships and Collaboration

Assumptions: The "smart" city involves partnerships and collaborations between city governments, communities, businesses, research institutions, etc. *within a framework that drives innovation.*

Other theorists view "smart" cities beyond "smart machine" analogies, shifting the focus from city functions to governance, especially from a liberal democratic perspective (Allwinkle and Cruickshank 2011). For example, Hollands (2008) adopts a critical view on self-proclaiming "smart" cities, highlighting that the use of ICTs is limited in the transformative capacity of cities without integrating human capital and shifts in the balance of power between government, businesses and communities. Haque (2012) critiques that 'smart' strategies should focus not on "the city as a single entity" but rather on 'the smartness of its citizens', who are idea generators rather than recipients. Townsend (2013) proposes the vision for "smart" cities to involve more social and inclusive processes of grassroots innovation. The social perspective is echoed in terms of empowering citizens without excessive emphasis on being a "machine city" (Sennett 2012) or "shallow" technical optimization (Greenfield 2013), and seeing "smart" efforts as a sociotechnical approach to solving urban wicked problems through collaborative planning (Goodspeed 2015). Hoornweg (2011) stresses that "smart" cities "ensure good communication between government and citizens", and "use all the local resources available in decision making and service delivery, e.g. universities, senior citizens, business community", thus underlining the need for cities to engage their communities and local organizations.

4.3 Learning, Relearning and Adapting

Assumptions: The "smart" city learns, relearns and adapts itself, through learning networks, and the use of metrics and feedback processes.

Under this theory, writers such as Campbell (2012) expand the engagement of "smart" cities beyond the involvement of communities to larger networks of cities, whereby cities learn from each other best practices in governance and management, and convert such learning to innovative application. City and institutional networks have been set up for this purpose, e.g. "Smart Cities" supported by the European Regional Development Fund (www.smartcities.info), "European Smart Cities" which benchmarks and outlines a "smart" cities model (www.smart-cities.eu), etc. Cities and their agencies have the capability to learn, and with the aid of ICTs, can incorporate feedback loops for re-learning and adaptation. The introduction of new knowledge and technologies, together with adapting traditional knowledge and practices, builds flexibility, a quality of city resilience (Arup 2014). This focus on trial-and-error learning draws on organizational learning theories developed by

Donald Schön in *The Reflective Practitioner* (Schön 1983) and in his two-volume set with Chris Argyris (Argyris and Schön 1978, 1996).

Implicit in the process of learning, re-learning and adaptation is the ability to assess performance, in particular, through metrics or performance indicators defined according to a city's goals. For example, as part of sustainable development and city resilience strategies, cities may have greenhouse gas emissions (GHGe) reduction targets or citizen health and well-being indicators. Cohen's (2012a) "Smart Cities Wheel" involves 100 indicators in six "smart" categories—economy, environment, governance, living, mobility and people—for cities to track their performance and adapt policies towards their goals. Walters (2012) describes the interconnected processes of monitoring, managing and using gathered data for future design as "virtuous cycles in city planning and operation" that lead to innovation.

4.4 Investing for the Future

Assumptions: The "smart" city is cognizant of its human, social and physical stocks of capital, and invests in technologies and functions that have the potential to reap greater economic, social and environmental benefits.

Another group of theorists frame "smart cities" from a resource or business perspective. Caragliu et al. (2009) emphasize stocks of capital, believing "a city to be smart when investments in human and social capital and traditional (transport) and modern (ICT) infrastructure fuel sustainable economic growth and a high quality of life, with a wise management of natural resources, through participatory governance". Recognizing cities' limited resources in the quest for sustainable development, Frenchman et al. (2011) argue that the "smart" use of technology could help cities transform and grow sustainably. For example, this includes the implementation of digital infrastructure and increasing access to information and the knowledge economy, managing long-term risks through monitoring and feedback systems, and facilitating new "city-making" processes such as urban prototyping, etc. "Smart" cities involve "savvy business and development decisions" as part of economic sustainability (Kotkin 2009), and developing a robust technological foundation on which to innovate city business processes (Kuk and Janssen 2011).

A report by The Climate Group et al. (2011) highlighted the potential for cities to realize savings and value through investments in "smart" applications. It urges cities to understand the value chain involved, so as to capture positive externalities and to explore new business models such as revenue streams from technology services. Thus, economic performance and sustainability, and perceived return on investment (ROI) are considerations in the formulation of "smart" initiatives.

On one level, the economic performance and long-term sustainability of cities may be a major consideration and driver in cities' conceptualization and adoption of "smart" applications. On another level, the business performance of "smart" applications, in relation to a city's technological capabilities and resources, its short and long-term objectives, as well as the perceived return on investment (ROI), are factors in determining the form and extent of these "smart" applications. These are also related to earlier points on organization and governance (e.g. city's capacity, available resources, etc.), formation of partnerships between city governments, businesses and communities (e.g. business model of "smart" applications, funding and implementation models, etc.), and the use of metrics that help to gauge business performance.

5 Case Studies: Concepts and 'Smart' Initiatives

Table 1 summarizes cities' initiatives in relation to the four theories and includes several websites related to the efforts. The "smart" initiatives presented below are non-exhaustive, and have been selected to illustrate "smart" and innovative practices. The governance and geography of the six initiatives vary widely from the municipality of Boston—comprising 617 K people (15 %) and 125 km^2 (1 %) of the 4.5 million people and 11.7 K km^2 within the metropolitan statistical area[2]—to the city-state of Singapore—comprising all 5.5 million people and 718 km^2 within the nation.[3] The public entities overseeing the initiatives also varied from Mayor's Office facilitators to newly established Authorities and public-private partnerships.

5.1 Boston, Massachusetts, U.S.A

Cohen (2012b) ranks Boston as North America's "smartest city", citing its entrepreneurial and innovation ecosystem, in particular, the Mayor's Office for New Urban Mechanics (MONUM). According to Osgood (2013), the city's "smart" efforts use technology and design to engage its citizens and address their concerns. MONUM is not a "skunkworks" laboratory, but is closely integrated with city departments, with whom it jointly develops, tests and implements prototypes. MONUM categorizes its "smart" initiatives as "participatory urbanism" which aims to engage citizens, "clicks and bricks" which focuses on technology infrastructure, and "education". The city's Department of Innovation and Technology (DoIT) has also implemented initiatives such as an open government portal and data cloud, and collaborated with technology providers. Boston's southern waterfront is also branded as an "Innovation District".

[2]US Census Bureau 2010 Quick Facts (http://quickfacts.census.gov/qfd/states/25/2507000.html) and 2000 statistics (http://www.census.gov/prod/2006pubs/smadb/smadb-06tableb.pdf).

[3]SingStat, Singapore (2014): http://www.singstat.gov.sg/statistics/latest-data#14.

Table 1 Examples of 'smart' initiatives and corresponding theories

"Smart machines" and informated organizations		
Boston: Data warehousing and integration (e.g. addresses, crime, public health, etc.) to allow city departments to identify hotspots and analyze problems	**Singapore**: Traffic prediction tool that predicts conditions and allow traffic managers to divert traffic, minimize disruptions and economic costs from delay	**Rio de Janeiro**: "Smart" integrated weather-prediction and emergency response functions, with some reorganization of city departments and processes (http://theinstitute.ieee.org/video/technology-focus/technology-topic/rio-de-janeiro-a-smart-city)
Partnerships and collaboration		
Boston: Citizen engagement initiatives include "Citizens Connect", "Street Bump" and "Community PlanIT"; "Boston About Results" web portal publishes city "scorecard" reports, providing greater transparency (http://www.cityofboston.gov/bar/scorecard/reader.html)	**San Francisco**: MOCI and "Chief Innovation Officer" position roles created to foster entrepreneurial innovation; idea-generating platforms implemented to enhance collaboration, solicit ideas and apps, e.g. hackathons and "ImproveSF"	**Amsterdam**: AIM created to foster entrepreneurial innovation within a "triple helix model" engaging technology providers, research institutions and community; ASC operates as an "open platform" that handles a wide variety of initiatives and does not rely on any single technology provider; "Climate Street" Utrechtsestraat initiative heavily involves local stakeholders and community
Learning, relearning and adapting		
Boston: Part of the "G7" network which exchanges ideas, e.g. public health app based on shared code from Chicago; MONUM evaluates its projects on performance and process of implementation	**Amsterdam**: ASC website with extensive "smart city"-related resources including research findings, and detailed reports (http://amsterdamsmartcity.com/); KPIs set for each project (e.g. CO_2e reduction, number of jobs created, number of citizens involved, etc.), and monitoring of partners involved and investments made; possible upward cascading of changes to tax policies to facilitate exchange of domestically produced energy, following community feedback on decentralized solar energy production in residential areas	**Stockholm**: Stockholm Royal Seaport will incorporate "smart" monitoring technology at the district, block and apartment levels, with a new sustainability unit to build up assessment models and follow up strategies; long-term approach for bi-directional feedback, where ICTs can influence behavior and plans, and evolving behavior and plans can in turn influence ICT

(continued)

Table 1 (continued)

Investing for the future		
Amsterdam: AIM spurs innovation and economic development—technology providers invest resources and stand to gain from commercializing it, while the city and communities benefit from its application	**Stockholm**: Fiber network has generated positive benefits to the local economy, and supports social programs and environmental initiatives; transportation-related initiatives such as the congestion management system and "Journey Planner" (http://reseplanerare.trafiken.nu/bin/query.exe/sn?) aim to reduce CO_2e	**Singapore**: Testing and development of initiatives for own needs, but can subsequently be "exported" commercially to other cities (e.g. water technologies)

"Smart" initiatives include "Citizens Connect", where citizens are engaged to provide feedback that help to enhance municipal services. Through mobile apps, a website, Twitter, or SMS, citizens report issues (e.g. potholes, graffiti, fallen trees, requests for snow plowing, etc.) which generate city work orders requiring action. Another initiative "Street Bump" uses mobile devices' sensors to record the location of uneven road surfaces. In 2011, MONUM and Boston Public Schools (BPS) used "Community PlanIT", an engagement game platform, to gather over 4600 online comments from the community on a school performance metrics and accountability system. DoIT's "Open Government Portal" provides citizens with better access to city data and city government performance indicators (e.g. crime, GIS, permits, Renew Boston Solar Map, etc.), while the "Data Boston" portal contains more than 50 datasets. In 2012, the city collaborated with IBM and Boston University to explore how data from city video cameras, street sensors, citizens' mobile phones, and social media can represent the real-time traffic situation (Dillow 2012), so as to optimize traffic flow and reduce vehicle emissions.

5.2 San Francisco, U.S.A

Cohen (2012b) ranks San Francisco as North America's second "smartest city", noting its environmental leadership and "thriving entrepreneurial economy". Miller (2013), from the Department of the Environment (SF Environment), cites Cohen's definition of "smart cities" as the concept and basis "to take (our) sustainability operations to the next level", achieving "cost and energy savings, improved service delivery and quality of life, and reduced environmental footprint", and to be carbon-free by 2030. Its "SF Energy Map" shows locations of buildings with solar installations and lets users calculate the photovoltaic potential for properties, while "Honest Buildings" is an online network that shares energy-efficiency building strategies. The "ChargePoint" app monitors 110 public electric vehicle (EV) charging stations, while "SF Park" distributes real-time information on parking availability, adjusting parking rates to match demand. The city recently passed its "Existing Commercial Building Ordinance", which requires commercial buildings with more than 10,000 square feet to report energy usage data to the city, with the intention that owners and managers will address inefficiencies (Nutter 2012).

The city encourages entrepreneurial innovation and the use of open data through the Mayor's Office of Civic Innovation (MOCI) and Department of Technology. Mayor Lee (2012) describes the government's role as a "convener" in encouraging experimentation, declaring October as the city's "innovation month". Initiatives include an open data policy and legislation, with the "DataSF" portal containing more than 500 datasets and a showcase of apps. The city's "living innovation zones", part of its "CleantechSF" initiative, aims to encourage businesses to use city-owned properties and public assets to pilot clean technology, products and design concepts. Idea-generating platforms have also been initiated by private organizations, community groups and city agencies. For example, through the "ImproveSF" online collaboration platform, the Planning Department issued the "Green Connections Challenge", soliciting ideas on making walking and cycling easier and safer, and ideas and suggestions for routes and activities. Other idea-generating platforms include hackathons[4] such as "Unhackathon" and "Summer of Smart", which produced the "Smart Muni" app that tracks city buses in real time and identifies transit system incidents.

5.3 Amsterdam, Netherlands

The Economist's (2012) "smart" cities report contrasted "problems" of "top-down" projects (e.g. Masdar, Songdo City, etc.) with Amsterdam's "bottom-up" approach, which relies on a collaborative platform rather than a master plan. The Amsterdam Smart City (ASC) platform was initiated by the Amsterdam Innovation Motor (AIM),[5] the City of Amsterdam and technology providers. ASC provides test-beds for initiatives that contribute to CO_2e reductions, economic development and improving the quality of life. There are more than 30 initiatives implemented by over 70 partners, categorized under five themes, living, working, mobility, public facilities and open data.

Energy-focused initiatives include "Geuzenveld sustainable neighborhood" where more than 500 households received smart meters to raise energy awareness

[4]Intensive one or multi-day workshops during which programmers write code to address pre-specified 'challenges' using newly available data and resources.

[5]In 2013, after the time of the study, AIM merged with the Amsterdam Economic Board.

and influence consumption behavior, and the "ReloadIT" smart electric vehicle(EV) grid where photovoltaic supply is matched with EV energy demand to optimize the use of renewable energy. Under the "Zuid Oost laws and regulations" initiative, the city is considering the implementation of a "freezone" for the testing of sustainability ideas, where rules and regulations are minimized. Amsterdam's "Open Data" portal contains datasets from 19 categories, while "Apps for Amsterdam" encourages app development along the themes of safety, mobility, vacancy, energy, tourism and culture, and democracy. The Utrechsestraat "Climate Street" initiative is a collaboration between local entrepreneurs, the city and technology providers. The group mapped out the base measurements of energy consumption and CO_2 and NO_2 emissions along the street, and introduced initiatives such as smart meters, displays on energy consumption, smart plugs to automatically dim or shut down appliances, dimmable lamps and tram stop lighting, solar-powered BigBelly waste bins, centrally located reverse osmosis water sources for cleaning vehicles, clustering and optimization of logistics and deliveries, etc.

"Almere smart society" is a collaboration between the Almere Economic Development Board and technology providers providing digital infrastructure to facilitate interaction between citizens and public organizations. Car navigation devices are used to analyze traffic flows and internet video communication is used in health care. CISCO implemented the first "smart work center" (SWC) in an Almere residential community to reduce transportation demands and increase productivity, as an office center equipped with workstations, telepresence teleconference equipment, a childcare center, restaurant, etc.

5.4 Stockholm, Sweden

The 2012 Smart Cities Expose report featured Stockholm as a city which "is doing a few things right" (Smart + Connected Communities Institute 2012). Stockholm's Vision 2030 is "…to become one of the world's cleanest, safest and most beautiful cities where Stockholm is a world leader in information technology and in the development, commercialization and application of new environmental and energy related technology" (City of Stockholm 2011). Stokab, a city-owned infrastructure company, implemented an extensive fiber network that extended from the financial center to the region, serving as backbone infrastructure to support the city's innovation efforts. Stockholm's Green IT strategy aims to use IT to reduce its environmental impact, addressing transportation, energy use, land use, water and waste management, GHGe reduction, etc.

Recognizing that the transportation sector contributes to 31 % of the city's CO_2e, the city aims to "create a long-term sustainable transport system, based on new technology, non-fossil fuels, and more information" (City of Stockholm 2011). In 2006, it developed a variable toll rate traffic and congestion management system with IBM, involving in-car transponders and optical character recognition to identify license plates. IBM (2010) reported that the system had reduced traffic by 20 %, average travel times by 50 %, and the amount of emissions by 10 %. Other initiatives include providing comprehensive transportation information (e.g. timetables, routes, real-time traffic speeds, incidents, road works, etc.) and a journey planner that estimates the length and cost of the journey and the estimated CO_2e per month weighted by vehicle fuel type.

"Smart" efforts will be extended to sustainable development projects such as Stockholm Royal Seaport, which aims to reduce its per capita CO_2e to 1.5 tons by 2020, and to be "free of fossil fuels" and "climate-neutral" by 2030. The district will comprise a "smart ICT" open and shared communications infrastructure, integrated city management system, street lighting, transport, education, health services, etc. (Stockholm Royal Seaport 2013). The district will also incorporate a smart grid and a vacuum waste collection system that includes user-level waste weighing, a single kitchen sink food waste chute, and energy recovery from the collection system.

5.5 Singapore

Singapore's "smart" efforts center on the application of ICTs in government, society and the economy. The Infocomm Development Authority's (IDA) Intelligent Nation 2015 master plan aims to harness ICTs as "enabling infrastructure" (Tay 2013) and to achieve targets of being "number one in the world in harnessing infocomm to add value to the economy and society", a "twofold increase in value-added of infocomm industry to S$26 billion", a "threefold increase in infocomm export revenue to S$60 billion", the creation of "80,000 additional jobs", "90 % of homes using broadband", and "100 % computer ownership in homes with school-going children". The recently announced "smart nation" platform includes initiatives such as "creating standards for Internet of Things@Home" (IDA 2014). The Economic Development Board (EDB) envisions Singapore as a 'living lab' that tests and develops innovative solutions for city management and the built environment.

"Smart" efforts include the "e-Citizen" web portal with over 385 e-services from 60 ministries and statutory boards (e.g. online filing of income and property taxes, paying mortgage loans, making medical appointments in public health centres, etc.), the "data.gov.sg" portal with over 5000 datasets, "OneMap" geo-located data, etc. For example, users can access data on all property transactions by location, property type, transaction period, price range, etc., visualize residential address locations to schools, or search for rental of government-owned property and space. Mobile apps that provide information include "MyENV" on stormwater levels, air quality, and dengue fever occurrences, "TransportSG" on traffic speed, incidents, electronic road pricing charges, and number of parking lots, and "Police@SG" on crime statistics. Government agencies support private hackathons (e.g. "UP Singapore") to seek new ideas and apps.

In terms of infrastructure, a 1Gbps fiber network is planned to connect 60 % of households, "Wireless @SG" provides free wi-fi in public areas, while near field communication e-payment solutions are being studied. The "intelligent energy system" initiative is a collaborative pilot project by the Energy Market Authority and Singapore Power, an energy provider, to test new smart grid applications. These include equipping households with a "smart meter" to view electricity consumption data, and in future phases to develop advanced management applications (e.g. time-of-use tariff information, demand response and energy management, outage management, integration of EV charging and vehicle-to-grid functions, etc.). The Land Transport Authority's (LTA) "smart" initiatives include the "e-Symphony" integrated fare card system and the "traffic prediction tool" that anticipates and helps the management of traffic flow to minimize congestion. New pilot "smart" initiatives will be tested at the Jurong Lake District, including autonomous buggy transportation in parks, real-time microclimate monitors, a smartphone platform that enables users to contribute data on the quality of public transportation trips, park lighting that responds to motion and natural conditions, etc. (IDA 2014).

5.6 Rio de Janeiro, Brazil

Rio de Janeiro's "smart" program is synonymous with its Rio Operations Center (COR). Developed with IBM, COR integrates the functions of over 30 city agencies, private transportation and utility companies, including emergency response (BBC 2013; IEEE 2014). According to Mayor Eduardo Paes, COR and ICTs are instrumentsthat benefit citizens, allow quick and reliable communication, and improve city operations (Sterling 2011). Carlos Osorio, Secretary for Conservation and Public Services, describes COR as a "collaborative tool" and "catalyst to make the broader metropolitan area function better" (Smart + Connected Communities Institute 2012: 24). Hamm (2012) describes COR as the "first such facility in the world" that coordinates the complex "human-made and natural systems of a city in a holistic way".

COR involves the use of real-time data (e.g. video from 560 traffic cameras) to manage traffic and crowd-intensive events, e.g. the Carnival, 2014 FIFA World Cup, etc. COR integrates a situation room where city leaders and emergency response officials communicate and make decisions. In 2010, floods and mudslides caused 200 deaths and made 15,000 homeless (Heim 2011). With COR, the city can forecast the weather 48 h ahead, predict wind speeds, rainfall and runoff intensities and the impacts of floods and landslides, coordinate its emergency response agencies and deploy resources such as emergency shelters. COR's citizen warning and communication system includes sirens, SMS, email, its web portal, Facebook, Twitter, etc.

6 Findings and Discussion

The cities' concepts and perceptions of "smart cities" and different approaches were assessed against the four theories, and key initiatives are summarized in Table 1.

6.1 "Smart Machines" and Informated Organizations

6.1.1 "Smart" Machines

Several cities used ICTs for automation and intelligent functions—i.e. "smart machines". For example, Rio's COR uses data on soil composition, topography, population, land use, and hydrology models to predict the weather and impacts. The rain forecasts, which assist decision-making, are reportedly 91.8–93.6 % accurate (Treinish et al. 2012). According to the LTA official, Singapore's "Traffic Prediction Tool", developed with IBM, uses algorithms to predict traffic conditions 30 min in advance, allowing traffic managers to divert traffic, minimizing disruptions and economic costs from delay. Boston's data integration efforts (e.g. address points, crime, Constituent Relationship Management System, code violations, public health records, etc.) allowed city departments "to easily identify and analyze problem(s)" and "hotspots to be identified and mapped" (Lane 2013). A number of observations were drawn. First, cities need infrastructure and processes to collect and organize data (e.g. sensors, data integration and warehousing), and algorithms and visualization tools for analyses. Second, these functions tend to be immediate-term city management operations with behavioral rules, e.g. emergency responses linked to prediction threat levels. Third, integrating data from multiple sources may lead to new insight, as seen in Boston's case, or lead to enhanced city functions, as seen in Rio's case.

6.1.2 "Informating" and Organizational Change

While most 'smart city' efforts involve some level of automation, the efforts have different extents of "informating" and re-organization. For Rio's COR, reorganization is inseparable from its "smart" functions; predicting landslides will be incomplete unless information is relayed to emergency response agencies close at hand. Osgood (2013) revealed that given the prototypical nature of "Street Bump" and the challenges faced in changes in processes and management practices, Boston's Public Works Department and MONUM took an incremental approach to change. According to the Singapore LTA official interviewed, there were no organizational changes arising from the use of the "Traffic Prediction Tool", as existing traffic management staff took on the trial predictive functions (Table 1).

6.2 *Beyond "Smart Machines": Partnerships and Collaboration*

6.2.1 New Innovation Agencies

Many of the examined cities consciously created new agencies to lead "smart" efforts. Some play a facilitator role, e.g. Amsterdam's AIM, and Singapore's "Smart Nation Program Office". Others have been charged to spearhead innovation, such as Boston's MONUM, and San Francisco's MOCI and "Chief Innovation Officer" position.

6.2.2 "Smart", Citizen-Focused Governance

The motivation for cities to achieve better governance—e.g. improved delivery of city services, engagement of citizens, creating transparency, etc.—through "smart" initiatives was evident. For Boston, this can be seen from the "Boston About Results" scorecard web portal that publishes city reports (e.g. number of permits issued online, percentage of streetlight outages addressed in 10 business days, etc.), "Citizens Connect" and "Community PlanIT", and for Singapore the government e-services. More investigations will be needed to evaluate the impacts of new service channels against traditional methods of delivery, as well as combined impacts.

6.2.3 Collaboration and Leveraging Local Human Capital

Cities were cognizant and took a collaborative approach in implementing their "smart" initiatives. According to a city official interviewed, Amsterdam based its efforts on the triple-helix model, which taps the "intellectual capital of universities, the wealth creation of industries, and the democratic government of civil society" (Leydesdorff and Deakin 2011). The development of Stockholm's Royal Seaport involves the city, developers, stakeholders, technology providers and research institutions. For Singapore's JLD, agencies fund and implement prerequisite infrastructure (e.g. fiber networks, data sensors) upon which technology providers build solutions. Cities (e.g. San Francisco, Boston, Amsterdam, Singapore, Stockholm) also commonly use hackathons to engage partners and seek innovative ideas.

Several factors influence cities' structuring of partnerships. First, the choice of technology provider(s) depends on their levels of expertise and business models. For example, few technology providers may be able to implement COR, integrating analytical capabilities, software and hardware like IBM. Technology providers may choose to collaborate as consortia, or not, due to their business interests. Second, the nature of each project matters. For example, cities may structure less public involvement for high-risk initiatives such as COR emergency response, in contrast to pothole repairs. Third, cities with a clear picture of objectives and technologies (e.g. setting up COR, a prediction-response function) may handpick specific partners, instead of crowd-sourcing ideas.

6.2.4 Avoiding Lock-In

Some cities set out to avoid technological "lock-in" (e.g. proprietary data formats, inflexible partnerships with technology providers), to ensure long-term sustainability. For example, Boston's MONUM takes an open source ideas approach, ASC is an open platform not relying on any single technology provider, and Stockholm Royal Seaport adopts an open and generic ICT infrastructure.

6.2.5 Overcoming Challenges in Collaboration

A challenge faced in collaboration is overcoming friction and expectations between partners. For Singapore, the EDB official interviewed cited agencies' different "level(s) of ambition" as a challenge; for instance, an economic development agency could be interested in "disruptive solutions", whereas a line agency could prefer "tried and tested low-cost solutions". To close the gap with operational agencies, MONUM's approach is to "broker the partnerships with the thought leaders within these agencies" (Osgood 2013), establishing buy-in and ensuring practical initiatives. Tratz-Ryan (2011), analyzing Rio's COR, highlighted that integration across agencies "will not happen overnight", and stressed the important role of people in uniting different functions. Hence, the human element that establishes the middle ground is a success factor.

6.3 Learning, Relearning and Adapting

6.3.1 Continual Learning

Most of the cities examined learn and share their experiences, for example, hosting delegation visits (e.g. Stockholm's Professional Study Visits on city governance and green efforts), organizing and attending conferences (e.g. Amsterdam's "Smart City Event", Singapore's "World Cities Summit"), and websites (e.g. Amsterdam's ASC website with extensive research findings and detailed "Smart Stories" reports). Ideas are also exchanged through city networks such as "G7" in North America. For example, based on shared code obtained from Chicago, Boston released its own public health app during the 2013 flu epidemic (Osgood 2013).

6.3.2 Use of Metrics

While this study could not obtain specific data, cities were generally aware of the use of metrics in assessing projects' performance. For Amsterdam, key performance indicators (KPIs) are set (e.g. CO_2e reduction, number of jobs created, etc.) and AIM monitors the partners involved and investments made for each project.

Boston's MONUM evaluates its projects on two dimensions, in terms of projects' performance (e.g. how "Citizens Connect" changed behavior) and the process of implementation (e.g. effectiveness in sourcing ideas) (Osgood 2013). Stockholm's Royal Seaport plans to incorporate "smart" monitoring technology at the district, block and apartment levels, and a new sustainability unit will focus on "building up assessment models and follow up strategies" (Claeson 2013).

6.3.3 Use of Feedback Loops

Some cities incorporate feedback loops from their "smart" initiatives. For San Francisco, data from EV "Charge Point" stations are collected to help determine future strategies, e.g. new charging station locations (Nutter 2012). For Singapore's traffic prediction, according to the LTA official interviewed, information is used for traffic management (e.g. relayed to road users) in the immediate term, operational improvements (e.g. changes to road junction geometry and signaling) in the medium term, and for planning purposes (e.g. traffic modeling) in the long term. For Amsterdam, arising from a pilot project involving decentralized solar energy production in residential buildings, the community provided feedback and advocated changes to tax policies to facilitate the exchange of domestically produced energy. According to the AIM official interviewed policies like this, if adopted, may cascade to the national level. For Stockholm's Royal Seaport, Bylund et al. (2011) outline a long-term approach for bi-directional feedback, where ICTs can influence behavior and plans, and evolving behavior and plans can in turn influence ICTs, to the extent of "discarding… outdated technologies".

Feedback mechanisms, when strategically incorporated through bootstrapping, allow cities to address long-term issues, yield wider benefits, and build up capabilities. For example, to address Rio's flooding and landslide hotspots, preemptive actions can be taken in the form of infrastructure (e.g. retaining structures, stormwater management systems, etc.) and policy adjustments (e.g. land use policies around high-risk zones).

6.4 *Investing for the Future*

6.4.1 Returns-On-Investment and Funding

From the cities examined, there was no evidence of cities adopting a strong return-on-investment (ROI) business perspective for their "smart" initiatives. For Singapore's "Traffic Prediction Tool", the LTA official interviewed highlighted that overall benefits were difficult to assess for the trial project; notwithstanding that, negative externalities resulting from traffic delays were considered to justify expenditure for the initiative. Some cities do not have dedicated budgets for their "smart" initiatives, for example, Amsterdam's AIM does not fund or own initiatives but provides manpower and organizing resources for ASC. Boston's MONUM receives funding from non-profit foundations (e.g. MacArthur Foundation, Bloomberg Foundation), the State, and line agencies for various initiatives. More investigation will be needed to analyze interactions between city objectives, perceived costs-benefits and ROI, and funding models.

6.4.2 Directly Monetizing "Smart" Initiatives

There was also no evidence that cities directly monetized their "smart" initiatives. For example, instead of the sale of data, many cities (i.e. Boston, San Francisco, Amsterdam, Stockholm, Singapore) have taken an open data approach, being aware of the longer-term benefits to spur innovation, improve the delivery of city services and enhance community engagement.

6.4.3 Longer-Term Wider Benefits

Some cities adopt a longer-term perspective on the wider economic, environmental and social benefits. For Amsterdam, AIM's interest is to spur innovation and economic development. Under their win-win model, technology providers invest resources as business decisions, own the product and stand to gain from commercializing it, while the city and communities benefit from its application. Stockholm's fiber network generated "significant positive benefits" to the economy through enhancing the city's attractiveness as a technology and innovation hub with excellent infrastructure and high administrative efficiency (Felten 2012). According to Broberg (Smart + Connected Communities Institute 2012), the network contributed to the city's knowledge economy, buffered it from the economic crisis, and supports social programs and environmental initiatives. In Singapore's case, the test-bedding and development of initiatives primarily meet its own needs, but can be "exported" commercially to other cities, as seen from its water technologies industry. Hamm (2012) sees Rio's implementation of COR as "investing for the long term," a way to mitigate the risks arising from severe weather and flooding.

7 Cities' Different Approaches

Cities have taken different approaches in relation to their characteristics, the nature of their "smart" initiatives, and the business models of their technology providers. Here, cities' approaches are also analyzed in terms of whether they are "top-down", where initiatives are determined and implemented through cities' directives, "bottom-up", where initiatives are grassroots driven, or "middle-out", where city objectives are addressed through a less-deterministic approach involving government initiation combined with the efforts of technology providers and the community.

7.1 Cities' Characteristics

With its small agencies (e.g. MONUM, DoIT), Boston takes a "middle-out"/"bottom-up" and incremental approach that relies on partnerships. It capitalizes on ideas from various sources (e.g. "Community PlanIT" from Emerson College, "Citizens Connect" from MONUM) (Osgood 2013), and matches them to specific needs. San Francisco takes a "bottom-up" approach with small agencies engaging local technology entrepreneurs to stimulate innovation. Amsterdam, through its collaborative ASC platform, typically use "middle-out"/"bottom-up" approaches to tackle different types and scales of initiatives. Stockholm's initiatives range from "top-down" (e.g. network infrastructure) to "middle-out" (e.g. Stockholm Royal Seaport), where its partnership framework and feedback mechanism is designed to support the long-term development of the district. Some of Singapore's initiatives are "top-down"(e.g. traffic prediction), while others, like JLD, are "middle-out". For Rio, COR was implemented as a "top-down" initiative through a mayoral decision (Singer 2012).

7.2 Nature of "Smart" Initiatives

The "smart" initiatives examined range from community engagement platforms (e.g. "Community PlanIT"), programs to meet environmental, economic and social goals (e.g. "Energy Map"), to city-scale infrastructure (e.g. COR, Stokab network). Initiatives of similar nature share common characteristics, for example the "middle-out" approach to develop new sustainability districts (e.g. Stockholm Royal Seaport, Singapore's JLD, Almere Smart Society) through partnerships. "Middle-out"/"bottom-up" initiatives (e.g. Boston's "Citizens Connect", San Francisco's "Unhackathon") involve community engagement and enhancements to city services. Compared to infrastructure, they require less capital resources, and involve communities in their conceptualization and use. The approaches taken for different initiatives do not appear to be "interchangeable". For instance, Rio's COR cannot be implemented through a "bottom-up" approach due to the high degree of expertise needed, and it will be incongruent to implement "Citizens Connect" through a "top-down" approach.

7.3 Technology Providers' Business Models

The nature of technology and technology providers' business models are factors influencing cities' approaches. For example, IBM's partnership with Rio can be described as a "turnkey" model, where the vendor brings forth its technology, innovations and systems integration capabilities. In this model, the city receives a fully packaged solution that meets its objectives without significant change in the skill set and personnel of the client (i.e., the city). Hence, this model requires shared goals to be established between the technology provider and city, which must also be willing to make capital investments and accept limited partnerships. Cisco's TelePresence technology rethinks the nature of work and commuting. Its model brings forth technology and new solutions to issues that may not already be recognized as problems. By doing so, it is investing in potential new areas where cities may reap benefits (e.g. greater productivity, reduced negative impacts from commuting), creating markets in which it will have a headstart. Stokab, a city-created agency-cum-technology provider, focuses on implementing prerequisite infrastructure that underpins other "smart" initiatives, while building new technical capacity within government rather than adopting a more turnkey approach. For the city, this model encompasses the idea of "sequencing" or "layering" of technologies, while from the business perspective, the base infrastructure may support retail services to be developed by non-governmental third-parties, in this case the delivery of telecommunication services.

8 Lessons for Planners and "Pathways" for Cities

There is no single "smart" model for cities. Table 2 summarizes lessons and best practices for city planners to consider different pathways in their conceptualization and implementation strategies. For example, a city may lean towards a "top-down" approach and position itself to implement large-scale infrastructure in partnership with an expert technology provider. If such an effort is treated as a 'turnkey' project by the city, there may be limited opportunity for capacity building and discovery of new ways to build on the new data infrastructure through 'informating.' However, the city need not follow a traditional "top-down" approach that narrowly seeks solutions to specific problems. Instead, it can invest "smartly" by deliberately creating and harnessing positive spillover effects, and shaping its efforts to be strategic enablers. The city might also incorporate multi-scale feedback mechanisms, taking a long-term view towards reaping maximum benefits through knowledge transfers, and sequential infrastructure development, while avoiding lock-in by remaining flexible in terms of technology and partnership structures.

Another city may lean towards a "bottom-up" approach, e.g. having limited in-house expertise and resources to engage in turnkey projects, and/or having an entrepreneurial grassroots. Yet some grassroots-driven efforts, while innovative, may be unsustainable if robust partnerships with city agencies are not established, or if a longer-term framework is not set up (e.g. city data management, knowledge sharing, data regulatory environment, etc.). The city can focus on developing the collaborative platforms, upon which processes of innovation, knowledge-sharing and implementation are made sustainable for the long-term (Table 2).

Finally, a city may be inclined towards the "middle-out" approach of maintaining a degree of openness and test-bedding initiatives through an innovation-

Table 2 Lessons for planners: best practices to consider

"Smart machines" and informated organizations
Identify "smart machine" functions that provide speedier and more accurate results within and across city agencies, and establish infrastructure and processes for data collection, integration, interpretation and analysis, etc.
Identify complementary informating functions and reorganize city agencies to harness automation, e.g. incorporating data analytics in planning support systems and evolving planning processes that inform, collaborate and co-create with the community
Partnerships and collaboration
Identify appropriate "smart" approach/"pathway" appropriate to the context of the city and initiative, i.e. "top-down", "middle-out" or "bottom-up"
Create innovation-fostering agencies that understand technologies, organization and processes of and across agencies, to form partnerships and initiate projects that support city economic, environmental and social objectives
Identify partners including city agencies, technology providers, research institutions, and the community
Implement citizen-focused initiatives to expand available city resources, provide innovative ideas, engage citizens and allow greater transparency
Anticipate and address possible challenges in collaboration, whether arising between partners or from technology gaps
Learning, relearning and adapting
Establish and use metrics to assess the effectiveness of "smart" initiatives against environmental, economic or social sustainability targets, as well as to improve their processes of implementation
Incorporate multiple, multi-scale feedback loops to assess the effectiveness of initiatives using metrics, and apply feedback inputs to refine cross-agency processes and long-term strategies
Avoid lock-in through the use of open data and open-source platforms instead of proprietary data formats, and through the structuring of partnerships where not a single technology provider or technology dominates
Establish avenues for continual learning and city knowledge management, e.g. through conferences, city networks, learning visits, publications, etc.
Recognize the cost, focus, and speed differences **between turnkey systems** delivering new capability, **and capacity-building efforts** that assist agency staff in discovering new ways to harness digital data and information technologies
Investing for the future
Define and pursue city goals and long-term strategies that reap wider benefits to ensure that there are net environmental, economic or social benefits in the long-term, beyond short-term objectives
Establish clear objectives, and assess ROI (return-on-investment) from both a financial and non-financial perspective, to evaluate "smart" initiatives and to justify investments
Sequence development and investment to carefully plan the implementation of prerequisite backbone infrastructure, or the use or testing of specific technologies

focused agency and/or collaborative platform. While it may not achieve the technological benefits of a "top-down" turnkey approach, or the level of innovation of a "bottom-up" grassroots approach, the "middle-out" approach may provide more flexibility in matching technological innovation to organizational capacity.

This approach may also be able to address initiatives ranging from smaller-scale community engagement to larger-scale infrastructure projects, sustain partnerships, and reap wider long-term benefits. In this way, the city can concentrate on developing a strong collaborative framework, and incorporate systematic learning and feedback loops to ensure long-term improvements.

9 Conclusion

The study of six "smart" cities found that overall, planners' concepts were supportive of, and included elements of the four theories of being "smart". The four theories were complementary and not mutually exclusive; the cities adopted various combinations of elements according to their specific contexts, and had different approaches and partnership frameworks depending on the nature of the initiatives.

These findings suggest multi-dimensionality in being a "smart" city. On one dimension, being "smart" involves harnessing ICTs, for example through automation, to achieve optimum results and to meet economic, environmental and social objectives. But there is enormous potential for cities to consider how they can "informate", where new digital information generated through automation can be used to exploit and innovate organizational structures and processes within and across city agencies, as well as with technology providers, research institutions and the community through sustainable partnerships. For example, through planning support systems that generate insight based on analyzing data obtained from city sensors, planners could evolve planning processes to inform, collaborate and co-create with the community. Being "smart" also involves prudent decision-making that matches a city's resources and capabilities with its objectives, maximizing long-term benefits, and maintaining a flexible approach that fosters innovation. Being "smart" involves continual learning and feedback monitoring, for cities to remain aware and nimble. Could these dimensions conflict with one another? Yes. For example, a city that focuses on automation to achieve efficiencies, without adequate attention to longer-term city objectives and ROI, or building up robust partnerships, may be missing the big picture. City leaders and planners need to be "smart" about being "smart".

The multi-dimensionality of "smart" cities may add to the diversity in definition, and discourse will likely continue on what being "smart" or "smarter" entails. Yet this multi-dimensionality reflects the complex nature of city planning and management and difficult issues that "smart" initiatives attempt to address. This study set out to uncover concepts and perceptions of "smart" cities. Rather than focus on a definition of the term, we have tried to identify viable pathways and learning in order to provide an applicable guide for city planners to consider the various theories and best practices, as they embark on their own "smart" initiatives.

Acknowledgments This chapter is primarily based on the Master's thesis of Tuan-Yee Ching submitted to the Massachusetts Institute of Technology (Ching 2013). We also acknowledge the helpful comments of Rob Goodspeed and anonymous reviewers and the partial support of the Singapore National Research Foundation through the "Future Urban Mobility" program of the Singapore-MIT Alliance for Research and Technology.

References

ABI Research. (2012). *Smart city technologies will grow fivefold to exceed $39 billion in 2016*. http://www.abiresearch.com/press/smart-city-technologies-will-grow-fivefold-to-exce. Accessed November 3, 2012.

Allwinkle, S., & Cruickshank, P. (2011). Creating smart-er cities: An overview. *Journal of Urban Technology, 18*(2), 1–16.

Argyris, C., & Schön, D. (1978, 1996). *Organizational learning* (Vol. 2), Addison-Wesley ODseries. Reading, Massachusetts: Addison-Wesley.

Arup. (2014). *City resilience framework*. http://publications.arup.com/Publications/C/City_Resilience_Framework.aspx. Accessed January 28, 2015.

Batty, M., Axhausen, K. W., Giannotti, F., Pozdnoukhov, A., Bazzani, A., Wachowicz, M., & Portugali, Y. (2012). Smart cities of the future. *European Physical Journal-Special Topics, 214* (1), 481.

BBC. (2013, September 8). Tomorrow's cities: Rio de Janeiro's bid to become a smart city. *News Article*. http://www.bbc.com/news/technology-22546490/. Accessed January 30, 2015.

Belissent, J. (2011). *The key to being a smart city is good governance: Smart governance*. http://blogs.forrester.com/jennifer_belissent_phd/11-05-15-the_key_to_being_a_smart_city_is_good_governance_smart_governance. Accessed November 1, 2012.

Brynjolfsson, E., & McAfee, A. (2011). *Race against the machine: How the digital revolution is accelerating innovation, driving productivity, and irreversibly transforming employment and the economy*. Lexington, MA: Digital Frontier Press.

Bylund, M., Andersson, P. O., & Olofsson, G. (2011). *Stockholm royal seaport: Smart communication*. Final Report. http://smartict.swedish-ict.se/files/2012/06/SRS-SC-pre-study-final-report-2011-05-25-1.00.pdf. Accessed March 23, 2013.

Campbell, T. (2012). *Beyond smart cities*. New York: Earthscan.

Caragliu, A., Del Bo, C., & Nijkamp, P. (2009). Smart cities in Europe. In *3rd Central European Conference in Regional Science—CERS*, 2009 (pp. 45–59).

Ching, T. (2013). *Smart cities: Concepts, perceptions and lessons for planners*. Master's Thesis, MIT. http://hdl.handle.net/1721.1/81146. Accessed November 23, 2014.

City of Stockholm (2011). *GreenIT: Green IT strategy for the City of Stockholm*. http://international.stockholm.se/InternationalGlobal/Stockholmbytheme/GrnIT-strategi_eng.pdf.. Accessed February 21, 2013.

Claeson, J. (2013). *Email interview*. City of Stockholm Planning Department, March 24, 2013.

Cohen, B. (2012a). What exactly is a smart city. *Fast Company*. http://www.fastcoexist.com/1680538/what-exactly-is-a-smart-city. Accessed September 29, 2012.

Cohen, B. (2012b). *The top 10 smartest cities in North America*. Co.Exist. http://www.fastcoexist.com/1680967/the-top-10-smartest-cities-in-north-america#1. Accessed December 9, 2012.

Dillow, C. (2012). *IBM tackles boston traffic, merging multiple data streams to predict, ease congestion*. http://www.popsci.com/technology/article/2012-07/bostons-ibm-built-traffic-app-merges-multiple-data-streams-predict-ease-congestion. Accessed November 17, 2012.

Felten, B. (2012). *Stockholm's stokab: A blueprint for ubiquitous fiber connectivity?* Report by Diffraction Analysis.

Frenchman, D., Joroff, M., & Albericci, A. (2011). *Smart cities as engines of sustainable growth*. Washington, DC: World Bank Institute. https://wbi.worldbank.org/wbi/Data/wbi/wbicms/files/

drupal-acquia/wbi/WBI_Challenge%20Paper_Final_20110614-2-3%20COMPRESSED.pdf. Accessed November 21, 2014.

Goodspeed, R. (2015). Smart cities: Moving beyond urban cybernetics to tackle wicked problems. *Cambridge Journal of Regions, Economy and Society, 8*(1), 79–92. doi:10.1093/cjres/rsu013.

Greenfield, A. (2013). *Against the smart city*. New York: Do Publications.

Hamm, S. (2012). *Smarter leadership: How Rio de Janeiro created an intelligent operations center*. A Smarter Planet Blog. http://asmarterplanet.com/blog/2012/03/smarter-leadership-how-rio-de-janeiro-created-an-intelligent-operations-center.html. Accessed March 5, 2013.

Haque, U. (2012). Surely there's a smarter approach to smart cities? *Wired.co.uk*. http://www.wired.co.uk/news/archive/2012-04/17/potential-of-smarter-cities-beyond-ibm-and-cisco. Accessed January 13, 2013.

Heim, A. (2011). *How data is making Rio de Janeiro a smarter city*. The Next Web. http://thenextweb.com/la/2011/07/13/how-data-is-making-rio-de-janeiro-a-smarter-city/. Accessed March 14, 2013.

Hollands, R. G. (2008). Will the real smart city please stand up? *City, 12*(3), 37–41.

Hoornweg, D. (2011). *Smart cities for dummies*. Sustainable Cities Blog. http://blogs.worldbank.org/sustainablecities/smart-cities-for-dummies. Accessed 13 September 2012.

IBM. (2010). IBM helps city of stockholm predict better commuting options. *News Release*. http://www-03.ibm.com/press/us/en/pressrelease/29903.wss. Accessed February 23, 2013.

Infocomm Development Authority of Singapore IDA. (2014). Singapore unveils building blocks of smart nation vision. *News Release*. https://www.ida.gov.sg/~/media/Files/About%20Us/Newsroom/Media%20Releases/2014/0617_smartnation/AnnexC_sn.pdf. Accessed February 28, 2015.

IEEE Newsletter Article. (2014, June 18) *Rio de Janeiro: A smart city—Protecting its residents using big data and advanced technology*. Accessed January 30, 2015.

Kotkin, J. (2009). *The world's smartest cities*. Forbes. http://www.forbes.com/2009/12/03/infrastructure-economy-urban-opinions-columnists-smart-cities-09-joel-kotkin.html. Accessed November 17, 2012.

Kuk, G., & Janssen, M. (2011). The business models and information architectures of smart cities. *Journal of Urban Technology, 18*(2), 39–52.

Lane, C. (2013). *Interview*. Boston Department of Innovation and Technology (DoIT), February 11, 2013.

Lee, E. (2012). *San Francisco's secret sauce—Innovation drives livability and sustainability*. Presented at meeting of the minds conference October 9–11, 2012. http://cityminded.org/talk/san-franciscos-secret-sauce-innovation-drives-livability-and-sustainability. Accessed March 18, 2013.

Leydesdorff, L., & Deakin, M. (2011). The triple-helix model of smart cities: A neo-evolutionary perspective. *Journal of Urban Technology, 18*(2), 53–63.

Miller, P. (2013). *Email interview*. San Francisco: San Francisco Department of the Environment (SF Environment), January 11, 2013.

Morier, R. (2012). *Who needs smart cities for sustainable development?* The World Bank. http://web.worldbank.org/WBSITE/EXTERNAL/TOPICS/EXTSDNET/0,,contentMDK:23146568~menuPK:64885113~pagePK:7278667~piPK:64911824~theSitePK:5929282,00.html. Accessed September 25, 2012.

Neirotti, P., De Marco, A., Cagliano, A. C., Mangano, G., & Scorrano, F. (2014). Current trends in smart city initiatives: Some stylised facts. *Cities, 38*, 25–36.

Nutter, M. (2012). *Smart and sustainable solutions in San Francisco*. Presented at CaFFEET 2012. http://www.youtube.com/watch?v=bHAArfyGH9I&list=SPYTiwx6hV33sHzO2G8MfqZxz4H6NLmWUE&index=5. Accessed March 18, 2013.

Osgood, C. (2013). *Interview*. Boston Mayor's Office of New Urban Mechanics (MONUM), January 23, 2013.

Pike Research. (2011). *Global investment in smart city technology infrastructure to total $108 billion by 2020*. http://www.pikeresearch.com/newsroom/global-investment-in-smart-city-technology-infrastructure-to-total-108-billion-by-2020. Accessed November 3, 2012.

Schön, D. (1983). *The Reflective practitioner: how professionals think in action*. New York: Basic Books.

Sennett, R. (2012). No one likes a city that's too smart. *The Guardian*. http://www.guardian.co.uk/commentisfree/2012/dec/04/smart-city-rio-songdo-masdar. Accessed December 6, 2012.

Singer, N. (2012). Mission control, built for cities. *The New York Times*. http://www.nytimes.com/2012/03/04/business/ibm-takes-smarter-cities-concept-to-rio-de-janeiro.html?_r=1. Accessed March 5, 2013.

Smart + Connected Communities Institute. (2012). *Smart cities expose: 10 cities in transition 2012*.

Sterling, B. (2011, November 9). IBM smart cities in Rio de Janeiro. *Wired.com*. http://www.wired.com/beyond_the_beyond/2011/11/ibm-smart-cities-in-rio-de-janeiro/. Accessed March 5, 2013.

Stockholm Royal Seaport. (2013). *Stockholm royal seaport*. http://stockholmroyalseaport.com/. Accessed February 23, 2013.

Tay, R. (2013). *Interview*. Infocomm Development Authority of Singapore (IDA), January 29, 2013.

The Climate Group et al. (2011). *Information marketplaces the new economics of cities*. http://publications.arup.com/Publications/I/Information_Marketplaces_The_new_economics_of_cities.aspx. Accessed November 21, 2012.

The Economist. (2012). *Mining the urban data*, November 21, 2012. http://www.economist.com/news/21566408-cities-will-become-smarter-different-ways-many-people-expected-mining-urban-data. Accessed February 16, 2013.

Townsend, A. (2013). *Smart cities: Big data, civic hackers, and the quest for a new utopia*. New York: WW Norton & Company.

Tratz-Ryan, B. (2011). *Observations from a 'smarter' operations center in Rio de Janeiro*. The Gartner Blog Network. http://blogs.gartner.com/bettina-tratz-ryan/2011/11/17/observations-from-a-%E2%80%9Csmarter%E2%80%9D-operations-center-in-rio-de-janeiro/. Accessed April 2, 2013.

Treinish, L., et al. (2012). Enabling an advanced numerical weather prediction model for operational forecasting in Rio de Janeiro. In *92nd American Meteorological Society Annual Meeting—Conference on Transition of Research to Operations: Successes, Plans, and Challenges*. https://ams.confex.com/ams/92Annual/webprogram/Paper200773.html. Accessed April 2, 2013.

Vanolo, A. (2013). Smartmentality: The smart city as disciplinary strategy. *Urban Studies*, 0042098013494427.

Walters, J. (2012). *Smarter urban mobility systems around the Pacific rim*. Urban Systems Collaborative Fall 2012 Meeting. http://urbansystemscollaborative.org/meetings/usc-fall-2012-university-of-california-berkeley/. Accessed March 11, 2014.

Zuboff, S. (1988). *In the age of the smart machine: The future of work and power*. New York: Basic Books.

Chapter 9
Who's Smart? Whose City? The Sociopolitics of Urban Intelligence

Kian Goh

Abstract Visions of the "smart city" are becoming reality, translated from the realm of concepts into actual urban space. Proponents of smart city technologies invoke their potential to free us from the drudgery of urban life and solve our environmental problems. But can "smart cities" save us? There has been long-standing resistance to the scientific, positivist basis for planning. What happens when intelligent plans encounter messy politics, social systems, and divergent scales of urban governance? This paper explores the promises of "smart cities" and their stated rationale, and grounds a review of theoretical paradigms with new empirical research in Singapore and London. I present two key findings: First, there is no one "smart city," even *within* a city. Second, differences in scales and ideologies of urban governance *across* cities have significant impact on the way that actors frame their priorities and objectives around the role of urban technologies. Finally, I speculate on the ways that urban networked systems might enable and empower a transformative planning.

1 Introduction

Visions of a kind of technology-infused "smart city" are becoming reality, translated from the realm of concepts into actual urban space. In Singapore, real-time flood sensors give updates on water levels in rivers and reservoirs; in London, an "intelligent" video system automates congestion toll collection; and in Rio de Janeiro, an integrated city management system is being set up to monitor and predict everything from landslides to traffic. Scholars, engineers, urban designers, city leaders, and corporate executives alike invoke the promise of digitally networked urban technologies such as these to free us from the drudgery of urban life

K. Goh (✉)
Department of Urban Studies and Planning,
Massachusetts Institute of Technology, Cambridge, MA, USA
e-mail: kiangoh@mit.edu

S. Geertman et al. (eds.), *Planning Support Systems and Smart Cities*,
Lecture Notes in Geoinformation and Cartography,
DOI 10.1007/978-3-319-18368-8_9

and solve our environmental problems (see, for example, Batty et al. 2012; Mitchell 1995, 2003; Ratti and Townsend 2011; Townsend 2013).

And yet, alongside the rhetoric of real-time, intelligent measurement, control, and management of this urban future, we confront daily reminders of the *lack* of control in cities. Cities increasingly exhibit economic, social, and spatial stratification (Goldsmith and Blakely 2010), with more and more urban residents living and working in substandard conditions (Garau et al. 2005; UN Habitat 2003). Disasters, "natural" or otherwise, expose the dysfunction of urban infrastructure—levees in New Orleans, flood channels in Jakarta, factories in Bangladesh, nuclear power plants in Fukushima, and commuter trains in New York City. In 2011 urban centers around the world erupted both in organized protest and violent riots (Castells 2012; Wasik 2012).

To what extent has the implementation of such "smart city" technologies fulfilled their promises? And what can we learn from the competing views of such urban interventions, especially as they hit the ground? In this chapter I explore the promises of "smart cities" and their stated rationale—specifically the vision of digitally networked urban space that has captured the attention of scholars and urban managers alike. I review two competing paradigms on scientific measurement of cities and the potential of urban technological systems, and situate the theoretical discussion with new empirical research in Singapore and London. The two cases explore the space between the advocacy of technology companies, and the reality checks of urban managers. I close by speculating on a way in which such technologies might prove transformative.

2 Corporate Digital Urbanism

Technology and infrastructure companies like IBM, Cisco, and Siemens vie for the attention of city governments. They promise similar results, with important distinctions. IBM emphasizes its extensive history in computing systems. The company embraces the idea of the city as a "system of systems" in its "Smarter Cities" initiative, and urges cities to pay attention to what IBM considers to be the six interconnected core systems of a city—people, business, transport, communication, water, and energy (Dirks and Keeling 2009). IBM's Rio Operations Center epitomizes this focus on systems, attempting to monitor and analyze information on weather, hydrology, and traffic in real-time (Fig. 1). Cisco's "Smart+Connected Communities" program builds on its core expertise in networking equipment and emphasizes the company's history of "translating" across different systems and networks, enabling the extensive interconnectivity we experience today. "Everything will be connected, intelligent, and green," proclaims (Cisco 2010, 2), a vision it is trying to realize in Songdo, Korea's self-proclaimed "City of the Future" (Fig. 2). Siemens (n.d.), an engineering and electronics company, stresses that it is not just the digital network but also its connection to the city's physical infrastructure, actual places and things, that is critical. The company touts its expertise in

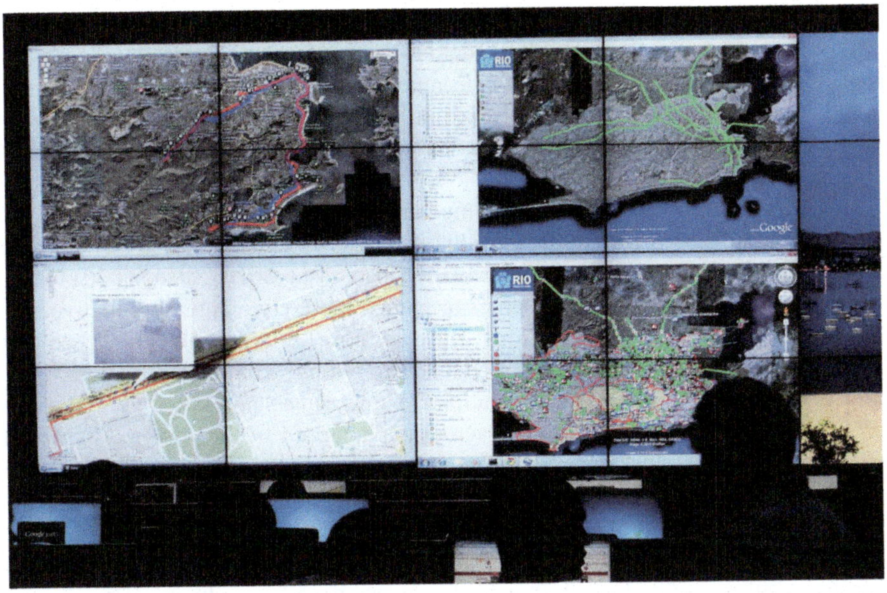

Fig. 1 IBM control room in Rio de Janeiro. (Photograph by IBM)

Fig. 2 Songdo, a flagship "smart city" project involving Cisco, is Korea's self-proclaimed "city of the future." (KOREA Magazine)

building automation, transportation infrastructure, and utility grids. In Siemens' Crystal center in London, exhibits demonstrate physical changes in cities in response to desired economic and environmental outcomes (Figs. 3 and 4).

While emphasizing their specific core business focus, each of these companies brings up almost identical themes about the contemporary global and urban condition. Each cites the well-worn expression that "more than half the world's population lives in urban areas," as well as burgeoning environmental crises. They all proclaim the arrival of advanced technologies that now enable measurement of a city's "exact conditions" (Dirks and Keeling 2009, 12; see also Cisco 2010; Siemens 2011).

The motivations of these companies to launch these city-centered initiatives are evidently centered on potential profits. But why now? It has long been argued that information and communications technologies, along with transportation infrastructure, have propelled decentralization, leading to a "post-city age" (Webber 1968; see also Fishman 1987). But, even as explanations of the post-World War II shift out of (U.S.) cities was solidifying in both the sprawling suburbs and in scholars' thinking, others like Harvey (1989) and Zukin (1982) have noted the continuation of increasingly privatized and selective investment in specific places in central city areas in the 1970s. The shift seems not so much in a singular spatial direction, but in methods and arenas of expansion and investment. The shifting grounds and modes of urban growth and capital investment reflect Harvey's (1985)

Fig. 3 Exploring interdependent elements of a sustainable city at the Siemens Crystal in London. (Photograph by author)

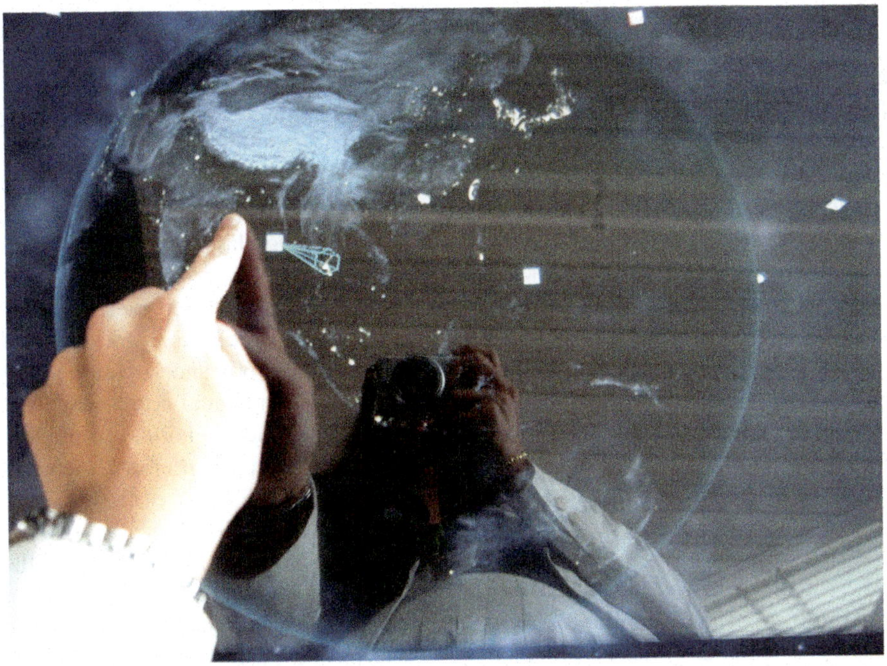

Fig. 4 Controlling the world and its cities with a flick of a finger at the Siemens Crystal in London. (Photograph by author)

assertion of the relationship between processes of urbanization and capitalist accumulation. The idea of the relationship between technology companies' urban focus and the search for new markets is concretely illustrated in Paroutis et al. (2014) study of IBM's business strategy during the 2008 recession.

Today, industry analysts, economists, and planners share optimism about the aligned future of markets and cities. A New York Times article on IBM's activities in Rio cites an estimate that the "smart" urban systems market will reach $57 billion by 2014 (Singer 2012). Economist Glaeser's (2012) embrace of urban density and height proposes that a de-regulated city would enable wealth, sustainability, health, happiness, and *intelligence*. Planner and architect Chakrabarti's (2013) "manifesto for urban America" details how design for "hyperdensification" can result in prosperity and sustainability. These themes are consistent with what is increasingly being viewed as the "urban age" (Burdett and Sudjic 2007; Brenner and Schmid 2014), when discourses of global social and environmental challenges *and* the opportunities for solutions are channeled by and through the continued growth of large urban centers.

While "smart cities" may be relatively new, the premise of being able to measure, know, and plan societal advancement extends a long lineage of justifications behind urban planning. Friedmann (1987, 67) details the evolution of a scientifically based notion of planning, beginning with Saint-Simon's vision of the scientists and engineers who would "observe and measure" the laws behind society in order to plan its progress. This positivist worldview underpins what Hall (2002) charts as the embracement of scientific analysis, monitoring, and control in professional planning in the 1960s. During this time, technological advances, including computerized data processing, and the idea of cities as complex systems, brought on new methods of planning, including modeling and predictions. A recent theoretical paper on "smart cities," written by IBM employees (Harrison and Donnelly 2011), appeals to these notions of the city as a system, invoking classic works by Jacobs (1961), Forrester (1974), and Alexander (1965). In fact, the references to the not-so-distant past may reflect lessons not learned. Goodspeed (2015) exposes the similarities between the current wave of technology companies' efforts to measure and optimize the city and the urban cybernetic theories of the 60s and 70s.

And today, confronting global urbanization and environmental crises (precisely the themes cited by technology companies), the call for the scientific, quantitative measuring of everything from the scale of the city to its impacts on the planet is as strong as ever (see, for example, Rosenzweig et al. 2010; Solecki et al. 2013).

It is important to note that there has been long-standing resistance to this scientific, positivist basis for planning. Rittel and Webber (1973, 158) warn of the inherent problems of an "idealized" planning system always on the search for "instruments of perfectability." What happens when the most intelligent plans and systems encounter "wicked problems" characterized by messy politics, stubborn social systems, and divergent scales of urban governance, geography, and ecology? And Friedmann (1987, 60) asserts the "illusion" of planners attempting to "build" a society like engineers build a bridge, and contrasts this scientific mode of thought with the planning traditions of "social learning" and "social mobilization." The decades since have been witness to both the explosion and fragmentation of post-positivist planning theories (Allmendinger 2002). Considerations of the role of technology in planning have followed alongside. Klosterman (1997), for example, builds on the communications view of planning to detail a vision of collective planning via information technologies.

3 Two Paradigms for the Smart City

The definition of the "smart city" is much contested (Hollands 2008). One prevalent definition—embraced by a broad constituency of technology companies, urban planners and designers, engineers, and city managers—is premised on the notion of an urban space threaded with digitally networked infrastructures, services, and devices, brought on by the pervasive increase in information and communication technologies (ICT) in the last thirty years. In this paper I explore the potential and drawbacks of these systems by looking to two specific diverging viewpoints—one a critical view concerned about aspects of power, epitomized by Manuel Castells' research on technology and society, and the other a more hopeful idea of digitally-supported liberation, perhaps best attributed to William Mitchell's writings and projects. These two theorists provide the bounds of possibilities, defined by macro level critique on one end, and a largely acritical embrace of technological potential on the other. Considering both together illuminates the conceptual terrain on which new "smart city" projects are enacted.

Castells (1989/1991) provides a critical analysis of the interrelationship of new technologies and the social structure in which such technologies arise, in particular, the spatial and social reorganization that accompanies the "informational mode" of development. In explaining the patterns of socioeconomic concentrating and dispersal that accompanies new information technologies, Castells coins the "space of flows," the "placeless" organizational space that enables the interaction between placed-based command operations and distributed services and production activities (Castells 1989/1991, 170). For Castells, the increasingly dominant nature of the space of flows brings on urban spatial and social differentiation, an unequal "dual city" (Castells 1989/1991, 172).

Castells (1996/2000) then exposes the contradictions in a "network society" between globally and instrumentally integrated informational technologies, and the emerging of locally specific actions based on primary sociocultural identities, creating oppositions between the internet and the "self." For Castells, "when the Net switches off the self, the self… constructs its meaning without global, instrumental reference" (Castells 1996/2000, 24). In elaborating on the "dual city," Castells traces differentiation not only within urban spaces but also across them, a global, striated system of connection and disconnection.

Mitchell (1995), in contrast, offers a prescient and generally optimistic view of how digital networks change the city. He relates structures and spatial arrangements of the digital age with economic opportunities and public services, public discourse, cultural activity, and urban experiences. "Traditionally, you needed to *go* someplace to do this sort of thing—to the agora, the forum, the piazza, the café, the bar, the pub, Main Street…" etc. (Mitchell 1995, 7, italics in original). Mitchell forecasts a technological reconfiguration of human habitat, a "bitspace" that will overlay traditional urban and rural landscapes (Mitchell 1995, 167). Ultimately, Mitchell envisions a new "global village" when all scales of people and objects are networked in "one densely interwoven system," human body and various scales of infrastructure interfaced with each other (Mitchell 1995, 173).

Mitchell (2003) further posits how technology has softened the physical boundaries of the city, "connectivity" supplanting "enclosure" as being the definitive urban condition. "My biological body meshes with the city," he enthuses (Mitchell 2003, 19). Mitchell describes a kind of two-prong sensory augmentation, where technologies have increasingly sculpted themselves to our bodies (for example, the miniaturization of handsets), and our own senses have been amplified or reproduced into space (with, say, cameras and scanners). In envisioning the multiplication of human sensory facets beyond the body into various realms of the city, Mitchell brings forth an urban cyborg fantasy, in which the breakdown of definitions between body and machine is further extended into the boundlessness of urban space.

These two concepts—Castells' dominating "space of flows," and Mitchell's embracing "cyborg self and networked city"—underlie two significant strands of scholarly thinking about the "smart city." Together these paradigmatic concepts present two poles against which to assess ongoing "smart city" initiatives.

In the one instance, in tune with Castells, new urban networked technologies enable corporations and political elites to create hierarchical spaces of physical and informational connectivity and settlement. It results in a "splintering urbanism," in which socioeconomic inequality is even further consolidated into the physical environment of the city (Graham and Marvin 2001; Graham 2002a, b). Technological fixes to presupposed urban vulnerabilities are also seen as potential tools to legitimize the continued growth, privatization, and securitization of urban centers (Hodson and Marvin 2010a, b). And the pro-business stance of both technology companies and city governments have reinforced criticisms that these solutions are harnessed towards the spatially and socially selective development of neoliberal capitalism, the "corporate smart city" (Hollands 2008, 2014).

In the other instance, extending Mitchell's vision, the increasing ubiquity of networked urban infrastructures and mobile devices like cellular phones and RFID chips allows a distributed intelligence to emerge between places, things, and people. The socio-technologically positivist view conceives this urban intelligence as bottom-up, "sociable," driven by the admittedly wondrous vision of our cities like "computers in open air" (Ratti and Townsend 2011, 44; see also Roche et al. 2012; Townsend 2013). The Copenhagen Wheel, a "smart bicycle" project by MIT Senseable Lab, for example, promises real-time mapping of a city's air quality (Fig. 5). While, in many cases, the projects promoted by the proponents of this view

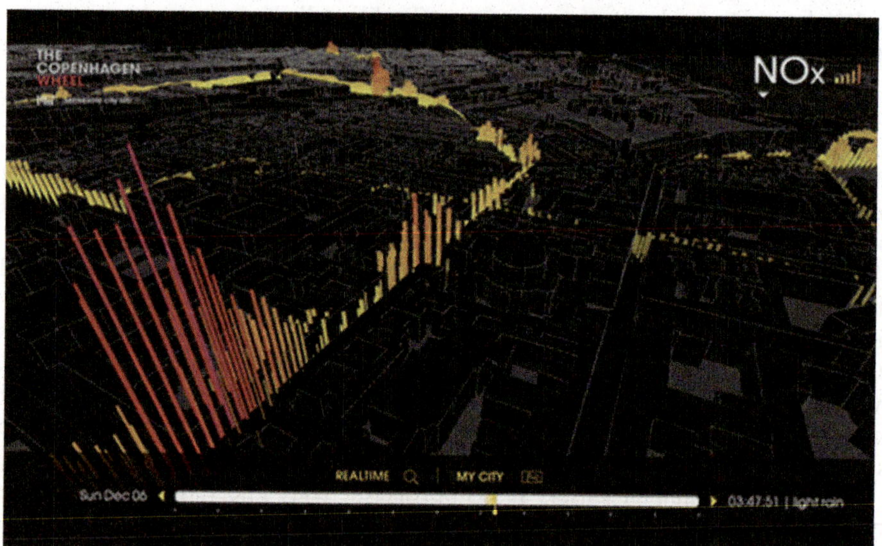

Fig. 5 Levels of atmospheric nitrogen oxides sensed and transmitted by the Copenhagen Wheel. (Image by MIT Senseable Lab)

include partnerships with the technology companies previously mentioned, they also tend to reject the state-controlled, centralized vision that many corporate-municipal initiatives hold (see, for example, Ratti and Townsend's (2011), Sassen's (2011), and Sennett's (2012) critiques of ground-up projects like Masdar in the United Arab Emirates and Songdo in Korea). Still, while these accounts invoke the use of technology in sociopolitical struggles, they often avoid a sustained analysis of power. In prioritizing the promises of technology over the structural reasons for contestation, they conflate the means and logics through which people mobilize with the tactics of those in power. Political change is treated like urban management, even though the objective of the former is transformation, the latter often continuity and the perpetuation of power structures. These arguments end up placing struggles for democracy on the same plane as traffic problems and the provision of health care and education (see, for example, Ratti and Townsend 2011).

4 Singapore and London

In probing these questions further, I look at "smart city" initiatives in two cities, Singapore and London. In this fast-moving context, it is understandable to point out the limitations of one or two case studies. However, in-depth studies of actual cases of "smart city" governance and implementation are critical in order to ascertain what really happens beyond the concepts and hype that is prevalent in the discourse (see, for example, Ching 2013; Shelton et al. 2015; Yigitcanlar and Lee 2014). These two cities hold particular lessons, separately and considered together. Both cities have active, established "smart city" initiatives and, as well, built examples of "smart" technologies in urban space. Both city governments have openly declared their support for such techno-urban futures (IDA n.d.; GLA 2013).

Given that both Singapore and London hold privileged positions as historical and present-day centers of trade—primary hubs in the global space of flows—with cohesive governance structures and ample wealth, one might ask whether and how these two cities offer lessons for others. In my view, they serve as two related yet differentiated paragons. Two ideal conditions: one, Singapore, the epitome of top-down planning, with enviable ability to implement social and spatial policies; the other, London, the archetypal Global City, still a powerful center of finance and culture. In comparison, they offer differences in political structure and administrative scales that illuminate the possibilities and challenges of governing and implementing smart urban technologies.

In my research, I focus on physical, spatial manifestations and sociopolitical systems. I conducted 14 in-depth, semi-structured interviews with city managers and planners, engineers and technology specialists, economic development officials, representative of a transnational corporation working on urban technologies, and privacy and technology advocates in both cities. I also made field visits to existing smart technology sites and city services, including transportation and water infrastructure, and reviewed documents by city agencies, technology companies, and media outlets, including local and national government "smart city" reports, various corporate white papers, primarily by IBM, Cisco, Siemens, and Veolia, and news accounts of "smart city" projects and investments.

4.1 Singapore

Small in spatial extent (less than half the size of London), Singapore boasts impressive achievements. It measures in the top ten in the world by purchasing power parity, twenty-seventh by human development index. Urban development in Singapore is generally top-down, a strong-state-led capitalism. Adhering to a model of centralized planning, the state strictly controls land use, environmental policies, and as well social relationships, evident in its reliance on ethnic quotas in public housing developments. As a historic port city, it was developed during British occupation and became a city-state after being invited to leave the Federation of Malaysia after disagreement over national social policies. Nevertheless, Singapore's urban and economic development is often invoked as an example for rapidly urbanizing cities of the Global South. Cities like Bangalore and Dalian look to Singapore as a model of technology-centered economic progress and a "livable" city (Chua 2011; Hoffman 2011), and Singaporean and Indian authorities have recently been in talks about potential partnerships in Indian Prime Minister Narendra Modi's stated plans to build 100 "smart cities" there (Tolan 2014).

Several initiatives in Singapore's built environment have been held up as "smart." These include the island's transportation network—with its extensive and smoothly running public bus and rail system, augmented with its "smart" fare card, and the central congestion pricing zone with automated estimated travel time notifications (Tan and Subramaniam 2012) (Fig. 6)—and real-time flood warning sensors in drainage canals. Beyond the realities on the ground, the city-state has also encouraged pilot initiatives in collaboration with technology companies, including winning a IBM Smarter Cities Challenge grant for the Jurong Lake District in 2012 (IBM n.d.), and announcing a partnership between the Singapore Housing Development Board (HDB) and French firms EDF and Veolia, that same year, to develop an urban modeling tool for Singapore public housing developments (EDF 2012).

Recently, Singapore's Infocomm Development Authority (IDA) has exploited its city-state status towards a new catchphrase, a "smart nation" (IDA 2014), or, according to Executive Deputy Chairman Steve Leonard, presenting the keynote at CommunicAsia 2014, "Singapore as one giant dashboarded entity." It is a vision in which an entire country is in sync, where "policy, people, and technology come together." Urban data analytics and management is seen as a "new frontier" for the nation, confirms Goh Chee Kiong, a senior economic development official, (2013, personal communication, 11 July). In this light, Singapore's overall approach to "smart cities" engagements appears to reflect a broader approach to urban development and nation building. As I've argued elsewhere, Singapore's development as

Fig. 6 Singapore traffic speed and travel time monitors. (Photograph by author)

a city has been wrought alongside its development as a nation, post-independence. Based on discourses of scarcity and survival, the country stakes its future on global links and the continued relevance of its development model (Goh, K. 2013). The current "smart nation" approach is consistent with the city-state's historical focus on establishing and maintaining ties to flows of global capital.

This vision of the "smart nation" is carried in lockstep within the upper levels of Singapore's government agencies and its private partners. Inquiring further into conceptualizations of the "smart city" across the various agencies reveals shades and differing outlooks and priorities. One issue concerns the role of corporations in "smart city" planning. Corporate technology companies like IBM and Siemens have enthusiastically touted their presence and embeddedness on the island. In the view of Singaporean economic development planners, this is because the corporations see the island as a "reference" market for Asia, a "leader for urban solutions" (Goh Chee Kiong 2013, personal communication, 11 July). And the government is confident about defining the terms of these relationships.

Another official involved in informational technology policy and planning is more direct in characterizing the government-corporate relationships as simply one of pragmatic commercial decision-making between (government) buyer and technology vendor. Because of the complexities of urban policy making and planning implementation, the companies, in his view, are not yet capable of providing much beyond the technology itself. They cannot yet "imagine the product." According to this official, the rhetoric of "smart cities" is simply that, rhetoric—"smoke and mirrors" (Henry Quek 2013, personal communication, 9 July). Indeed, the advocacy role of those in government information technology agencies, like IDA in Singapore, is critical in this aspect. They take the rhetoric in brochures and white papers, sift through the possibilities and opportunities in relation to their knowledge of the realities of urban governance processes, and explain them to those responsible for the "traditional" planning realms of transportation, housing, etc.

Whether speaking with government technology experts, officials in "traditional" municipal services, or key figures in economic development, one receives a significantly different response in terms of understanding the potential of the "smart city" and its place in current and future urban governance. This in itself is not a startling finding. More importantly, it suggests that the cohesive vision of a smooth urban digital space envisioned by many "smart city" proponents is difficult, perhaps impossible, to find. Instead, one finds a somewhat exploratory space where promised technologies are pondered over, tinkered with (tires kicked, so to speak), and literally bought and sold in bits and pieces as they are tentatively inserted into the workings of the city. The vagaries of urban governance, at the end of the road (sometimes literally), in large part define the impact of even the most comprehensive, ambitious, and forward-thinking technological vision.

Government officials hold conviction about the role of the state in making key—and good—decisions for citizens. This reinforces the leverage they believe they hold in public-private partnerships—evident, for example, in the planning and implementation of transportation. On the other hand, this context also hints at the darker side of pervasive urban sensing. Singapore's strict regulation of public behavior is well known—including laws curbing free speech. Government officials express no qualms about further securitization of public spaces, envisioning surveillance cameras smart enough to detect littering (Goh Chee Kiong 2013, personal communication, 11 July). Active government regulation, however, has not yet managed to quell increasing socioeconomic inequality in the city-state. This condition threatens to undermine basic societal balances that have maintained economic growth and political stability on the island in the recent decades (Economist 2014).

4.2 London

London, on one level, might be an odd place to find an aspiring global "smart city." A historic European urban center, it might be known more for monuments and a sense of propriety rather than broadband and automation. However, as the erstwhile center of empire, it has retained its position in international commerce and finance. Early efforts to digitally control city services and urban space were notable, including the iBus system that enables tracking of public transportation and easing of traffic flow; automated video cameras monitoring congestion pricing zones in central London (Fig. 7); and the oft-mentioned "Boris Bikes," a bike share program

Fig. 7 London's video-automated congestion pricing zone. (Photograph by author)

with centralized monitoring of bikes and stations (now widely replicated). Already considered one of the most surveilled cities in the world, the result of the so-called "Ring of Steel" developed during the Irish Republican Army insurgency period, the city now moves to install even more cameras and sensors.

London's historical role as a global center of commerce is evident in its embrace of "smart city" initiatives. City officials expect continued population growth, and are at pains to deliver services and an environment conducive to traditional banking and insurance sectors, as well as the growing technology sector. In 2013 the Smart London Board (2013) was formed to advise the Greater London Authority on visions for a smarter London. It released its first Smart London Plan late that year. The plan attempts to build off of the previous work, including the efforts to reform transportation and security during the 2012 Olympics and open data initiatives such as the CityDashboard. It ardently promotes London's existing digital infrastructure, and references the city's position as a center of culture and creativity. London's efforts are couched within a broader United Kingdom-wide initiative to encourage cities to pursue the economic potential of "smart urban systems" (BIS 2013).

Many aspects of the Smart London Plan are just beginning to be implemented. Speaking to local authorities, a key challenge in "smart city" planning is the relationship between territorial boundaries and urban governance. Cities often, if not always, comprise a set of administrative entities tied to specific territories—either nested, or in series, or both. Depending on the structures of governance, such entities may or may not be inclined to cooperate. London poses a particularly striking example of this. Greater London comprises the City of London and 32 boroughs, each with its own responsibilities for certain city services. The creation of the Greater London Authority in 1999 (and with it the first Mayor of London), was an attempt to bring greater regional governance and strategic urban planning. Within this structure, the City of London, the "Square Mile," presents a highly distinct sociopolitical space as a center of commercial activity—home of names like Rothschild and Lloyds—with relatively few residents. In contrast, this urban fragmentation is much less of a problem in Singapore due to its status as a city *and* nation and strong top-down national government.

This mismatch between priorities and governance scales is apparent in a number of ways. The "Boris Bikes," generally much loved and touted, was announced by the first Mayor of London Ken Livingstone and implemented by (and colloquially named after) his successor and current mayor Boris Johnson. The City of London acquiesced to this plan, and provided space for bike stations. But City of London planner Peter Wynne Rees brings up the contradiction of people getting off public transportation and onto bicycles, which in turn get in the way of the buses. He also notes the issue of the "loads and loads of vehicles moving the bikes around" (2013, personal communication, 30 July). To a broader point about "smart city" planning, Rees points out that some so-called "smart" aspects of London—including its transportation network and early digital communications infrastructure, were put in a "piecemeal way," because people wanted it, not in search of a "technopole."

On another level, Janet Laban, a senior planner in the City of London focused on sustainability, notes too the reluctance of large, high-profile companies in the Square Mile to sign on to smart grid initiatives, because of potential interference with high-velocity, algorithmic trading, or to the existing combined cooling, heat and power (CCHP) system (2013, personal communication, 30 July). Such reluctance poses roadblocks to the connectivity of "smart" infrastructure in the places it is arguably needed most.

London, like Singapore, is witnessing increasing socioeconomic inequality. Rapid and uneven economic growth—both within the city and globally—have had clear impact on the skylines and streets of both cities. Recent developments, including luxury towers and the massive Olympics effort, have been derided for their exclusionary nature. London's plan explicitly states the challenge of inequality, the task of addressing the "digital divide," and making access inclusive (Smart London Board 2013, 21). And yet, the plan—and its proponents—seems reticent to fully embrace the politics behind this concept. How do the principles of the "smart city" plan alter processes of urbanization and economic growth—processes that are themselves not in contradiction with the thrust for "smart city" initiatives?

The foregrounding of open data in London's plan offers some potential in this regard, something less evident in parallel discussions in Singapore. Open data advocate Gavin Starks, who sits on the Smart London Board, proclaims, "What would be the impact, for example, of having free Wi-Fi everywhere? At speed, for consumers and for businesses. That could be a very disruptive play for a telco (telecommunications company)" (2013, personal communication, 1 August). Starks' view of open data betrays an idealism about the possible community benefits of such initiatives. Of course, it remains to be seen if the open data initiatives in the Smart London Plan will achieve the objectives of socioeconomic inclusion.

4.3 Learning From…

Two key observations are evident from investigations in each city, and across them. First, echoing Hollands (2008), there is no one "smart city." But, additionally, there is no one "smart city" even *within* a city. Depending on whether I spoke to urban planners, technology consultants, infrastructure engineers, city economic development officials, industry experts, or technology company officials, I received a very different story of motivations, possibilities, practices, and end goals of the so-called "smart city." Differences in disciplinary expertise, job descriptions, implementation time-scales, and scales of urban governance are as important to the understanding of urban technologies as a common vision. This notwithstanding, some so-called "smart" technologies are already being implemented—in transportation and water infrastructure systems in Singapore; and in transportation and security in London. It is not that plans are not being realized; there seems to be a wide gap between what these various actors might envision today and what is and will be realized in the future.

Second, differences in scales and ideologies of urban governance *across* cities seem to have significant impact on the way that the many actors involved in "smart city" visioning frame their work and their priorities. In London I found distinctly stronger dialogues on open-source systems, and the ways that these might benefit smaller groups of users in a way that is, interestingly, both market and community friendly. In Singapore I found greater emphasis on government actions—simultaneously inviting private partnerships and investment and affirming government autonomy. These findings might appear unsurprising and even stereotypical. But they do demonstrate the uncertainty and intractability that transnational technology companies take on when partnering with local governments on long-term, large-scale projects.

5 "Smart" "Urban" Movement Building?

Perhaps neither of the paradigms offered by Castells and Mitchell finds its place wholly on the ground in Singapore or in London. But the threat of Castells' striated and unequal "dual city" is increasingly a reality. One might wait for either an equally dominating government structure or a more diffuse one, characterized by a commitment to openness, to ameliorate the worst effects of the growing inequality. Or, as planners tasked with envisioning urban futures and making them real, we might look to the technology itself for another way.

Given (1) the contesting scholarship on the threats and opportunities of urban network technologies, (2) the competing visions of what a "smart city" is, both in corporate literature and within urban governance, (3) the profit motive of corporations and its uncertain relationship with modes and objectives of city governance —and further, given that (4) technologies arise out of and are part of dominating socioeconomic structures (Castells 1989/1991)—can these technologies and strategies be harnessed towards a more just, inclusive, socially and environmentally sustainable city?

Castells himself provides a valuable and timely intervention. In 1989 he warned of the impending hegemony of the new "techno-economic paradigm," the superseding of spaces of places by the space of flows. He asserted that the response would involve knowing "how to articulate the meaning of places to this new functional space" (the space of flows) (Castells 1989/1991, 350). Ironically, in 2011 he finds such meaning, and space for resistance, in the flows themselves. Castells traces the origins and growth of social movements during the Arab Spring, European austerity protests, and Occupy Wall Street. He notes that movements may start on digital social networks, but they coalesce as movements by occupying urban space. But it's also not that simple. Movements then endure through interactions between cyberspace and urban space, in what Castells calls the "space of autonomy"—merging the globally-connected "free" space of networks with transformative power of actually claiming space in the city and openly challenging institutional structures (Castells 2012, 222).

This "space of autonomy," carved out by protestors and occupiers from both the emblematic urban spaces of oppressive regimes and the spaces of a now more mature Internet, offers an opening. The implementation of any part of the rhetorical and promised "smart city" is at best in its infancy at the moment. But, alongside the rapid growth of technologies and platforms such as social media, smartphones, and apps, it seems likely that many aspects of these smart urban systems will fast become part of our lives. As I write, infrastructure is being threaded with cable and sensors, public parks get open hotspots, people are being increasingly connected in less and less visible ways, the objects around us, from cars, to houses, to refrigerators, to our eye glasses, made to talk to each other. Planners, like many others, will have too many connections, too much data, and our problems will remain wicked. We could choose to look past the rhetoric of measurement and control, and explore the possibilities of this permeation of urban space (our realm, after all) with sensors and devices that can help *communicate, network, strategize, and organize,* to find our own space of autonomy. We might find, then, the spaces for a new wave of transformative planning.

Acknowledgments This study was partly funded by the Harvard Program on the Study of Capitalism. I am grateful to the anonymous reviewers and the editors of this publication for their invaluable criticism and suggestions.

References

Alexander, C. (1965). A city is not a tree. *Architectural Forum, 122*(1), 58–62, *122*(2), 58–61.

Allmendinger, P. (2002). Towards a post-positivist typology of planning theory. *Planning Theory, 1*(1), 77–99.

Batty, M., Axhausen, K. W., Giannotti, F., Pozdnoukhov, F., Bazzani, A., & Wachowicz, M. (2012). Smart cities of the future. *The European Physical Journal Special Topics, 214*(1), 481–518.

Brenner, N., & Schmid, C. (2014). The "urban age" in question. *International Journal of Urban and Regional Research, 38*(3), 731–755.

Burdett, R., & Sudjic, D. (Eds.). (2007). *The endless city: The urban age project by the London School of Economics and Deutsche Bank's Alfred Herrhausen Society*. London: Phaidon.

Business, Innovation & Skills (BIS), UK Department for. (2013). *Smart cities: Background paper*. London: UK National Archives. https://www.gov.uk/government/uploads/system/uploads/attachment_data/file/246019/bis-13-1209-smart-cities-background-paper-digital.pdf. Accessed January 28, 2015.

Castells, M. (1989/1991). *The informational city: information technology, economic restructuring, and the urban-regional process*. Oxford, UK; Cambridge, MA: Blackwell.

Castells, M. (1996/2000). *The rise of the network society*. Cambridge, MA: Blackwell.

Castells, M. (2012). *Networks of outrage and hope: Social movements in the Internet age*. Cambridge, UK; Malden, MA: Polity Press.

Chakrabarti, V. (2013). *A country of cities: a manifesto for an urban America*. New York: Metropolis Books.

Ching, T.-Y. (2013). *Smart cities: concepts, perceptions and lessons for planners. Master in city planning*, Department of Urban Studies and Planning, MIT.

Chua, B. H. (2011). Singapore as model: Planning innovations, knowledge experts. In A. Roy & A. Ong (Eds.), *Worlding cities: Asian experiments and the art of being global* (pp. 29–54). Chichester, West Sussex; Malden, MA: Wiley-Blackwell.

Cisco. (2010). *Smart+connected communities: Changing a city, a country, the world*. http://www.cisco.com/web/strategy/docs/scc/09CS2326_SCC_BrochureForWest_r3_112409.pdf. Accessed December 4, 2013.

Dirks, S., & Keeling, M. (2009). *A vision of smarter cities: How cities can lead the way to a prosperous and sustainable future*. Somers, NY: IBM Corporation.

Economist, The. (2014, August 23). *Singapore's politics: Cracks in the façade*. http://www.economist.com/news/asia/21613346-strains-pension-fund-hint-broader-problems-cracks-fa-ade?fsrc=scn/tw/te/pe/ed/cracksinthefacad. Accessed December 2, 2014.

EDF. (2012). *EDF and Singapore, in partnership with Veolia Environnement, innovating to make the future of the cities sustainable using advanced modelling approach*. http://press.edf.com/press-releases/all-press-releases/2012/edf-and-singapore-in-partnership-with-veolia-environnement-innovating-to-make-the-future-of-the-cities-sustainable-using-advanced-modelling-approach-91265.html&return=42680. Accessed December 2, 2014.

Fishman, R. (1987). *Bourgeois utopias: The rise and fall of suburbia*. New York: Basic Books.

Forrester, J. W. (1974). System analysis as a tool for urban planning. In N. J. Mass (Ed.), *Readings in urban dynamics* (pp. 13–28). Cambridge, MA: Wright-Allen Press.

Friedmann, J. (1987). *Planning in the public domain: from knowledge to action*. Princeton, NJ: Princeton University Press.

Garau, P., Sclar, E. D., & Carolini, G. Y. (2005). *A home in the city (report from the UN Millennium project)*. London: Earthscan Publications.

Glaeser, E. L. (2012). *Triumph of the city: how our greatest invention makes us richer, smarter, greener, healthier, and happier*. New York: Penguin Books.

Goh, K. (2013). Global garden city: Trans-territorial ecologies in Singapore. In I. Berman & E. Mitchell (Eds.), *ACSA 101: New constellations new ecologies*. ACSA.

Goldsmith, W. W., & Blakely, E. J. (2010). *Separate societies: poverty and inequality in U.S. cities*. Philadelphia, Pa: Temple University Press.

Goodspeed, R. (2015). Smart cities: Moving beyond urban cybernetics to tackle wicked problems. *Cambridge Journal of Regions, Economy and Society, 8*(1), 79–92.

Graham, S., & Marvin, S. (2001). *Splintering urbanism: Networked infrastructures, technological mobilities and the urban condition*. London; New York: Routledge.

Graham, S. (2002a). Bridging urban digital divides? Urban polarisation and information and communications technologies (ICTs). *Urban Studies, 39*(1), 33–56.

Graham, S. (2002b). FlowCity: Networked mobilities and the contemporary metropolis. *Journal of Urban Technology, 9*(1), 1–20.

Greater London Authority (GLA). (2013, March 22). "*Smart London*" vision. http://www.london.gov.uk/media/mayor-press-releases/2013/03/mayor-announces-smart-london-board-to-realise-london-s-ambition.

Hall, P. (2002). *Cities of tomorrow: An intellectual history of urban planning and design in the twentieth century*. Oxford, UK ; Malden, MA: Blackwell Publication.

Harvey, D. (1985). *The urbanization of capital: Studies in the history and theory of capitalist urbanization*. Baltimore, MD: John Hopkins University Press.

Harvey, D. (1989). From managerialism to entrepreneurialism: The transformation in urban governance in late capitalism. *Geografiska Annaler. Series B. Human Geography*, pp.3–17.

Harrison, C., Donnelly, I. A., IBM Corporation. (2011). A theory of smart cities. In *Proceedings of the 55th Annual Meeting of the ISSS* (Vol. 55). http://journals.isss.org/index.php/proceedings55th/article/view/1703.

Hodson, M., & Marvin, S. (2010a). Urbanism in the anthropocene: Ecological urbanism or premium ecological enclaves? *City, 14*(3), 298–313.

Hodson, M., & Marvin, S. (2010b). *World cities and climate change: producing urban ecological security*. Maidenhead, UK; New York: Open University Press.

Hoffman, L. (2011). Urban modeling and contemporary technologies of city-building in China: The production of regimes of green urbanisms. In A. Roy & A. Ong (Eds.), *Worlding cities: Asian experiments and the art of being global* (pp. 29–54). Chichester, West Sussex; Malden, MA: Wiley-Blackwell.

Hollands, R. G. (2008). Will the real smart city please stand up? Intelligent, progressive or entrepreneurial? *City, 12*(3), 303–320.

Hollands, R. G. (2014). Critical interventions into the corporate smart city. *Cambridge Journal of Regions, Economy and Society*. doi:10.1093/cjres/rsu011.

IBM. (n.d.). IBM smarter cities challenge: Singapore, Jurong Lake District. http://smartercitieschallenge.org/city_singapore_singapore.html. Accessed December 2, 2014.

Info-communications Development Authority of Singapore (IDA). (n.d.). *Smart city programme office*. http://www.ida.gov.sg/Infocomm-Landscape/Infrastructure/Smart-City-Programme-Office. Accessed December 5, 2013.

Info-communications Development Authority of Singapore (IDA). (2014, December 4). *Smart nation vision for Singapore*. http://www.ida.gov.sg/blog/insg/featured/smart-nation-vision-for-singapore/. Accessed January 29, 2015.

Jacobs, J. (1961). *The death and life of great American cities*. New York: Random House.

Klosterman, R. E. (1997). Planning support systems: a new perspective on computer-aided planning. *Journal of Planning Education and Research, 17*(1), 45–54.

Mitchell, W. J. (1995). *City of bits: Space, place, and the infobahn*. Cambridge, MA: MIT Press.

Mitchell, W. J. (2003). *Me++: The cyborg self and the networked city*. Cambridge, MA: MIT Press.

Paroutis, S., Bennett, M., & Heracleous, L. (2014). A strategic view on smart city technology: The case of IBM smarter cities during a recession. *Technological Forecasting and Social Change, 89*, 262–272.

Ratti, C., & Townsend, A. (2011). The social nexus. *Scientific American, 305*(3), 42–48.

Rittel, H. W., & Webber, M. M. (1973). Dilemmas in a general theory of planning. *Policy Sciences, 4*(2), 155–169.

Roche, S., Nabian, N., Kloeckl, K., & Ratti, C. (2012). Are "Smart Cities" smart enough. In *Global geospatial conference*. http://www.gsdi.org/gsdiconf/gsdi13/papers/182.pdf.

Rosenzweig, C., Solecki, W., Hammer, S. A., & Mehrotra, S. (2010). Cities lead the way in climate-change action. *Nature, 467*(7318), 909–911.

Sassen, S. (2011). Open source urbanism. *The New City Reader: A Newspaper of Public Space*, (15). http://www.newcityreader.net/issue15.html.

Sennett, R. (2012, December 4). No one likes a city that's too smart. *Guardian*.

Shelton, T., Zook, M., & Wiig, A. (2015). The "actually existing smart city". *Cambridge Journal of Regions, Economy and Society, 8*(1), 13–25.

Siemens. (n.d.). *Sustainable cities*. http://w3.usa.siemens.com/topics/us/en/sustainable-cities/pages/home.aspx. Accessed December 6, 2013.

Siemens. (2011). *Smarter neighborhoods smarter city: Solutions for a more sustainable New York - today*. Siemens Corporation. http://www.usa.siemens.com/sustainable-cities/pdf/smarter-neighborhoods-smarter-city.pdf. Accessed May 27, 2013.

Singer, N. (2012, March 3). IBM. Takes "smarter cities" concept to Rio de Janeiro. *New York Times*.

Smart London Board. (2013). *Smart London Plan: Using the creative power of new technologies to serve London and improve Londoners' lives*. London: Smart London Board.

Solecki, W., Seto, K. C., & Marcotullio, P. J. (2013). It's time for an urbanization science. *Environment: Science and Policy for Sustainable Development, 55*(1), 12–17.

Tan, L. W. H., & Subramaniam, R. (2012). Creating smart cities with intelligent transportation solutions: Experiences from Singapore. In O. Y. Ercoskun (Ed.), *Green and ecological technologies for urban planning: Creating smart cities* (pp. 174–189). Hershey, PA: IGI Global.

Tolan, C. (2014, July 18). Indian PM Modi pushes plan for 100 "smart cities." *CNN*.

Townsend, A. M. (2013). *Smart cities: Big data, civic hackers, and the quest for a new utopia*. New York: W.W. Norton & Company.

UN-Habitat. (2003). *The challenge of slums: global report on human settlements 2003*. London: Sterling, VA: Earthscan.

Wasik, B. (2012, January). #Riot: self-organized, hyper-networked revolts—coming to a City Near You. *Wired Magazine*, 20(1). http://www.wired.com/magazine/2011/12/ff_riots.

Webber, M. M. (1968). The post-city age. *Daedalus, 97*(4), 1091–1110.

Yigitcanlar, T., & Lee, S. H. (2014). Korean ubiquitous-eco-city: A smart-sustainable urban form or a branding hoax? *Technological Forecasting and Social Change, 89*, 100–114.

Zukin, S. (1982). *Loft living: culture and capital in urban change*. Baltimore: Johns Hopkins University Press.

Chapter 10
Knowledge-Mining the Australian Smart Grid Smart City Data: A Statistical-Neural Approach to Demand-Response Analysis

Omid Motlagh, Greg Foliente and George Grozev

Abstract Large scale field trials of smart grid technologies provide important insights as they capture the complex interdependencies of all the key variables, including consumer behaviours, which are needed for their effective evaluation. We present the Australian Smart Grid Smart City program and describe its big data using a narrative approach to hasten understanding and further analyses by others. Then we present a novel statistical-neural approach to maximise knowledge extraction from large datasets of diurnal load profiles, and demonstrate its use in evaluating the effectiveness of two cost-reflective product offerings, a Network-type and a Retail-type product bundle. The methods of analyses include Principal Component Analysis and Self-Organising Mapping. The results for the mid-winter electricity consumption profiles of participating households in July 2013 in New South Wales showed consumption behaviour changes with up to 12 % reduction in relative peak demand at 700 households who accepted the offerings compared to the control group. The resultant load factor of the high consuming outliers improved by about 18 % under demand-response compared to the control group. The feature-based classifier also revealed which behavioural components change due to users' demand-response activities; results compared favourably with third party consumer survey results.

O. Motlagh (✉) · G. Foliente · G. Grozev
CSIRO Land and Water, 37 Graham Road, Highett, Vic 3190, Australia
e-mail: omid.motlagh@csiro.au

G. Foliente
Faculty of Architecture Building and Planning, The University of Melbourne, Melbourne, Vic 3010, Australia
e-mail: greg.foliente@csiro.au

G. Grozev
e-mail: george.grozev@csiro.au

S. Geertman et al. (eds.), *Planning Support Systems and Smart Cities*,
Lecture Notes in Geoinformation and Cartography,
DOI 10.1007/978-3-319-18368-8_10

1 Introduction

With more than a decade of smart grid development in different parts of the world, the electricity sector is one of the transition leaders towards smart cities. A smart grid (Marris 2008; Gellings and Samotyj 2013) is distinguished from conventional grids by enhanced information and communication technologies and systems to manage in more intelligent ways the generation, delivery, storage and end-use consumption of electricity, to save energy, shave peak electricity usage, reduce risks of power interruptions and blackouts, and reduce greenhouse gas emissions. Ideally, like other concepts of smart city service provisions, these are supposed to be achieved in a cost-effective way for industry and businesses, consumers and society as a whole.

In Australia, the federal government and fourteen partners, electricity distribution and transmission companies, technology companies, universities and CSIRO, invested AUD 490 million in the *Smart Grid Smart City* (SGSC) field trial. One of the largest commercial-scale smart grid technology assessment projects worldwide, SGSC is implemented across eight local government areas in New South Wales (NSW), covering 30,000 dwellings since 2009 (Fig. 1). The coverage includes Upper Hunter, Muswellbrook, Cessnock, Lake Macquarie, Newcastle, Ku-Ring-Gai, Auburn, and Sydney. The segmentation is based on climate zone, household income, dwelling type, electricity consumption and gas consumption level, which —considering all possible combinations—makes 108 socio-demographic statistical cells.

Large commercial-scale field trials are important to understand the critical variables and relationships in the context of a selected region and community that could affect the success or failure of specific smart grid technology and product-service deployment to achieve its intended purpose. Trials that have been designed well allow stakeholders to observe and or discover the nature and scope of informational, technological, socio-economic and governance issues and challenges ahead. But the massive amount of data generated in such an undertaking can be *big* and overwhelming. Utility companies have some ideas what data they need and how they may use them to inform their investment and product offering options, but as Beyea (2010) has pointed out, there has been little discussion or exploration "of

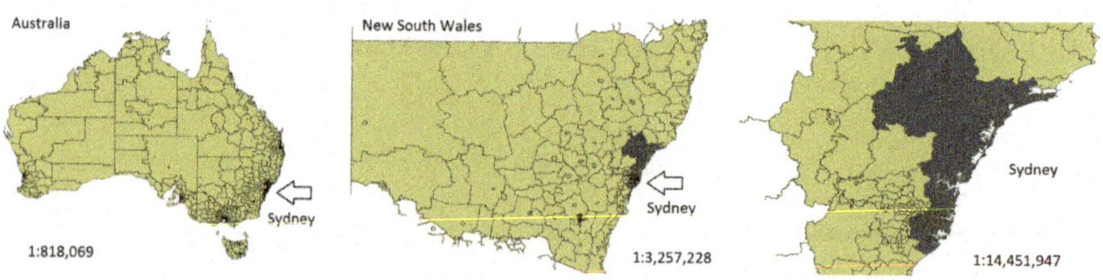

Fig. 1 The SGSC field trial location (*from left*): Australia, the state of NSW, and the areas in and near Sydney

the full scientific, economic, and historic potential of these data, whose usefulness may extend beyond the original purposes for which the data will be collected and stored". Third-party researchers (i.e., neither customers nor utilities) have the potential to investigate and analyse such data in various ways to contribute to a greater understanding of their potential for environmental and societal good, and for guiding policy, legislation, and regulations.

Towards this goal, this chapter aims to: (a) present the SGSC database and its key features and contents, and (b) demonstrate a novel approach to better understand customers' response to two popular smart grid "products" in the Australian context. The former is served by a narrative approach to describe the data sets (Sect. 2). To achieve the latter, we present a statistical-neural approach to maximise knowledge extraction from load profile data (Sect. 3). In electricity industry, household *load profile* is described as variations of electricity demand often expressed in kilowatt (kW) given the rate of sampling. A common representation consists of 48 samples per day at half hour rate which is known as the diurnal load profile. Samples could be further detailed to monitor power usage by individual electrical appliances. However, in this chapter diurnal load profiles refer to variations of the main import from the grid.

2 The Smart Grid Smart City (SGSC) Field Trial Data

One of the main aims of the SGSC Program is to quantify the costs and benefits of the deployment of smart grid technologies. Their applications are made possible by smart grid protocols that facilitate the interactions between customers and suppliers to actively manage demand on different parts of the supply-demand value chain. The interests and responsibility of key stakeholders are maintained: the customer or user on Demand Side Management (DSM) (Gellings and Samotyj 2013; Barbato and Capone 2014) the supplier on resolving peak demand issues, assuring service reliability and managing asset investments, and government on planning and enacting efficient and effective energy policies.

2.1 Overview of the SGSC Program

A range of cost-reflective smart-grid product bundles, demand-response protocols, and feedback technologies have been made available to customers. Accordingly, their impacts are measured in terms of electricity bill saving, and more importantly resolving peak demand issues and optimality of load factors. In electrical engineering, the diurnal load factor is defined as the average daily load (in kW) divided by the maximum peak load. Therefore, electricity network providers prefer load factor close to unity on the user side (i.e., "best" load factor) as it provides better network utilisation, which is often attained by load shifting-levelling. Accordingly, two of the project partners, the retailer *Energy Australia*, and the distributor *Ausgrid*, tailored different product bundles, referred to as the customer treatments.

Users are divided to treatment groups whose demand response behaviours towards cost-reflective products would be compared against a control group. There are twenty product bundles including different combinations of pricing options (e.g., dynamic peak rebate, pre-payment, seasonal time-of-use, dynamic peak price, etc.), usage feedback options (e.g., in-home displays, online portal, text messaging, etc.), and a home area network option. The characteristics of the product offerings are summarised in Table 1. The SGSC trials were divided into two streams, the *Network* trial managed by *Ausgrid*, and the *Retail* trial managed by *Energy Australia*.

The Network trial (products N01-N08 in Table 1) measured the effectiveness of smart meter based products without changing the customer's retail tariffs. Therefore, it mainly tested feedback technologies, financial incentives (rebates), and a lifestyle audit. The products consisted of online portal, in-home display, appliance control and sub-metering devices, interruptible load (for air conditioning) control rebate, and dynamic peak rebate. In contrast, the Retail trial measured the effectiveness of alternative electricity tariffs either as standalone products or bundled with feedback technologies. Accordingly, the retail trial tested smart meter based tariffs, feedback technologies and a rebate. There were twelve retail products (R01-R012) offered to customers each including a tariff (e.g., dynamic peak pricing, seasonal time-of-use or top-up plan), rebate (e.g., interruptible load) and one or more optional feedback technologies (e.g., online portal, in-home display, or sub-metering devices).

2.2 The SGSC Database

The SGSC database is publicly available through the Information Clearing House (ICH) (Ausgrid 2014). It is a massive database covering most physical aspects of the smart grid, as well as, aspects of users' demographics (Table 2).

To make the most of what the database offers, a good understanding of its features and contents is essential. Further analyses require careful identification and pre-processing. We describe herein the resultant database using a narrative approach for describing its contents. The specific names of various data objects and fields are made clear using [*italicised*] font within square brackets. The database is well organised and supported with an interactive data analysis portal for populating targeted queries. Table 2 provides a quick insight into the published databases at both physical and survey layers.

The data models, user guides and instructions are available on the portal (the ICH). The two main datasets which are needed for effective analysis of the program objectives are the data showing adoption of product bundles by the customers

Table 1 List of the products trialled throughout the SGSC program (Frontier Economics 2014)

Product	Trial	Marketing name of product bundle	Retail tariff	Feedback technologies[b]
N01	Network	Home energy online	–	Online portal
N02	Network	Home energy monitor	–	In-home display
N03	Network	Home energy assessment[c]	–	Smart meter only
N04	Network	Home energy rebate (PeakRebate)	DPR[d]	Smart meter only
N05	Network	Home energy network + home energy online	–	HAN[e]-smart plug, online
N06[a]	Network	Home energy rebate + home energy monitor	DPR	In-home display
N07	Network	Home energy online + home energy monitor	–	Online, in-home display
N08	Network	Air[f]	–	DR[g] device, online portal
R01	Retail	BudgetSmart	PPP [h]	Smart meter only
R02	Retail	BudgetSmart with PowerSmart monitor	PPP	In-home display
R03	Retail	BudgetSmart with PowerSmart online	PPP	Online portal
R04	Retail	BudgetSmart with PowerSmart online & home control	PPP	Online, HAN-smart plug
R05	Retail	FlowSmart with PowerSmart online	Interruptible	DR device, Online portal
R06	Retail	PriceSmart	DPP[i]	Smart meter only
R07[a]	Retail	PriceSmart with PowerSmart monitor	DPP	In-home display
R08	Retail	PriceSmart with PowerSmart online	DPP	Online portal
R09	Retail	PriceSmart with PowerSmart online & home control	DPP	Online, HAN-smart plug
R10	Retail	SeasonSmart	STOU[j]	Smart meter only

(continued)

Table 1 (continued)

Product	Trial	Marketing name of product bundle	Retail tariff	Feedback technologies[b]
R11	Retail	SeasonSmart with PowerSmart monitor	STOU	In-home display
R12	Retail	SeasonSmart with PowerSmart online	STOU	Online portal

[a]Products analysed in this chapter
[b]All technologies have the smart meter infrastructure
[c]Life style assessment
[d]DPR: Dynamic Peak Rebate
[e]HAN: Home Area Network
[f]Air: Direct control air-conditioner with minimal acceptance
[g]DR: Demand response enabling device
[h]PPP: Pre-payment plan
[i]DPP: Dynamic peak pricing
[j]STOU: Seasonal time-of-use

Table 2 Various aspects of the smart grid smart city database, derived from Ausgrid (2014)

Subject area	Application	Description
Customer sale (application)	Survey, customer segmentation, product uptake	Survey data of the sales cycle including the number of customers and date they attempted, subscribed or quitted each of the products offered. It also provides wide range of customers' demographic data
Half-hour power consumption-generation	Load profile representation, clustering	Provides electricity consumption and generation data for all customers including Retail and Network (trial customers) and the control group customers
Network model	System level	Provides the data of the electrical connections across the network and their physical properties
Network state history	Physics-based model	Data of voltage, current, etc. related to many points in the network monitored at points in time together with physical properties such as temperature, humidity, etc.
Distributed temperature sensing	Regression-correlation modelling	Four feeder cables were equipped with temperature sensing devices to record temperature along the cable
Home area network plug readings	Demand side management (demand response)	Data from smart plugs (and therefore individual appliances) that consumers can view in real-time. Customers can then decide on how to respond to impending peak events, etc.
Peak event response	Demand response modelling	Data showing how customer consumption behaviours were modified over peak events. For Network events it shows the applied rebates
Electric vehicles	Disruptive technology	Data from a fleet of 20 electric vehicles run by Ausgrid

[*Customer Sale*] and the customers' electricity consumption behaviours data [*Half-Hour Power Consumption and Generation*]. In later sections, we provide an insight into the physical layer and an example of a disruptive technology, i.e., electric vehicles, respectively. More details of the database could be found in a series of reports generated by different SGSC partners (AEFI 2014).

2.3 Customer Sale and Half-Hour Power Consumption and Generation Datasets

This is a big portion of the data from about 5900 households who accepted a product out of about 24,000 offers made to the target population. Spatial resolution down to anonymous individual dwellings [*Customer Sales*: *Households*: *Customer_Key*] is also given based on {[*Postcode*], [*Local Government Area*], or [*Suburb Name*]}. In order to make these datasets easily understandable, here we present a new narrative approach to explain the available data fields and their interdependencies in a brief yet logical and memorisable way as follows:

Under the SGSC trial, the stakeholders ran a campaign [*Campaign*] to enrol some of their customers into a demand-response program. Datasets are ordered to reflect the objectives of the trial and the sequences which make up the demand-response cycle. Customers [*Household*] of different characteristics [*Household Demographics*] and located in different geographical locations [*Locality*] signed-up for different product bundles [*Offered Product Bundle*: *Tariff Code*] on different dates [*Product Installation Date*]; from that date onwards the customers were notified in advance about every impending [*Peak Event*], via a feedback technology [*Product Bundle*: *Primary Feedback Technology*] and were reminded to reduce their electricity use during the event. Customers were said to have participated in the event if their electricity use dropped during the event relative to other times.

Before customers respond to peak signals, their responses could be anticipated based on their known consumption characteristics [*Power Usage Factors*], e.g., from whether or not they have solar panels, pool pumps, gas cooker, etc. A customer's actual response [*Peak Event Response*] would be due to any reason for example their interest in the offered incentives or a psychological factor, etc. The Network monitored customers import and export of electricity [*Half-Hour Power Consumption and Generation*] throughout the trial [*Consumption Date and Time*]. Customers themselves were also able to monitor their own consumption at circuit level [*Plug Readings*] using the feedback technology provided to them [*Secondary Feedback Technology*].

The [*Half-Hour Consumption and Generation*] data provide kWh readings of the actual flow of energy. This includes the electricity supply imported from the grid [*General Supply*] which is the main source of energy under any of the applied tariffs, import during off-peak or for a control load [*Off Peak*], gross or net values of domestic (often solar) electricity that is exported back to the grid [*Generation*]. The [*Peak Events*] data field gives the details of every event [*Event Key*] including its [*Type*], [*Date*], [*Start Time*], and [*End Time*]. Accordingly, the entire sequence of user demand responses to any of the cost-reflective products is available in the dataset to evaluate the project objectives as previously discussed.

2.4 The Physical Network Datasets

This data records a measurement of some electrical or environmental property at a point in the network at a point in time. On SGSC interactive data portal, the different electrical and environmental properties recorded in the subject area are provided under [*Network State History*] while information about the conductive network that SGSC devices are connected to is found in [*Network Model*]. The type of data includes: monitored load data at devices connected to conductive network (in kWh), and measurements of environmental property (and meteorological variables) where the device is located. Data is sourced at 1 s for peak events, and down to 10 min for device readings. The data is presented at both feeder level and local area level so that could be used for both physical and statistical applications.

Devices on which the measurements are taken [*All Measured Devices*] are identified by their names, types, and locations. More specifically their locations on the grid from the grid hierarchy point of view are identified in [*GridApps Device*]. That is for example about which feeder and substation a device is attached to, and which company monitors it. The devices data are recorded frequently [*Readings Date-Time*] and [*Time of Day*], while the peak events are also recorded [*Readings*]. The type of measurements, e.g., analogue versus digital, and units, e.g., kWh, are indicated as well [*Measurement Types*]. There are partner companies who monitor [*Partner Devices*] that are often measuring devices for meteorological variables, e.g., air temperature, etc.

The SGSC also trialled an Electric Vehicle (EV) research to bring a better understanding of the time, location and the scale of EV charging impact on the electricity grid. The project trialled 20 Mitsubishi i-MiEV electric battery-only vehicles assigned to residential and fleet customers. Additionally, fast and standard EV charging infrastructure was deployed in the trial regions. The data provides EV charging patterns and grid impacts. The project was also informed by uptake and grid impact modelling. Other significant EV trials were conducted in Victoria and Western Australia. Several organisations investigated different aspects of integration of EVs, including the CSIRO Electric Driveway project (Paevere et al. 2014).

3 Demand-Response Analysis by a Statistical-Neural Approach

The datasets described in Sect. 2 have been used by several stakeholders and research institutes to infer its knowledge using statistical and survey analysis in a number of significant research areas. In this section, we first discuss a summary of their current outcomes (Sect. 3.1); then, present a new analytical approach to knowledge-mining from a specific example case (Sect. 3.2). The results and analysis are given in Sect. 3.3. This method could also be effectively used for the other data described in Sect. 2.

3.1 Household Participants' Survey

The SGSC activities and data priorities were designed to prove or challenge a number of trial hypotheses and assumptions. The hypotheses were focussed on determining whether different smart grid technologies could achieve an economic (or other) benefit for Australian electricity consumers. Therefore, the [*Customer Application*] program aimed to complete a commercial-scale trial of customer products by installing and operating customer-side applications at around 14,000 residential dwellings. The applications included price incentives and or consumption information feedback technologies, as supported by the installed smart meter infrastructure.

Progress reports have been published by the project team over the years to allow the Australian government to oversee and assess the trial progress against the contractual obligations for the duration of the project. In March 2013, a consortium of Australian-based consultancy firms was commissioned to utilise the trial results to develop an integrated business case for smart grid technologies in a national context. The consulting consortium, called AEFI, includes Arup, Energeia, Frontier Economics, and the Institute for Sustainable Futures at the University of Technology Sydney. In line with the main objectives of the project, the AEFI released a survey (ISF and UTS 2014) aiming at analysing the experience that SGSC trial participants had with the products they were trialling. However, this analysis lacks the critical insight about how the users' actual electricity load profiles changed after participating in the program, i.e., subscribing to any of the SGSC products.

A customer research survey was commissioned for qualitative analysis of the users' data. The key findings and conclusions have been reported and are available from the Information Clearing House (Ausgrid 2014). The survey data fields are *italicised* for clarity. In terms of [*Increased Control*], the products had a very positive impact on overall usage awareness. Almost 70 % of participants reported better bill-saving. As for [*Energy Efficiency*], 83 % of participants took action to change behavior or reduce their usage in order to reduce overall consumption.

Table 3 Proportion of customers (1798 samples) taking action towards consumption reduction (ISF and UTS 2014)

Action	Any action[a]	Reduced usage	Shifted usage	Reduced standby	Upgraded
Participation %	85	65	58	33	25

[a]Includes reduced or shifted usage, or reduced standby time of use or upgraded appliance(s)

Around 67 % of participants reported on reduced usage, 58 % reported on changing time of use, and around 25 % reported on upgrading one or more appliances for higher efficiency. In the comprehensive reports, each of the efficiency aspects is presented in details. For instance, it is reported how the demand response protocol assisted with appliance control (ISF and UTS 2014). The survey showed 36 % found it very useful, against 5 % who found it not useful.

Regarding [*Peak Event Products*], almost 90 % of households reported participating in peak events. The incentive-based Dynamic Peak Rebate (DPR) products inspired 10 % higher participation than the tariff-based Dynamic Peak Pricing (DPP) products. Additionally, people with peak rebate were more likely to report on bill-saving than people with DPP and non-peak event products. Overall, the surveys were mainly focused on the following products from Table 1: DPR (or PeakRebate), DPP (or PriceSmart), STOU (or SeasonSmart), and PPP (or BudgetSmart). From [*Frequency of Engagement*] perspective, customers with peak event products, i.e., incentive-based (PeakRebate) or tariff-based (PriceSmart), engaged more frequently with their feedback devices as compared to households with the other tariff types. For [*Perception of Financial Savings*] users with peak event products reported the largest savings, with the PeakRebate being the strongest performer. Finally, [*Product Satisfaction and Likelihood to Recommend*] is high for peak event products compared to the conventional time of use products.

To summarise the survey based results, Tables 3 and 4 show the impact of the demand-response bundles on overall energy efficiency behaviours, and details of appliances usage, respectively. For instance, 85 % of users reported taking at least one action towards efficiency improvement on at least one electrical appliance (Table 3). The specific type of action as well as percentages of the appliances subject to at least one action is presented as well. The survey based results provide an insight into overall impact of demand response program on consumption behaviours. However, these results do not present a quantitative analysis of the users' load profiles. We need a systematic computational model to automatically extract the changes to the load profiles for different households, including comparisons with the control group. Understanding the impacts on the households' actual load profiles is significant when it comes to supply-demand optimization.

Table 4 Households appliances subject to energy saving and time shifting actions (ISF and UTS 2014)

Appliance	Air-conditioner	Dryer	Washing machine	Lighting	Dish washer	Pool pump-filter	Heater	TV	Computer	Entertainment	Shower or bath	Oven or stove	Iron	Microwave	Kettle	Fridge
Participation %	56	54	53	52	51	50	43	34	33	31	24	22	21	19	16	10

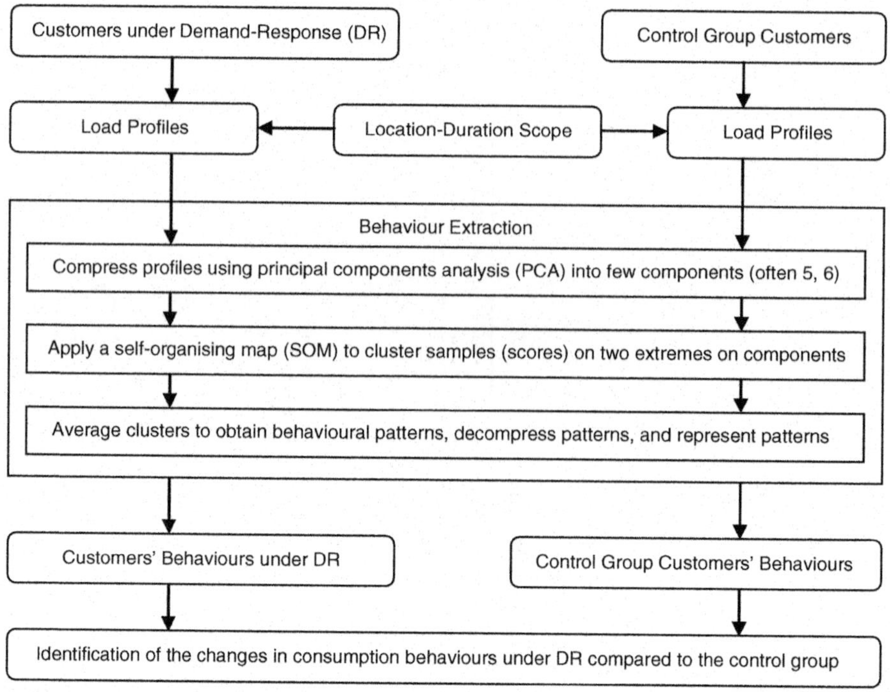

Fig. 2 Quantitative analysis of electricity demand behaviours with and without demand-response

3.2 The Method of Clustering

The objective of this methodology is to distinguish what exact changes occur in load profiles due to demand-response behaviours. In order to limit the scope of the modelling, we conduct herein a stochastic analysis looking at the households' load profiles rather than their typologies and the underlying properties. To demonstrate the technique, the analysis is presented on the aggregated scale rather than for individual demographic or geographic cells, covering all sample households (around 30,000) during the entire month of July 2013 (mid-winter in NSW).

Figure 2 provides a summary of the methodology and the employed techniques, namely Principal Component Analysis (PCA) and Self-Organising Map (SOM). Given the big data nature of the load profiles (i.e., kW readings at half-hour resolution), the first step involves compression of the load profiles from 48 half-hours to a few principal components (PCs). Preliminary results showed that a bunch of diurnal profiles (P) can often be presented into no more than six principal components (from an elbow on the 6th eigenvalues on the scree test) which reflect much of the correlation information (around $\simeq$75 %). Equation 1 presents the PCA for all $p \in P$ profiles; where P minus its empirical mean make rows of matrix X. There are 48 variables representing half hourly diurnal demand levels in kW (columns of X). A feature vector F is then built from the order of the principal eigenvectors of X, to reveal the sample scores Y (Eq. 2). Accordingly, the samples are mapped into PCs-dimension space, where PCs equals the number columns in Y. Now let us cluster the samples in the PCs-space. The goal is to cluster compressed samples $y \in Y$ into k subregions around the positive and negative extremes on every PC axis showing pairs of opposite behaviours which are also distinct from component to component. To cluster the sample profiles into $\{k\}$ clusters, SOM winner-takes-all strategy is adopted where winner is chosen based on minimal Euclidean distance (e) in Eq. 3. The algorithm is repeated iteratively until convergence. Next, the scores around each extreme on each PC are averaged to return a pattern. Both clusters and patterns are then decompressed and presented as time series, i.e., representative diurnal profiles.

$$p \in P \Leftrightarrow x \in X: \quad x = p - E(P) \tag{1}$$

$$Y = X \times F, \quad y \in Y \tag{2}$$

$$\begin{aligned} |e|_{min\{k\}} &\Rightarrow y_{winner\{k\}} : |e|_{min\{k\}} \\ &= \min\left\{ \left(\sum_{c=1}^{PCs} \left[Y_{\{k\}}(c) - y(c) \right]^2 \right)^{\frac{1}{2}} \right\}, \quad y \in Y \end{aligned} \tag{3}$$

3.3 Results and Analysis

With the analysis done once for the control group households and once for the trial households, a comparison could be made between their respective behaviours to infer the impact of the trialled products. Since the survey results on [*Frequency of Engagement*] with feedback devices (i.e., deemed equivalent to the extent of demand response engagement) showed higher frequency under DPR (PeakRebate) and DPP (PriceSmart) products, we selectively present our results only from these two products and in comparison with the result from the control group. The candidate products represent N06 and R07 from Table 1. The samples include 700 households subscribed to the Network's DPR plan, 700 households subscribed to Retailer's DPP plan, and 400 households from the control group. Each household is represented by its 31 diurnal load profiles during July 2013. There are therefore 21,700 profiles under DPR, and similarly under DPP, against 12,400 control profiles. We then conduct the PCA on each set which reveals 5 components to present ≃64 % of the correlation information.

Applying the SOM on the PC-space, a number of distinct behavioural components are revealed as the SOM untangles the scores by attracting them towards maximum (positive) and minimum (negative) extremes on the PCs axes. Table 5 shows a summary of the findings and the observed behaviours. As it was expected, there is a pair of opposite behaviours on the two extremes of every PC which are named according to their representative profile shapes, e.g., high demand versus low demand on PC1, evening demand versus morning demand on PC2, etc. The significance of each pair of opposite behaviours in the analysis (shown by percentage) is determined by the ratio of the eigenvalue of their respective PC over the total of the principal components.

With the major behavioural components given in Table 5, each household may exhibit one or a combination of such behaviours, e.g., HC and PP that is high consumption during both morning and evening peaks. The centre of the coordinates (i.e., zero on all PCs) is indeed a big cluster of households that exhibit no extreme feature compared to the outliers with extreme features that populate around the ± extremes on the PC axes. We call this the natural cluster for the particular product they are trialling. The plots in Fig. 3a–c show the profiles of the natural clusters and their variability under flat tariff, DPR, and DPP, respectively.

The control group shows a relatively steady diurnal demand (averaged across samples and over days) at around 0.33 kW. However, with the existing samples design, it is apparent that both averaged DPR and averaged DPP profiles show higher variability, and slightly higher average demand (around 0.37 kW).

Figure 3d shows the average wholesale electricity price during the same period in NSW. Apparently with the highest price at around 7 and 8 a.m. (half-hours 14 and 16) and 6–7 p.m. (half-hours 36–38), DPR and DPP variability with lowered evening consumption could be regarded as favourable bill saving activities. The downward arrows indicate the demand deliberatively lowered. The lowered demand

Table 5 Summary of the behaviours and their significance percentage under DPR and DPP versus control group

Product	Control (%)	Control group behaviours (abbreviated name)	DPR (%)	DPR behaviours	DPP (%)	DPP behaviours
PC1	24.8	High consumption (HC) versus low consumption (LC)	30.1	HC LC	29.8	HC LC
PC2	13.0	Only evening (E) versus only morning (M)	9.8	E M	9.3	D N
PC3	9.8	Only day time (D) versus only night time (N)	9.1	D N	9.0	E M
PC4	8.5	Early morning and early evening (EMEE) versus late morning and late evening (LMLE)	8.6	EMEE LMLE	8.1	EMEE LMLE
PC5	7.4	Midday and midnight (MM) versus morning peak and evening peak (PP)	7.6	MM PP	7.4	MM PP

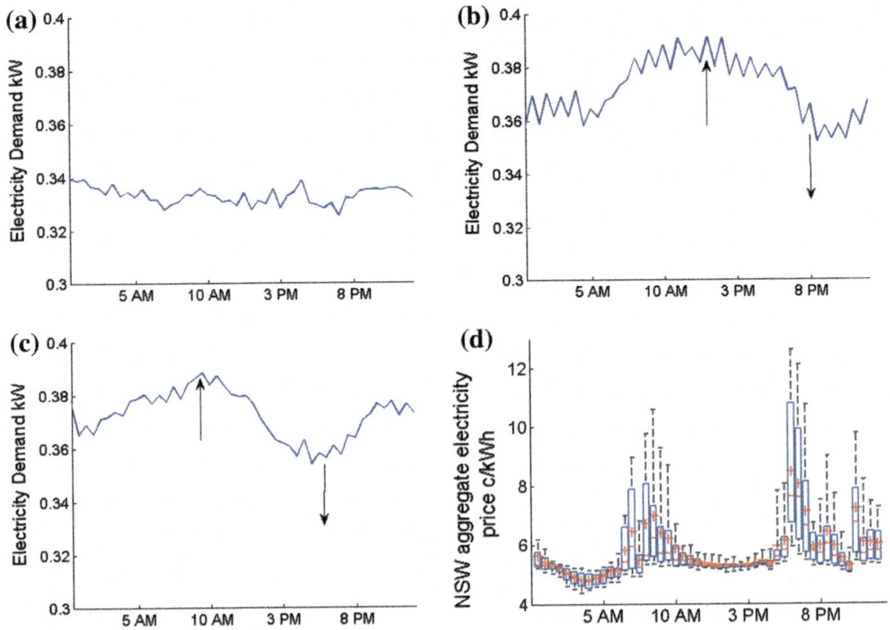

Fig. 3 Representative profiles of the natural clusters, **a** averaged profile of the control group with almost no variability, **b** average of the DPR group with deliberatively lowered evening consumption and increased midday consumption, **c** average of the DPP group with deliberatively lowered evening consumption, **d** average wholesale electricity price during July 2013 in NSW (AEMO 2014) varying between 5 and 10 c/kWh

is however compensated with increased demand deliberatively at around low-rate and shoulder-rate hours, i.e., midday hours under DPR and post morning peak hours under DPP. The upward arrows show the increased demand. The overall increase in level of consumption could be due to skewness in sample design, or other factors such as lavish consumption due to higher household confidence in bill saving under either of the DR plans (i.e., a rebound effect).

It is also critical to discover which of the behavioural elements could be the root cause of the variability. Accordingly, we shrink the clusters' boundaries for a focused observation of the outliers corresponding to each of the behavioural components in Table 5. Figures 4, 5 and 6 show the findings under the three situations. Under both DPR and DPP, the high consuming outliers (HC behavioural component in Figs. 5a and 6a) have become more level (compared to Fig. 4a) signifying the effectiveness of the demand response plans. With PC1 ≃30 % this could be a considerable achievement to have high consumers approach load factors (L_f) closer to unity. Equation 4 reveals respective load factors 0.84 and 0.73 for the HC group under DPR and DPP respectively, against 0.71 for the HC control group. In a similar fashion, better load factors are observed in the LC group under DPR

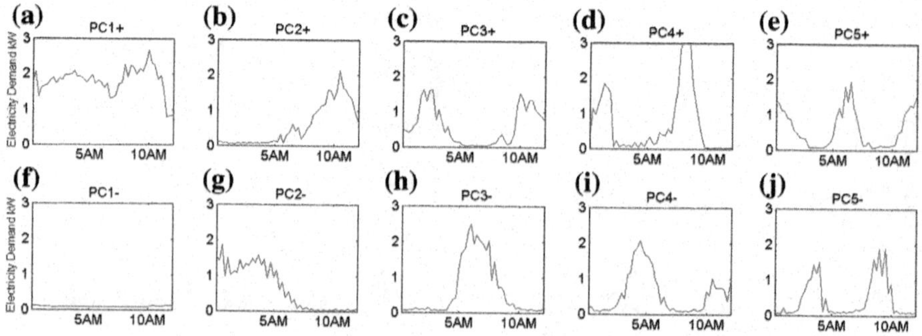

Fig. 4 Flat tariff (control): clusters of outliers formed at the two extremes on every individual PC axis, **a**, **f** PC1, **b**, **g** PC2, **c**, **h** PC3, **d**, **i** PC4, **e**, **j** PC5

and DPP against the control group (i.e., 0.86 and 0.91 respectively in Figs. 5f and 6f, vs. 0.72 in Fig. 4f).

Best load factor (i.e., L_f approaching unity) among high consuming groups is therefore achieved under DPR which is relatively about 18 % better than the L_f of the high consuming outliers among the control group. In terms of the behavioural components the difference between DPR and DPP is that under DPR households' behaviours are of the same significance order as the control households' behaviours. Under DPP however, the second and third behavioural components are opposite to that of the control group. Table 5 best shows the differences. A quick justification of the similarity between behavioural significance order in the DPR and the control group is that the households under no demand-response plan (here the control group) more easily adapt to the DPR plan than to the DPP plan. This however requires further validation.

The fact that the significance percentages for Day-Night consumption differ just slightly between DPR and DPP (9.1 % vs. 9.3 %) reflects that the two groups are not considerably different in terms of diversity of average midday and average midnight consumption. However, the relatively more significant difference for Evening-Morning consumption (9.8 % vs. 9.0 %) shows that the respective diversity is higher under DPR plan which could indicate wider range of deliberative behavioural choices.

Relative reduction of evening peak demand over the average demand is also higher under DPR than DPP as could be observed on the natural DPR and DPP profiles in Fig. 3b–c, respectively. These ratios are 12.04 % under DPR and 8.71 % under DPP. The underlying cause could be more instances of peak shifting and peak shaving under DPR as could also be observed on PC2 and subsequent components in Fig. 5. In fact peak demand reduction of 0.37–3.22 % has been verified on the behavioural components on PC2 to PC5 under DPR. These reductions plus those on the subsequent components (i.e., PC6, etc.) collectively reflect similar extent of reduction as seen among the natural DPR group. The peak shaving behaviours are indicated with downward arrows on both Figs. 5 and 6.

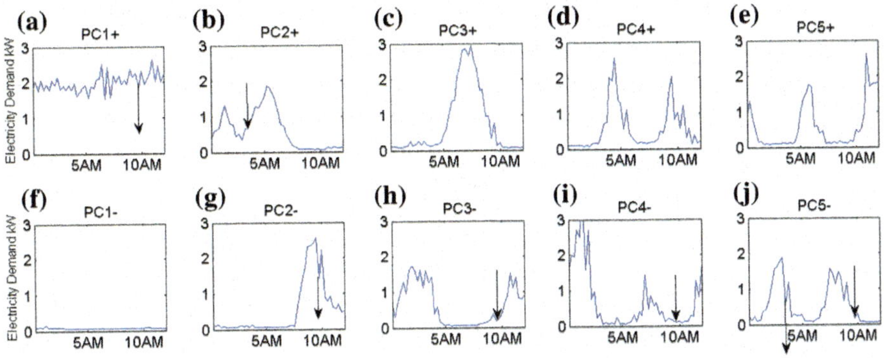

Fig. 5 DPR (PeakRebate): clusters of outliers formed at the two extremes on every individual PC axis, **a**, **f** PC1, **b**, **g** PC2, **c**, **h** PC3, **d**, **i** PC4, **e**, **j** PC5

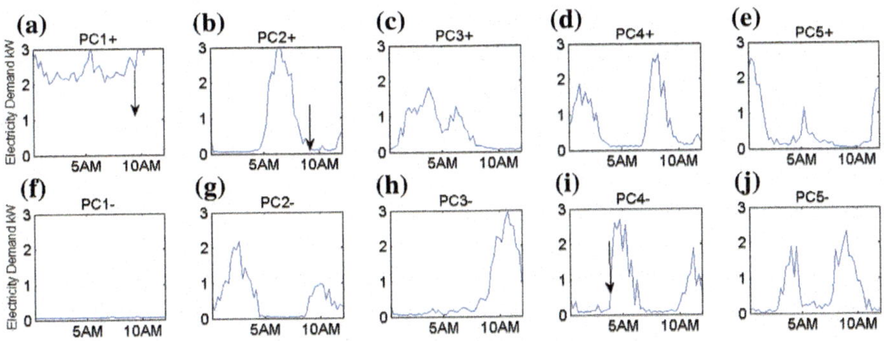

Fig. 6 DPP (PriceSmart): clusters of outliers formed at the two extremes on every individual PC axis, **a**, **f** PC1, **b**, **g** PC2, **c**, **h** PC3, **d**, **i** PC4, **e**, **j** PC5

Similar findings have been confirmed via unsupervised Hebbian algorithm and other neural and statistical models, e.g. in (Zhou et al. 2013).

$$Y_{\{k\}} \Leftrightarrow P_{\{k\}} : L_f = E\big(E\big(P_{\{k\}}\big)\big)/\max\big(E\big(P_{\{k\}}\big)\big) \tag{4}$$

The variations observed on representative diurnal profiles under DPR and DPP compare well with the survey findings that 58 % of SGSC participant reported on changing their time of use behaviours upon subscription to these products. More importantly, our analysis revealed more inclination towards the DPR plan which is also seen in the survey results, i.e., the DPR products inspired 10 % higher participation than the DPP products. Another common result with the survey is that DPR users were more likely to report on bill-saving than DPP users, which could be associated with the higher reduction of peak demand under DPR compared to DPP.

4 Conclusions and Future Work

This chapter presented an insight into the nature and content of the Smart Grid Smart City big data, and the household participants' research survey results. Two of the most popular trial products representing the Network-type (PeakRebate or DPR) and the Retail-type (PriceSmart or DPP) boundless were analysed in more details using a novel computational methodology that use Principal Component Analysis combined with Self-Organising Mapping as the feature-based classifier. Users' consumption behaviours were detected and compared against the control group showing effectiveness of demand response with up to 12 % reduction in relative peak demand in evening, as well as up to 18 % improvement of the load factor among the high consuming outliers.

Analysis results—which indicate that many participants changed their time of use behaviours after subscribing to these products, customers have greater inclination towards the DPR plan compared to the DPP plan, and that DPR users were more likely to report on bill-saving than DPP users—compared well with the customer survey results. The method described in this chapter could also be extended to analyse the impacts of the other products in Table 1, in comparison to one another, and or against the control group.

Apart from user side applications, the future directions of this research also include the analysis of behaviours among distribution and retail industries towards market adaptation, e.g., how retailers sustain profitability in line with the government's policies as well as market transitions such as increased uptake of solar photovoltaic panels. For understanding the underlying socio-demographic factors that impact on consumers' demand-response behaviours towards specific smart grid product offerings, a bottom-up modelling approach could be employed to complement and or extend the insights from the type of analyses presented herein.

Finally, on top of demand-response analysis for the offered smart grid product bundles, users' reactions towards controlled load strategies and advanced dynamic demand technologies could be investigated as well. The FlowSmart product in this project (R05 in Table 1) could be a good example case where the SGSC data would include air-conditioners control load data, duty cycle data, as well as the demographic data of the subscribing customers. This data could be linked with other extensive user consumption datasets such as the Residential Building Energy Efficiency (RBEE) project (Ambrose et al. 2013).

Acknowledgments The authors are grateful to Mr Daniel Burke, Mr Shayne Baird, and Mr Victor Lambe from Ausgrid, for their help and constructive discussions in relation to SGSC datasets.

References

AEFI: Arup, Energeia, Frontier Economics, Institute for Sustainable Future. (2014). Smart Grid Smart City: shaping Australia's energy future, executive report.

AEMO: Australian Energy Market Operator. (2014). *Aggregated price and demand data files*. http://www.aemo.com.au/Electricity/Data/Price-and-Demand/Aggregated-Price-and-Demand-Data-Files/. Accessed 20 July 2014.

Ambrose, M. D., James, M., Law, A., Osman, P., & White, S. (2013). *The evaluation of the 5-star energy efficiency standard for residential buildings*. Australia: CSIRO.

Ausgrid. (2014). *SGSC information clearing house*. https://ich.smartgridsmartcity.com.au/Main. Accessed online 22 November 2014.

Barbato, A., & Capone, A. (2014). Optimization models and methods for demand-side management of residential users: A survey. *Energies, 7*(9), 5787–5824.

Beyea, J. (2010). The smart electricity grid and scientific research. *Science, 328*, 979–980.

Frontier Economics. (2014). *Smart grid smart city customer applications trials: Data analysis methodology report*. https://ich.smartgridsmartcity.com.au/Main. Accessed 10 July 2014.

Gellings, C. W., & Samotyj, M. (2013). Smart grid as advanced technology enabler of demand response. *Energy Efficiency, 6*(4), 685–694.

ISF & UTS: Institute for Sustainable Future, University of Technology Sydney. (2014). *Smart grid, smart city: Customer research report*. https://ich.smartgridsmartcity.com.au/Main. Accessed 26 July 2014.

Marris, E. (2008). Upgrading the grid. *Nature, 454*, 570–573.

Paevere, P., Higgins, A., Ren, Z., Horn, M., Grozev, G., & McNamara, C. (2014). Spatio-temporal modelling of electric vehicle charging demand and impacts on peak household electrical load. *Sustainability Science, 9*(1), 61–76.

Zhou, K. L., Yang, S. L., & Shen, C. (2013). A review of electric load classification in smart grid environment. *Renewable and Sustainable Energy Reviews, 24*, 103–110.

Chapter 11
Urban Emotions: Benefits and Risks in Using Human Sensory Assessment for the Extraction of Contextual Emotion Information in Urban Planning

Peter Zeile, Bernd Resch, Jan-Philipp Exner and Günther Sagl

Abstract This chapter introduces the 'Urban Emotions' approach. It focuses on integrating humans' emotional responses to the urban environment into planning processes. The approach is interdisciplinary and anthropocentric, i.e. citizens and citizens' perceptions are highlighted in this concept. To detect these emotions/perceptions, it combines methods from spatial planning, geoinformatics and computer linguistics to give a better understanding of how people perceive and respond to static and dynamic urban contexts in both time and geographical space. For collecting and analyzing data on the emotional perception to urban space, we use technical and human sensors as well as georeferenced social media posts, and extract contextual emotion information from them. The resulting novel information layer provides an additional, citizen-centric perspective for urban planners. In addition to technical and methodological aspects, data privacy issues and the potential of wearables are discussed in this chapter. Two case studies demonstrate the transferability of the

P. Zeile (✉) · J.-P. Exner
Department of CAD and Planning Methods in Urban Planning and Architecture—CPE, University of Kaiserslautern, Kaiserslautern, Germany
e-mail: zeile@rhrk.uni-kl.de

J.-P. Exner
e-mail: exner@rhrk.uni-kl.de

B. Resch · G. Sagl
Department of Geoinformatics—Z_GIS, University of Salzburg, Salzburg, Austria
e-mail: bernd.resch@sbg.ac.at; g.sagl@cuas.at

B. Resch · G. Sagl
Chair of GIScience, University of Heidelberg, Heidelberg, Germany

B. Resch
Center for Geographic Analysis, Harvard University, Cambridge, USA

G. Sagl
Department of Geoinformation and Environmental Technologies, Carinthia University of Applied Sciences, Salzburg, Austria

S. Geertman et al. (eds.), *Planning Support Systems and Smart Cities*, Lecture Notes in Geoinformation and Cartography, DOI 10.1007/978-3-319-18368-8_11

approach into planning processes. This approach will potentially reveal new insights for the perception of geographical spaces in spatial planning.

1 Introduction

The question of how people perceive a city and how they feel about it has always been an important issue in urban planning. Typically, citizens/residents as well as visitors classify existing urban situations or new planning projects with the attributes ‘good’ or ‘bad’. “From a nonprofessional perspective, visual quality may be the most important influence on how people experience and respond to urban areas and planning initiatives” (Kaiser et al. 1995, p. 223). For instance, pedestrians or cyclists may emotionally respond differently to the city depending on a variety of context factors including personal mood, environmental conditions like the weather, traffic density or road conditions. Methods for an objective measurement of the quality of architecture and the resulting urban space last boomed in the late 1950s–1970s (Debord 1957; Lynch 1960; Cullen 1961; Franke and Bortz 1972; Krause 1974; Trieb 1974). However, in contrast to landscape design (Bishop and Hull 1991) or especially architectural design, there is still a scientific lack in the discussion of emotional aspects in urban planning. This is relevant because the amenity value or perceived safety are important aspects for making decisions in many urban design proposals. In recent years, the integration of technical and human sensors in combination with a direct feedback from citizens to stakeholders via real-time participatory communication channels (Burke et al. 2006), such as social media, enabled new insights into urban patterns, and dynamic and static contexts like traffic or architecture.

Research fields including computer science, geoinformatics, computational linguistics, sensor technology, citizen science, architecture and spatial planning are overlapping impacts in this interdisciplinary topic, which we call ‘Urban Emotions’. Concisely, the Urban Emotions approach focuses on a new and human-centric perspective onto the city, in which humans as ‘users of a city’ represent the main sensing element. Therefore, we combine and merge objective elements of sensor technology with subjective measurement methods to create a ‘human sensor network’. The long-term goal is to develop a new information layer for planners, in which a visualization of the measured spatial perception is possible. These visualizations allow conclusions about human behavior in an urban environment and enable a new citizen-centered perspective in planning processes.

2 Scientific Background

The research field of human perception of the city originated from the 1970s and deals with the perception of the natural and the built environment (Downs and Meyer 1978). Main components and the chosen cartographic representation are so called 'mental maps' or 'cognitive maps'. These maps reflect the subjective perception of a person in (urban) space segments (Downs and Stea 1974). 'Cognitive Maps and Spatial behavior' (Downs and Stea 1974) or 'Image of the City' (Lynch 1960) describe the concepts of cognitive representation of space. "We are not simply observers of this spectacle, but are ourselves a part of it, on the stage with the other participants. […] Nearly every sense is in operation, and the image is the composite of them all" (Lynch 1960, p. 2).

These days, up-to-date methods like sensing or crowdsourcing offer a valuable supplement in creating these maps and using them in spatial planning. Analog techniques like sketches or descriptions are combined with digital methods like the above mentioned. Emotions and space are linked closely. Each situation creates its own specific atmosphere and triggers an emotional reaction to the respective observer (Mody et al. 2009). The first to combine global positioning systems (GPS) data with biometric human sensor data and investigate the question of collecting cartographically referenced emotional data was Christian Nold in 2004 with his project 'Biomapping'. Therefore, it was possible to visualize user gathered psychophysiological data and share with the community. In his collection of essays entitled 'Emotional Cartography', he presents this concept and other works around the topic of collecting data for a better understanding of people's perception in urban areas (Nold 2009).

Besides the emotion aspects, the perception of humans' urban surroundings and how they were created or triggered is another important point/perspective. Feelings or perceptions are important key facts and not negligible aspects in urban planning, especially for supporting recommendations in design processes and in some infrastructure projects like trespassing of streets, cycle lanes etc. The Urban Emotions concept refers closely to new methods of a 'new understanding of cities', beginning with Baudelaire's 'flaneur', Patrick Geddes' 'outlook tower' from 1892, Walter Benjamin's 'Passagen-Werk' from the late 1920s/40s (Benjamin and Tiedemann 1983), concept of 'dérive' (Debord 1956) and 'guide psychogéographique de Paris' (Debord 1957) up to Lynch's 'mental maps'.

2.1 New Kinds of Participation—Urban Emotions as New Information Layers Within Planning Processes

"Cities have the capability of providing something for everybody, only because, and only when, they are created by everybody" (Jacobs 1961, p. 238). Spatial planning is cross-sectional and interdisciplinary and takes into account all spatial and social structures within the city. In an ideal case, all public and private interests are evaluated to minimize conflicts within planning processes. The quote from Jacobs highlights the need to develop an ideal type of participatory spatial planning approach for a 'good' and citizen-centric planning. This includes theoretical visions as well as human centered implementation strategies.

However, the question is: how can relevant and objective issues be measured and included in the process of planning and evaluation? And why is the use and knowledge about citizens' emotion so interesting for urban designers and planners? Moreover, how can planners recognize which problems citizens are really confronted with in a spatial context?

The use of emotions in urban design is not well-established and far grown. One of the few research efforts is presented by Hoch (2006), yet describing the use of emotional intelligence in communication in planning processes rather than psychophysiological measurements. In landscape design, the discussion of an acceptance of a project is much more integrated in the scientific discourse. Bishop and Hull (1991) discussed 'visual quality' of landscapes as an important emotional issue, even for planners and stakeholders. In the context of 'mental health', they state: "Emotional and mental well-being may be one benefit of a high-quality visual environment. (…) On the negative side, the cluttered visual environment may be distracting and cause stress or emotions which hinder the accomplishment of one's objectives." (Bishop and Hull 1991, p. 298). Good visual quality is, according to the authors, good for residential satisfaction and an overall confidence in land management. These aspects "should be the concern of landscape management and planning because they are socially relevant" (Bishop and Hull 1991, p. 298). These findings will be adopted in Urban Emotions and transferred to other application areas in the domain of urban planning.

The Urban Emotions concept also tries to give the answer by developing a new set of methods for urban planning and spatial planning. Taken into account are the concepts of a network society (Castells 1996), which will join up today's fundamental change of the understanding of planning (Streich 2014). So-called bottom–up processes of participation, a proactive involvement of citizens, are the core elements in this strategy. So, the repertoire of methods in spatial and urban planning has to be extended significantly by using this 'sensor technologies'; traditional deductive planning will be supplemented by inductive planning approaches, like the above mentioned crowdsourcing processes in bottom-up mode (Streich 2014).

2.2 *Techniques, Wearable Computing and 'Quantified-Self' Movement*

There are several techniques for detecting emotions. One is to combine several technical sensors into a new network. A simple configuration and a communication hub for detecting emotions are enabled by the new generation of smartphones. GPS, microphones and cameras are the basic requirements of participatory sensing (Burke et al. 2006). Smartphones are probably the most efficient standard sensing devices in daily life, because they are mobile, embedded and, thus in situ (Martino et al. 2010). Further, the smartphone's integrated sensors for physical activities will come into vogue more and more. "The mobile phone is the new gateway to people-centric urban sensing, a new sensor-networking paradigm that leverages humans as part of the sensing infrastructure" (Martino et al. 2010, p. 2).

Wearable devices are portable mini computers equipped with various sensors, which are always on and always ready and accessible (Mann 1998) and constitute a supplement for smartphones (Schumacher 2013). They should serve as an "intelligent daily assistant", in which the user's role is passive without any explicit data input (Kotrotsios and Luprano 2011, p. 279). State-of-the-art wearables allow us to measure physiological data like skin conductance, skin temperature or electrocardiography (ECG) and it is possible to derive indicators of negative arousal from these datasets. Hence, wearable technology has an increasingly important role in supporting everyday life and is essential equipment in the Quantified-Self and Life Logging movement.

Kevin Kelly and Gary Wolf initiated the Quantified-Self movement in 2007. "Self-knowledge through self-observation" is the slogan of the movement. People measure their bio data digitally and want to learn from these data, to live better, or at least to be able to understand certain physical and psychological processes better (Klausnitzer 2013, p. 29).

3 Methodology—The Urban Emotions Concept

The Urban Emotions concept proposes a new human-centered approach in order to extract contextual emotion information from human and technical sensor data. As illustrated in Fig. 1, the methodology comprises four main components: first, detecting emotions using wristband sensors to measure the human's bio-feedback in the urban context; second, 'ground-truthing' these measurements (assigning a

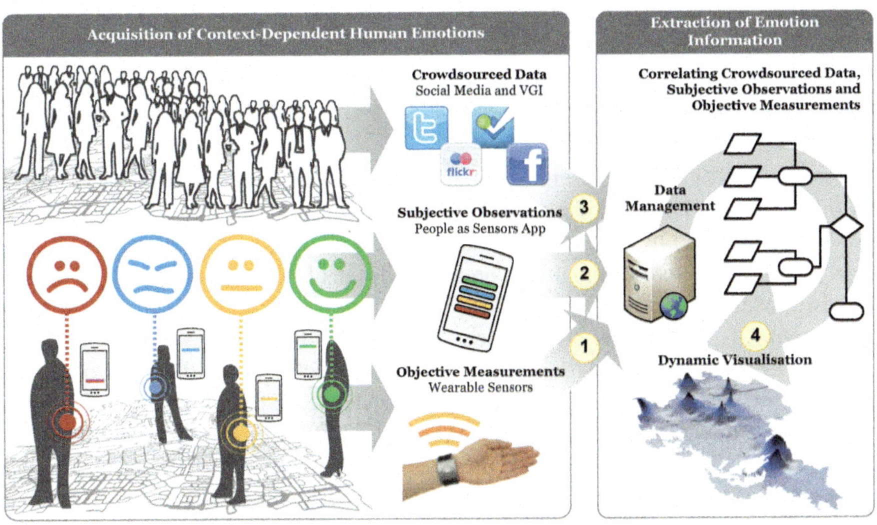

Fig. 1 Urban emotions concept (Zeile et al. 2014; Resch et al. 2015a)

formal emotion category to each measurement) using a 'people as sensors' location-aware mobile application; third, extracting emotion information from crowdsourced data and geo-social media like Twitter; and fourth, correlating the measured and extracted emotions in order to provide urban planners with additional insights into the complex human-city relationships. However, the Urban Emotions concept's major innovation is its trans-disciplinary nature as it consolidates the know-how and perspectives of at least four additional scientific disciplines, namely GIScience, computational linguistics, sociology and computer science. Hence, the contextual emotion information extracted can serve as the citizens' direct feedback for urban planning and decision support for ongoing planning and design processes.

The most important factor in this approach is that Urban Emotions is not conceived as a general tool of solving all planning tasks, but it can help to create another view and better understanding of 'the body of the city'. A big success would be, if this new knowledge can be integrated as a fact or indicator system in weighting in official planning processes. Urban Emotions can provide valuable information not in every case, but in special tasks like the above mentioned design processes or in the discussion of personal perceived safety.

3.1 Modes of Measuring Emotions

There are several modes to measure the emotions of people within physical spaces. These range from simple online questionnaires to a basic localization 'tagging' of the attributes 'good' or 'bad' using a smartphone app, over automated text-extraction from social media channels up to psychophysiological measurements. These different modes are considered in the Urban Emotions concepts and explained in detail below.

3.1.1 Tagging

An easy method is to 'tag' annotations with the help of a digital place mark within the urban space. This is for example possible by using the smartphone RADAR SENSING app, with which users can locate explicitly GPS-based positive or negative impressions within urban areas (Zeile et al. 2012). The RADAR infrastructure (Memmel and Groß 2011) can generate 'on the fly heat maps' or 'density maps' out of the attributed dataset. The RADAR SENSING app runs on AndroidOS devices and saves all the information such as user data and user contributions in an appropriate database in RADAR SENSING back-end. Simply a registration and the installation of the RADAR SENSING app are required. Then users can give both negative and positive ratings, with the help of predefined categories, all relevant to planning issues. If there is a missing category, it is also possible to define a free text for a new classification. In order to ensure easy and fast handling, the app locates the user's position directly with the help of several sensors

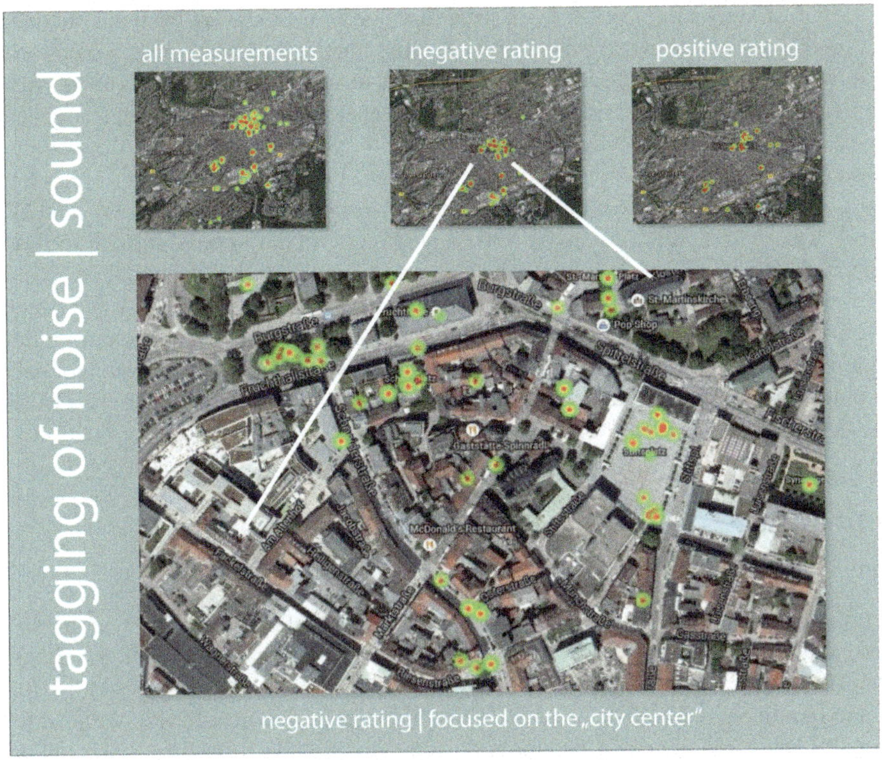

Fig. 2 Tagged classification of negative or positive rating concerning 'sound' into the city of Kaiserslautern (Zeile et al. 2012)

(GPS and WiFi) and displays them on a map. If the position is not accurate enough, it can be manually edited. The aforementioned interactive heat map of the marked points is interesting for spatial planning (Fig. 2). In addition, it is also possible to display only a certain period of time or the real time to allow the generation of the heat map (Zeile et al. 2013).

The above-mentioned technology is easy to use and useful for a quick examination and first overview for planning facts, but it is not to collect 'emotional data'. Therefore, the following modes were developed.

3.1.2 Extraction of Emotion Information from User-Generated Data

Previous methods for extracting information from crowdsourced data have relied on methods from a single discipline; for instance, geographers tried to detect hot spots in geographic space, whereas computational linguists analyzed crowdsourced data with respect to their semantic content. The Urban Emotions approach extracts information by means of a trans-disciplinary algorithm, in this case, integrating methods from computational linguistics, GIScience and urban planning.

Figure 3 shows the workflow for the extraction of information from unstructured datasets like Twitter tweets. Generally speaking, the algorithm assigns an emotional quality (joy, anger, fear, sadness, surprise and disgust) to each tweet. In practice, this means that we produce a set of labelled tweets from initially unlabeled tweets in four steps. *First*, the tweet text is pre-processed by applying basic text processing algorithms (part-of-speech [POS] tagging to detect emoticons, negations, etc.) and by comparing their content to the Affective Norms for English Words (ANEW) word list. *Second*, the similarity between all tweets is computed according to three dimensions: linguistic and semantic content, temporal distance, and geo-spatial distance. *Third*, a tweet graph is constructed, where "thicker" edges between tweets indicate higher similarity scores. *Fourth*, a graph-based label propagation algorithm is applied to the graph, resulting in a probability distribution over all labels for each previously unlabeled node. The methodological details of the approach are described in Resch et al. (2015a).

3.1.3 "People as Sensors" for Ground-Truthing Emotion Measurements

Current emotion sensors are able to detect spikes in different biometric parameters like additional heart rate, skin conductance or body temperature, but they are not capable of identifying the emotional quality, as mentioned above. To address this shortcoming, a 'people as sensors' smartphone app was developed, with which persons can manually enter the emotional category and the according context in the case of an emotional spike (Resch 2013). So, the app's inputs are twofold, namely objective sensor measurements, and subjective observations—both are send to the back-end (refer to Fig. 1). In the design of the app, the main challenge was to account for technological requirements for web development and at the same time integrate psychological theories for interaction design and emotion detection. More information about the design can be found in Resch et al. (2015a). The technical design considerations including the standardized communication with the back-end will be presented in Resch et al. (2015b).

3.1.4 Psychophysiological Monitoring

The so-called psychophysiological monitoring is the technology to measure the arousal of a person. In principle, the measurement occurs in nearly real time and provides data about the change in body reactions (body physiology) of a participant. There is a variety of assessment methods to identify emotions (Schumacher 2014). Using 'emotional stimuli' with IAPS (International Affective Picture System), in which a collection of pictures is shown representing the range of human experience and the users rate them with the attributes 'positive', 'negative' or 'neutral' (Bradley and Lang 2007). Another method is to measure the 'startle reflex', a peripheral physiological parameter. Thereby, negative stimuli can be measured over

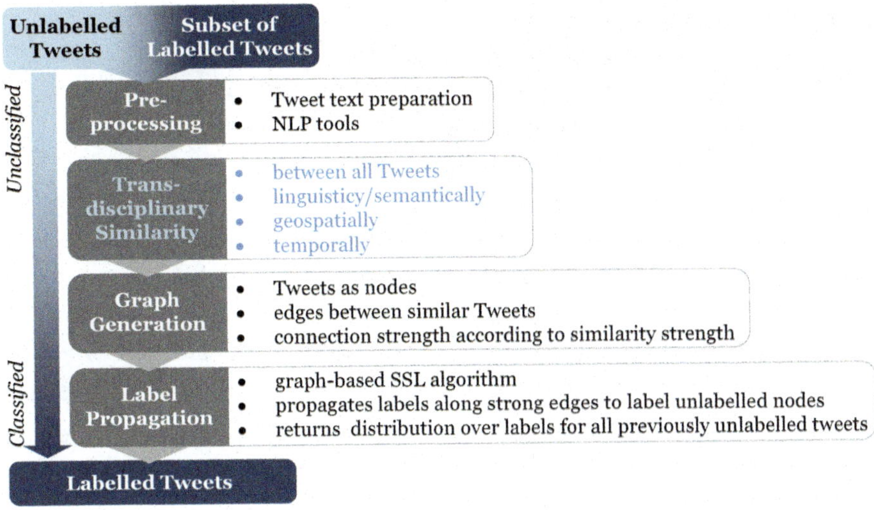

Fig. 3 Labelling tweets in a semi-supervised learning approach

a reflex of the neck or the eye (eyeblink). With the help of two electrodes, placed under the eye, minimal muscle tension can be recorded and is an indicator of negative stimuli (Geyer and Swerdlow 1998). For a mobile use in field research in urban environment, only the following two methods are suitable: A combination of body temperature and skin conductance (Stern et al. 2001; Boucsein 2012) or by using the additional heart rate (Fahrenberg and Myrtek 2001; Myrtek et al. 2005; Schächinger 2003).

Currently, only negative emotions are clearly identifiable based on these physiological parameters. In a popular scientific manner, this negative emotion can also be classified as 'stress'. "According to emotion researchers, when a negative experience occurs, the skin conductivity increases and the measured skin temperature decreases" (da Silva et al. 2014, p. 97). For instance, a test person has the experience of anger or fear—a negative emotion—skin conductance (the difference between sweat production and absorption of the skin) increases and skin temperature in the extremities decreases. A well-known example in extreme mental stress situations is known as 'cold sweat' (Bergner and Zeile 2012). These identified and georeferenced points can then be visualized as exemplarily shown in Fig. 4.

4 Results

The research in this topic is ongoing; the following first results show the potential of the new approach.

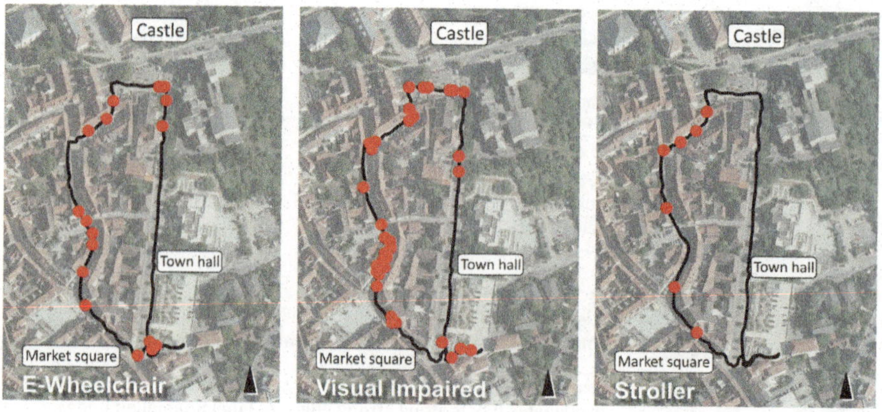

Fig. 4 Identified stress points of the test persons within the city of Kirchheimbolanden (Bergner and Zeile 2012)

4.1 Barrier-Free Planning

The topic of accessibility plays an increasingly important role in urban planning processes. Due to demographic changes and high expectations and demands on mobility, especially in old ages, the accessibility of infrastructures in new planning projects is now more and more important. A requirement for accessibility is that people can use objects, media and infrastructure fully independently, regardless of possibly having a disability or mobility limitation. Especially in the design of public space, various problems can thereby arise concerning accessibility.

At the core of Urban Emotions, georeferenced and time stamped psychophysiological measurements from the combination of human and technical sensors enrich traditional planning methods with a new layer of contextual emotion information. This is a new and objective method of rating the accessibility of infrastructures. Besides the known experiments (da Silva et al. 2014), the example shows the potential benefits of using this technology in planning processes. This study is set-up in the small village of Kirchheimbolanden, which is characterized by having a high age structure and a correspondingly high number of senior residences. The four participants included a woman with an electric wheelchair, a visual impaired woman, a mother with a stroller and a non-handicapped reference person (Bergner and Zeile 2012).

In Kirchheimbolanden, an urban renewal project to renovate old infrastructure like pavements, streets, stairs and crossing sections was planned. The announcement of the project included a part to implement 'barrier free planning'. After termination, the accessibility towards special industry standards like DIN 18024 (surface indicators in public space) and DIN 32984 (barrier-free construction) was checked again. This served as a reference to compare these results to the psychophysiological monitoring, where the participants delivered data to identify 'stress points' within the urban area.

As shown in Fig. 4, a number of individual stress points were identified, which were compared with the DIN survey on-site. With the help of the DIN survey, recorded movies of the walk and the stress points, it was possible to interpret if there really is a planning deficiency, a temporal barrier or only a personal negative impression. For example, it was noticeable in this study, that in the pedestrian zone, located in the southwest, different cobblestones and temporary barriers of the retailers had a considerable impact on the handicapped participants, especially to the blind person. The results were presented to the public, stakeholder participants and the representative of people with disabilities. The subsequent debate discussed a limitation of mobile billboards within the pedestrian zone and if it was possible to remove the identified deficiencies.

4.2 Emocycling

The topic of cycling in urban areas emerged as another application field in the Urban Emotions approach. The possibility to detect urban development problems for bicycle traffic can give hints for better planning and a resulting acceptance of using cycles, especially against the background of the need to conserve energy and the search for alternatives in modal choice. 'Emocycling' examines negative arousals of a cyclist while riding a bike in the city (Höffken et al. 2014). Again, participants were equipped with a variety of 'wearables' to measure psychophysiological data. A GoPro-camera and a GPS-tracker to geolocate the measurement made it possible to detect areas of negative arousal in the city. Because of the analysis, we identified hotspots of 'stress' in the city, where an increased change of vital data was detected. Even nonprofessionals can clearly identify potential danger spots with this technology. Consequently, this kind of survey helps in public discussions of traffic safety to communicate between all parties concerned.

In addition to cycling, this study also compared whether or not there are differences in the subjective feeling of safety by using a traditional bicycle or a 'pedelec' (pedal electric cycle), especially if the drivers ride on tracks with exhausting topography like ramps or streets with more than 6 % gradient. All the records were aggregated to identify stress hotspots, and the trigger points were extracted and condensed into a heat map, as shown in Fig. 5. Subsequently, particular areas of increased stress reactions can be viewed by video material and deliver hints for optimization in traffic planning. A lower stress density is noticeable when riding up the hill during the test with a pedelec. One reason for this is the reduced physical effort of the drivers. They are not deflected and thereby attentive in detecting obstacles. The subjective feeling of security is increased (Höffken et al. 2014). The simultaneously recorded movie material proves this assumption, the traffic conditions and obstacles didn't change in comparison to the test rides with normal bicycles.

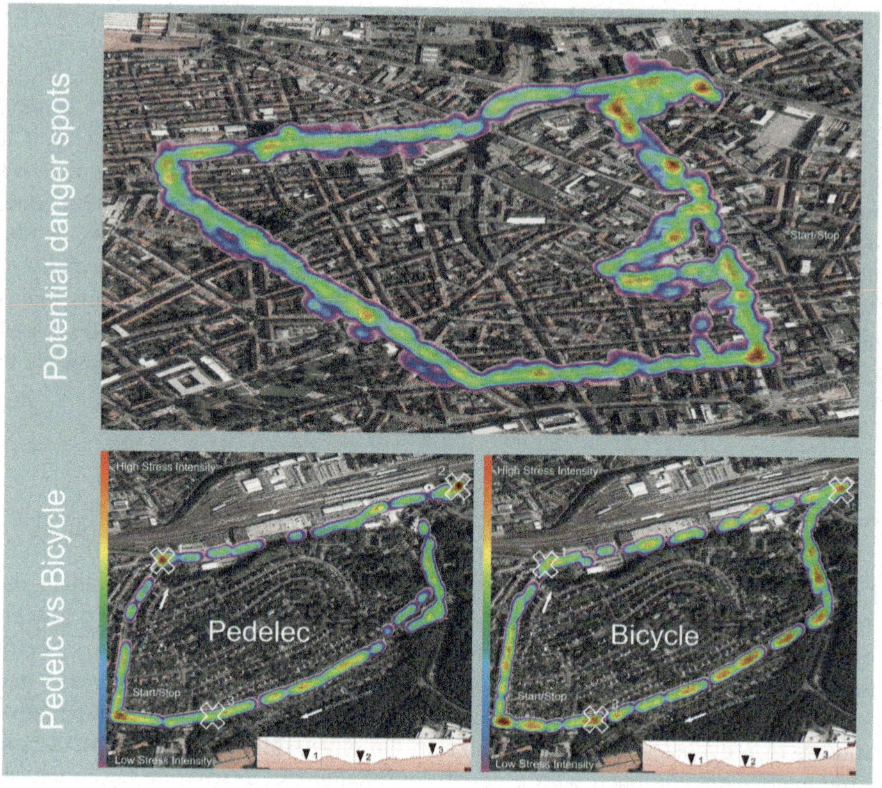

Fig. 5 Potential danger spots for cyclists, identified by psychophysiological monitoring (*top*) and different stress points by using a pedelec (*bottom left*) or a bike (*bottom right*) (Höffken et al. 2014)

5 Risks and Limits of Using Physiological Data

As elaborated above, the results of the various studies show that the use of technical and human sensor data brings an enormous potential for a variety of applications, specifically in urban areas. However, an issue like data privacy has to be borne in mind, because people as mobile users produce the most sensitive data. If there are additional parameters involved, such as private messages or the physical sense, the demands for data protection are even higher. Hence, the smartphone itself is in focus of various privacy concerns […] "Could mobile phones become the most wide spread surveillance tools embedded in history" (Shilton 2009, p. 48).

Apple's privacy policy in terms of location-based services reveals, for instance, that the company and its partners use and share precise location data in real time. Also, installed apps can access a variety of relevant data from the GPS sensor, calendar, browser history, etc. (Kersten and Klett 2012). There are no reliable facts, but it is assumed that about one third of all free apps access the location of its mobile device. Furthermore, about 50 % of free Android apps have a so-called Third Party Code and "support the sending and showing of mobile advertising and the analysis and tracking of user behavior" (Kersten and Klett 2012, p. 88). This can bring users into permanent conflict between participative motivation and concerns for their own privacy. The postulated freedom in the internet is equally compromised by commercial companies, which use the data for their own purposes (Caesar 2012). This dilemma can also be seen in the Internet, because "lack of clarity, commercialization and information overload must be monitored continuously. Nevertheless, it can be stated that the advantages of using social media in terms of strengthened legitimacy, transparency and democracy prevail through citizens' participation" (Caesar 2012, p. 84).

Some approaches which consider four geographic locations as sufficient to reveal a person's identity through the analysis of movement patterns (Montjoye et al. 2013), although the requirements for an effective anonymization have to be discussed continuously in the light of the technological development because the possibility of geographical traceability will increase. Zang and Bolot explored in their study for example, that the location of call via call details records (CDR) together with publicly available data will make it possible to re-identify users despite anonymized data records (Zang and Bolot 2011). The concerns about data protection and privacy have to be dealt with wisely in the context of the ongoing development. Hence, the problems are less about technical hurdles, but are rather to raise awareness of the multi-dimensional connections. A biased attitude against the use of sensor data for planning purposes is not beneficial in this case. As Shilton (2009, p. 53) explains: "participant engagement in privacy decision making must therefore be fortified by supporting social structures". Thus, it should be noted, that as well as technical protection mechanisms, people's awareness for transparency, openness and especially control over the data are all important. For instance, every new or edited 'End User License Agreements' (EULA) of Facebook opens a public debate. Following campaigns, public debates and blogs can contribute to social acceptance as well as clear and easy to understand EULAs (Shilton 2009). Or, if the usage of personal data is not well documented, people quit the service or won't take part of a survey.

6 Conclusion

Urban Emotions is an interdisciplinary approach at the interface between the research domains of geoinformatics/GIScience and spatial planning, between language processing, sensor technology, citizen science and architecture. It can be seen as a novel anthropocentric approach for understanding the complex spatio-temporal dynamics and interactions in the human-space framework. Thus, the approach focuses on the human being with their perceptions and sensations in the urban context.

The objective and subjective measurements of human feelings and perceptions in terms of urban circumstances like architecture or traffic represent the basis for the extraction of contextual emotion information in a fine-grained spatial and temporal resolution. The extracted information can be used and implemented in planning processes and provide innovative opportunities for citizen-centered urban and spatial planning. Moreover, such contextual information as emotion data enables new forms of validation to make planning more sustainable; for example—do the newly designed traffic management systems reveal the desired effect, or how is the newly created city park perceived at different times of the day by citizens?

Technological advances in sensor technology, smart-phones and networks as well as the evolution of web 2.0 and social media enable new opportunities for networking and the collaboration of different research domains—these possibilities are not limited to geoinformatics and spatial planning. It is rather a question of joining the paradigms of engineering, applied science, natural sciences, social sciences and humanities to identify synergies and potential benefits.

Measuring emotions in a complex microstructure what we know as a city is a core element in the Urban Emotions approach. The methods of detecting emotions are manifold, ranging from a simple localization of opinions ('good' or 'bad') up to psychophysiological measurements. By using such sensitive data, a prognosis of the acceptance of the above-mentioned approach is difficult to estimate. Accessibility or emocycling are only a two examples of a range of relevant issues in spatial planning, in which emotions and human beings are increasingly being focused on in the shift towards human-centric planning.

Privacy issues have to be discussed in an open and transparent way to inform and clarify the methods. However, a tendency in sharing personal information through online platforms, even sensitive physiological data or personal emotions can be observed in the above mentioned 'Quantified-self' movement. How these datasets can influence or be used in urban planning processes remains to be determined. It is positive to note that 'life loggers', with their personal motivation and initiative, can play an important factor in the collection of emotion data. Spatial planning should recognize this potential and use them to enhance participation processes.

Acknowledgments The authors would like to express their gratitude to the German Research Foundation (DFG—Deutsche Forschungsgemeinschaft) for supporting the project Urban Emotions, reference number ZE 1018/1-1 and RE 3612/1-1. This research has been supported by the Klaus Tschira Stiftung GmbH. We would also like to thank Linda Dörrzapf, Anja Summa, Martin Sudmanns, Daniel Broschart, Johann Wilhelm, Claire Dodd and Dennis J. Groß for their support.

References

Benjamin, W., & Tiedemann, R. (1983). Das Passagen-Werk (1st edn, Edition Suhrkamp). Frankfurt: Suhrkamp.

Bergner, B. S., & Zeile, P. (2012). Ist Barrierefreiheit messbar? *Planerin, 2012*(2), 20–24.

Bishop, I. D., & Hull, B. R. (1991). Integrating technologies for visual resource management. *Journal of Environmental Management, 32*(4), 295–312. doi:10.1016/S0301-4797(05)80068-4.

Boucsein, W. (2012). *Electrodermal activity* (2nd ed.). New York: Springer Science + Business Media LLC.

Bradley, M. M., & Lang, P. J. (2007). The International Affective Picture System (IAPS) in the study of emotion and attention. In J. A. Coan, & J. J. B. Allen (Eds.), *Handbook of emotion elicitation and assessment* (Series in affective science). Oxford, New York: Oxford University Press.

Burke, J., Estrin, D., Hansen, M., Parker, A., Ramanathan, N., Reddy, S., et al. (2006). Participatory sensing. *World Sensor Web 2006 Proceedings*, 1–5.

Caesar, I. (2012). *Social Web-politische und gesellschaftliche Partizipation im Netz: Beobachtungen und Prognosen*. Berlin: Simon Verlag für Bibliothekswissen.

Castells, M. (1996). *The rise of the network society* (Information age, v. 1). Malden, Mass: Blackwell Publishers.

Cullen, G. (1961). *Townscape*. New York: Reinhold Publishing Corporation.

da Silva, R., Zeile, A. N., Aguiar, P., de Oliveira, F., Papastefanou, G., & Bergner, B. S. (2014). Smart sensoring and barrier free planning—project outcomes and recent developments. In N. N. Pinto, J. A. Tenedório, A. P. Antunes, & J. R. Cladera (Eds.), *Technologies for urban and spatial planning: Virtual cities and territories* (pp. 93–112). Hershey, PA: IGI Global.

Debord, G. (1956). Theory of the Dérive. Les Lèvres Nues (9). Paris. Reprinted in Situationistische Internationale (1976), Band 1. Körle: SI-Revue.

Debord, G. (1957). *Guide psychogéographique de Paris: Discours sur les passions de l'amour*. Copenhagen: Permild & Rosengreen.

Downs, R. M., & Meyer, J. T. (1978). Geography and the mind: An exploration of perceptual geography. *American Behavioral Scientist*. doi: 10.1177/000276427802200104.

Downs, R. M., & Stea, D. (1974). *Image and environment: Cognitive mapping and spatial behavior*. Piscataway, NJ: Transaction Publishers.

Fahrenberg, J., & Myrtek, M. (2001). *Progress in ambulatory assessment: Computer-assisted psychological and psychophysiological methods in monitoring and field studies*. Seattle: Hogrefe & Huber Publishers.

Franke, J., & Bortz, J. (1972). Beiträge zur Anwendung der Psychologie auf den Städtebau: 1. Vorüberlegungen und erste Erkundungsuntersuchung zur Beziehung zwischen Siedlungsgestaltung und Erleben der Wohnumgebung. *Zeitschrift für experimentelle und angewandte Psychologie, 19*(1), 76–108.

Geyer, M. A., & Swerdlow, N. R. (1998) Measurement of startle response, prepulse inhibition, and habituation. In C. Gerfen, A. Holmes, D. Sibley, P. Skolnick, S. Wray (Eds.), *Current protocols in neuroscience* (pp. 8.7.1–8.7.15). New York: Wiley.

Hoch, C. (2006). Emotions and planning. *Planning Theory & Practice, 7*(4), 367–382. doi:10.1080/14649350600984436.

Höffken, S., Wilhelm, J., Groß, D., Bergner, B. S., & Zeile, P. (2014). EmoCycling—Analysen von Radwegen mittels Humansensorik und Wearable Computing. In M. Schrenk, V. V. Popovich, P. Zeile, & P. Elisei (Eds.), *Real CORP 2014* (pp. 851–860). Wien.

Jacobs, J. (1961). *The death and life of great American cities*. New York: Random House Vintage Books.

Kaiser, E. J., Chapin, F. S., & Godschalk, D. R. (1995). *Urban land use planning* (4th ed.). Chicago: University of Illinois Press.

Kersten, H., & Klett, G. (2012). *Mobile device management* (1st edn). Heidelberg u.a.: mitp.

Klausnitzer, R. (2013). *Das Ende des Zufalls: Wie Big Data uns und unser Leben vorhersagbar macht* (1st ed.). Salzburg: Ecowin-Verl.

Kotrotsios, G., & Luprano, J. (2011). The commercialization of smart fabrics: Intelligent textiles. In A. Bonfiglio & D. de Rossi (Eds.), *Wearable monitoring systems* (pp. 277–294). US: Springer.

Krause, K.-J. (1974). *Stadtgestalt und Stadterneuerung*. Bonn: Bundesvereinigung Deutscher Heimstätten e.V.

Lynch, K. (1960). *The Image of the city*: Cambridge Mass: MIT Press.

Mann, S. (1998). Wearable computing as means for personal empowerment: Keynote address. In *Keynote address for the first international conference on wearable computing, ICWC-98*. Fairfax, VA, 12 May.

Martino, M., Britter, R., Outram, C., Zacharias, C., Bidermann, A., Ratti, C. (2010). Senseable city —digital urban modelling and simulation. Working paper. MIT senseable city lab. http://senseable.mit.edu/papers/pdf/2010_Martino_et_al_DUMS.pdf. Accessed 25 Feb 2015.

Memmel, M., & Groß, F. (2011). RADAR—potentials for supporting urban development with a social geocontent hub. In M. Schrenk, V. V. Popovich, & P. Zeile (Eds.), *REAL CORP 2011* (pp. 777–784). Wien: Essen.

Mody, R. N., Willis, K. S., & Kerstein, R. (2009). WiMo: location-based emotion tagging. In N. Milic-Frayling, J. Häkkilä, J. Crowcroft, C. Mascolo, & E. O'Neill (Eds.): *Proceedings of the 8th International Conference on Mobile and Ubiquitous Multimedia* (pp. 1–4). New York: ACM.

Montjoye, Y.-A. de, Hidalgo, C. A., Verleysen, M., Blondel, V. D. (2013). Unique in the Crowd: The privacy bounds of human mobility. Scientific reports, doi: 10.1038/srep01376

Myrtek, M., Aschenbrenner, E., & Brügner, G. (2005). Emotions in everyday life: an ambulatory monitoring study with female students. *Biological Psychology*,. doi:10.1016/j.biopsycho.2004.06.001.

Nold, C. (2009). Emotional cartography: Technologies of the self. http://emotionalcartography.net. Accessed 15 Jan 2015.

Resch, B. (2013). People as sensors and collective sensing-contextual observations complementing geo-sensor network measurements. In J. M. Krisp (Ed.), *Lecture notes in geoinformation and cartography* (pp. 391–406). Berlin, Heidelberg: Springer.

Resch, B., Summa, A., Sagl, G., Zeile, P., & Exner, J. P. (2015a). Urban emotions—geo-semantic emotion extraction from technical sensors, human sensors and crowdsourced data. In G. Gartner & H. Huang (Eds.), *Progress in location-based services 2014* (pp. 199–212, Lecture Notes in Geoinformation and Cartography). Berlin: Springer International Publishing.

Resch, B., Sudmanns, M., Sagl, G., Summa, A., Zeile, P., & Exner, J. P. (2015b). Crowdsourcing physiological conditions and subjective emotions by coupling technical and human mobile sensors. In T. Jekel, A. Car, J. Strobl & G. Griesebner (Eds.), GI_Forum 2015—geospatial minds for society (in press).

Schächinger, H. (2003). Herz-Kreislauf-Erkrankungen. In U. Ehlert (Ed.), *Verhaltensmedizin* (pp. 225–263, Springer-Lehrbuch): Berlin, Heidelberg: Springer.

Schumacher, F. (2013). Angewachsen—Wie Wearables unseren Alltag verändern werden. *c't*(25), 86ff.

Schumacher, S. (2014). *Psychophysiological responses to emotional stimuli and their alterations in stress—related mental disorders*: Universität Konstanz.

Shilton, K. (2009). Four billion little brothers? *Communications of the ACM, 52*(11), 48–53. doi:10.1145/1592761.1592778.

Stern, R. M., Ray, W. J., & Quigley, K. S. (2001). *Psychophysiological recording* (2nd edn). Oxford [England], New York: Oxford University Press.

Streich, B. (2014). *Subversive Stadtplanung*. Wiesbaden: Springer VS.

Trieb, M. (1974). *Stadtgestaltung: Theorie und Praxis (Bauwelt-Fundamente, 43: Städtebau, Architektur)*. Düsseldorf: Bertelsmann-Fachverlag.

Zang, H., & Bolot, J. (2011). Anonymization of location data does not work. In P. Ramanathan, T. Nandagopal, & B. Levine (Eds.), *Las Vegas, Nevada, USA, 19.9.-23.9.2011* (pp. 145–156), New York, NY: ACM.

Zeile, P., Exner, J. P., Bergner, B. S., & Streich, B. (2013). Humansensorik und Kartierung von Emotionen in der räumlichen Planung. In E. Buhmann, S. Ervin, & M. Pietsch (Eds.), *DLA Conference 2013* (pp. 129–141). Berlin: Wichmann Verlag.

Zeile, P., Memmel, M., & Exner, J. P. (2012). A new urban sensing and monitoring approach: Tagging the city with the RADAR SENSING app. In M. Schrenk, V. V. Popovich, P. Zeile & P. Elisei (Eds.), *REAL CORP 2012* (pp. 17–25). Wien.

Zeile, P., Resch, B., Exner, J. P., Sagl, G., & Summa, A. (2014). Urban Emotions—Kontextuelle Emotionsinformationen für die räumliche Planung auf Basis von Echtzeit- Humansensorik und Crowdsourcing-Ansätzen. In J. Strobl, T. Blaschke, G. Griesebner, & B. Zagel (Eds.), *Angewandte Geoinformatik: Beiträge zum AGIT-Symposium Salzburg* (pp. 664–669). Berlin: Wichmann.

Chapter 12
Leveraging Cellphones for Wayfinding and Journey Planning in Semi-formal Bus Systems: Lessons from Digital Matatus in Nairobi

Jacqueline Klopp, Sarah Williams, Peter Waiganjo, Daniel Orwa and Adam White

Abstract For many cities in the developing world, public transit consists mainly of semi-formal mini-buses (paratransit). However, little to no digital information is typically available on routes, bus stops, passenger boarding, service frequency or scheduled trip times. Cities that rely on these bus systems can benefit from the generation of digital data on these systems for planning and passenger information purposes. Perhaps more importantly, this data can provide the ability to generate citizen-based information tools, such as transit routing applications for mobile devices widely discussed in the smart cities dialog (Townsend in Re-Programming mobility: the digital transformation of transportation in the United States, 2014). Through our work in Nairobi, this paper shows that cell-phone technology that is ubiquitous in most countries can be used to generate a dataset in an open standard, GTFS. Citizens can then leverage open source tools made for that standard, enhancing access to information about the transit system. We argue that one of the

J. Klopp
Center for Sustainable Development, Columbia University, 475 Riverside Dr. Suite 520, New York, NY 10115, USA
e-mail: jk2002@columbia.edu

S. Williams (✉)
Department of Urban Studies and Planning, Massachusetts Institute of Technology, 77 Massachustts Ave, Cambridge, MA 02139, USA
e-mail: sew@mit.edu

P. Waiganjo · D. Orwa
School of Computing and Informatics, University of Nairobi, 30197 Nairobi GPO 00100, Kenya
e-mail: waiganjo@uonbi.ac.ke

D. Orwa
e-mail: dorwa@uonbi.ac.ke

A. White
Groupshot, 157 Pearl Street, Cambridge, MA 02139, USA
e-mail: adam@groupshot.org

S. Geertman et al. (eds.), *Planning Support Systems and Smart Cities*, Lecture Notes in Geoinformation and Cartography,
DOI 10.1007/978-3-319-18368-8_12

most important components of our work in Nairobi was the engagement process that created trust in the data and knowledge of its existence for the development of civic technologies. The lessons learned in Nairobi can be translated to other areas with the potential to use mobile applications to develop data on essential urban infrastructure and to extend the use of that data by sharing it with a larger community.

1 Introduction

In this chapter, we describe a research effort in Nairobi, Kenya to collect basic data on the semi-formal bus system that is the basis of the public transit system in this city of over 3.5 million people. These buses of all sizes are called "matatus", and they are ubiquitous in Nairobi, forming the bulk of the mass transit system. We show that data collection using GPS-enabled cell phone applications can generate valuable data that allows for the development of wayfinding tools, both digital and analog, for transit users in the matatu system. These tools include trip planning applications on mobile devices and a stylized paper map of bus routes produced from the data. Our research illustrates how by actively engaging the Nairobi technology and transit community during the data collection process we were able to extend the benefits of the data to multiple actors in the city, thereby extending the benefit of the data collection process itself.

Increasingly, formal transit authorities are releasing data to facilitate the development of tools to help transit users make more informed choices, which, in turn, can improve transit systems as a whole (Roth 2010; McHugh 2013). Access to data is part of a broader agenda of leveraging technology and analytics to make cities including their transit systems "smart" (Townsend 2013). However, in many rapidly developing cities where people rely on semi-formal bus systems operated by diverse actors, access to information about these systems is not readily available. Often regulation and planning is poor which means operators are not required to collect information similar to more formal transit agencies. Even if these operators were required to collect data, the task of gathering this data from many small operators would be challenging. Each operator would have to collect data in the same standard and would have to ensure the accuracy and quality control (Williams et al. forthcoming).

Lack of data has made it hard to include semi-formal transit systems in transportation models. However, these systems often make up the majority of the vehicles on the roadway and, therefore, are essential for accurately representing traffic patterns in these cities. The limited access to information also makes it hard for users to know how to navigate the systems, and they often must rely on word of mouth for information about stops, routes, and schedules.

Given that data is essential for planning and citizen knowledge of semi-formal transit systems, it is important to develop technologies that allow for ease in data collection. The rapid growth of mobile phone use along with the increased accuracy of GPS transceivers in mobile devices allow them to be more easily used for data collection (Hein et al. 2008). A number of recent experiments have demonstrated that using cell phone transaction logs, which include location information, can assist in mapping traffic flows (Caceres et al. 2012; Herrara et al. 2010; Wang 2010; Ratti et al. 2006) and empower transit users in many ways "exemplifying the smart city vision" (Zegras et al. 2014). However, telecommunications companies, who collect these records are often unwilling to release cell phone data, even when anonymized, citing privacy concerns. One recent exception is the analysis of data from 2.5 billion call records of 5 million cell phone users in the Ivory Coast which helped analyze and suggest improvements to traffic movement in and around Abidjan, Ivory Coast (Talbot 2013; Wakefield 2013).

While access to cell phone use records can be difficult, the information can also be collected through direct user interaction with cell phones. In our research in Nairobi, we developed cell phone based mobile applications that made it easier to collect location data on routes, trips, fares, stops and schedules using a small group of dedicated people. After we had begun our work in Nairobi, we found several other projects that have attempted the same process, in Dhaka, Manila, and Mexico City (Eros et al. 2014; Klopp et al. 2014; Zegras et al. 2014). However, with the exception of Manila these projects did not open the semi-formal transit data for anyone to use.

In our project in Nairobi, we developed and released the data on the semi-formal transit system in an open transit standard, GTFS. Openly sharing data in this way allows it to be interpreted by multiple groups and extends the knowledge production that can be created from it (Craig 2005; Williams et al. 2014). The data collection process included continuous communications with the local technology and transit community in Nairobi. Through this process, the technology community learned about GTFS and that it could be used with open source tools for trip routing, such as Open Trip Planner among others. GTFS is the standard used for most routing applications. It was originally developed by Google and Portland's Tri-met transit agency to allow for transit routing in Google maps (McHugh 2013). Beyond routing applications, there are also a number of open source analysis packages that use GTFS as the base (Wong 2013). We distributed the data on the GTFS exchange, a website repository for open transit data, as well as our website www.digitalmatatus.com, allowing anyone to access it and develop projects using it.

By informing Nairobi's transit community about the development of the GTFS data throughout the data collection process, we were able to build trust in the data set created. Public Participation GIS (PPGIS) scholarship has shown this trust is essential for the adoption and use of data (Elwood 2006). We held a series of workshops where we invited actors from the local technology community, governments, NGOs, academia, as well as operators and drivers of the semi-formal buses. At these workshops, we explained the data collection process. Participants also learned how they could access and use the information. By developing this inclusive process, we broke down the knowledge barriers often established by the control and dissemination of information (Harvey and Christman 1998; Sieber 2006; Williams et al. 2014).

Mobile technology can serve as a powerful tool for data collection and feedback for next generation PPGIS projects, given that processes for collecting data can create new forms of knowledge production (Elwood et al. 2012). Smartphones now come with several sensors including GPS receivers; this means that smartphone users can contribute GPS tagged data with relative ease, which, can in turn, be used to develop data beneficial for everyone. Data collection on mobile devices can be exploited using a controlled collection process to assure quality data for a multiplicity of actors and users, especially if it is made open (Ching 2013; Klopp et al. 2014; Williams et al. 2014; Zegras et al. 2014). The cellphone can thus be thought of as the "quintessential smart city instrument" (Zegras 2014) although, as always, how this instrument will be used, just like PPGIS as a whole, is heavily linked to context and politics (Goodspeed 2015; Ghose and Elwood 2006; Kyem 2004).

In the last decade, citizen participation in the development of crowdsourced data collection tools has shown the power of data collection that can often be performed outside formal governmental channels. The push for open data, such as GTFS for transit, as well as open software means that the private sector can more easily be involved in the development of smart technologies that can enhance citizen interaction with the city (Townsend 2013). This is particularly true for the development of mobile phone technologies which can democratize access to essential information about city resources.

2 Methodology

Our team, consisting of a consortium of universities (University of Nairobi, MIT and Columbia University) and one small technology firm (*Groupshot*), mapped the city's bus (matatu) routes, primarily using easily available smartphone apps. Data collection took place by having students directly riding a matatu route, either in the car or matatu, using a data collection tool to generate latitude and longitude points along the route. A record of all of the stops as well as specific metadata about each stop, such as the stop name, stop signage and stop infrastructure (shelter and bench) was also recorded. While some semi-formal bus systems do not utilize fixed stops, in Nairobi the matatu system features regular stops and large terminals. We identified a stop based on local knowledge of data collectors, information from frequent users of these routes, visual notation (signs, shelters etc.), and if necessary, confirmation from discussion with an operator or group of commuters on a route.

Prior to beginning of data collection, we surveyed existing bus data and found that no complete up to date publicly accessible data existed. The city had MS Word document files for some routes dated 1997 and some app developers had tried to collect data using GPS devices but this was for individual routes and did not represent a comprehensive data set. Other groups like the weekly magazine of events in Nairobi, Kenya *Buzz*, created a stylized matatu map from scratch but it had many geographical inaccuracies and was not based on latitude and longitude points, making it hard for the information to be used with digital technologies. However, these attempts showed demand for this data from different constituencies.

In the fall of 2012, we tested various software developed for data collection on cell phones to determine one suitable for collecting route and stop data of Nairobi's matatu system (see Table 1). We compared the latitude and longitude data collected on these systems to data collected by hand-held GPS units to determine accuracy. Many of the devices captured data comparable to GPS units. However, some lost connection with the satellites which created incomplete data. On the basis of this review, the team decided to use the *MyTracks* App developed by Google for Android phones. *MyTracks* was chosen because it allows users to enter the name of stops digitally. It also snaps latitude and longitude points to the Google base dataset that is used as the map interface for the program. The ability to snap the GPS data to roads allowed for a cleaner data set as the GPS receivers often have errors of 10 m. Snapping to the roads allowed us to ensure our data on stops and routes was in the actual roadway making for a much cleaner dataset. The *MyTracks* application also made it easy to add our own metadata to the data, including information about whether stops were designated or undesignated.

TransitWand, an open source mobile data collection software for Android phones developed by the firm Conveyal with support from the World Bank, was released in the last weeks of our data collection and had features similar to *MyTracks* in that it allows for more metadata collection and the ability to snap GPS points to the roadway. The team investigated the use of *TransitWand* for the project. However, the tool was not suitable for our collection process as it needed modification for use in Nairobi. These modifications included: developing a script that would allow the program to generate unique ids using the structure we had created specifically for our Nairobi GTFS data, extending the code to create data columns and input forms to accommodate this way we extended the GTFS data format for Nairobi, and modifying the underlying base interaction as in our preliminary test the latitude and longitude points were not snapping to the roads properly. These modifications were not extensive. However given the fact that we were at the end of our data collection process and the export to GTFS function was not working properly in *TransitWand*, we did not shift to using this data collection device. Future research should test the modification and adaptation of this tool, because its interface and expandable metadata collection in the field holds much potential.

Formal data collection began in February of 2013 using a modified version of *MyTracks* App for Android smart phones developed specifically for the project to allow for the entry of stop names, as well as identifying if a stop was designated or undesignated. (see Fig. 1) GPS units were used as backup so that we could compare the accuracy of the two forms of data collection. Through testing we found that GPS units and data collected through *MyTracks* had similar accuracy. Some of the challenges with using this mobile based application was the limited battery life, slow speeds of affordable Android phones, the risk of losing more high value Android phones to theft in matatus where security could be a challenge (that, did happen, unfortunately), and the small screen size that made digital data entry more time consuming—particularly in situations where bus stops were very close to one

Table 1 Comparison of various mobile collection software tested for project

Tool tested	Pro	Con
MyTracks	• Snapped latitude and longitude data to the roadways allowing for cleaner data. • Easy to modify for additional data elements. • Easy to visualize the data captured immediately.	• Needed to develop tool to post process data collected in the field to GTFS data standard.
TransitWand	• Snapped latitude and longitude data to the roadways allowing for cleaner data.	• Needed programming modifications for use in the field which was already developed for MyTracks. However had this tool been found earlier in the process these modifications might have been very similar to what was needed to be performed by MyTracks.
	• Developed specifically for the collection of GTFS data. • Allowed for modification of questions asked in the field. • Good user interface.	• GTFS export was not working properly making post processing to GTFS necessary and cumbersome.
FlockTracker	• Allowed for latitude and longitude data collection on phones.	• Needed many modifications to be used in Nairobi. A new interface would need to be developed.
	• Could be modified to allow for survey questions.	• As this tool was in the early phase of development when we used it, programming communication errors causes it to lose connection with GPS satellites. We understand some of these bugs have been fixed in more recent versions of the software. • Did not snap data to the roadways - which meant that much post-processing would be needed.
Fulcrum	• Allowed for latitude and longitude data collection on phones. • Could be modified to allow for survey questions.	• Designed for development on iPhones which are not common in Kenya as they are too expensive. Most smartphone users in Kenya use Android phones.
GPS surveyor	• Allowed for latitude and longitude data collection on phones.	• Works much like a GPS unit and does not allow for additional data (such as stop names to be collected in the field) no advantage over GPS units.
Open data kit	• Allowed for latitude and longitude data collection on phones.	• Needed significant programming development to allow for use in the field.
App inventor		• Needed significant programming development to allow for use in the

(continued)

Table 1 (continued)

Tool tested	Pro	Con
	• Allowed for latitude and longitude data collection on phones.	field. Was the basis for the FlockTraker program previously mentioned.

It should be noted that for all the products we tested, the interaction between the phone and GPS satellites needed for data collection drained the battery life of the phones, making it difficult to complete whole routes without extra battery packs for the phones themselves

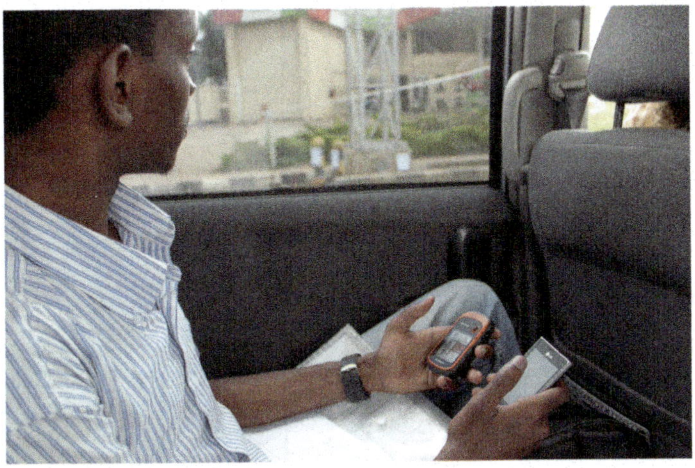

Fig. 1 University of Nairobi Student collecting GTFS data using *MyTracks* software and GPS unit. Image credit Adam White

another. A clear opportunity exists to continue to explore, compare and improve this transit data collection technology.

Once the data was collected, it needed to be cleaned and formatted into the standard we had chosen-the General Transit Feed Specification (GTFS) used by Google (see Fig. 2). The GTFS standard was chosen as it is the most common form of openly shared transit data. There are several open source applications, from routing to accessibility analysis, that use this format. Levering the existing technology developed for this format would extend its use in Nairobi. Translating the data into GTFS included the development of a unique identification system for routes and stops. There were also a number of challenges including the fact that GTFS required scheduling information, which matatus loosely follow, so a hypothetical schedule had to be generated in order for the data to be used with routing software. We also needed to develop new fields to include information we collected about designated and undesignated routes (Williams et al. forthcoming).

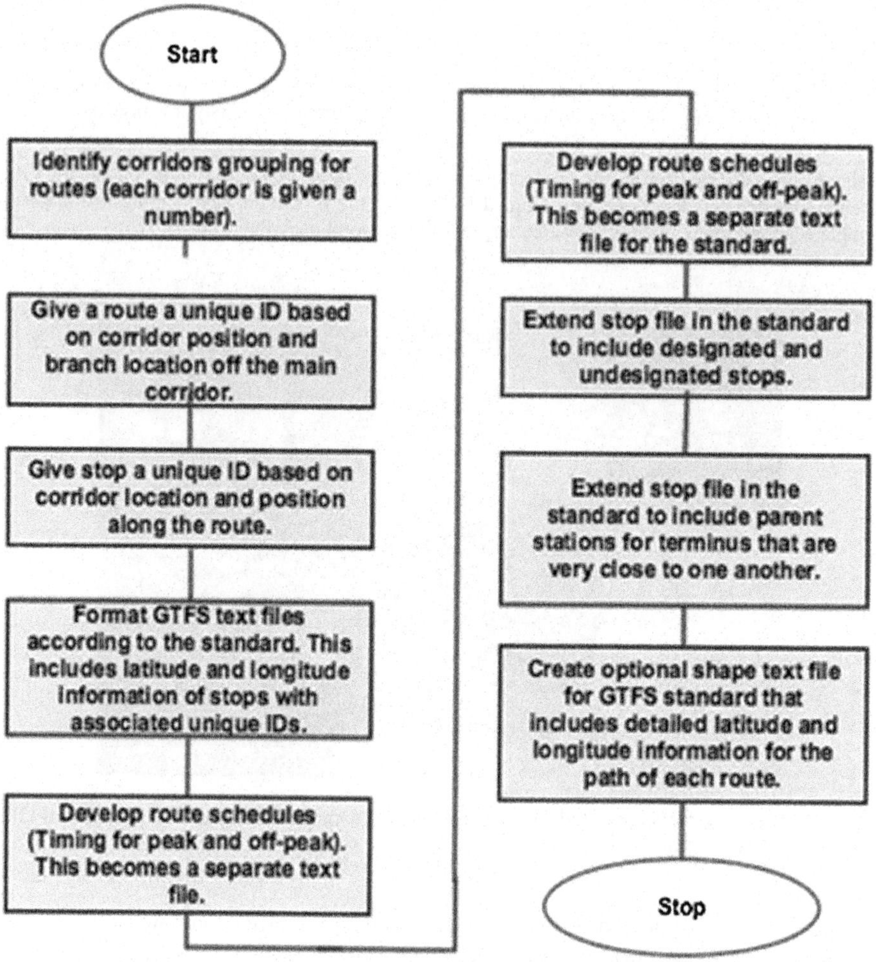

Fig. 2 Flowchart for the conversation of data collected into GTFS format

As we collected this data, we held a series of conversations with potential users in the technology and transit community to help them become aware of the data, its structure and format, and how they might be able to use it. We also used these discussions to get feedback on our process. In all, we collected 136 routes and 6 rail lines and made this data open on our website (www.digitalmatatus.com), Facebook page, and the GTFS Exchange. Making this data open was part of our strategy to encourage technology to be developed on top of the data and hence to generate innovation in wayfinding, journey planning for users, and more analytical applications for the transportation community.

In order to encourage the use and adoption of the data to users inside and outside the transit community, it was necessary to show community members what was

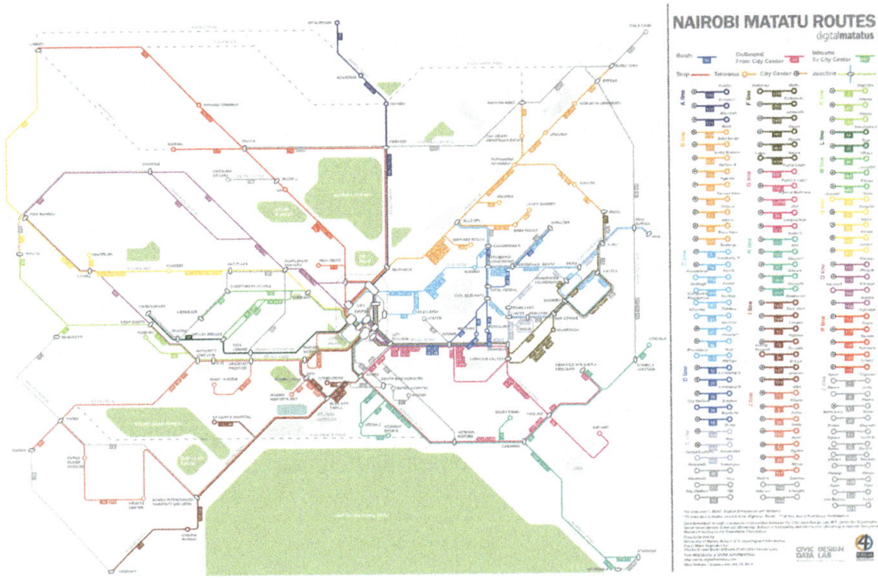

Fig. 3 Stylized Nairobi Matatu route map created through the project

Fig. 4 Image of team during an editing session. Image Credit: Adam White

collected through visualization of the data. We thus created a paper version of the matatu data as a way to orient the public. It borrowed its style from traditional transit maps such as those found in London, Paris, New York, and Washington DC (Fig. 3). University of Nairobi students helped us edit the map to correct for errors, identify missing routes and review overall legibility (see Fig. 4). The maps transformed how Nairobi's transit community understood the matatu system, because, for the first time, they could see it as one comprehensive network. Matatu owners used the map to illustrate their strategies for relieving congestion. The launch of the data and map, where the governor of Nairobi made it the official Matatu map, helped start a meaningful dialogue about the role of matatus in Nairobi's transit landscape. The downloadable maps went viral on social media, and versions were printed in local newspapers allowing anyone to access the information we created.

After the release of the data and map, we held two focus groups to solicit feedback on the map's accuracy, usability and value. One focus group was with twenty-five university students who were regular transit users, and another focus group was with sixteen matatu drivers and owners who worked on one of the Southern routes. We also conducted a more formal survey of first-year planning students at the Technical University of Kenya. A total of sixty-six surveys were collected after respondents had the opportunity to review and comment on the paper version of the matatu map. Respondents were asked a total of nineteen questions which included demographic information on participants as well as overall familiarity with matatus.

3 Results

Our Nairobi study illustrates that by modifying standard software developed for Android mobile phones, it is possible to develop a comprehensive data set for semi-formal bus systems. The data collection requires careful groundwork identifying and riding all possible routes. The team also repeated travel in some routes to locate the most common path, as these are prone to some variability. Once developed, the dataset provides base data that can be built upon and updated. We are currently researching and developing methods including crowdsourcing and new tools to easily update changes to routes and stops.

The release of the data showed the demand for the data by planners and researchers working in Nairobi. We received multiple requests for the data from transportation consultants including the Institute for Transportation and Development Policy (ITDP) which was creating a bus service plan for a proposed Bus Rapid Transit (BRT) corridor, and required knowledge of existing routes. Providing the data saved the organization much time as a lot of work went into investigating the various matatu routes along the corridor which was the primary emphasis for their study. They used the data and map to select routes and locations on these routes to conduct ridership capacity and frequency surveys (ITDP and UN-Habitat 2014). The matatu owners and drivers instantly saw the potential of the map and used it to demonstrate the logic of new routes that would help improve traffic circulation. The consultants at Mott McDonald who were advising the Ministry of Transport and Infrastructure in setting up a new transit authority also found the data valuable for their purposes. Numerous researchers used the data and maps for their work. Providing the data openly allowed the transit community to share ideas from a common knowledge base creating more dialogue, improved trust and enhanced collaboration between some of the various stakeholders (Williams et al. 2014).

Our strategy of developing open data for innovation was very successful. A number of software developers including those we included in the early discussions about the project used the data to develop trip planning applications for cell phones. These included the popular trip planning application *Ma3route* which also gives traffic alerts and allows you to rate your matatu driver. *FlashCast Sonar* was another trip planning application that also gave some real time information for a subset of the matatus in Nairobi. The *Digital Matatu* trip planning app was created for Windows, named presumably after this research project and another trip planning app called *Matatu Map* was created. More recently a Canadian company used our data to produce *Transit App*. Research remains to be done on how many users these apps attract and the impacts of their use. *Ma3route* also crowdsources information about traffic conditions, creating potentially useful data for planning. Google queries may also be interesting to analyze once the data is live on their system.

The stylized matatu map was extremely well-received when launched in Nairobi and the city intended to adopt it as its official map. The map was used in a number of transit planning initiatives. Citizens were generally very positive about the map including the 66 undergraduate planning students from the Technical University of Kenya who took our formal survey. The survey sample did not reflect the makeup of the city- a more representative sample would be needed. However, the results are still helpful to gauge the response of a young better educated male demographic. About two-thirds of respondents were under the age of 22 years of age, and nearly 88 % were males. Most respondents (48 %) had lived in Nairobi for between two and ten years and most rode matatus four or more times per week, (42 %). Almost 29 % claimed only to ride matatus a few times per month reflecting the fact that many of the students walk.

More than 80 % of respondents reported never having seen a public transit map like the Digital Matatu map for their city. More than 90 % claimed that they were able to find where they were at the time and also the route they typically rode. Almost all, 86 %, believed the map made it easier to find the way around the city. About 83 % reported that with the map they are more likely to take a new matatu route and 77 % reported that they were more likely to take a matatu to a part of Nairobi they usually did not travel to. In focus group discussions, it was clear that finding information about new areas of the city was a problem with people needing to rely on word of mouth. In addition, respondents agreed that the matatu map would be useful for a variety of users: visitors and tourists (19 %), newcomers to the city (14 %), residents (12 %), and matatu users/passengers (11 %). Nearly 45 % of respondents reported that posting the map at matatu stops would be the most convenient way to access the map, and this corresponded to what we learned in the focus groups with drivers who had a similar request. One respondent suggested posting the map in public places like supermarkets and shops. Overall, the results showed an interest in having public transit information in map form.

Several respondents hoped that real-time information could be incorporated in the data some way, which the team would like to investigate in the future. As the cost of GPS devices lowers further, more vehicles could carry such devices and give real time information to passengers although many vehicles do not wish to be tracked when they deviate from designated routes or make more trips than they might report to the matatu owner. The *FlashCast* sonar app already provides real-time data for a several matatu routes.

It is interesting that almost 50 % of survey respondents claimed that after seeing the map, they believed that Nairobi's matatu system was better than they previously thought. The stylized representation of the map helps give that impression as it copies the style of more formalized systems but also shows the expanse and network structure of the system which was previously hard to understand. About 86 % reported that the Digital Matatu Map was more useful than their available sources of matatu information, the most typical of which for 42 % of respondents was to ask a friend. Nearly two-thirds of respondents claimed that they would be willing to pay between KES 20 (US $0.22) and KES 100 (US $1.10) for the map. While we need to survey a representative sample of the Nairobi population, these initial results of young primarily male university students suggest that a demand exists for better transit information within this demographic category. Further research should explore how other ways of disseminating the map data might provide better information for users and the public to understand and relate to their transit system (Gosselin 2011).

Finally, our focus group with drivers and owners of matatus revealed a willingness to participate in improving data on landmarks and stops on the map. Like the university respondents, they thought the map was generally useful and wanted to have parts of it at bus terminals and certain route information available on their vehicles so people knew the stops. They said that they spent time giving directions to passengers, and they did not always have good information on routes in other parts of the city. They took a certain pride in creating and running part of the extensive system depicted on the map.

4 Conclusions

Overall, this work in Nairobi gives preliminary evidence that specialized data recording phone apps, utilizing the integrated GPS tracking function of mobile phones, can effectively and efficiently capture important transit information. Our research so far also demonstrates demand exists for both the open data and information in both digital and paper forms. More importantly, we illustrate the importance of engaging the transportation and technology community in data collection strategies. In the era of "big data" and the prevalence of crowdsourcing devices, our project demonstrates how valuable data can be collected and disseminated in a way that extends the possibilities of innovation in civic technologies by creating networks of trust and collaboration.

Our research also suggests that semi-formal bus systems that are fragmented and operated by many vehicles, could benefit from a more centralized data collection effort by a university or transit authority (which currently does not exist in Nairobi but is now being created). This would allow for data quality control and the development of standard such as GTFS. We have also shown that universities in these cities can help with these efforts and can help spur innovation by linking to the tech community. However, substantial challenges remain in institutionalizing data collection and supporting better official city efforts to provide needed transit information to citizens in varied forms that cater to different categories of people including the poor, the elderly and children. Currently, such an effort does not exist because officials in Nairobi still have negative attitudes to their "semi-formal" transit system (Klopp and Mitullah 2015), and generally do not have a culture of providing information to citizens (Klopp et al. 2013; Williams et al. 2014) although this is changing.

In the movement to develop cities that are "smarter" and more responsive to citizens' needs, the development of data on essential infrastructure such as the matatus is critical. Better wayfinding and journey planning along with new planning applications can help Nairobi's transit system become more responsive to citizen needs. Yet there is much work to do to link this information up with transportation planning processes. Wayfinding, for example, is highly dependent not only on technology and information but on the structure of the environment and the transportation network (Woyciechowicz and Shliselberg 2005). However, we have shown that cell phone technology is making it easier to help users and operators even in these highly flexible and dynamic semi-formal bus systems. The kinds of digital data we collected, the openness in which it was released and shared, and the visualization of the network of routes are important steps but not the only ones necessary for better transit information.

5 Recommendations

Overall, this work suggests the fruitfulness of transit data collection efforts using GPS enabled cell phone devices. However, more research needs to be done to improve transit data applications, data collection, visualization, and dissemination efforts to make transit information widely available in appropriate forms. This in turn, will be critical for improving wayfinding and journey planning but also analysis and modeling to develop interventions to improve existing transit systems. More effective phone applications for easy, user-friendly data collection might be developed that streamline information processing while in the field. With good quality tools and continued engagement with residents, government, and civic "hacktivists" (Townsend 2013) there are opportunities to crowdsource important data from citizens. More research needs to be performed on what additional transit information is helpful to users, operators, and planners. Further research should also investigate the potential impacts that providing this data and information has on the relationships between the various actors of the transportation network.

Acknowledgments This work would not have been possible with the critical thinking and hard work of the University of Nairobi, School of Computing & Informatics students: Ikamar Ekessa, Peter Kamiri, Samuel Kariu, Maureen Mbinya, Jackson Mutua, Mureri Ntwiga. Researchers at MIT's Civic Data Design Lab also contributed significantly to the work including; Jonathan Andrew Campbell, Emily Eros, Alexis Howland, Lindiwe Rennert, Alicia Rouault, Christopher Van Alstyne, Catherine Vanderwaart. Special thanks to Wenfei Xu from the Civic Data Design lab who was instrumental in the development of the Matatu Map. Thanks also to Professor Lawrence Esho and his students at the Technical University of Kenya for their valuable feedback on the Digital Matatu map. We also gratefully acknowledge the support of the Rockefeller Foundation which provided a grant for this work. Special thanks to Benjamin de la Pena at the Rockefeller Foundation for his support of and critical insights on this work. We also benefitted greatly from conversations that compared projects with Prof. Chris Zegras, Albert Ching, Stephen Kennedy, Neil Taylor, Kevin Webb and Emily Jean Eros. Last but not least, we thank James Gachanja and Dr. Zachary Gariy at KIPPRA for hosting the workshops and the Kenya Alliance of Resident Associations for their work on the launch of the data and map for this work helped our thinking, made connections and showed us a way forward in sustaining data collection work.

References

Caceres, N., Romero, L. M., Benitez, F. G., & Del Castillo, J. M. (2012). Traffic flow estimation models using cellular phone data. *Intelligent Transportation Systems, IEEE Transactions on, 13*(3), 1430–1441.

Ching, A., Zegras, C., Kennedy, S., Mamun, M. (2013). A user-flocksourced bus experiment in Dhaka: New data collection technique with smartphones. *Transportation Research Record: Journal of the Transportation Research Board.*

Craig, W. (2005). White knights of spatial data infrastructure: The role and motivation of key individuals. *URISA Journal, 16*(2), 5–13.

Elwood, S. (2006). Negotiating knowledge production: The everyday inclusions, exclusions, and contradictions of participatory GIS research. *The Professional Geographer, 58*(2), 197–208.

Elwood, S., Goodchild, M. F., & Sui, D. Z. (2012). Researching volunteered geographic information: Spatial data, geographic research, and new social practice. *Annals of the Association of American Geographers, 102*(3), 571–590.

Eros, E., Mehndiratta, S., Zegras, C., Webb, K., Ochoa, M. C. (2014). Applying the general transit feed specification (GTFS) to the global south: Experiences in Mexico City and beyond. *Transportation Research Record.*

Goodspeed, R. (2015). Smart cities: Moving beyond urban cybernetics to tackle wicked problems. *Cambridge Journal of Regions, Economy and Society, 8*(1), 79–92.

Gosselin, K. (2011). Study finds access to real-time mobile information could raise the status of public transit. *Next American City.* http://nextcity.org/daily/entry/study-finds-access-to-real-time-mobile-information-could-raise-the-status-o. Accessed March 6, 2015.

Harvey, F., & Chrisman, N. (1998). Boundary objects and the social construction of GIS technology. *Environment and Planning A, 30*(9), 1683–1694.

Hein, J. R., Evans, J., & Jones, P. (2008). Mobile methodologies: Theory, technology and practice. *Geography Compass, 2*(5), 1266–1285.

Herrera, J. C., Work, D. B., Herring, R., Ban, X. J., Jacobson, Q., & Bayen, A. M. (2010). Evaluation of traffic data obtained via GPS-enabled mobile phones: The mobile century field experiment. *Transportation Research Part C: Emerging Technologies, 18*(4), 568–583.

Klopp, J. M., Marcello, E. M., Kirui, G., Wing, N., & Mwangi, H. (2013). Can the internet improve local governance? The case of Ruiru, Kenya. *Information Polity, 18*(1), 21–42.

Klopp, J., Mutua, J., Orwa, D., Waiganjo, P., White, A., Williams, S. (2014). Towards a standard for paratransit data: Lessons from developing GTFS data for Nairobi's Matatu system. In *Transportation Research Board 93rd Annual Meeting* (No. 14-5280).

Klopp J., & Mitullah, W. (2015). Politics, policy and paratransit: A view from Nairobi. In R. Behrens, D. McCormick & D. Mfinanga (Ed.), *Paratransit for African Cities.* Routledge.

Kyem, P. A. K. (2004). Of intractable conflicts and participatory GIS applications: The search for consensus amidst competing claims and institutional demands. *Annals of the Association of American Geographers, 94*(1), 37–57.

McHugh, B. (2013). Pioneering open data standards: The GTFS story. *Edited by Brett Goldstein with Lauren Dyson*, 125.

Ratti, C., Williams, S., Frenchman, D., & Pulselli, R. M. (2006). Mobile landscapes: using location data from cell phones for urban analysis. *Environment and Planning B: Planning and Design, 33*(5), 727.

Roth (2010). How google and Portland's TriMet set the standard for open transit data. Streetsblog. http://sf.streetsblog.org/2010/01/05/how-google-and-portlands-trimet-set-the-standard-for-open-transit-data/. Accessed January 5, 2014.

Sieber, R. (2006). Public participation geographic information systems: A literature review and framework. *Annals of the Association of American Geographers, 96*(3), 491–507.

Talbot, D. (2013). African bus routes redrawn using cell-phone data. *MIT Technology Review.* http://www.technologyreview.com/news/514211/african-bus-routes-redrawn-using-cell-phonedata/. Accessed October 6, 2014.

Townsend, A. (2013). *Smart cities: Big data, civic hackers, and the quest for a New Utopia.* New York: W.W Norton.

Townsend, A. (2014). Re-programming mobility: The digital transformation of transportation in the United States. http://reprogrammingmobility.org/wp-content/uploads/2014/09/Re-Programming-Mobility-Report.pdf . Accessed March 4, 2015.

Wakefield, J. (2013). Mobile phone data redraws bus routes in Africa. *BBC*, http://www.bbc.com/news/technology-22357748. Accessed September 30, 2014.

Wang, H., Calabrese, F., Di Lorenzo, G., Ratti, C. (2010). Transportation mode inference from anonymized and aggregated mobile phone call detail records.In *Intelligent Transportation Systems Conference* (ITSC 2010), pp. 318–323.

Williams, S., Marcello, E., & Klopp, J. M. (2014). Toward open source Kenya: Creating and sharing a GIS database of Nairobi. *Annals of the Association of American Geographers, 104* (1), 114–130.

Williams, S., White, A., Waiganjo, P., Orwa, D., Klopp, J. (Forthcoming). The digital Matatu project: Using cell phones to create an open source data for Nairobi's semi-formal bus system. *Journal of Transportation Geography* (Working Paper Submitted October 2014).

Wong, J. (2013). Leveraging the general transit feed specification for efficient transit analysis. *Transportation Research Record: Journal of the Transportation Research Board, 2338*(1), 11–19.

Woyciechowicz, A., & Shliselberg, R. (2005). Wayfinding in public transportation. *Transportation Research Record: Journal of the Transportation Research Board, 1903*(1), 35–42.

Zegras, P. C., Eros, E., Butts, K., Resor, E., Kennedy, S., Ching, A., Mamun, M. (2014). Tracing a path to knowledge? Indicative user impacts of introducing a public transport map in Dhaka, Bangladesh. *Cambridge Journal of Regions, Economy and Society*, rsu028.

Chapter 13
The Australian Urban Intelligence Network Supporting Smart Cities

Christopher J. Pettit, John Barton, Xavier Goldie, Richard Sinnott, Robert Stimson and Tom Kvan

Abstract As the global population continues to grow and an increasing number of people move to cities, there is need for ambitious approaches to provide urban information infrastructures and analytical tools to support smart urban design and planning. This chapter introduces the Australian Urban Intelligence Network, which brings together a network of researchers, planners and policy-makers from across Australia who have access to an online workbench of data and tools. The work-bench comprises over 1100 datasets and 100 spatial statistical routines, and a select number of planning support systems and geodesign tools. In this chapter, we outline the urban data and analytical capability the online workbench; introduce a couple of the PSS tools and spatial statistical capabilities through a case study approach. We also discuss the user outreach and capacity building capability program which is a critical component to assist with user adoption. We conclude the chapter with some reflections on the lessons learnt and next steps in the project.

Keywords Planning support systems · Spatial statistics · Scenario planning · Walkability

---

C.J. Pettit (✉) · J. Barton · X. Goldie · R. Stimson · T. Kvan
Faculty of Architecture Building and Planning, University of Melbourne, Victoria 3010, Australia
e-mail: cpettit@unimelb.edu.au

J. Barton
e-mail: jack.barton@unimelb.edu.au

X. Goldie
e-mail: xavier.goldie@unimelb.edu.au

R. Stimson
e-mail: rstimson@unimelb.edu.au

T. Kvan
e-mail: tkvan@unimelb.edu.au

R. Sinnott
Department of Computing and Information Systems, University of Melbourne, Victoria 3010, Australia
e-mail: rsinnott@unimelb.edu.au

S. Geertman et al. (eds.), *Planning Support Systems and Smart Cities*, Lecture Notes in Geoinformation and Cartography, DOI 10.1007/978-3-319-18368-8_13

1 Introduction

At a time when more people are moving to living in urban areas and there is significant global population growth, there are significant challenges for the sustainable development of cities and their resilience. At the same time we are seeing the rise of cloud computing and the compilation of many data sets across all aspects of urban activities and life. Subsequently, there are increasing expectations by planners and decision-makers on what information and communications technologies can deliver to improve urban quality of life. Concepts such as smart cities, big data, information visualisation and predictive analytics are leading to the formation of new tools and techniques for analysing and understanding cities and their future paths of growth and development. Yet for these concepts to gain traction and to make a difference in planning and urban design, there is a need for smart people to be asking the right questions using existing and emerging new data, tools and techniques appropriately.

The Australian Urban Research Infrastructure Network (AURIN) is one such initiative which is providing a secure, online platform to access big data, to facilitate the integration of diverse types and sources of data, and to enable the real-time interrogation of data using a sophisticated suite of open source statistical, spatial analytical, modelling and visualisation tools. AURIN is also fostering and connecting a network of smart end-users, combining both the technological and human dimensions. We refer to this as the Australian Urban Intelligence Network. In short, AURIN is becoming Australia's Urban Intelligence Network connecting urban researchers, planners and policy-makers with an online workbench of databases and decision support tools.

In this chapter we also provide an overview of the technologies and capabilities being developed by AURIN. This includes an online workbench, which comprises an extensive and diverse range of federated datasets (currently over 1100) from autonomous providers, more than 100 spatial-statistical tools, and a suite of planning support systems (PSS) and geodesign tools. The online workbench is available to all researchers from all universities across Australia and also policy and decision-makers from all levels of government. By providing the workbench to a broad range of stakeholders who design and plan our cities, it is anticipated that new collaborative outcomes will foster multidisciplinary research to inform the future of our cities.

In this chapter, we provide the next instalment of research and development occurring as part of the AURIN project. In previous work (Pettit et al. 2013a, b) reported on the technical development and first application of the AURIN Online What if? (OWI) as applied in Hervey Bay. In this chapter we focus on a selection of case studies applying AURIN data and tools being used across Australia to support urban settlements research and policy analysis. We discuss the networks of data hubs that have been established across Australia and how these are being used by researchers and by government policy and decision-makers. We introduce a walkability tool that is being used to plan for more walkable neighbourhoods and cities. We briefly revisit the OWI PSS and how this is being used for medium to long term strategic planning in the City of Perth. We introduce the comprehensive suite of spatial-statistical modelling routines available via the online workbench and how these can be used on big data, such as the Australian Property Monitors residential property transactions database.

Such an online workbench is of little value if it is not being used to support research, planning, policy and decision-making. In this chapter we therefore discuss some of the capacity building activities being undertaken to up-skill the workforce in being able to appropriately use the suite of data and tools currently available via the online workbench.

The chapter concludes by reflecting on some of the lessons learnt in developing the online workbench, and the challenges and opportunities in connecting urban researchers, policy and decision-makers through an urban intelligence network.

2 Background

The Australian government made a commitment to supporting urban research by allocating $20 million in the 2009 budget and noting the need for:

> … the establishment of facilities to enhance understanding of urban resource use and management, involving a collaboration between several universities.

The need was also recognised in the *Strategic Roadmap for Australian Research Infrastructure* 2008, under the 'Built environments' capability area (and identifying links to other capabilities such as humanities and general social sciences) that identified the need for

> *….facilities that enable the collection and integration of datasets in energy and water consumption, resource management, and social and environmental interactions.*

The Roadmap recommended integrated approaches, linkage across disciplines, jurisdictions and sectors, and the importance of social and environmental aspects of "built environments".

The allocation was made under the Education Infrastructure Funding (EIF) framework that mandated that the money be only used to create a national infrastructure to support researchers across the nation. Within these guidelines yet with considerable resources, it became clear that a powerful infrastructure could be created that provided portal access to a wide range of data held by many owners, governmental, institutional and private sources. Since the initial investment an additional $4 million has been made available through the Australian Government National Collaboration Research Infrastructure Strategy (NCRIS) program.

The initial proposal recognised that the opportunity lay in creating a web portal to the wide range of data created and gathered across the country. These data reside in institutional, governmental and corporate systems, taking the forms of archival data, transactional data that will at some point be archived, and ephemeral data such as locational data of telephone handsets at a particular moment or social media data that may not be captured or, if fleetingly captured, not archived. All of these are of relevance to urban research.

From this opportunity, the Australian Urban Research Infrastructure Network (AURIN) was created. The University of Melbourne submitted the initial project plan in 2010 and funding approved by midyear. In particular, the proposal was framed to respond to the Communiqué issued in December 2009 by the Council of Australian Governments (COAG) that set as a goal to:

> …..ensure our cities have strong, transparent and long term plans in place to manage population and economic growth, plans which will address climate change, improve housing affordability and tackle urban congestion.

This identified the national capacity for strong urban management was to be underpinned by effective research that informed policy-making. The proposal noted that, without good systems by which to bring reliable data, current and historical, to support analysis and the development of models with which to inform and test policy, this goal could not be achieved. It was recognised immediately that the opportunity in Australia was not to collect or create the data but to provide secure facilitated access to this wide range of data and to create a workbench of tools for users to work with these data. A major portion of research funding was being deployed by researchers to identify and obtain access to relevant data. The value of AURIN therefore would lie in reducing significantly these access issues, reducing too the related data management costs for the owners, and thus leverage available funding for increased research activity. This chapter describes how the system has been developed to achieve this and illustrates its application in select case studies.

3 An Urban Data and Analytics Online Workbench

The suite of tools and data that comprises AURIN is collectively referred to as the workbench. It is realised through an extensive online platform providing urban researchers, policy and decision-makers with seamless and secure (where required) access to a wide range of distributed (and heterogeneous) data with supporting analytical and visualisation tools. The workbench offers (at present) over 1100 data sets from 37 major agencies across Australia. Key to the solution is web-based, programmatic access to distributed data in situ, and i.e. urban research requires access to the definitive and most up-to-date from the major data providers across Australia with accompanying tools reflecting best practice across the urban research domain. Collectively these data sets and tools are available through one website (http://aurin.org.au/) which provides access to a data search and discovery tool and a number of facilities with the portal (https://portal.aurin.org.au) being the flagship application shown in Fig. 1.

Each data provider has its own demands and protocols on access to and use of their data and typically their own data/metadata formats. A major demand that the

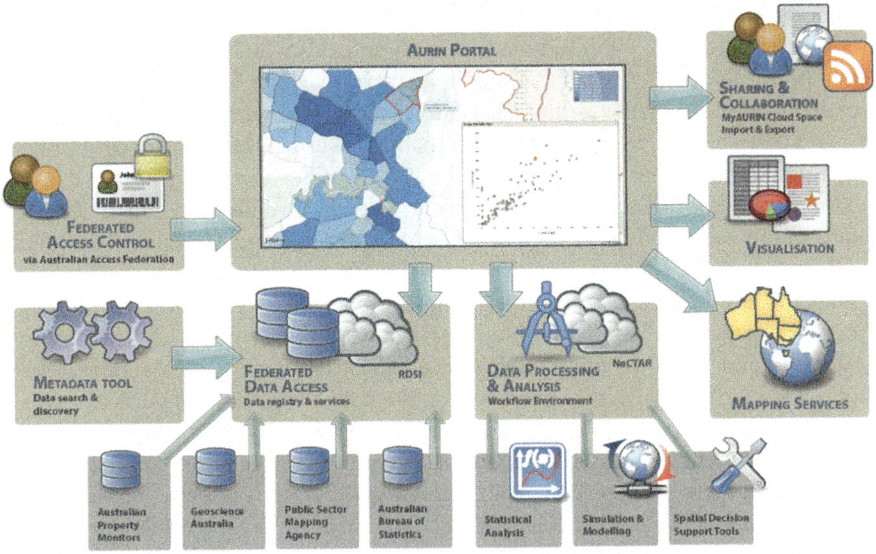

Fig. 1 AURIN system architecture for workbench adapted from Sinnott et al. (2014)

e-Infrastructure was to meet was making access and use of distributed and heterogeneous data transparent to the end users and supporting the notion of single sign-on, i.e. users log in once to the portal and thereafter are able to access remote information from remote providers without the need for further usernames and passwords. This approach allows the systems to scale when new data sets are added to these providers and importantly it allows data providers to define their own access and usage conditions since each data provider is completely autonomous. Indeed in many cases this autonomy is demanded since the data is of a commercial or sensitive nature since it deals with individual-level information.

To provide common access to all researchers across Australia, the portal is delivered through the Australian Access Federation (AAF—www.aaf.edu.au). The AAF provides federated authentication whereby users from different academic organisations (that is, those users with an @edu.au University login) across Australia can access resources, e.g. the AURIN portal, by logging in (authenticating) at their own research organisation. This means that usernames and passwords specific to AURIN are not required. Whilst the focus has been primarily based on supporting academic researchers (approximately 50 % of the portal's current user base), for example those at Australian Universities, many others involved in urban environments are increasingly requesting access to the portal including those from government, industry and indeed researchers from outside of Australia.

Access to these other users is also enabled through the AAF through what is known as the Virtual Home Organisation (VHO). With the VHO model a user account can be created through and is associated with an email address provided by the non-academic collaborator. At present there are approximately 400 government users (~17 % federal, ~58 % state and ~25 % local) who have been given access to the portal through the VHO. As at January 2015, there have been over 25,000 user sessions utilising the portal. This information is captured and made available by the AAF. The trend in utilisation information is shown for the last year in Fig. 2. These users come from over 30 academic organisations across Australia.

Considerable work has been undertaken to gain an understanding of both the nature of the users and of their experiences within the portal, to drive the iterative development of the portal within an agile software development environment. This is undertaken through regular surveying of both new and experienced users alongside an established monitoring and evaluation framework. Master classes, workshops and tutorials with students, academics and professionals are surveyed after hands on training sessions, asking them a series of standardised closed questions relating to their experience with the portal and their backgrounds, together with open-ended questions asking for positive and negative aspects of their experience, and suggestions for future features or improvements. These are fed directly into the software release cycle, with a number of new portal features having evolved directly from this user feedback (Barton et al. 2015). For example, the way users navigate through the hierarchy of geographies to search data was identified as a significant usability issue. This was taken on board by the system developers and the hierarchy was refactored in the next portal release to make it easier to navigate up and down, and across geographies. Also, a bounding box selection feature was added to further improve the user experience.

The organisations that provide data primarily do so through offering programmatic web services. Typically these are either geospatial services such as Open Geospatial

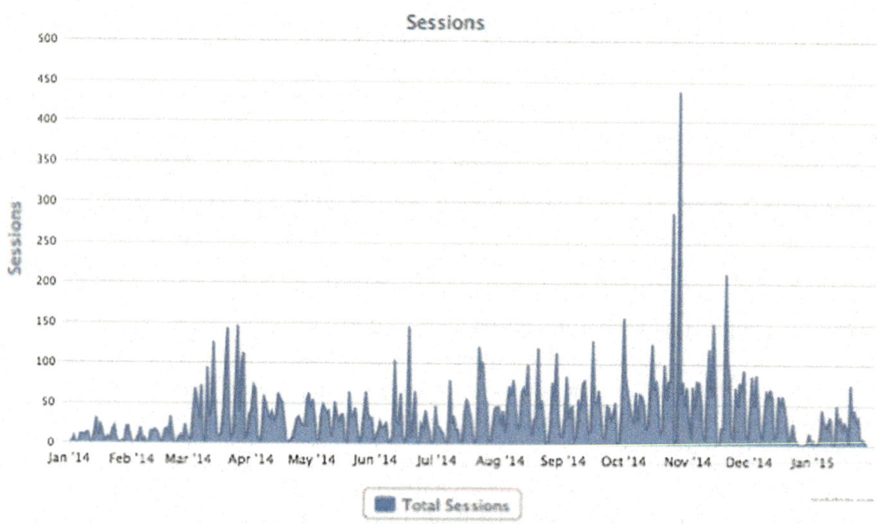

Fig. 2 AURIN usage patterns for 12-month period, from Jan 2014 to Jan 2015

Consortium (OGC) Web Feature Services (WFS), web services such as Representational State Transfer Services (ReST) or other targeted flavours of web service aligned with the organisation demands and existing technology base, e.g. in the case of the Australian Bureau of Statistics (ABS—www.abs.gov.au) the Statistical Data and Metadata Exchange (SDMX—www.sdmx.org) format was required.

3.1 A Network of Data Hubs

As a concept 'data hubs' have been created to address issues arising from the discovery, access and format diversity of research data (Delaney and Pettit 2014) Through the federated technical architecture of the workbench (as illustrated in Fig. 1) a number of data hubs have been established across Australia including in Melbourne, Sydney, Brisbane, Perth, Adelaide, Canberra, Townsville and Illawarra. Each of these data hubs is making available hundreds of direct programmatic data feeds into the portal that end users can analyse visualise or download directly to use in their software of choice.

One of the significant advantages of such a federated data approach is that when datasets are updated users to the AURIN workbench automatically receive access to the updated data. One such example can be given in reference to the property sales transaction data, which is available for every land parcel across Australia through the Australian Property Monitors (APM) (Pettit et al. 2014). This data service is updated quarterly by the data custodian APM. AURIN users automatically gain access to additional quarterly snapshots of transaction along with the previous sales transaction dating back approximately 20 years.

All of the data hubs and services are driven by metadata enriched using the AURIN metadata tool, which is based on Geonetwork (Delaney and Pettit 2014). The metadata tool is semi-automated in that a number of attributes can be harvested from each of the data hubs. However, the system does require the data custodian to have their metadata available in a version similar to ISO19115. This is an important quality assurance mechanism to ensure the data meets the AURIN e-infrastructure minimum standard and requires the data custodian to ensure their metadata is of sufficient quality.

Once the metadata has been enriched with the necessary attributes it is registered in the data registry and finally made discoverable via the portal. AURIN has also implemented an instance of CKAN as part of the workbench so that the metadata can be searched and discovered without having to log into the system, which is a necessary step when accessing the portal. Through the implementation of the data hubs concept the portal is enabling researchers, policy and decision-makers to collaborate on multi-disciplinary research endeavours which cut across a range of domains, referred to as lenses (Pettit et al. 2012). These include for example urban health, transport, housing, water and energy supply and consumption and innovative urban design.

The current version of the AURIN e-infrastructure is not semantic web enabled, nor is it driven by ontologies. The data is structured using a hierarchy of geographies, predominantly based on the Australian Statistical Geography Standard (ASGS) and Australian Spatial Geographic Classification (ASGC) developed by the Australian Bureau of Statistics (ABS). Cadastre level data is based upon the Public Sector Mapping Agency (PSMA) National Cadastre dataset which comprises approximately 11 million parcels. By using such nationally consistent geographical boundaries, spatial analysis operations can be performed across datasets of similar geographies. Also, where possible nationally agreed upon nomenclature has been used for fields and attributes. However, the user needs to explore the enriched metadata when undertaking data analysis across datasets from different organisations to ensure analytical integrity.

3.2 Planning Support System Tools

PSS have been described as systems that support those involved in planning to explore, represent, analyse, visualise, predict, prescribe, design, implement, monitor and discuss issues associated with the need to plan (Batty 1995). In this section we will introduce a number of PSS tools that have been implemented or are under development as part of the AURIN workbench. This includes a brief summary of the walkability and What If? scenario planning tools which can support the geodesign and planning of our cities. These PSS tools are a sub-category of spatial decision support system tools which comprise the AURIN workbench—see Fig. 1.

3.2.1 Creating More Walkable Neighbourhoods

The AURIN workbench includes a walkability tool used for assessing and comparing the walkability of neighbourhoods (Giles-Corti et al. 2014; Badland et al. 2013) which has been demonstrated to have a significant impact on health outcomes (McCormack et al. 2008). Input data for these tools are sourced from several sources—The Australian Bureau of Statistics (ABS), the Public Sector Mapping Agency (PSMA) and Points of Interest (PoI) from either user input, or existing datasets such as train stations or bus stops.

Factors influencing the walkability of a neighbourhood include the proximity to a destination, density of people, street connectivity and land use mix (LUM) (Frank et al. 2010). These factors can be measured to provide a Z-score indicating if an identified neighbourhood is more or less walkable than the average of the other neighbourhoods in a study. The Sum-Z-score allows the decision maker to review a range of alternatives and discern options based on a comparative score. The resulting vector-based walkability pedestrian catchment coverage (ped-shed) is exportable in .shp/GeoJSON format for integration into other geospatial tools.

This tool processes data from multiple sources, and offers an intuitive interface to facilitate an evidence-based approach to designing the built environment. There is a balance in offering either a heavily automated system offering ease-of-use or a system that offers manual control and expert-level flexibility. The walkability tool is designed to process routine analyses in the background, such as density calculations and land-use mix. This has a benefit of maintaining a consistent method and reduces the need for the user to undertake specialised analyses. At the front end, the user can specify which datasets service as inputs and what analyses is performed- this adds a degree of freedom and lets the user tailor the system to a specific task. Importantly, the results of the analysis are explicitly outputted into a new data table that can be openly interrogated, further processed and shared. This data-driven approach to geodesign is valuable for engaging with other stakeholders. Furthermore, the map-based output helps present complex scenarios as a legible map and, importantly, facilitates discussion amongst stakeholders based on an explicit method.

Transit oriented development (TOD) can be supported by revealing not only the network-generated proximity to a node, but the current density of dwellings and persons within that catchment. Further assessments can be made as to if this catchment is appropriate for increased density, if changes to the street network improve walkability or if increased LUM is warranted. Adding further demographic data can assist in generating new layers of information, for instance, are there areas with a high instance of sedentary behaviour that are under-serviced by access open space? Are facilities appropriately located, or might local residents be car-dependent?

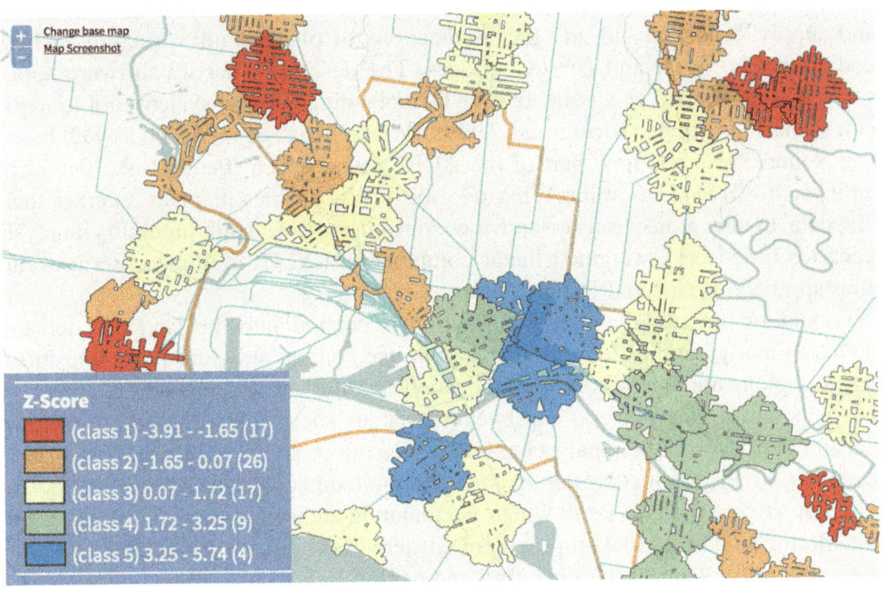

Fig. 3 Walkability analysis supporting the Geodesign of transport oriented development in Melbourne. Z-scores in *red* indicate relatively low walkability catchments around trains stations, while Z-scores in *blue* indicate relatively high walkability catchments

Figure 3 provides an example of a walkability analysis undertaken for the Inner-Melbourne area. Pedestrian catchment clouds (known as ped-sheds) have been formed by following the street network away from the train station using a distance of 800 m. This is a standard distance used in the planning community assuming a typical person is willing to walk 800 m to a train station. It applies especially well in the context of Melbourne where a policy exists to cluster increasingly dense new developments around existing public transport infrastructure (The State of Victoria 2014). Within the resulting polygon, attributes are calculated for population density, the number of street connections, and LUM by area. Having assessed these factors for each ped-shed, a relative Z-score is calculated for the study areas with the average is set to zero. A Z-Score above zero indicates that the area ranks better than the average; less than zero, below average.

A choropleth map was generated with the deeper blue areas in the centre of Fig. 3 illustrating the catchments scoring a relatively better overall score, and the peripheral stations indicating a relatively weaker score, as indicated by a red fill. As these attributes are encoded into a corresponding data table, further analysis into the subcomponents factoring into the score can be undertaken either by interactively brushing the mouse over the polygons or exporting the dataset into a standard GIS environment.

3.2.2 What if? Scenario Planning for Future Urban Growth

When undertaking strategic planning it is important to understand the dimensions of land supply, land demand and the implications of planned infrastructure such as roads, schools, water and energy services. The desktop What if? software application was developed as a collaborative GIS planning support system tool to assist with scenario planning (Klosterman 1999). What if? has been recoded as web based open source, which is now part of the AURIN workbench (Pettit et al. 2013a, b); Pettit et al. 2015). The online What if? tool (OWI) enables planners to enter their GIS data in the project set-up and then run customised land suitability analysis scenarios based on a weighted linear combination (WLC) multiple criteria evaluation approach (Pettit and Pullar 1999).

Population and employment information is entered into the land use demand module. Finally, control layers such as land use zoning, structure plans and infrastructure plans are entered into the allocation module. At which time the end users (planners and policy-makers) can tweak variables such as population projections, housing density, and boundary constraints to explore an envelop of urban growth scenarios. Figure 4 illustrates an example of the land suitability map generated for Joonulup area in north of Perth where the planning authority was exploring the land suitable for an educational employment cluster. This is the first time that OWI has been used with planners in generating and exploring land use suitability scenarios. In Pettit et al. (2013a, b) we reported on the first demonstration of OWI but this was using the Hervey Bay data acquired by Pettit (2005) purely to demonstrate the tool could work and to validate the underlying algorithms.

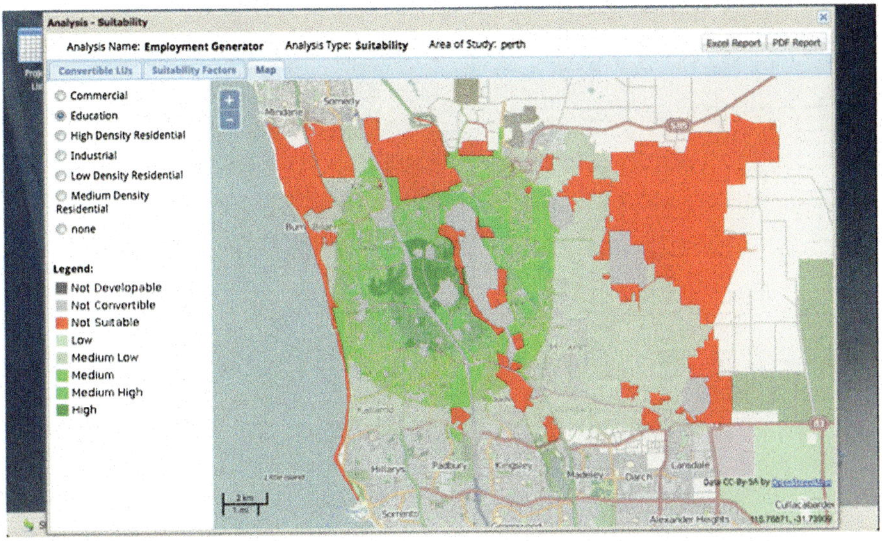

Fig. 4 Land suitability analysis map produced using the online What if? tool as applied in Perth (Pettit et al. 2015)

3.3 Visualisation and Analytical Tools

A central feature of the AURIN workbench is the implementation of a comprehensive suite of statistical tools for the interrogation and analysis of spatial data (both from the data custodians, and the users' own data)—see Fig. 1. These tools range from basic visualisation tools (e.g., choropleth mapping, charts, graphs), to classical statistical techniques (correlation, regression, ANOVA), through to more complex spatial statistical routines (e.g. spatial autocorrelation and geographic weighted regression). This breadth of tools ensures the utility and value of the workbench is maximised for a diverse range of users with different competencies, requirements and goals.

These tools have been developed in a collaborative process between AURIN's project partners and the core technical development team, and in many instances represent the implementation of a number of statistical packages and functions written in the open-source R statistical language (R Development Core Team 2008). The inclusion of these tools in the workbench in an easily navigable graphical user interfaced (GUI) format represents a significant strength, where the user has access to complex and robust statistical routines without having to learn the R language.

The tools themselves are grouped together according to their likely utilisation pattern by users, within intuitive clusters such as "chart tools", "data manipulation tools" and "spatial statistical tools". Each tool presents as a parameter input pane within the portal, allowing the user to enter both mandatory and optional components of the workflow. The outputs of the tools vary in their formats, comprising interactive map and chart visualisations, static images, derived datasets or simple

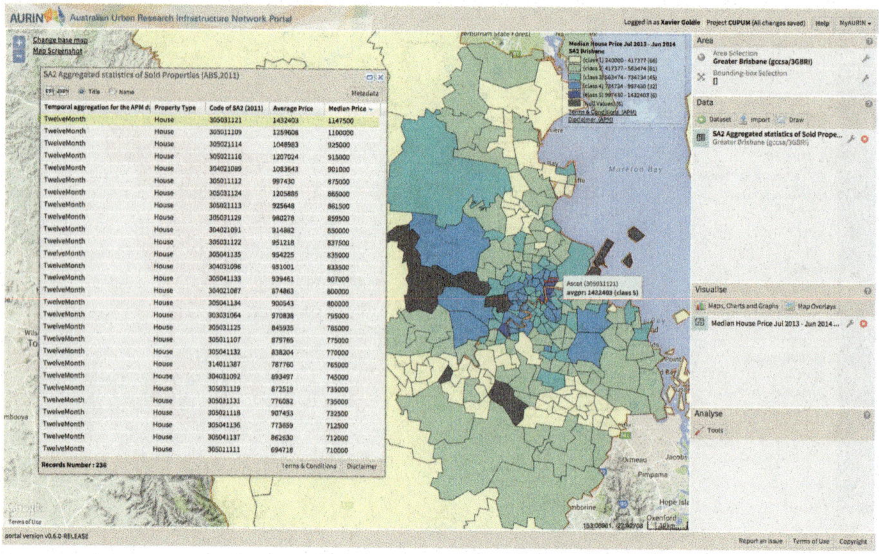

Fig. 5 Map visualisation tool in the AURIN Portal: Choropleth map output for APM property data across Brisbane showing data and interactive brushing capabilities

text outputs, allowing users to either continue interrogating their data within the portal, or to download the outputs for their own purposes. Each of the tools has a companion documentation page, explaining the use of the tool, the required inputs and the likely outputs for each workflow.

As an example of the utility of the online spatial statistical tools, we present here a simple worked example, focussing on property prices and socio-demographic factors in Brisbane. To begin, we load the Australian Property Monitors (APM) aggregated property sale statistics for Brisbane for the 2013–2014 financial year at SA2 (suburb) level. Using the map visualisation tools (Fig. 5) we are able to map the average price for sold properties across the study area.

A strength of the online workbench is its ability to compare datasets from disparate data sources. For this example, first we load selected averages and medians for populations at an SA2 level for Queensland. After bringing this data into the workbench, we are able to use data joining tools to merge the datasets into combined derivative datasets. These outputs can either be downloaded directly to the users desktop, or processed further online. With a combined dataset, we are able to investigate the relationship between different variables. In this instance, we create an interactive scatterplot, comparing the relationship between average price and median household income across Brisbane (Fig. 6).

We can start to engage in more sophisticated spatial statistical routines, such as investigating spatial autocorrelation and clustering. In this instance, we use the APM data to generate a contiguous spatial weights matrix, and then estimate the Moran's I statistic for property prices across Brisbane, which indicates a high and significant level of clustering of average property price (Moran's I = 0.6022,

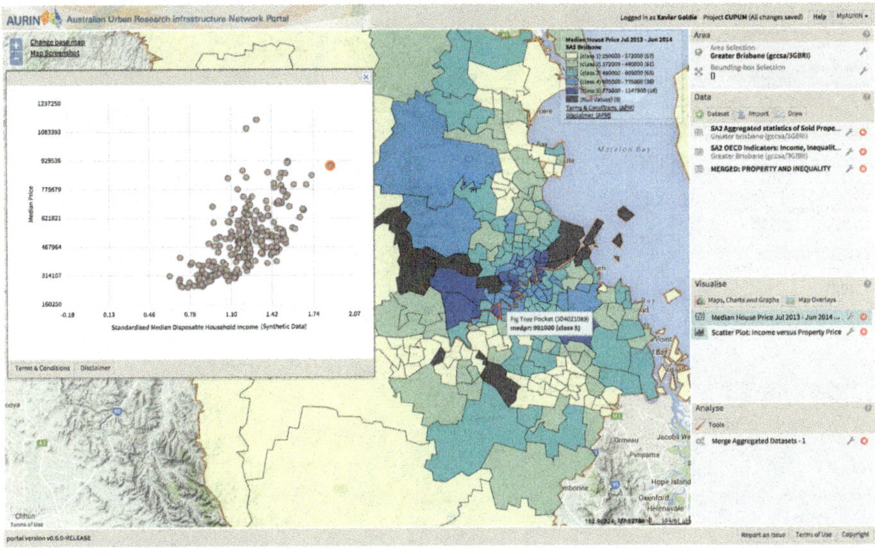

Fig. 6 Data comparison capabilities in the AURIN portal: interactive scatterplot of merged datasets, showing relationship between median household income and average property price by suburb across Brisbane

$P < 0.0001$). The results from these (and other statistical) analyses are presented in simple-text format within the workbench, allowing users to copy and paste the results into their own report or document (Fig. 7).

4 Training the Next Generation of Planners and Decision-Makers

The workbench caters to all members of the Australian university community and Australian government analysts and policy-makers by providing access to data and tools via a one-stop shop web-based environment. However, the user uptake of such tools as PSSs has not been as widespread as tool designers and developers would like (Vonk et al. 2005). Therefore, a critical component of the project has been to develop a user outreach and capacity building capability to up skill the existing and future work force in being able to use the workbench tools and available datasets.

As part of this teaching and learning program a series of workshops and master classes, and on demand on-site user training sessions are run across Australia. The six hour master classes (Fig. 8) contain a series of decision-support tasks for participants to work together on exploring data and testing ideas. The class concludes with the participants creating a series of graphical outputs to convey their findings to their peers.

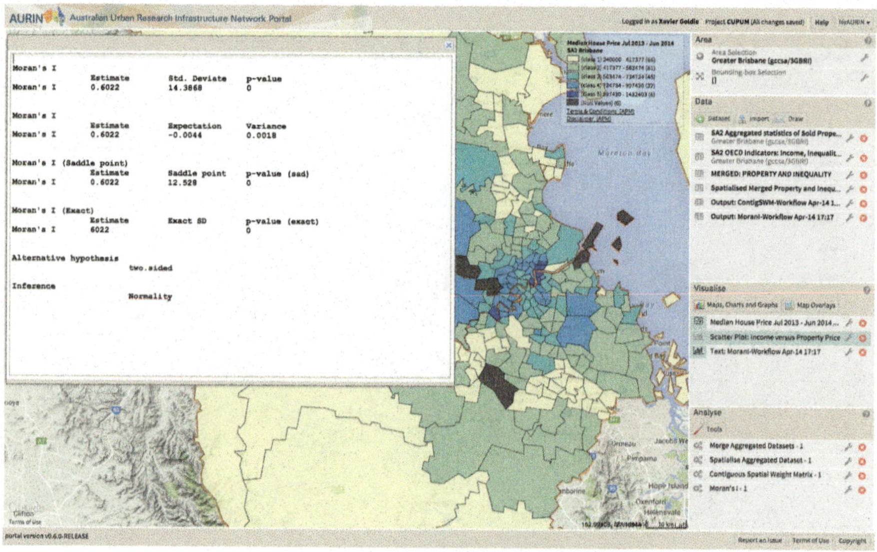

Fig. 7 Moran's I workflow output in simple text format for easy inclusion in user reports and documents

Fig. 8 AURIN master classes in action across Australia

As a companion to the workbench, a documentation repository is maintained (docs.aurin.org.au). This site contains a help page for each item in the workbench, plus a series of tutorials and use cases. These are structured to introduce a new user to a particular concept, guiding them through a series of steps to create a typical use case scenario. This resource is important for online end users and participants to the training courses to orientate themselves in using the workbench. To address this concern, the help pages have taken on a pedagogical role, introducing concepts with language designed to be easily comprehended by a user who might be completely unfamiliar with the subject matter. Once concepts and context are clearly presented, more detailed technical information is provided, where possible, with accompanying formulas, downloads and links to references.

User engagement is critically important on several levels. Firstly, with regard to the technical build, the technical build team are learning how people are using the system and how best to tailor the system for the end users. As previously discussed the feedback from workshops and master classes is itemised, prioritised and fed back into the successive phase of the workbench's Agile development (Barton et al. 2015).

The ease-of-use of the system is important for first-time users and experienced operators alike. User testing sessions at each Beta release of the workbench not only serve to identify areas for refinement, but permit a closer understanding of particular user groups. This underpins the capacity of performing decision-support tasks in the real world which requires community engagement and the generation of consensus-based policy outcomes.

With regard to up-skilling users from many different disciplines, a geospatial approach helps this diverse user group come together. Accessing shared data, common tools and adapting standard methods allows both local focus and national collaboration. Some examples of uptake include: (i) the Australian Commonwealth Department of Infrastructure, using the AURIN workbench to source data for the State of Australian Cities (SOAC) report, (ii) the NSW Department of Premier and Cabinet, using the AURIN workbench to rapidly acquire data from agencies across the state to better inform and test policy outcomes in relation to urban productivity, and (iii) local government using the AURIN workbench to search and discover population health data for their municipality, such as the rate of type II diabetes.

At the University of Melbourne, we have established the first "Geo for All" laboratory in Australia (http://osgeolab.unimelb.edu.au/). The OSGeo-Live software stack, developed by the Open Source Geospatial Foundation (OSGeo), aggregates the majority of open-source geospatial tools onto a bootable virtual machine. Students can readily extract urban data from the AURIN workbench and import it into specialised geospatial tools contained in the OSGeo stack. As this software is open-source, significant economies of time and cost savings are possible. Geospatial software can freely be loaded on student's personal computers at home or in the university lab. The OSGeo-Live software tools allow the training and outreach team to reach an international audience. This has begun to be realised in running two hands-on international workshops and a webinar.

5 Conclusions

In this chapter we have discussed both the technology component and the people interactions occurring with the AURIN workbench. We have introduced a select number of the datasets and tools and provide examples on how these can be put to use in supporting evidenced based planning and decision-making across Australia. Whilst much has been achieved there are significant lessons in developing such a comprehensive e-infrastructure. One of the main challenges concerns the design of such a system that is usable for a range of end users to access, from high-level policy-makers all the way to expert urban researchers who have a deep knowledge of spatial statistical algorithms. Given the workbench caters for a diverse ranges of end users ranging from naïve to power users it continues to prove a challenge in designing a toolkit which is simple, intuitive yet has advanced functionality for those who wish to use it. One way we have endeavoured to address this is through a dedicated training and outreach program which provides training to both naïve and expert users alike.

Another important learning concerns the combination of open data with sensitive data and how to develop tools that can bring these two worlds together. Work is currently underway in AURIN in developing a demonstration project which exemplifies how unit record health data can be interrogated through enabling queries to be performed at the unit record level but the results are only given to the end user in acceptable aggregate geographies. The results of this work are to be reported in a forthcoming research paper.

An important next step is the internationalisation of such a comprehensive workbench. In many ways, it is the first of its kind in the world. It has been developed on open source software and there is an opportunity to develop an international developer and end user community around the workbench, or components thereof. Finally, there is work in progress to extend the training and outreach program through a massive online open course (MOOC) where the workbench will be used as an exemplar on how to access evidenced based data and tool to understand how to plan for an ageing population.

Acknowledgments The authors would like to thank the AURIN technical team who are building the AURIN infrastructure and external collaborators including Professor Richard Klosterman for working with AURIN in migrating his desktop tool into the workbench and Professor Billie Giles-Corti and her team for developing the walkability tool. The authors also acknowledge the Australian Government who are funding the AURIN project through the NCRIS Program.

References

Barton, J., Goldie, X., & Pettit, C. (2015). Introducing a usability framework to support urban information discovery and analytics. *Journal of Spatial Science.* doi:10.1080/14498596.2015.997315.

Badland, H., White, M., Macaulay, G., Eagleson, S., Mavoa, S., Pettit, C. J., & Giles-Corti, B. (2013). Using simple agent-based modeling to inform and enhance neighborhood walkability. *International Journal of Health Geographics, 12*, 58.

Batty, M. (1995). Planning support systems and the new logic of computation. *Regional Development Dialogue, 16*, 1–17.

Delaney, P., & Pettit, C. J. (2014). Urban data hubs supporting smart cities. In S. Winter & C. Rizos (Eds.), *Research@Locate'14* (pp 13–25). Canberra, Australia, April 07–09, 2014. Published at http://ceur-ws.org.

Frank, L. D., Sallis, J. F., Saelens, B. E., Leary, L., Cain, K., Conway, T. L., & Hess, P. M. (2010). The development of a walkability index: Application to the neighborhood quality of life study. *British Journal of Sports Medicine, 44*, 924–933.

Giles-Corti, B., Macaulay, G., Middleton, N., Boruff, B., Bull, F., Butterworth, I., Badland, H., Mavoa, S., Roberts, R. & Christian, H. (2014). Developing a research and practice tool to measure walkability: a demonstration project. *Health Promotion Journal of Australia*, 25(3), 160–166. doi:dx.doi.org/10.1071/HE14050.

Klosterman, R. E. (1999). The What if? Collaborative planning support system. *Environment and Planning B: Planning and Design, 26*, 393–408.

McCormack, G. R., Giles-Corti, B., & Bulsara, M. (2008). The relationship between destination proximity, destination mix and physical activity behaviors. *Preventive Medicine, 46*(1), 33–40.

Pettit, C. J. (2005). Use of a collaborative GIS-based planning support system to assist in formulating a sustainable-development scenario for Hervey Bay, Australia. *Environment and Planning B, 32*(4), 523–546.

Pettit, C., & Pullar, D. (1999). An integrated planning tool based upon multiple criteria evaluation of spatial information. *Computers, Environment and Urban Systems, 23*, 339–357.

Pettit, C., Stimson, R., Tomko, M., & Sinnott, R. (2012). Building an e-infrastructure to support urban and built environment research in Australia: A Lens-centric view. In *Proceedings of the Spatial and Surveying Sciences Conference 2013*. Australia: Canberra.

Pettit, C. J., Klosterman, R. E., Nino-Ruiz, M., Widjaja, I., Tomko, M., Sinnott, R. (2013a). The online What if? Planning support system. In S. Geertman & J. Stillwell (Eds.), *Planning support systems for sustainable urban development* (pp. 349–362). Berlin: Springer Publishers.

Pettit, C. J., Tomko, M., Sinnott, R., Stimson, R. (2013b). Building an e-infrastructure to support urban research in Australia: A lens centred view. *At Surveying and Spatial Sciences Conference 2013*, April 15–19, 2013, Canberra, Australia.

Pettit, C. J., Stimson, R., Nino-Ruiz, M., Morandini, L., Widjaja, I., Delaney, P., Tomko, M., Sinnott, R., Randolph, B., Kvan, T. (2014). Supporting urban informatics through a big data analytics online workbench. *At National Science Foundation Workshop, Big Data and Urban Informatics Conference*. Chicago: University of Illinios, August 11–12, 2014.

Pettit, C. J., Klosterman, R. E., Delaney, P., Whitehead, A. L., Kujala, H., Bromage, A., Nino-Ruiz, M. (2015). The online what if? Planning support system: A land suitability application in Western Australia. *Applied Spatial Analysis and Policy*. doi:10.1007/s12061-015-9133-7.

R Development Core Team (2008). R: A language and environment for statistical computing. R Foundation for Statistical Computing, Vienna, Austria. ISBN 3-900051-07-0, URL http://www.R-project.org.

Sinnott, R. O., Bayliss, C., Bromage, A., Galang, G., Grazioli, G., Greenwood, P., Macauley, G., Mannix, D., Morandini, L., Nino-Ruiz, M., Pettit, C., Tomko, M., Sarwar, M., Stimson, R., Voorsluys, W., Widjaj, I. (2014). The Australia urban research gateway. *Concurrency and Computation: Practice and Experience*. doi:10.1002/cpe.3282.

The State of Victoria. (2014). *Plan Melbourne: Metropolitan planning strategy*. Available from: http://www.planmelbourne.vic.gov.au/. Accessed on January 19, 2015.

Vonk, G., Geertman, S., & Schot, P. (2005). Bottlenecks blocking widespread usage of planning support systems. *Environment and Planning A, 37*, 909–924.

Chapter 14
Smart Governance, Collaborative Planning and Planning Support Systems: A Fruitful Triangle?

Yanliu Lin and Stan Geertman

Abstract The scientific literature on smart cities has focused on innovative developments in information and communication technology (ICT) and on its consequences for urban life and policy making. In line with these, Batty et al. (Eur Phys J Spec Top 214:481–518, 2012) state that new technological developments are providing for new types of analysis, public participation and multi-actor collaboration, blurring the boundaries between smart cities and urban planning. We take this statement as a starting point for our discussion and put attention on the interplay between new ICT, smart cities and spatial planning. We focus in particular on the triangular relationship between smart governance as one of the areas of smart cities (Giffinger et al. in Ranking of European medium-sized cities, 2007), collaborative planning as a present form of spatial planning, and planning support systems (PSS) as a specific form of ICT dedicated to planning tasks. Collaborative planning is characterized by consensus building among distinctive stakeholders in participatory processes. Smart governance adds ICT-related components (e.g. efficient communication; data exchange) to the concept of collaborative planning. Finally, PSS involves the creation and use of tools to support professional planners' tasks, including introducing relevant geoinformation and facilitating participation. To illustrate this triangular relationship, we examine some practical case studies from China, Finland, and the USA that suggest how web-based and model-based PSS can fulfil a supportive role, to realize smart governance in spatial planning by promoting effective communication and collaboration between various actors, and by strengthening the transparency of the decision-making process. We conclude that collaborative planning can become a form of smart governance under two basic conditions, namely of recent developments in the applicable ICT and an appropriate institutional design.

Y. Lin (✉) · S. Geertman
Department of Human Geography and Planning, Utrecht University, Utrecht, The Netherlands
e-mail: y.lin@uu.nl

S. Geertman
e-mail: S.C.M.Geertman@uu.nl

S. Geertman et al. (eds.), *Planning Support Systems and Smart Cities*,
Lecture Notes in Geoinformation and Cartography,
DOI 10.1007/978-3-319-18368-8_14

1 Introduction

In cities around the world, residents and government officials are increasingly inundated with the concept of the 'smart city' as they attempt to make decisions about how to utilize information and communication technologies (ICT) (Hollands 2008; Goodspeed 2015). The main focus of the concept of smart city is on ICT infrastructure, although increasingly research also examines the role of human capital, social and relational capital, and environmental issues as important drivers of sustainable urban growth (Caragliu et al. 2009). Batty et al. (2012) argue that new technological developments are providing us with new ways of public participation and collaboration in the decision-making processes, blurring the boundaries between smart cities and urban planning. They point out that the new technologies provide ways in which citizen groups, governments, businesses and various agencies can interact in augmenting their understanding of the city and providing essential engagement in the design and planning process (Batty et al. 2012).

This chapter explores the potential connections between collaborative forms of planning, smart governance and PSS, the latter as a component of the wider group of ICT dedicated to planning. However, until now, the existing literature has paid very little attention to this interrelationship. Remarkably, the focus still is put foremost on the achievements of upcoming ICT, leaving behind its more precise relationship to smart governance and with spatial or collaborative planning. It is the purpose of this chapter to shed more light on the triparted relationship between smart governance as one of the areas of smart cities (Giffinger et al. 2007), collaborative planning as a present form of spatial planning, and PSS as a specific form of ICT dedicated to planning.

After a brief discussion of these three areas, the chapter examines some case studies in China, Finland, and the USA. We find that the case studies have some unexpected similarities, although they are different in terms of political, social and institutional contexts. These similarities include the specific relationship between collaborative planning and smart governance. Under certain conditions, collaborative planning can become a form of smart governance. The first condition is appropriate institutional design, which encourages collaborative and inclusionary consensus building and makes clear ground rules to ensure the transparency of the planning process. This second condition is the introduction of ICT in general, and PSS in particular, in collaborative planning, which can help to form the link with smart governance. In the final part, the main findings are discussed, where we call for greater investigation of the use of ICT to foster connections between smart governance and planning.

2 Smart Governance, Collaborative Planning and PSS

Smart governance is a key component of a smart city. A city is smart when "investments in human and social capital and traditional (transport) and modern (ICT) communication infrastructure fuel sustainable economic growth and a high quality of life, with a wise management of natural resources, through participatory governance" (Caragliu et al. 2009, p. 70). The inclusion of participatory governance in this definition triggered our understanding of the links between the concepts of smart city and smart governance. According to Giffinger et al. (2007), there are six main areas under the umbrella of smart city, namely smart living, smart governance, smart economy, smart environment, smart people, and smart mobility. Therein, smart governance entails participation in decision-making processes, transparency of governance systems, the availability of public services and the quality of political strategies.

Governance is the process of interaction and decision-making among the stakeholders involved in a collective issue (Hufty 2011). Smart governance involves the organizational and institutional embedding of ICT into urban governance (Komninos 2009). On the one hand, big data, data warehousing and monitoring tools are used to strengthen the information base of the government (Leydesdorff and Deakin 2011; Batty et al. 2012). On the other hand, social media and the Internet are used to enable various actors to collaborate (Hoon et al. 2013). Chourabi et al. (2012) summarize a range of characteristics of smart governance, including effective collaboration, leadership, public participation, transparency in decision making, public-private partnerships, efficient communication, data exchange and service and application integration. Since smart governance focuses on enabling—with the help of ICT—the participation and collaboration of various actors in the decision-making process and on supporting transparency in governance, it has an obvious connection with collaborative planning.

The notion of collaborative planning, a term popularized by Healey (1997), has become the dominant theoretical approach for understanding spatial planning processes. Spatial planning can be understood as the process of decision making in a society on the use of land and supporting infrastructure and facilities, based on assessing and balancing competing demands (Nuissl and Heinrichs 2011). The label "collaborative planning" has been used to describe a range of inclusive and participative governance processes in spatial planning over the last few years. According to Healey (1997), collaborative planning can be viewed as an activity which generates networks across society by engaging diverse stakeholders in consultation. The involvement and interaction of three spheres (the economy, civil society, the state) in governance process can generate ways of thinking and acting that may be carried forward into new relationships and new forms of governance (Healey 1997). To the traditional focus on deliberation, Healey argues for the importance of considering institutional structures that shape decision-making (Healey 2003). According to Innes and Booher (1999), consensus building among stakeholders can be considered a more systematic and sophisticated version of collaborative planning. It refers to an array of practices in which stakeholders, selected to represent different interests, come together for face-to-face, long-term dialogue to address a policy issue of common concern (Innes 2004). However, collaborative planning processes may be more practically understood as fully interactive processes from which the outcomes may be fragile, incomplete and contestable (Healey 2003; Cheng 2013). Ansell and Gash (2008) identify several critical variables that determine the success of collaborative planning, including the prior history of conflict or cooperation, the incentives for stakeholders to participate, power and resource imbalances, leadership and institutional design.

Collaborative planning is often treated as a desirable mode of governance because it provides the necessary flexibility that was lacking in the traditional blueprint planning approaches. However, it is criticized as well: "Ironically the progressive credentials of spatial planning in terms of consensus building, policy integration, and the search for 'win–win–win' solutions may have helped script out oppositional voices" (Allmendinger and Haughton 2010, p. 803). Such an exclusionary effect will reduce the democratic legitimacy of the planning process. Democratic legitimacy may also be undermined by the incorporation of strong private interest groups into collaborative planning process as this may threaten attention to the public interest (Hartmann and Barrie 2012). Another challenge is identified by Rydin, who acknowledges that communication and collaboration are critical characteristics of planning, but who is also concerned about the validity of different knowledge claims posed by different stakeholders in the planning process (Rydin 2007). Conflicts are not automatically solved simply by bringing all stakeholders to a roundtable and expecting them to discuss the problem until it is resolved (Billé 2008; Davy 1997). Collaborative planning needs institutions to make its planning results robust and enforceable (Huxley 2000, p. 371). Healey (1997) elaborates on the notion of institutional design and identifies two distinctive levels. The first level is the soft infrastructure of individual efforts in planning, with a focus on stakes, arenas, routines and styles. The second level is the hard infrastructure. It is related to the design of the planning system, which needs to be critiqued and invented by a careful assessment of the constitution of rights and duties, of resource allocation mechanisms, of performance criteria and of competencies. Based on the distinction and the observation that consensus-building processes are not always effective and inclusive, Healey (1997) concludes that a proper institutional design is needed that has the capacity to encourage collaborative and inclusionary consensus building.

In recent years, with the development of ICT, the internet and social media, new forms of collaborative planning have emerged. Rather than face-to-face stakeholder meetings, online communication and interaction between different online actors (e.g. civic organizations and citizens) become crucial for consensus building (Cheng 2013; Deng et al. 2014). Batty et al. (2012, p. 19) identify in the concept of smart governance at least four modes in which ICT supports interactive collaboration and participation: "first, portals and other access points to [add] useful information about any aspect of routine living and working in cities, second, ways in which citizens can interact with software that enables them to learn more about the city by engaging with other users online and actually creatively manipulating information, third, citizens engaging with crowd-sourced systems in which they are responding to queries and uploading information, and fourth, fully fledged decision-support systems which enable citizens to engage in actual design and planning itself in terms of the future city ".

From this it can be seen that ICT can play a key role in linking collaborative planning and smart governance. According to Epp (2012), ICT, including PSS, can play an important role in supporting higher levels of public participation. Therein, PSS are envisioned as a subset of geo-information-based instruments that incorporate a suite of components (including data, information, GIS, statistical tools, and models) that collectively support a unique professional planning task (Geertman 2006). Epp (2012) points out that new types of GIS, Web 2.0 and mobile phone apps, as well as their precursors, rely on user generated content and are community based and are designed to harness and communicate a collective wisdom. For instance, web-based PSS can allow citizens, as either individuals or members of civil society organizations, to participate in public debates, to express their opinions and to hear about or develop new solutions to urban problems (Poplin et al. 2013). The application of PSS in collaborative planning can be seen to assist stakeholder participation and can be considered an alternative method of dealing with the wicked problems of planning practice (Goodspeed 2015). This coincides with the reflections by Klosterman (1997) in which he considers the increasing application of PSS in practice as continuing the planning trends from applied science to communication and collaboration, including broader concerns with intelligence and collective design. In fact, PSS are going more and more online. Online PSS can be used by citizens in social media on their smartphones, and make its wider application possible. Notwithstanding these positive signs, it should be noted that the actual application of PSS in planning practice is still lagging behind its potential (see Vonk et al. 2005; Brommelstroet 2011; Pelzer et al. 2014).

In retrospect, the well-known notion of collaborative planning, with its focus on public participation and on the collaboration of distinctive actors in decision-making processes, remains at the heart of many planning discussions. The related concept of smart governance puts its focus on the ICT-supported participation and collaboration of various actors in the decision-making process and on supporting transparency in governance processes. Furthermore, smart governance argues for the incorporation of a wide variety of ICT instruments (e.g. Batty et al. 2012), while collaborative planning highlights the need for proper institutional design to encourage collaborative and inclusionary consensus building (Healey 1997; Ansell and Gash 2008). Within the mentioned diversity of ICT instruments, the group of so-called PSS stands out and is expected to fulfil a distinctive supporting role in the establishment of smart governance. Given the debate about how to improve the adoption and performance of PSS, an examination of the relationship among collaborative planning, smart governance and PSS seems especially needed to inform developments in all three areas (see Fig. 1).

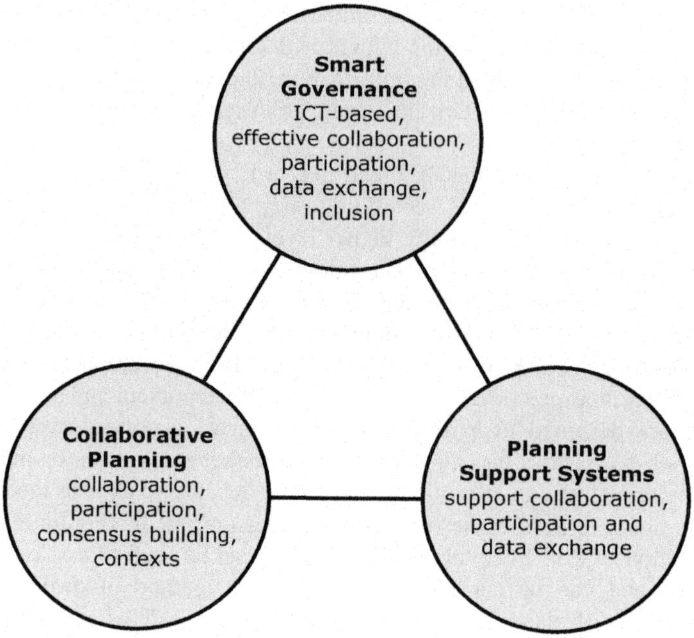

Fig. 1 The triangular relationship between smart governance, collaborative planning and planning support systems

In the next section, we illustrate our case through cases taken from three continents that differ substantially in political and cultural context and in spatial planning tradition.

3 Case Studies

To illustrate our argument, we present examples illustrating this relationship from three case countries—China, Finland, and the USA—that differ substantially in political and cultural context and in their urban planning traditions. Collaborative planning has been advocated in Finland and the USA since the 1970s but has emerged much more recently in China, where it exerts an increasing influence on both planning practice and theoretical planning research. In particular, the rise of collaborative planning in China has a strong relationship with the extremely fast development and implementation of the concept of smart cities. In Finland, web-based collaborative planning and e-participation have been promoted by the extensive investment in broadband Internet, which has resulted in the highest broadband coverage in Europe. Finally, a growing number of cities in the USA are offering smart city infrastructures that blur the boundary between collaborative planning and smart cities.

The data for this section were mainly collected by searching the Internet and from semi-structured interviews with planners, PSS experts and citizens. It should be noted that these data are not the result of an extensive and elaborate case study research project. As a consequence, we are using this information simply as a means of exploring the relationship between collaborative planning, smart governance and ICT including PSS, in order to challenge existing thinking and guide further research on the topic.

3.1 China

In the last few years, the smart city concept has been widely advocated in many Chinese cities, which has resulted in strategic cooperation with big ICT firms like IBM and Cisco. The development of smart cities in China has largely focused on technological issues and improving ICT infrastructure. Collaborative planning is emerging together with the rapid development of the Internet and social media, which act as new platforms for public participation and communication among various actors.

For centuries, the institutional system of government in China was hierarchical and characterized by top–down approaches. In 2008, however, the Urban–Rural Planning Law was introduced, clearly stating the basic requirements for public participation in the planning process. As a consequence, Chinese planners needed to develop their skills in communicating with the public (Sun and Yin 2008). Recently, the rapid development of social media and the Internet have offered new participatory platforms for marginal social groups, citizens, and civic organizations to express their interests and take collective action (Cheng 2013; Deng et al. 2014). For example, citizens in Nanjing used microblogs and city forums to oppose the felling of old trees to make way for a new subway project and forced the local government to communicate with them. The Internet and microblogs became the communicative and cooperative platforms where consensus was built concerning the conservation of the old trees and the adjustment of the project (Yan and Zhu 2013). In other cases, the combination of web-based PSS, social media and the Internet has recently impacted on several planning practices in China (e.g. Xu 2013).

The "Bell and Drum Tower" neighborhood regeneration project in Beijing illustrates this development. In 2010, the city initiated a project to regenerate the Bell and Drum Tower neighborhood. The project would lead to the demolishment of a number of valuable buildings along several historical streets in the city center of Beijing, near this famous, historic tower. However, the plans aroused a wide range of criticism from experts, planners, and citizens. A civic group (the Bell and Drum Tower Neighborhood Team) comprising experts, planners, and students was established and created an account on Weibo, China's Twitter-like microblogging service. This civic group took photos of historical buildings, mapped the neighborhood, interviewed local residents, and posted research reports on the Internet. The group also asked a PSS expert to create a web-based PSS called "The protection of the north axis: community participation and communication website for Beijing's 'Bell and Drum Tower' neighborhood regeneration project" (Fig. 2). The web-based PSS shows the map and pictures of the neighborhood and offers users the possibility to add comments to specific locations. Besides this, many citizens criticized the original project plans via the local government's Weibo and via the civic group's Weibo. As such, the local government was faced with popular pressure, and as a result revised the project (Xu 2013). However, with reference to the discussion above concerning power imbalances in collaborative planning, although the local government revised the project, it did not fully consider the opinions of the participants and still demolished some of the historical buildings. Experts mapped the demolished buildings in the neighborhood and uploaded them in the civic group's Weibo. They pointed out that while these actions were legal, they produced undesirable results. This case illustrates the hierarchical Chinese planning system, which lacks an effective mechanism to guarantee public participation and require decision-makers to take input into account in the decision-making process (Lin et al. 2014).

This brief example illustrates how planning practices in China are evolving. New forms of collaborative planning are emerging, fueled by the rapid development of the Internet and social media, as well as by the development of new types of PSS. They are also facilitated by the recent development of smart cities which largely improve ICT infrastructure and support ICT-based public participation and governance in China. In other words, as a consequence of new technologies, projects like the Bell and Drum Tower neighborhood regeneration project feature a combinations of elements typically viewed as unrelated in the existing literature: collaborative planning through the Internet and social media, the emergence of new types of PSS, and the development of smart cities (particular smart governance).

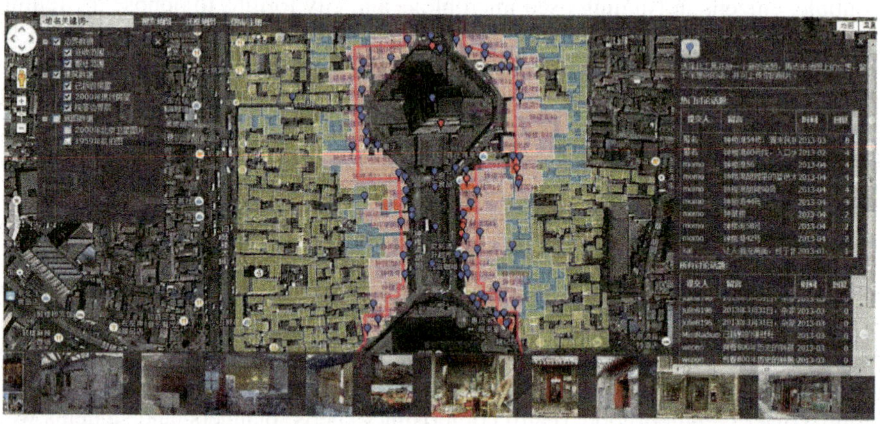

Fig. 2 A web-based PSS: protection of the north axis: community participation and discussion website for Beijing's "Bell and Drum Tower" neighborhood regeneration project (http://archlabs.hnu.cn/bj/index.php)

In summary, the case study of Beijing's Bell and Drum Tower neighborhood regeneration project suggests that there is a need to have a close look at the relationship between three conceptual categories (collaborative planning, smart governance and PSS) which are separated in literature, and analyze how institutions should be adapted for this new planning context.

3.2 Finland

Within Europe, Finland possesses a very high level of broadband coverage, and partly as a result is noted for ICT-based collaborative planning. As of 2011, 98 % of homes in Finland have access to basic broadband and almost 68 % can have access to superfast broadband, also known as for Next Generation Access (European Commission 2011). This is the highest total superfast coverage in Europe and one of the highest levels for rural areas in Europe. The extensive coverage of broadband Internet in Finland makes the web a natural platform for public participation and communication between various actors in the planning process. As a parliamentary democracy, citizens can vote and run in parliamentary, municipal and presidential elections. Collaboration in Finland is supported by European Union (EU) directives, and the Finnish Building and Land Use Act (1999) which aims to ensure wide participation and support open and high-quality planning decisions and processes.

Nuojua et al. (2008) examined web-based participation methods by taking Pyhäjärvi (a town in the south of Northern Ostrobothnia region, Finland) as a case study. A web mapping application called WebMapMedia (WMM), which was based on Google Maps, was developed to help planners to acquire local knowledge from the citizens in the form of comments and pictures. WMM can be considered a virtual combination of two traditional participation methods: (1) PhotoVoice, a technique where which citizens identify, represent and enhance their community through photography (Wang and Burris 1997); and (2) sticker-map method that enables citizens to mark locations with personal significance by placing colored symbols on the map. In the WMM tool, when citizens click a marker on the map, a bubble opens with a thumbnail of the picture and hyperlink to the discussion about the place. Moreover, RSS feeds are provided by WMM for the planners in order to monitor the continuous flow of local data. Nuojua et al. (2008) found that such a web-based application was especially suitable for acquiring local knowledge and that it was an easy and inexpensive way to enlarge and diversify the group of participants. However, with reference to the discussion above on the pitfalls of collaborative planning, Nuojua et al. (2008) also identified a big challenge for web-based participatory planning was combining local knowledge with the knowledge of planning professionals. The interviews with the planners showed that the proposals made by citizens were often considered to be too general to be useful in the specific planning work (Nuojua et al. 2008). In addition, the planners appeared to be

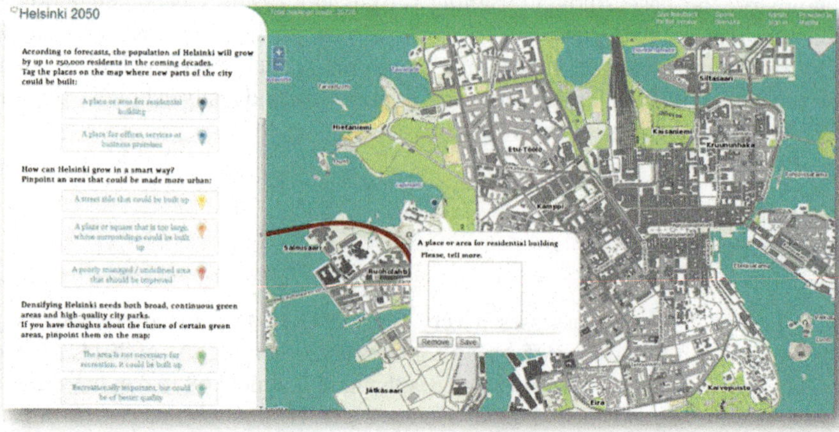

Fig. 3 SoftGIS tool created for Helsinki 2050 master planning process (Kahila et al. 2014)

unwilling to participate directly in the discussions on the WMM forum. In other words, this web-based collaborative planning hasn't led to effective and equitable collaboration between citizens and planners. On the one hand, clear ground rules may be needed to facilitate more equitable relationships and conversation between planners and citizens. On the other hand, the improvement of the PSS from a technological perspective (such as the visualization of the planning sketches) may be required to facilitate truly two-way communication between the planners and local people, and further integrate the professional knowledge of the planners and local knowledge of citizens.

As another example from Finland, professionals have adopted the PSS SoftGIS to facilitate online participation (Kahila and Kyttä 2009). It is an example of a Public Participation GIS (PPGIS) with its main purpose being to enhance participation and collaboration by allowing residents to share their knowledge about their living environment with that of urban planners and researchers. SoftGIS methods build a bridge between residents and urban planners by allowing residents to produce and distribute local knowledge through user-friendly internet-based applications and not by attending in-person events. Citizens can use SoftGIS tools to fill in online questionnaires about topics like everyday modes of transport, and preferences for future living environment, as well as make planning proposals by pinpointing appropriate locations for new building sites or identifying areas in need of improvement (see Fig. 3). Kahila et al. (2014) built upon and examined the application of these SoftGIS methods in supporting the Helsinki 2050 master planning process, documenting how 3745 citizens participated in this planning process through SoftGIS.

3.3 USA

There is a growing number of cities in the USA making efforts to offer smart and more efficient city infrastructures (Michelle 2014). For instance, New York City has initiated ‘City 24/7’, an interactive platform that integrates information from government programs, local businesses, and the city’s citizens to provide knowledge to anyone, anywhere (Frazier and Touchet 2012).

In addition to smart infrastructure, some cities are also using advanced PSS tools. Calthorpe Associates (2012) has created two novel PSS, which are mainly used by government agencies within specific projects to support collaboration between government agencies and other stakeholders in the planning process. These two tools for scenario development and modeling, named RapidFire and Urban Footprint, are able to express the varying impacts of developments and infrastructure investments on a variety of spatial scales (http://www.calthorpe.com/scenario_modeling_tools). The first one, RapidFire, is a user-friendly, spreadsheet-based tool that is used to produce and evaluate scenarios at different scale levels (statewide, regional, county, and jurisdiction-level). It emerged out of the need for a comprehensive modeling tool that can inform state, regional and local agencies and policy makers when evaluating climate, land use and infrastructure investments. In the “Envision Bay Area” project, RapidFire is used to build regional scenarios that depict the land use and transportation choices the future of the region faces, and the consequences of those choices for a range of critical indicators. With the support of RapidFire, residents and community leaders can make informed decisions about building activities that will shape the future environment. The second open-source geo-spatial model mentioned, Urban Footprint, was first developed and deployed across California’s major regions as part of the Vision California process (Fig. 4). It serves as a scenario development and analysis system, which includes powerful data organization, scenario building and analytical capabilities. This model is designed to work via a web-based interface, allowing for detailed mapping and

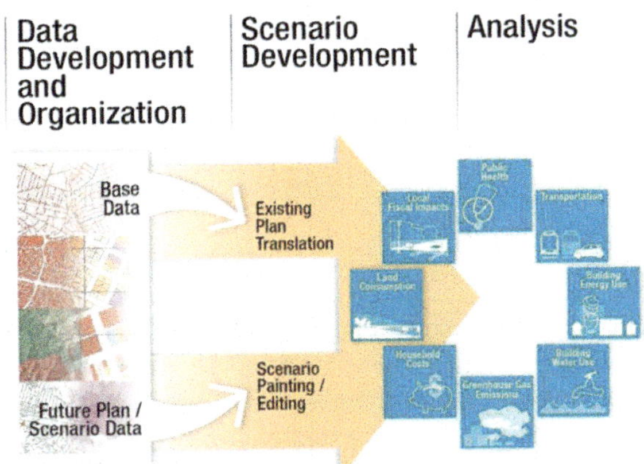

Fig. 4 UrbanFootprint painting tool screenshot (*Source* http://www.calthorpe.com/scenario_modeling_tools)

"painting" of land use and transport futures and for working at regional, sub-regional and local planning levels. Urban Footprint includes the ability to analyze scenarios based on a full range of fiscal, environmental and public health metrics. However, the previous version of Urban Footprint was mainly operated by experts and difficult to be used by other stakeholders who lacks technical knowledge. Calthorpe Associates (2012) completed the first fully-operational version of the model in 2012 and is now working to advance the model for use by a broad range of public agencies and organizations.

Besides model-based PSS for collaboration between government agencies and other stakeholders, there is also web-based PSS for supporting public participation. One of the case studies is CitySourced (http://www.citysourced.com/), which is a real time mobile civic engagement platform. CitySourced can be downloaded to mobile phones (e.g. iPhone, Android, Windows Phone7 and BlackBerry). The mobile phone applications allow data from a range of sources, including social networking sites, to be tied to the physical world, allowing devices such as mobile phones to instantly overlay information about a location or object (Epp 2012). The primary applicability of CitySourced is the crowd-sourcing of place-based issue identification. Through having CitySourced on their mobile phones, citizens can identify civic issues and report them to city hall for a quick resolution. For instance, citizens can report a problem with a sidewalk by pinpointing the location and adding text to explain the issue. CitySourced supports "citizens engaging with crowd-sourced systems in which they are responding to queries and uploading information" (Batty et al. 2012, p.19). With the help of this app, local government can identify problems in the city in a timely fashion and find prompt solutions, improving accountability to the citizens. Therefore, CitySourced has supported collaborative process and promoted smart governance by facilitating a more effective exchange of information between citizens and the government and can become a cooperative platform for real action. However, to date the focus has been on automating customer service requests to municipalities, although the software can also be applied to record planning issues (Epp 2012). The effort and support of the government seems necessary to further use and integrate the outcomes of CitySourced in specific planning tasks. A remaining question is how to integrate the knowledge of planners with the input of citizens obtained through Citysourced. This problem is related to the improvement of PSS and quite similar to that of the mentioned Web mapping application (WMM) in Finland.

4 Discussion

The case studies in China, Finland and the USA illustrate interesting relationships between collaborative planning, smart governance and ICT including PSS. Although they differ in term of context, they show some interesting similarities. In general, PSS play an important role in supporting collaborative planning and facilitating smart governance. However, collaborative planning cannot become smart governance without appropriate institutional design and ICT supports (PSS, social media, smartphone, ICT infrastructure, etc.).

The Chinese case studies show that collaborative planning is emerging together with the rapid development of smart cities and ICT including PSS. The recent development of smart cities has largely improved overall ICT infrastructure. The Internet and social media have become two of the most influential collaborative platforms through which urban planning conflicts are anticipated to be resolved (Cheng 2013; Deng et al. 2014). Grassroots participants have had an impact on planning practices by using the Internet and enlightened the public's engagement in planning participation (Cheng 2013). The emergence of web-based PSS, together with the rapid development of ICT and social networking sites, assist community participation and the communication between the local government, civic organizations and citizens, with the Bell and Drum Tower neighborhood regeneration project being a good example. Therefore, ICT can improve transparency in decision-making processes and can lead to more efficient communication between citizens, planners and public authorities.

However, the case study shows that an appropriate institutional design is required too. The associated question is that: how do you develop an effective mechanism to guarantee an appropriate synthesis of bottom–up and top–down approaches to public participation, to improve transparency in decision-making processes, and to take into account the outcomes of such participation and processes in the actual decision-making process? We believe the Chinese case studies show that in addition to new PSS, new forms of collaborative planning and smart governance require appropriate "institutional design" (Healey 1997) that ensures transparency and collaboration.

ICT-based collaborative planning in Finland has been supported by a very high standard of broadband coverage, the development of web-based PSS, and the widespread use of social media and smartphones. Furthermore, formal national regulations ensure wide participation in planning decisions and processes. In the case study of Pyhäjärvi, a web mapping application was developed to help planners to acquire local knowledge from the citizens in the form of comments and pictures. However, this web-based collaborative planning hasn't led to effective and equitable collaboration between citizens and planners. To achieve smart governance, clear ground rules which is a key component of institutional design (Ansell and Gash 2008) may be needed to facilitate more equitable relationships between planners and citizens, and the improvement of the PSS from a technological perspective is required to visualize the planning sketches and thus facilitate a truly two-way communication. In response to this last mentioned requirement, a new type of web-based PSS called SoftGIS was recently developed and applied in supporting online public participation in the Helsinki 2050 master planning process. Toolsets that could easily be used on smartphones were designed to support citizen participation and their outcomes were then used by planners to (re-)formulate the plan. This form of collaborative planning which is based on SoftGIS is much closer to smart governance, since it features transparency, effective uses the outcomes of public participation in decision-making, and leads to effective cooperation between citizens, planners and other stakeholders.

The case studies in the USA show that the development of improved PSS are on its way to support the effective collaboration among government agencies and institution, and crowdsourcing-based PSS has become a new cooperative platform between the government and citizens. RapidFire and Urban Footprint are two scenario building tools that show to support the communication and collaboration between government agencies, organizations and other stakeholders. These tools contribute to the communication and collaboration between different key stakeholders through scenario building, modelling and visualizing, and thereby linking collaborative planning with smart governance. The form of collaborative planning based on this model-based PSS however is difficult to be used by stakeholders who were lacking the technical knowledge. How to advance the model for use by a broad range of public agencies and organizations is a key concern in achieving smart governance which requires inclusive and effective cooperation. CitySourced in the USA supports citizens to engage with crowd-sourced systems in which they identify civic issues in the city and help the local government to become aware of problems and improve its accountability. Although CitySourced shows the potential to promote inclusive collaborative planning and smart governance by supporting effective information exchange between citizens and the government and acting as a cooperative platform for real action, its focus has still been on automating customer service requests to municipalities. The further application of this type of web-based PSS in assisting collaborative processes and smart governance requires not only continued efforts by cities, but also technical changes to enable the integration of participation and planners' expertise.

5 Conclusions

In the introduction to this chapter, we referred to the argument by Batty et al. (2012) that technical developments are eroding the distinction between planning and smart cities. According to these authors, new technologies provide ways in which citizen groups, governments, businesses and other stakeholders can interact in order to augment their understanding of the city and provide essential engagement in the planning process. Although, this all sounds exciting, many of these authors fail to elaborate more closely on the relationships between smart cities, planning and ICT due to a lack of empirical evidences. This chapter is a first attempt to take a closer look at the triangular relationships between smart governance, collaborative planning and ICT, in particular by studying case studies in China, Finland and USA. We find that the case studies have some similar aspects, although they have different political, institutional and social contexts. The similarities include the special relationship between collaborative planning and smart governance. Under certain conditions, collaborative planning can become a form of smart governance. These conditions are twofold.

First, the introduction of ICT in general, and PSS in particular, to collaborative planning can help form a link with smart governance. The Internet, mobile communication, cartography and visualization are converging rapidly to develop new means of collaboration and of displaying geographic information (Epp 2012). New forms of ICT (portals, crowd-source websites, online decision support systems, etc.) can assist in formulating new forms of participation that can help with developing smart cities (Batty et al. 2012). Web-based PSS, in particular, can play a crucial role in assisting with choice processes and enabling citizens to negotiate and engage in trade-offs with traditional power holders and in assisting the cooperation between various stakeholders (Epp 2012). Besides web-based PSS for supporting public participation, the case studies also showed that there are emerging model-based PSS for scenario building and spatial modelling. In the USA, the successful application of RapidFire and Urban Footprint is largely due to the integration of resources and data exchange at all levels and the collaboration between government agencies and institutions. This model-based PSS contributes to the communication and collaboration between key stakeholders through scenario building, modelling and visualizing, and thereby linking collaborative planning with smart governance.

Second, the institutional design of the specific planning practice needs to have the capacity to encourage collaborative and inclusionary consensus building, and has to ensure the transparency of the planning process. It should be pointed out that the introduction of ICT/PSS alone is not sufficient to arrive at smart governance. Institutional design is another important factor that helps to build this relationship. Collaboration and public participation in the planning process in Finland have been supported in the long term by policies and regulations at EU and national levels: "the Internet offers new possibilities for involving citizens in political decision-making through e-democracy" (Christensen 2012, p. 1). However, the case study of web-based participation methods in Pyhäjärvi shows that clear ground rules may be needed to facilitate more equitable relationships and conversation between planners and citizens. In the Chinese context, further changes to institutions seem necessary to arrive at smart governance. Effective collaborative governance requires a commitment to a positive strategy of empowerment and the representation of weaker or disadvantaged stakeholders (Ansell and Gash 2008).

In summary, drawing on our discussion of the literature and a reflection on the case material, it can be seen that at least two preconditions should first be fulfilled in order to arrive at smart governance. On the one hand, recent developments in ICT, such as the growing and extensive coverage of broadband Internet, expanding use of social media, widespread use of smartphone apps, growing possibilities for data exchange, and expanding cloud computing and online GIS, are important preconditions for smart governance. In addition, an appropriate institutional design is needed in order to arrive at truly smart governance. Further research can shed more light on the triangular relationship between smart governance, collaborative planning and ICT/PSS to enhance the potential role of the new technologies to arrive at smart governance and be of added value to planning practice.

Acknowledgments This project is supported by the State Key Laboratory of Subtropical Building Science, South China University of Technology (Refs: 2015ZC08, 2015ZB06). We thank the anonymous reviewers of this CUPUM book for their very helpful comments.

References

Allmendinger, P., & Haughton, G. (2010). Spatial planning, devolution, and new planning spaces. *Environment and Planning C: Government and Policy, 28*, 803–818.

Ansell, C., & Gash, A. (2008). Collaborative governance in theory and practice. *Journal of Public Administration Research and Theory, 18*(4), 543–571. doi:10.1093/jopart/mum032

Batty, M., Axhausen, K. W., Giannotti, F., Pozdnoukhov, A., Bazzani, A., Wachowics, M., et al. (2012). Smart cities of the future. *The European Physical Journal Special Topics, 214*(1), 481–518.

Billé, R. (2008). Integrated coastal zone management: Four entrenched illusions. *Surveys and Perspectives Integrating Environment and Society, 1*(2), 75–86.

Brommelstroet, T. M. (2011). Transparency, flexibility, simplicity: From buzzwords to strategies for real PSS improvement. *Computers, Environment and Urban Systems, 36*(1), 96–104.

Calthorpe Associates. (2012). Urban footprint technical summary model version 1.0. http://www.calthorpe.com/scenario_modeling_tools. Accessed 20 Jan 2014.

Caragliu, A., Del Bo, C., & Nijkamp, P. (2009). *Smart cities in Europe*. Paper presented at the 3rd central European conference in regional science—CERS.

Cheng, Y. (2013). Collaborative planning in the network: Consensus seeking in urban planning issues on the internet—the case of China. *Planning Theory, 12*(4), 351–368.

Christensen, H. S. (2012). E-democracy in Finnish municipalities. City of Turku, Urban Research and Statistics Unit. http://www.academia.edu/2162952/E-democracy_in_Finnish_municipalities.

Chourabi, H., Nam, T., Walker, S., Gil-Garcia, J. R., Mellouli, S., Nahon, K., et al. (2012). Understanding smart cities: An integrative framework. In *Proceedings of 2012 45th Hawaii International Conference on System Sciences*. IEEE (pp. 2289–2297).

Davy, B. (1997). *Essential injustice: When legal institutions cannot resolve environmental and land use disputes*. New York: Springer.

Deng, Z. H., Lin, Y. L., Zhao, M. X., & Wang, S. F. (2014). *Collaborative planning in the new media age: The Dafo Temple controversy*. Cities (accepted): China.

Epp, M. (2012). *Assessing incidence of and experiences with new information communications technologies in planning practice in Canada and the United States*. Master Thesis in Pratt Institute, School of Architecture.

European Commission. (2011). Broadband coverage in Europe in 2011: Mapping progress towards the coverage objectives of the digital agenda. Point Topic.

Frazier, J., & Touchet, T. (2012). Transforming the city of New York: New platform for public–private cooperation ushers in smart cities of the future. Cisco IBSG.

Geertman, S. (2006). Potentials for planning support: A planning-conceptual approach. *Environment and Planning B: Planning and Design, 33*(6), 863–881.

Giffinger, R., Fertner, C., Kramar, H., Meijers, E., & Pichler-Milanović, N. (2007). Ranking of European medium-sized cities. Final Report, Vienna.

Goodspeed, R. (2015). Smart cities: Moving beyond urban cybernetics to tackle wicked problems. *Cambridge Journal of Regions, Economy and Society, 8*(1), 79–92.

Hartmann, T., & Barrie, N. (2012). Introduction. In H. Thomas & N. Barrie (Eds.), *Planning by law and property rights reconsidered* (pp. 1–23). Farnham: Ashgate Surrey.

Healey, P. (1997). *Collaborative planning: Shaping places in fragmented societies*. London: Macmillan.

Healey, P. (2003). Collaborative planning in perspective. *Planning Theory, 2*(2), 101–123.

Hollands, R. G. (2008). Will the real smart city please stand up? *Cities, 12*(3), 303–320.

Hoon, L. J. H., Phaal, R., & Lee, S. H. (2013). An integrated service-device-technology roadmap for smart city development. *Technological Forecasting and Social Change, 80*(2), 286–306.

Hufty, M. (2011). Governance: Exploring four approaches and their relevance to research. In U. Wiesmann & H. Hurni (Eds.), *Research for sustainable development: Foundations, experiences, and perspectives*. Bern: Geographica Bernensia.

Huxley, M. (2000). The limits to communicative planning. *Journal of Planning Education and Research, 19*, 369–377.

Innes, J. (2004). Consensus building: Clarifications for the critics. *Planning Theory, 3*(1), 5–20.

Innes, J., & Booher, D. E. (1999). Consensus building and complex adaptive systems. *Journal of the American Planning Association, 65*, 412–423.

Kahila, M., Broberg, A., & Kyttä, M. (2014). Let the citizens map: public participation GIS as a planning support system in Helsinki 2050 master planning process. *Presentation in the Association of European Schools of Planning (AESOP)*, Utrecht University, 9–12 July 2014.

Kahila, M., & Kyttä, M. (2009). SoftGIS as a bridge-builder in collaborative urban planning. In S. Geertman & J. Stillwell (Eds.), *Planning support systems: Best practice and new methods*. Dordrecht: Springer.

Klosterman, R. E. (1997). Planning support systems: A new perspective on computer-aided planning. *Journal of Planning Education and Research, 17*(1), 45–54.

Komninos, N. (2009). Intelligent cities: Towards interactive and global innovation environments. *International Journal of Innovation and Regional Development, 1*(4), 337–355.

Leydesdorff, L., & Deakin, M. (2011). The triple-helix model of smart cities: A neo-evolutionary perspective. *Journal of Urban Technology, 18*(2), 53–63.

Lin, Y. L., Zhang, X. L., & Geertman, S. (2014). Towards smart governance and social sustainability for Chinese migrant communities. *Journal of Cleaner Production*. doi:10.1016/j.jclepro.2014.12.074

Michelle, R. (2014). 5 U.S. cities using technology to become smart and connected. http://www.forbes.com/sites/ptc/2014/08/15/5-u-s-cities-using-technology-to-become-smart-and-connected/. Accessed 15 Aug 2014.

Nuissl, H., & Heinrichs, D. (2011). Fresh wind or hot air—does the governance discourse have something to offer to spatial planning? *Journal of Planning Education and Research, 31*(1), 47–59.

Nuojua, J., Juustila, A., Räisänen, T., & Kuutti, K. (2008). Exploring web-based participation methods for urban planning. In *The Proceedings of the Tenth Anniversary Conference on Participatory Design* 2008 (pp. 274–277).

Pelzer, P., Geertman, S., van der Heijden, R., & Rouwette, E. (2014). The added value of planning support systems: A practitioner's perspective. *Computers, Environment and Urban Systems, 48*, 16–27.

Poplin, A., Pereira, C. G., & Rocha, F. C. M. (2013). The participatory cube: A framework for analysis of online participation platforms. In S. Geertman, F. Toppen, & J. Stillwell (Eds.), *Planning support systems for sustainable urban development*. Dordrecht: Springer.

Rydin, Y. (2007). Re-examining the role of knowledge within planning theory. *Planning Theory, 6* (1), 52–68.

Sun, S., & Yin, Y. (2008). Public participation based on urban-rural planning law. *Planners, 24*(5), 11–14. (in Chinese).

Vonk, G., Geertman, S., & Schot, P. (2005). Bottlenecks blocking widespread usage of planning support systems. *Environment and Planning A, 37*, 909–924.

Wang, C., & Burris, M. A. (1997). Photovoice: concept, methodology, and use for participatory needs assessment. *Health Education and Behavior, 24*(3), 369–387.

Xu, Y. X. (2013). *Research on public participation and decision making models in planning in the network era*. Master Thesis in Nanjing University.

Yan, H., & Zhu, X. G. (2013). Research on public participation in China's urban–rural planning in the new media age. *Shanxi Agricultural Sciences*, 211–213.

Part III
Planning Support Systems and Public Engagement

Chapter 15
Sentient PSS for Smart Cities

Brian Deal, Varkki Pallathucheril, Yong Wook Kim and Haozhi Pan

Abstract Being "smart" has become the Holy Grail for cities and city developers around the world. The idea of a smart city however, has been mostly limited to real-time data acquisition from ever expanding sensor networks. Utilizing these data in thinking about the future, planning, or decision-making has been largely overlooked. Developing a useful planning support system (PSS) for a smart city requires that the PSS possesses a degree of sentience—an awareness of application context and user needs—that few if any current PSSs currently possess. In this chapter we seek to make a case for sentient PSS by first briefly examining the notion of sentience from a computing perspective and by presenting case studies of emerging sensor-driven sentient computing applications. These case studies help us identify essential characteristics of a sentient PSS. We then consider how these characteristics might be manifested based on our experiences in PSS development. We argue, as we have elsewhere, that use-driven development—testing and refining the system in real-world applications—must be the signature of future work on a sentient PSS. We conclude with a discussion on potential challenges and paths forward.

B. Deal (✉) · Y.W. Kim · H. Pan
Department of Urban and Regional Planning, University of Illinois at Urbana-Champaign, Room 228 Temple Buell Hall 611 Taft Drive, Champaign, IL 61820, USA
e-mail: deal@illinois.edu

Y.W. Kim
e-mail: ywkim@illinois.edu

H. Pan
e-mail: hpan8@illinois.edu

V. Pallathucheril
College of Architecture Art and Design, American University of Sharjah, 26666, Sharjah, UAE
e-mail: varkki@aus.edu

S. Geertman et al. (eds.), *Planning Support Systems and Smart Cities*,
Lecture Notes in Geoinformation and Cartography,
DOI 10.1007/978-3-319-18368-8_15

1 Introduction

Although not new, the term Smart City has evolved over the last decade into a popular and perhaps overused phrase in the urban lexicon (Caragliu et al. 2011). Poole (2014) observes that the term generally implies a manifestation of "the internet of things" at an urban scale, predicated on "ubiquitous wireless broadband and the embedding of computerized sensors into the urban fabric". He also notes that "big technology, engineering and consulting companies" (most notably IBM and Cisco) have been the most energetic in promoting the term in the hope of profiting from big municipal contracts. Many consider the basic vision of a smart city as a *wired* (Kitchin 2014), *sensor-filled* (Hancke et al. 2013), *tech-heavy* (Lee et al. 2008), "*technolopolis*" (Gibson et al. 1992), that uses sophisticated computing techniques to transform the city into an *intelligent machine* (Hall et al. 2000; Humphries 2013) as fatally flawed. In this worldview, the *user* does not have a place in the system. This limits its relevance to urban planning and urban management which privileges consultation and participatory processes. Smart city proponents have not demonstrated if and how citizen voices can be heard in smart ways (Poole 2014). Greenfield (2013) agrees, "the notion of the smart city in its full contemporary form appears to have originated …(without) any party, group or individual recognized for their contributions to the theory or practice of urban planning".

In contrast, other new technologies appear to be paying *more* rather than *less* attention to end-users. Smartphones and other hand held devices with apps directed at the user experience are changing the computing landscape. At this writing, a majority of all internet traffic is now taking place on mobile devices (Lella and Lipsman 2014). Hopper (2000) describes this as a movement toward computational environments that are "more responsive and useful with more direct connections to users" within their own, physical world. He suggests these systems display a degree of 'sentience', a greater awareness of space, time, context, and user experience. Although sentience is not yet a part of the smart city discourse, an ability to gather information from a particular context in terms of space and time through sensor technologies is a key part of the smart city movement.

The computational basis for smart city approaches has been focused on urban data acquisition techniques, data structures and communication protocols, real-time analysis, and some short-term projection capabilities. We believe that this focus ignores the connection to longer time-frame analysis critical to urban planning. We argue that planning support systems can and must help bridge this gap. In this chapter, we examine the connection between smart cities, sentience, and the relationship to planning support systems (PSS). We do this by briefly examining the notion of sentience from a computing perspective. We then introduce PSS technologies and describe ways in which a sentient PSS might evolve. To describe the features might be included in such a system, we sketch how the current LEAM PSS (Deal and Pallathucheril 2009a; Deal et al. 2014) would need to evolve in order to manifest sentience. We conclude with a discussion on potential drawbacks and potential paths forward.

2 Sentient Computing Systems

Sentient computing systems are described as computational environments that respond to evolving or changing environmental conditions. It is considered one pathway toward ubiquitous or pervasive computing; an advanced concept where computing and computational services are available everywhere, anywhere, all of the time (Saha and Mukherjee 2003). Hopper (2000) uses the term "sentient computing" to suggest computational environments that are made more responsive and useful by more direct connections to users and to the physical world.

2.1 Context Awareness and Adaptability

Bill Schilit, the creator of one of the first system infrastructures for location-aware mobile computers while at Xerox PARC, suggests that "context-aware computing applications" do not occur at a single location in a single context, as in desktop computing, but rather span a multitude of situations and locations. He notes that this form of computing is much broader than mobile computing because it concerns mobile people not just mobile computers (Schilit et al. 1994). Harter et al. (1999) adds "a persistent distributed object system" as one of the most important features of a context-aware system, describing an early prototype of an application as 'sentient' because it knows the location of users and equipment, the capabilities of the equipment, and the networking infrastructure needed to perform certain tasks. More recently, Henricksen et al. (2006) argues that context-aware computing systems require infrastructure to gather, manage, and disseminate contextual information to applications.

Hewlett-Packard's *nomadic computing* is one early approach to developing a context-aware computing framework (Kindberg and Barton 2001). In the HP model, every physical entity has a web representation that includes both static attributes (such as names and locations) and dynamic attributes (in terms of its space and time context). The nomadic framework updates the information associated with the entity and its surroundings as the entity moves through space and time. It is generally intended to support users of wireless, handheld devices interacting with their environment. For example, a user's handheld device automatically displays information about the room into which the user has just entered or the projector in the room projects a presentation file that is on a device carried by the user. In HP's *Cooltown Project*, the nomadic framework has been applied to a virtual city (Kindberg and Barton 2001) in which entities in the model include people, places and things connected to an urban setting. The computing system "knows" real-time human perceptions of their physical environmental for the whole population (Abdelzaher et al. 2004). All the information is retrievable from a URL associated with the entity.

More common examples of context-aware computing include a tablet computer that switches its screen orientation when rotated, maps that orienting themselves according to the user's orientation and or adapting the zoom level in response to the user's speed of travel, and a smartphone adjusting the screen's backlight in response to changing ambient light levels. Although simple, these examples represent strides in context-aware applications. In contrast to traditional approaches, these applications are not designed for a single or a limited set of user contextual experiences; they are designed for a broad range of potential computer-user interactions. It is not just a matter of making sense of data but it is using the data to predict what a user is likely to want or need and being prepared to satisfy that need.

2.2 Interaction with Users

A sentient computing systems' ability to sense objects and adapt to change is not limited to its relation to the physical environment. The system is also expected to be capable of adjusting itself to user behaviors and facilitate smooth user experiences, as well as present easily interpreted information through advanced visualization techniques.

Addlesee et al. (2001) point out the importance of user perception in a sentient system: "a sentient computing system doesn't need to be intelligent or capable of forming new concepts about the world—it only needs to act as though its perceptions duplicate the users' [perceptions]." He also notes that one solution involves creating devices and applications that appear to cooperate with users, reacting as though they are aware of the context and manner in which they are being used, and reconfiguring themselves appropriately. Harter et al. (1999) suggest that the ultimate justification and test of sentient computing will be its capacity to deliver benefits to users, enabling them to interface directly to devices and expressing complex issues in a simple way.

As an early example, the University of Washington's Portolano project emphasized invisible, intent-based computing, which infers user intentions via their actions in the environment and their interactions with everyday objects (Esler et al. 1999). A user turns on a device (an e-reader) on one network and seamlessly picks up where she left off using another device (a different e-reader) on another network. Project devices are so highly optimized to particular tasks that they blend into the world and require little technical knowledge (Saha and Mukherjee 2003). The project was part of a DOD DARPA sponsored effort to make computing an integral part of manufacturing.

Another early adopter in sentient user interactions was the AT&T Laboratories Cambridge (2001). Their stated goal was to:

create devices and applications that appear to cooperate with users, reacting as though they are aware of the context and manner in which they are being used, and reconfiguring themselves appropriately.

The AT&T system used sensors to update a model of the real world in terms of object positions, descriptions, and state in a way familiar to the user—so that "the model describes the world much as users themselves would". They used the model to write programs that reacted to changes in the environment according to the user's preferences. They referred to this as sentient computing, "because the applications appear to share the user's perception of the environment".

A step closer to understanding the users' physical environment and their real-time emotion is a sentient computing system that can interact with users in a way that understands the social characteristics of the users. In his description of "social computing", Dourish (2001) calls for an interface design that "recognizes that the systems we use are embedded in systems of social meaning, fluid and negotiated between us and the other people around us".

Ubiquitous or ambient computing are more recent and related terms. They represent a vision for computing that is "everywhere all the time". One example is the Google House project which attempts to integrate ubiquitous sensing/computing networks with handheld user-interfaces (Cabitza et al. 2014). Google apps and technologies in general are moving in this direction. Google Now can exchange data with the Google platform in order to personalize the user experience. The recent purchase of the Nest thermostat system by Google begins to tie these entertainment-oriented systems to building controls and human comfort.

Proponents of smart cities have focused on real-time data acquisition from ever expanding sensor networks and the information and communication (ICT) infrastructure that facilitates use of these data (Giffinger and Gudrun 2010; Kloeckl et al. 2012; Hancke et al. 2013). The question is: how do we make use of the data available in ways that support planning and decision-making? Kitchin (2014) notes that "the production of sophisticated data analytics for understanding, monitoring, regulating and planning the city", is critical for realizing smart city goals. Currently however, the value of these data in thinking about future, planning, or using plans to make decisions has been largely overlooked.

3 Planning Support Systems and Sentience

The role of technology in support of planning has been the subject of a long-running and persistent debate (Klostermann 1997). Planning technologies assembled into planning support systems (PSS) according to Brail and Klostermann (2001) generally consist of base data, analysis engines (models), and information delivery systems (visualization interfaces). In terms of the planning process, Geertman and Stillwell (2003) identify the following as critical PSS components: integration into the process, focus on the planning problem at hand, meeting user needs, and accessibility. Updates to this work describe the current state of PSS technologies in general Geertman and Stillwell (2009), and in terms of large scale urban modeling systems (Brail 2008). More recently, the emphasis in PSS development has shifted to the management of information needs (Power and Sharda 2009), use-based systems (Te Brömmelstroet and Schrijnen 2010; Deal 2008), and web-based strategies of information retrieval and delivery (Budthimedhee et al. 2002; Deal and Pallathucheril 2009b).

We argue that in order to tackle the complexity of smart city visions, the next generation of PSS development should follow a path toward *sentience* as framed by the computational disciplines. Using this framework, a sentient planning system might be described as an interactive plan-making environment that has an evolving self-awareness about its context, underlying data and models that process this data, and is able to deliver useful and timely information to its user. We find that few existing PSSs possess these attributes. Generally current PSSs lack dynamic ways of updating their underlying data, and user interfaces tend to be static and cannot accommodate individual differences or multiple contexts. The following sketches out some of the challenges and potential approaches to dealing with more dynamic data, creating sentient user interfaces, and general sentient PSS system development.

3.1 Context Awareness and Adaptability

A sentient PSS should incorporate and automate data-update processes so that context depictions are constantly and transparently updated. This is important for maintaining context awareness. Sophisticated and automated manipulation procedures must be included in any raw data collection and update procedures. Local data sources pose a challenge to this continual update problem. Local data are typically not frequently updated due to cost and local administrative structure. Automation would also require changes in the way local organizations generate and share data.

Sensors, RFID and other related technologies present a great opportunity but perhaps an even greater challenge for data collection. The approach can produce vast quantities of data at very fine temporal resolutions. Masucci et al. (2013) recently studied 11 million records taken over a one-week period from transport for London's (the London tube subway system) electronic ticketing system. This data helped to reveal London as a polycentric system of up to 10 different centers that interlink in complex patterns. Batty (2013) argues that these kinds of data sets can help in planning resilient cities, but this type of analysis typically involves a retrospective examination of huge data dumps and not real-time or dynamic analysis. It is also unclear how it implicates the longer time horizon, and sometimes larger—or smaller spatial areas required for planning.

According to Paroutis et al. (2013) most 'smart' projects fall into 1 of 2 categories: those that focus on the features of the technology adopted to solve particular problems; or those that consider complex systems that are prone to particular behaviors such as adaptation and self-organization (Portugali 2000). A greater awareness of the broader contexts and potential implications of a given decision choice, would enable planning and design decisions that would be more readily adaptable to change. This capacity to adapt to and shape to change is an important component of planning. For example, those who make and use plans may not have facile access to the diverse body of information contained in climate simulations, especially as it relates to vulnerability and adaptation. The information is generally presented and communicated in ways that are difficult to understand and translate. This limits their ability shape and adapt to climate related stimulus and change.

3.2 Interaction with Users

PSS research has sought to incorporate innovative visualization tools (Brail and Klostermann 2001; Deal and Pallathucheril 2009b; Budthimedhee 2003). Even so, a PSS user is typically (although the state of the art is changing quickly) presented with a very limited range of options in terms of visualization approaches.

One area of improvement that will be needed in order to accomplish sentience is the accessibility and speed at which various bits of information are delivered and accessed. Accessible information is essential in engaging dialogue and in understanding the potential implications of the analysis (Deal and Schunk 2004). Effective visualizations catalyze discussion about divergent futures and provide a platform from which a set of stories about the future can be generated. There would also need to be a greater variety of media used to deliver content. While most systems use visual representations, there needs to be more effective way of integrating visual and verbal representations, particularly for those who are not spatially adept and rely more on verbal representations. One crucial factor for PSS viability is how it presents information to decision makers or how its users make sense of the information presented (Pallathucheril and Deal 2012).

In a study on perceptual speed (the speed at which different types of information are recognized), Cristina and McLaren (2008) found that perceptual speed varied greatly by individual. In terms of PSS development however, their work confirms that the effectiveness of any given visualization interface depends on the person using the interface. This supports our thesis that next generation PSSs must both understand the user and the context from which the user is seeking information. This means they must aware of who is seeking information and how they need it delivered.

4 Challenges and Opportunities

We argue that a sentient PSS should possess a basic self-awareness about itself, its data, and its user; be capable of learning and spatial and temporal reasoning; understand rules, potential conflicts and support conflict resolution; and be visually accessible and interactive. In this section, we use a PSS developed at the University of Illinois' Land-use Evolution and Impact Assessment Modeling laboratory (LEAM) to help illustrate the challenges in realizing a sentient PSS.

The LEAM PSS is intended to coordinate complex regional planning activities and aid in regionally-based thinking, decision support, and policy formulation (see Deal and Pallathucheril 2007, 2008, 2009b). It uses a content management system for handling data, managing model output, and for hosting a visualization interface for gathering and disseminating information from (and to) local planners and communities. It includes both tightly and loosely coupled models and modules (Fig. 1) for understanding land-use change, local regional economics, environmental impacts, and social implications of planning scenarios.

A typical LEAM PSS for a given region (see example in Fig. 2) includes: data storage and a manipulated data repository, scenarios described and articulated in images and maps, impact assessment analysis related to each scenario (examples form our work include traffic congestion changes, stresses on natural or cultural resources, water quality or quantity impacts, and green infrastructure implications, to name a few), a system of viewing existing plans (this is a searchable repository of all plans in the region so that cross check comparisons on what already exists in terms of plans can be quickly accessed (based on Hopkins and Zapata 2007), a

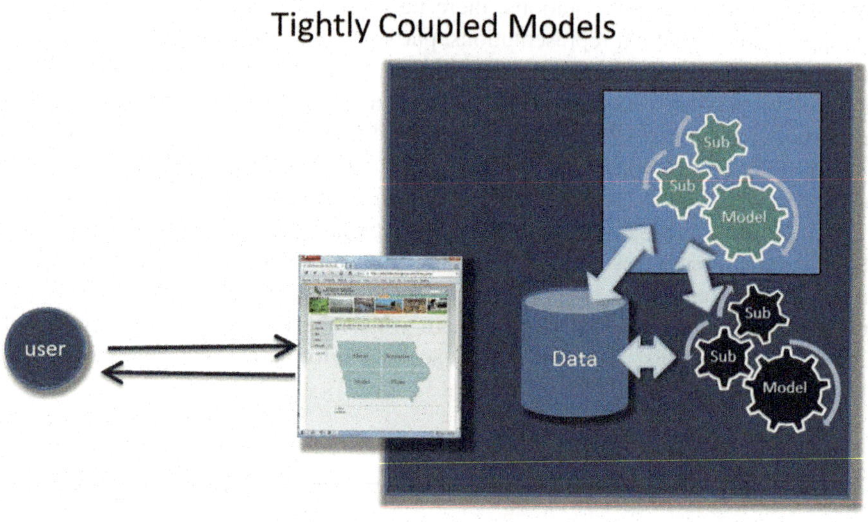

Fig. 1 The LEAM PSS. A two way user interaction (upload and download) with data, models, and sub models through a web-based CMS portal

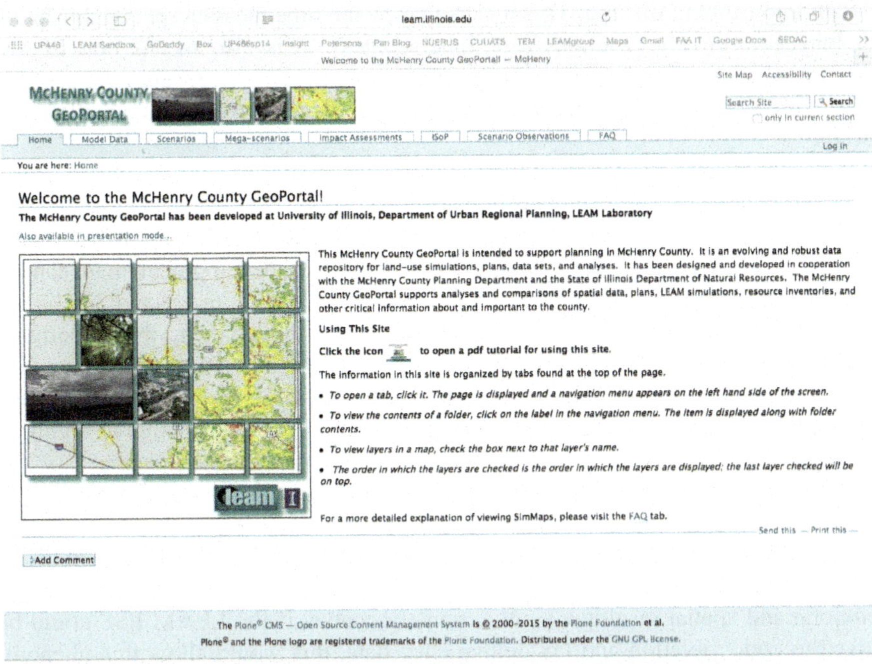

Fig. 2 The LEAM PSS for McHenry County (available at leam.illinois.edu/mchenry//scenarios-geoserver). Data, scenarios, impact assessments, a system of viewing existing plans, a summary view of scenarios, and an FAQ are the major 'tabbed' components of this system

summary view of scenarios for policy formation and dialogue, and an FAQ (in the use of the system).

4.1 Context Awareness and Adaptability

LEAM simulations require spatial data on land-use land-cover in the region of concern. These data describe the specific simulation context. They are also the basis for computing the likelihood that a particular land-use classification is associated with a specific underlying factor (say, distance from job centers). These likelihood values partially influence whether land uses can be expected to change at a particular location at a particular time in the future.

LEAM applications in the United States acquire land-use land-cover data from the National Land Cover Data (NLCD) produced by the United States Geological Survey (USGS). While the NLCD represents fairly good recent historic data (from 2001), it is not a very good description of the current context. NLCD only provides this data every 5 or so years (2001, 2006, 2011). It also uses an intensive collection and ground truthing method. This means that by the time the data set is published it no longer accurately describes a regional context. This also means that LEAM likelihood values are also not contextually current. While we believe that these values are not extremely volatile, it would help if the PSS were contextually more aware.

To provide better context awareness, a sentient version of the LEAM PSS could be linked to land-development permitting information in the region. As new permits are issued, the land-use data can be automatically updated. The permit data would provide more details about the type and intensity of development than can be extracted from the NLCD. This would, in turn, allow more refined likelihood values. With regularly updated information about the context, the underlying models can be recalibrated, and implications for a regions system of plans (see below) can be explicated. While the temporal resolution of the permit data is likely to be fine, the temporal resolution of implications must be large enough to highlight significant shifts. It would be too easy to miss the forest for the trees.

Another context awareness issue is how to capture obsolete land uses. NLCD captures physical properties without reference to the socio-physical landscape, i.e. whether properties are abandoned, vacant, or under utilized and accessible for redevelopment. Data on vacancy are available in the decennial census but the temporal and spatial resolutions are extremely coarse. If the LEAM PSS could be linked to code violation and tax delinquency data, this would allow this phenomenon to be tracked with much more precision. The PSS becomes more responsive or adaptable to change. We would not need massive sensor networks to achieve this valuable level of sentience.

The LEAM PSS for a particular region often has built into it a library of plans that affect that region. This library seeks to implement Hopkins et al.'s (2005) notion of an information system of plans (Finn and Hopkins 2007). It provides important context for thinking about the future of the region and is potentially a form of sentience. The library as currently implemented supports word searches of the text of plans but does not capture the logic of these plans. What is the spatial extent of the impact of the plan or one of its components? What are the actions and who are the actors affected by the plan? This kind of sentience would require implementation of a plan markup language (Hopkins et al. 2005) as well as ways of encoding the graphic content of plan (Li 2009). The overhead associated with marking up existing plans has thwarted our past efforts to realize the kind of sentience.

4.2 Interaction with Users

LEAM uses a customized version of the content management system Plone to store spatial and aspatial data, plans, scenarios, impact analysis, and other data. A map viewer is embedded to provide visual imagery of the stored spatial data sets so that users can easily pan, zoom, and move around the data. A lot of the information assemblage is automated so that when a scenario is simulated by the system, the different views of the results, including the underlying data and assumptions, are automatically generated and made available.

To become sentient, LEAM visualization interfaces should better support the use and processing of information by lay people, especially with respect to decision making and risk (Bostrom 2002, 2008). Interfaces should also be informed by an understanding of how communities of people (as opposed to the individuals in communities) think about issues and risks, and how communities acquire knowledge from a variety of sources. Budthimedhee (2003) provides some key insights into the characteristics of visualization devices that can effectively and efficiently support inferences from dynamic spatial data sets. First, because the speed at which inferences can be made is critically important with large data sets, she draws on the idea that we must pay attention to the ease and accuracy with which the pre-attentive visual system can assess relative magnitudes. Second, because of the amount of data needed to make inferences, she draws on the idea that graphic attributes of a visualization device may be more important than its efficiency in using ink to represent the data as is the conventional wisdom. Third, she reiterates Wickens' (1992) argument that if bits of information must be proximate mentally in order to support inferences, they must be proximate when visualized.

A sentient PSS, therefore, requires visualization interfaces that are context-relevant. That is, the system should adjust spatial and material configurations based on new information or individual user differences. There is very little in the literature on system intelligence of this type, but technologies are available for implementing this kind of flexibility. Users can indicate preferences for alternative visualization devices ("Show me these data in a table", or "Show me these data in a different format"). The ability to have user preferences persist over separate sessions is available in many Internet-based information systems and the PSS can appear to be learning and responsive to individual differences.

Another form of sentient interfaces involves how community comment and feedback are handled. In LEAM deployments, model builders invite collective public critique of the intelligibility of model results, and what assumptions must change to address unreasonable results. New model runs are informed by that feedback. A step towards this form of sentience could involve social media platforms for stakeholders to express concerns and critiques model assumptions and outcomes. The PSS could then aggregate these comments and infer from these new or different scenarios that are run and posted to the PSS. It would appear that the PSS is responding to community concerns.

Interfaces might also convey valuable information about the user to the system. Many of today's PSSs allow users to leave comments on the information being presented. Such information about users' interactions with the system can help drive system development. This type of system interaction and learning will be critical for PSS evolution in response to user preferences. For instance, an action like searching for the impact of a particular public policy or investment choice might cause the system to store that knowledge and remember that there is an interest in that topic area. If this same search happens repeatedly, the system might generate one or more new scenarios in response. PSS development activity can benefit from this kind of crowd-sourcing of questions to be addressed. Furthermore, at present, we rely on knowledge to identify data sets to help answer new questions. It would add to the sentience of PSSs if there were ways for the system to learn about and discover new and relevant data repositories that would add value to the generation of information needed.

Additional development is also required in spatial and temporal reasoning within the individual tools that make up a PSS, and there must be ways of generating and implementing rule-based procedures and addressing potential conflicts arising from these rules. This form of sentience could support planners as they attempt to resolve conflicts among different interests.

4.3 A Use-Based Approach to a Sentient PSS

Although PSS visualization interfaces are typically designed to deliver planning-related information and knowledge to users, an important question is how well these interfaces help in the act of making plans? Hopkins (2001) describes plan-making in terms of behaviors (things that people do), tasks (combinations of behaviors that accomplish particular functions), and processes (patterns of tasks that yield plans). Clearly tasks can be supplemented with planning tools, such as GIS systems for making maps. Combinations of tasks might also be supplemented by PSS. The critical question is how PSS help to support planning behaviors.

We find that an evolving, sentient PSS might be most effectively delivered through use-based development rather than solely based on a priori notions (Deal and Pallathucheril 2009b). This requires formative rather than summative assessments in system development. It requires working through specific sets of data sources and information with groups of real-world users, building a thorough understanding of the cognitive demands placed on users in a common virtual sandbox environment. The process of gaining acceptance and of becoming an integral part of local and regional planning evolves with time and familiarity. A use-driven, embedded approach to PSS development suits this evolutionary process.

5 Conclusions

In this chapter, we argue that smart cities require PSS that go beyond the ones we have today. We call for PSS that possess a greater degree of sentience—a greater awareness of application context and user needs. We need information and planning systems that are much more than sensor networks and data acquisition exercises. We suggest that in order to make use of big data and smart city technologies, we need sentient planning support systems that can supply timely and useful information in support of complex plan-making tasks.

We argue that a sentient PSS should: possess a basic self-awareness about itself, its data, and its user; be capable of learning, be capable of spatial and temporal reasoning, understand rules, potential conflicts and support conflict resolution, and is visually (and perhaps ultimately in the future verbally) accessible and interactive. We don't expect a well-constructed sentient PSS to replace human plan-making, but we do expect that will help us make better plans.

Our argument for a move toward sentient planning systems does have limitations however. For example, sentience implies self-awareness—what if that self-awareness is not self-assured? What if users don't have confidence in that awareness? Or have fears that it is misdirected? Users of sentient information systems must have confidence that the system is producing information objectively and in an unbiased way. Without this, users may not have confidence that the system will provide reliable support for difficult choices. Our experience with LEAM in regional planning processes suggests that participants in the process give the process legitimacy and assurance to others. Privacy is also a concern. Users are likely to need to disclose some personal information. Confidentiality is hard to maintain in today's 'cloud' based systems. Revealing any kind of personal information could cause a variety of problems.

Similarly, will users trust black box outcomes? In the past, opening the technology box to provide at least a moderate level of understanding of what's inside has been the hallmark of planning technology implementations. Without this understanding, there may be a hesitancy to trust outcomes. Is this possible given the complexity of the problem? Additionally, can technology based systems keep up with technology innovations themselves? Can sentient system designs adapt to technological innovations as they adapt to their contexts, surroundings and environments? What about radical changes to the environment or user demands.

These are important questions that need serious study in order to be adequately addressed. Clearly, a more detailed study on the role of sentience in smart city planning and support systems is necessary.

References

Abdelzaher, T., Blum, B., Chen, Y., Evans, D., George, J., George, S., et al. (2004). *EnviroTrack: Towards an environmental computing paradigm for distributed sensor network*. Proceedings of 24th International Conference on Distributed Computing Systems, pp. 582–589.

Addlesee, M., Curwen, R., Hodges, S., Newman, J., Steggles, P., Ward, A., & Hopper, A. (2001). Implementing a sentient computing system. *Computer, 34*(8), 50–56.

AT&T Laboratories Cambridge. (2001). Sentient Computing Project Home Page. http://www.cl.cam.ac.uk/research/dtg/attarchive/spirit/. Accessed February 26, 2015.

Batty, M. (2013). *Smart cities and big data: How we can make cities more resilient*. Dublin, Ireland. http://www.spatialcomplexity.info/files/2013/07/BATTY-AESOP-ACSP-20131.pdf.

Bostrom, A. (2002). *Risk communication: A mental models approach*. Cambridge, UK: Cambridge University Press.

Bostrom, A. (2008). *Risk assessment, modeling and decision support: Strategic directions*. Berlin, Germany: Springer.

Brail, R. K. (2008). *Planning support systems for cities and regions*. Lincoln Institute of Land Policy.

Brail, R. K., & Klostermann, R. E. (2001). *Planning support systems: Integrating geographic information systems, models, and visualization tools*. Redlands, CA: ESRI Press.

Budthimedhee, K. (2003). Effective visualization interfaces for planning support systems (PhD dissertation). University of Illinois at Urbana-Champaign.

Budthimedhee, K., Li, J., & George, R. (2002). ePlanning: A snapshot of the literature on using the world wide web in urban planning. *Journal of Planning Literature, 17*(2), 227–246.

Cabitza, F., Fogli, D., & Piccinno, A. (2014). Fostering participation and co-evolution in sentient multimedia systems. *Journal of Visual Languages and Computing, 25*(6), 684–694.

Caragliu, A., Del Bo, C., & Nijkamp, P. (2011). Smart cities in Europe. *Journal of Urban Technology, 18*(2), 65–82.

Cristina, C., & McLaren, H. (2008). *Exploring the role of individual differences in information visualization*. Presented at the Proceedings of the Working Conference on Advanced Visual Interfaces (AVI '08). New York, NY. doi: http://doi.acm.org/10.1145/1385569.1385602.

Deal, B. (2008). *Sustainable land-use planning: The integration of process and technology*. Saarbrücken, Germany: VDM Verlag.

Deal, B., & Pallathucheril, V. (2007). Developing and using scenarios. In L. D. Hopkins & M. A. Zapata (Eds.), *Engaging the future: Forecasts, scenarios, plans, and projects* (pp. 222–242). Cambridge, MA: Lincoln Institute for Land Policy.

Deal, B., & Pallathucheril, V. (2008). Simulating regional futures: The land-use evolution and impact assessment model (LEAM). In R. Brail (Ed.), *Planning support systems for cities and regions* (pp. 61–84). Cambridge, MA: Lincoln Institute for Land Policy.

Deal, B., & Pallathucheril, V. (2009a). Sustainability and urban dynamics: Assessing future impacts on ecosystem services. *Sustainability, 1*, 346–362. doi:10.3390/su1030346.

Deal, B., & Pallathucheril, V. (2009b). A use-driven approach to large-scale urban modelling and planning support. Planning support systems best practice and new methods. *The GeoJournal Library, 95*, 29–51.

Deal, B., Pallathucheril, G., & Pan, H. (2014). *Planning support systems in the rear view mirror*. Presented at the 49th Annual Conference of the Association of Collegiate Schools of Planning. Philadelphia, PA.

Deal, B., & Schunk, D. (2004). Spatial dynamic modeling and urban land use transformation: A simulation approach to assessing the costs of urban sprawl. *The Journal of Ecological Economics, 51*(1–2), 79–95.

Dourish, P. (2001). *Where the action is: The foundations of embodied interaction*. Cambridge, MA: MIT Press.

Esler, M., Hightower, J., Anderson, T., & Borriello, G. (1999). *Next century challenges: Data-centric networking for invisible computing: The Portolano project at the University of Washington*. Presented at the Proceeding MobiCom'99 Proceedings of the 5th Annual ACM/IEEE International Conference on Mobile Computing and Networking (pp. 256–262). New York, NY.

Finn, D., & Hopkins, L. (2007). The information system of plans approach: Using and making plans for landscape protection. *Landscape and Urban Planning, 81*(1), 132–145.

Geertman, S., & Stillwell, J. (2003). *Planning support systems in practice*. New York, NY: Springer.

Geertman, S., & Stillwell, J. (2009). *Planning support systems: An introduction*. Berlin, Germany: Springer.

Gibson, D. V., Kozmetsky, G., & Smilor, W. (1992). *The technopolis phenomenon—Smart cities, fast systems, global networks*. Savage, MD: Rowman & Littlefield.

Giffinger, R., & Gudrun, H. (2010). Smart cities ranking: An effective instrument for the positioning of cities? ACE: Architecture. *City and Environment, 4*(12), 7–25.

Greenfield, A. (2013). *Against the smart city*. New York, NY: Do projects.

Hall, R. E., Bowerman, B., Braverman, J., Taylor, J., Todosow, H., & Von Wimmersperg, U. (2000). *The vision of a smart city*. Presented at the 2nd International Life Extension Technology Workshop, Paris (FR), 09/28/2000–09/28/2000. Paris, France.

Hancke, G. P., Silva, B. C., & Hancke, G. P, Jr. (2013). The role of advanced sensing in smart cities. *Sensors, 13*, 393–425. doi:10.3390/s130100393.

Harter, A., Hopper, A., Steggles, P., Ward, A., Webster, P. (1999). *The anatomy of a context-aware application*. Proceedings of the 5th Annual ACM/IEEE International Conference on Mobile Computing and Networking (MobiCom'99). ACM. New York, NY.

Henricksen, K., Indulska, J., & Rakotonirainy, A. (2006). Using context and preferences to implement self-adapting pervasive computing applications. *Software: Practice and Experience, 36*, 11–12.

Hooper, A. (2000). The Royal Society Clifford Paterson Lecture, 1999-sentient computing. *Philosophical Transactions, Royal Society London, 358*, 2349–2358.

Hopkins, L. D. (2001). Planning as science engaging disagreement. *Journal of Planning Education and Research, 20*(4), 399–406.

Hopkins, L., Kaza, N., & Pallathucheril, V. (2005). Representing urban development plans and regulations as data: A planning data model. *Environment and Planning B: Planning and Design, 32*(4), 597–615.

Hopkins, L. D., & Zapata, M. A. (2007). *Forecasts, scenarios, plans, and projects*. Cambridge, MA: Lincoln Institute of Land Policy.

Humphries, C. (2013, May 19). The too-smart city. *The Boston Globe*. Boston, MA. http://www.bostonglobe.com/ideas/2013/05/18/the-too-smart-city/q87J17qCLwrN90amZ5CoLI/story.html . Accessed February, 9 2014.

Kindberg, T., & Barton, J. (2001). A web-based nomadic computing system. *Computer Networks, 35*(4), 443–456.

Kitchin, R. (2014). The real-time city? Big data and smart urbanism. *GeoJournal, 79*, 1–14. doi:10.1007/s10708-013-9516-8.

Kloeckl, K., Senn, O., & Ratti, C. (2012). Enabling the real-time city: LIVE Singapore! *Journal of Urban Technology, 19*(2), 89–112. doi:10.1080/10630732.2012.698068.

Klostermann, R. E. (1997). Planning support systems: A new perspective on computer-aided planning. *Journal of Planning Education and Research, 17*, 45–54.

Lee, S., Han, J., Leem, Y., & Yigitcanlar, T. (2008). Towards ubiquitous city: Concept, planning, and experiences in the Republic of Korea. In T. Yigitcanlar, K. Velibeyoglu, & S. Baum (Eds.), *Knowledge-based urban development : Planning and applications in the information era* (pp. 148–169). Hershey PA: IGI Global.

Lella, A., & Lipsman, A. (2014). *The U.S. Mobile App Report*. http://www.comscore.com/Insights/Presentations-and-Whitepapers/2014/The-US-Mobile-App-Report. Accessed February, 9 2015.

Li, J. (2009). Developing a markup language for encoding graphic content in plan documents. University of Illinois at Urbana-Champaign. Retrieved from https://www.ideals.illinois.edu/bitstream/handle/2142/14610/Li_Jinghuan.pdf?sequence=2.

Masucci, A. P., Stanilov, K., & Batty, M. (2013). Limited urban growth: London's street network dynamics since the 18th century. *PLoS ONE, 8*(8), e69469. doi:10.1371/journal.pone.0069469.

Pallathucheril, V., & Deal, B. (2012). *Communicative action and the challenge of discounting*. Presented at the 53rd Annual Conference of the Association of Collegiate Schools of Planning, Cincinnati, Ohio.

Paroutis, S., Bennett, M., & Heracleous, L. (2013). A strategic view on smart city technology: The case of IBM Smarter Cities during a recession. *Technological Forecasting and Social Change*,. doi:10.1016/j.techfore.2013.08.041.

Poole, S. (2014). *The truth about smart cities: "In the end, they will destroy democracy."* http://www.theguardian.com/cities/2014/dec/17/truth-smart-city-destroy-democracy-urban-thinkers-buzzphrase. Accessed February, 9 2015.

Portugali, J. (2000). *Self-organization and the city*. Heidelberg, Germany: Springer.

Power, D. J., & Sharda, D. (2009). Decision support systems. In S. Y. Nof (Ed.), *Springer handbook of automation*. Heidelbery, Germany: Springer.

Saha, D., & Mukherjee, A. (2003). Pervasive computing: a paradigm for the 21st century. *Computer, 36*(3), 25–31.

Schilit, B., Adams, N., & Want, R. (1994). *Context-aware computing applications* (pp. 85–90). Washington, DC: IEEE Computer Society.

Te Brömmelstroet, M., & Schrijnen, P. M. (2010). From planning support systems to mediated planning support: A structured dialogue to overcome the implementation gap. *Environment and Planning B: Planning and Design, 37*(1), 3–20.

Wickens, C. D. (1992). *Engineering psychology and human performance*. New York, NY: Harper Collins.

Chapter 16
Gaming, Urban Planning and Transportation Design Process

Jayanth Raghothama and Sebastiaan Meijer

Abstract In today's urban planning, two perspectives dominate the discourse: a technical-rational perspective and a communicative rational perspective. Bridging the dichotomy between the two perspectives and situating new planning support methods within the context of complexity theories leads to new structures for planning support systems. The implications of the inherent complex nature of planning when bridging these perspectives should be taken into account for new planning processes and support systems. The development of such methods requires an iterative cycle between methodological and technological aspects of tool development. The chapter presents a technical framework that enables the development of methods integrating both perspectives. The framework derives its requirements from the integration of the two perspectives, and is evaluated in the context of two design case studies in the cities of Stockholm and Paris. The development of the framework and method has implications for the design of tools in urban planning. The tools need to reflect the open nature of the complex systems they represent and operate in. Such methods also expand the boundaries of the design space, allowing for previously unknown configurations.

1 Introduction

Over the last couple of decades, mainstream discourse in urban planning has shifted to the view of urban systems as complex systems, accepting the fact that urban systems exhibit non-linear relationships between components and emergent behavior (Allen 2012). The tools for analysis and design have changed accordingly.

J. Raghothama (✉) · S. Meijer
School of Technology and Health, KTH Royal Institute of Technology, Stockholm, Sweden
e-mail: jayanthr@kth.se

S. Meijer
e-mail: smiejer@kth.se

S. Geertman et al. (eds.), *Planning Support Systems and Smart Cities*, Lecture Notes in Geoinformation and Cartography,
DOI 10.1007/978-3-319-18368-8_16

Bottom-up cellular automata or multi-agent models simulate urban systems based on individual preferences and actions, and group interactions, and represent the influences of actors on the system at a micro level (Batty 2007). Participatory methods were developed to make the planning process more democratic and to include the preferences of actors in the planning process (Healey 1997). The rise of computational power and techniques like visualization and geographical information systems (GIS) augmented both these streams (Batty et al. 1999; Dykes 2000; Ware 2000).

The methods and tools for analysis and design of urban systems can be broadly classified into two groups. The first group is a technical-rational approach using computational and mathematical methods rooted in systems engineering (Checkland 1978). The second approach derives from qualitative approaches which seek to build upon multiple narratives, values and perspectives (Burke 1979). There exists a significant gap between the two approaches. The mathematical approach ignores the multiple perspectives and values of different stakeholders, as well as the politics and process of planning, focusing on accurate representation of system dynamics. The participatory approach focuses on politics and process, and tends to neglect scientific facts as described by models. The best way forward are mixed methods which integrate advanced computer simulations with collaborative processes, referencing both the dynamics of underlying systems as well as the political dimensions of planning (Van Bueren 2009; Dearden and Wilson 2011). Collaborative decision support systems, participatory simulations and gaming simulations have all been used as such mixed methods (Devisch 2012).

While the use of gaming simulation in urban design is not new, the method has not been used significantly either (Cecchini and Rizzi 2001; Meier and Duke 1966). This could be because of a dissonance between the method and prevailing theory. By embracing the complexity and uncertainty inherent in urban planning, gaming contradicted the assumptions prevalent in the rational engineering oriented planning model dominant at the time (Hanzl 2007). It required a theoretical shift in urban planning practice towards a communicative and participatory planning paradigm for the use of gaming to increase.

The use of gaming in urban planning processes is mainly for pedagogical purposes, where games provide insights into complex processes, political and social dynamics. They facilitate communication among stakeholders and can be used to test the effects of different policies on the system. The focus of all these types of games is on the player, where the player has to translate the insights from the game into the real world (Duke and Geurts 2004; Klabbers 2006; Meijer 2009). The question of whether games can deliver operationalized policies and designs that can be implemented in the real world is very new and seldom attempted. This use for gaming places new requirements both on the game and the game designer.

In the following sections, the authors present a technical framework where an integrated method can be developed. We first review the evolution of discourse in planning theory and situate these theories and new technologies in the context of complexity theories. The technical framework is then described, in the context of two case studies in the cities of Paris and Stockholm. Preliminary process designs on the use of this framework in game sessions with stakeholders are then identified, followed by some preliminary results.

2 Methods for Planning in Complex Systems

As mentioned earlier, methodologies that address the analysis and design of complex systems can be broadly classified into two groups. The first group is rooted in computational and mathematical techniques, derived from systems analysis and/or systems engineering disciplines. The ability of computational models to process large sets of data between parts of a system linking to a systemic whole was seen as the key to solve complex problems, and 'systems thinking' was the approach to this kind of problem solving. As a purely reductionist approach, this technical-rational/rational-comprehensive perspective breaks a system down into component parts and tries to understand the structure of the system. It relies heavily on models and computer simulations, which serve as important toolkits for analysis, description and experimentation. Such technical-rational planning implied more control from the top-down through planners, who were expected to be able to foresee consequences (Roo and Rauws 2012). Doubts about this rational-comprehensive perspective started towards the end of the 1950s when people like Lindblom (1959) and Davidoff (1965) started to criticize and question this approach. These doubts followed the realization that planning is essentially a political process, and the scientific tools, at least then, were inadequate (Harvey 1973; Lee 1973). In response, Lindblom tried making the process more realistic and Davidoff tried making it more democratic.

Populist movements in urban planning started refuting the hegemony of the expert urban planner to make the process of planning more democratic and inclusive. Stemming from the need to address the specific needs of individuals, local communities started asserting their rights to organize their local environments participate in and influence the urban planning process. This led to the question of who had the power to design the urban, and a conflict between the traditional top-down model of urban planning and the self-organizing bottom-up influences of different actors. The solution was to envision cities as holistic systems consisting of multiple sub-systems organized under the influence of multiple networked actors, such as contractors, residents, planners, individuals and others (Hajer 2003; Healey 1997; Hughes and Sadler 2013; Innes 1995, 1996; Jacobs 1961).

This second group derives from this need for participation, and is rooted in qualitative approaches which seek to build upon multiple narratives, values and perspectives. This communicative approach accepted the uncertainties inherent in planning, and sought to build plans through collective and communicative processes, shifting the goals of planning to the process of planning. From the 1990s, planning shifted from the technical-rational perspective towards this communicative-rational perspective (Schoenwandt 2012).

A distinction can be drawn here between these two forms of planning. Scholars have suggested that this distinction needs to be bridged, and hybrid methods are the best way forward. Public participation in planning is a reaction to the notion that there exists only one form of planning: a global top-down approach. Public participation is a way of making planning more democratic and just, and giving people more say in the planning process. However, both this global form of planning and a local form of planning co-exist, the local in the form of self-organizing processes and dynamics that result from the actions of independent actors, and in many cases local planning can be more dominant and effective in the overall urban process. This implies that public participation needs to be perceived not just as a reactive force, but as an important source of ideas and initiatives (Portugali 1999, 2006, 2012). While public participation implies an engagement with citizens, scholars have also discovered that people participating in such processes need to have the requisite knowledge to engage with other stakeholders and make the process efficient. Stakeholders involved in these planning processes inevitably acquire or hire (through consultants or other experts) the knowledge required to participate (Alfasi 2003).

The technological changes that have dominated society over the last couple of decades have also given plenty of possibilities for planning. The networked society and the rise of computation have created new possibilities that did not exist before: GIS technologies for analysis, design and visualization, connected mobile phones that enable real time monitoring and data collection; virtual reality techniques that allow for real time exploration of future cities; and so on. Also dominant is the increasing sophistication of urban simulation models, backed by complexity theories, which allow for the study of the dynamics of cities. While each of these technologies support planning, integrating them together results in a more comprehensive planning support system (PSS).

This integrated system is advocated as the state-of-the-art in PSS (Brail 2006), which typically consists of three component parts (Fig. 1). It consists of a set of simulation models (SiMod), usually agent based (AB) or cellular automata (CA), a GIS system, and a set of 2D/3D virtual reality (VR) components. The AB/CA models allow for the simulation of future scenarios, GIS systems form the database for such scenarios and VR components are used to see the results. However, given the non-linearities that typically characterize cities, establishing causal relationships

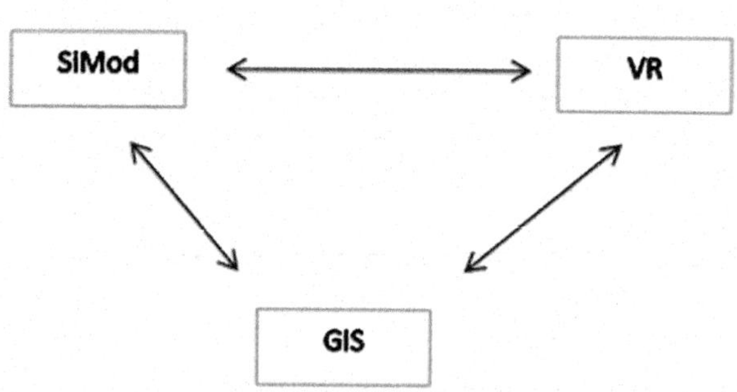

Fig. 1 A typical planning support system. *Source* Portugali (2012)

between variables to enable prediction of future states is difficult. The problem of collecting and analyzing enough data to validate and calibrate such models remains, rendering the ability of AB/CA models to accurately predict future scenarios suspect. Such systems therefore should enable planners to experience a planning scenario rather than trying to predict (Portugali 1999). The challenge then is to formulate processes by which such artificial exploration of simulated realities can be converted into operationalized designs for an inherently unpredictable open system such as the city.

3 Gaming Simulation in Urban Planning

Gaming simulation has been used in urban and regional planning, though sparingly, despite credible demonstrations of the utility of gaming in urban planning practice. In other areas, games have been used since the 1960s as a method to bring together policy makers and stakeholders in a participatory event. Games provide ways to collectively decide on the problem formulation, system boundaries and on the dynamics of the system that will be addressed. Then, policies can be formulated and tested in this simulated environment (Duke and Geurts 2004; Greenblat and Duke 1981).

Lately, games are usually constructed by combining sophisticated computer simulations with interaction and role play (Dearden and Wilson 2011). Simulations provide a realistic context to the policy being tested. In urban planning, there are a number of examples of games being used: in transportation planning, SimCity appears in planning curricula for learning (Raghothama and Meijer 2014). In actual practice, games are used mostly as playful participatory methods. Such methods sustain public engagement because they make it easy to participate, and facilitate learning-by-doing (Gaber 2007; Gordon et al. 2011; Minnery and Searle 2014; Poplin 2011, 2012).

As demonstrated by Tan and Portugali (2012), gaming satisfies many requirements for decision making in cities. Games functioning as PSS are collaborative, multi-agent, support open communication, evolve incrementally and run on rules which can reflect the rules governing the city, all of which are characteristics needed for decision support in complex systems. Combining both socio-political and technical-physical complexity, gaming creates a trans-disciplinary condition where spatial design, political governance, social and cultural structures can engage in problem-solving through an interactive dialogue that crosses scales, visions and fields of expertise. Such games, however, focus on the players, aiming to create learning or communication effects. The question of whether gaming can be used for design, producing operational plans that can be implemented, remains unanswered (Mayer 2010; Portugali 1999; Tan and Portugali 2012).

Such a function for gaming is new. The game will need to produce designs and solutions for a problem in a complex system, governed by the influence of multiple actors. Drawing upon the different paradigms of planning provides the requirements for game design. The game needs to bridge the two approaches in some fashion to be able to generate a viable design. From the systems thinking approach, the game needs a realistic representation of the system in a holistic fashion and allow for exploration for future states for design and problem solving. From the participatory approach, these realistic representations of the system should be open to debate, exploration and change. Incorporating elements from both perspectives, the game should allow for the description of the system through the use of formal models and simulations and allow for exploration through collaborative processes and interfaces. The rules of the game structure both the interaction and the mechanisms by which the models are changed. The design of the interfaces and the rules of the game facilitate the bridging of boundaries between disciplines, organizations and languages. Such structures also have the benefit of fitting into the traditional paradigms of planning.

In the following sections, we show how an integrated method could be constructed and present a technical framework where these different methods are integrated. The framework integrates multiple simulations representing different aspects of the system and presents the results in real time to multiple stakeholders in an interactive manner, creating a shared design space which all the stakeholders (players) can influence. The players can control the simulations and change scenario parameters during run time. The scenario design indicates the problems that the players need to address, and the design of interfaces, parameters, data representations, etc. are critical to the effectiveness of the game. Together with the integrated simulations, these other artifacts function as a boundary object bridging the multiple disciplines. The process design guides the scenario design, the structure of the sessions within which these scenarios are explored, facilitation and so on. The framework is integral too, but is meant mostly to support, these new process designs.

4 Framework

The framework creates a virtual environment within which different aspects of the system are brought together through integrated interactive simulations, creating a shared playable space where future scenarios can be explored in real time collectively by multiple stakeholders. The integrated simulations expand the scenario boundaries, and the results from these simulations are animated in real time in a game engine. The simulation (system) parameters can be changed in real time by the stakeholders, the effects of which can be experienced immediately. To make this possible, the framework combines a game engine, networked data interaction services, databases providing real data and multiple simulations to provide realistic representations of the urban space.

Conceptually, the (participatory) simulation game can be represented in Fig. 2. Different facets of the simulation are combined in a realistic virtual environment, which is then influenced collectively by multiple stakeholders. The system behavior

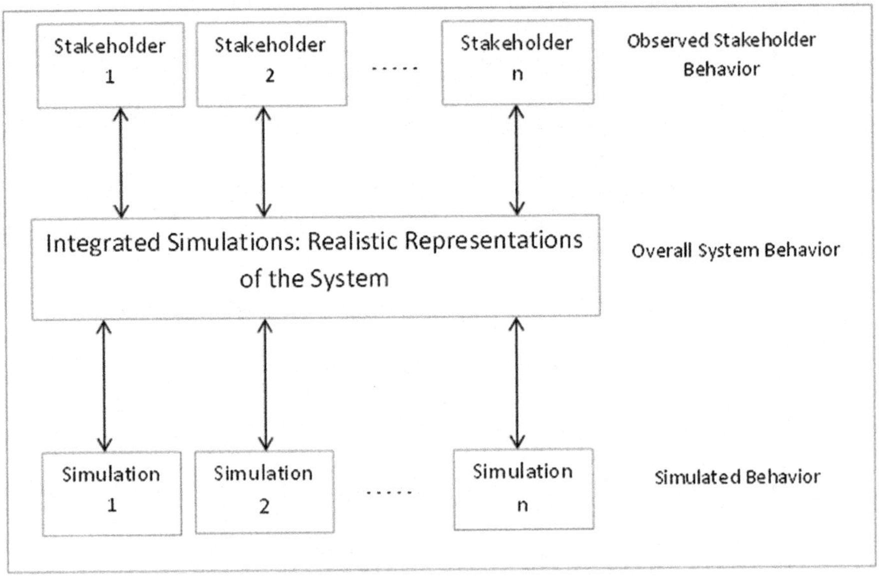

Fig. 2 Conceptual participatory simulation

is replicated through the simulations, and the inter-group dynamics, values and perspectives of different stakeholders are observed through their facilitated explorations of this environment.

The architecture of the framework is described in Fig. 3. The gaming layer is built in Unity3D, a commercial game engine. The gaming layer preprocesses OpenStreetMap (OSM) data and stores it in the database. 3D models of cities or sections of cities are then automatically generated. The section to be generated is decided based on user input. The logic for generating the city, controlling access to the database etc. resides in the web service, built using Microsoft's Web Communication Framework (WCF). The WCF exposes different functionalities as networked (local or web based) services, which can be accessed by any component that has the address.

Results from the different simulations are sent to the gaming layer either independently over the network, or through the web service. The gaming layer parses these results and animates them in the generated city. Results from the traffic simulation can be seen as cars moving around in the virtual city. The gaming layer also controls the clocks (where available) of the simulations so that the animations are synchronized with the simulation. This ensures that what players see is what is happening in the city. The players can pause, stop, restart, speed up (if the simulation can respond) and slow down the simulations. Players can also scroll backwards and forwards to the latest simulation data, allowing them to observe patterns.

The simulations interact with each other in the physical space generated in Unity. For example, the interactions between pedestrians and cars are achieved in

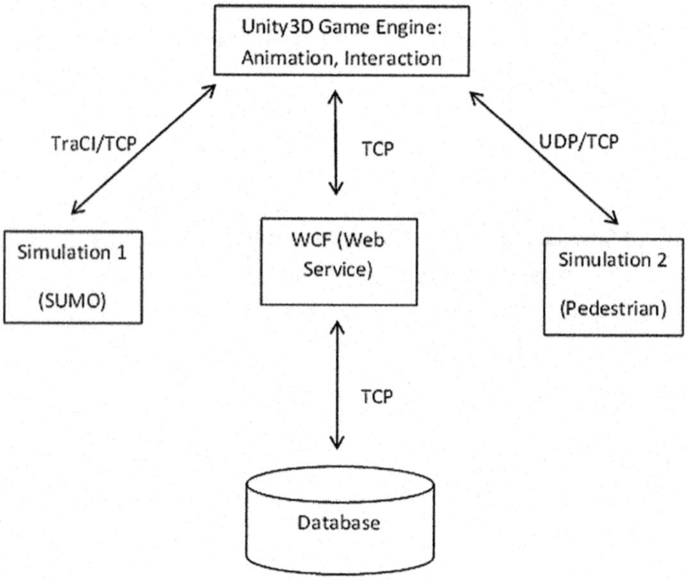

Fig. 3 High level architecture

Unity, where pedestrians and cars detect each other and respond accordingly. Neither the pedestrian simulation nor the traffic simulation is aware of each other. But the pedestrian and car game objects in Unity are aware of each other, and each tells their respective simulations to respond accordingly. For example, cars stop for pedestrians at crossings and pedestrians stop for cars at signal lights and other places and so on. Given that mostly pre-built commercial or off the shelf simulation engines are used, the only implementation effort is in building this higher order logic of interaction. A new modeling exercise where the combined pedestrian-traffic model needs to be built is not required. The technical interfaces for these interactions are modularized, and can be reused for any type of urban simulations. Apart from OSM data to generate the 3D city, the framework also includes interfaces to use operational data from public and open sources. Data feeds such as transport delays, schedules, origin destination matrices and so on are made use of in the simulations to make them realistic.

Crucial to game sessions are the options for design and the artifacts that represent them. The design of the game scenario is also through an interface in the game engine. An editor is available where players select the different options they want to experiment with. The scenario is constructed through a state machine which structures the logic of different machines, and players can change the state machine to formulate the model and edit the assumptions behind some of the models. Table 1 describes the requirements for the new design method and how the framework satisfies the requirements.

Table 1 Match between requirements and features

Requirement	Feature
Realism	Integrated simulations which expand the system boundaries, publicly available real data to enhance simulations, validated models
Participation	Multiplayer or single player games through Unity3D game engine. Process design (explained in later sections) to design the game sessions around the framework
Exploration	Run time interaction with integrated simulations. Changing the realistic simulations at run time in a session with other stakeholders to experience effects of changes
Design	For every design choice, there is a wide range of changes possible. In addition, stakeholders can (if possible) configure their own changes through the interface editor
Content	The use of open and public data and the expanded boundary of the integrated simulations ensure that the context for the problem is not too specific, but rather allows the players to design in a range of options
Cost	Reuse of commercial off-the-shelf, or existing simulations, and open data feeds ensure no expensive development efforts are needed. Also very specific models need not be designed. This also ensures validity since accepted models are reused
Openness	Models are open for exploration during run time. Also some assumptions behind the models can be changed, or the logic for the scenarios can be formulated by the players when the scenario is being designed
Infrastructure	Special computational or simulation infrastructure is required. All components work over a network through different interfaces, and if required distributed over different machines

The framework is tested with two case studies, one in the city of Paris and one in Stockholm. The use cases and the implementation are described in the following sections.

4.1 Stockholm

The problem for the city of Stockholm is to change the scheduling of university classes to reduce congestion in traffic and public transport. Three major universities in the city of Stockholm are located close to each other, and leads to a lot of congestion. The challenge is to provide appropriate information to students so that they leave classes at times when congestion is less likely to occur.

The 3D geography for the case study application is a small area around the technical university. To simulate traffic and public transport, Simulation of Urban MObility (SUMO) (Behrisch et al. 2011) an open source microscopic traffic simulator is used. The SUMO network file is obtained by converting OSM data, and the geography is the same as that of the 3D geography. The routes and trips for the cars are derived using origin/destination (O/D) matrices, which are output from the

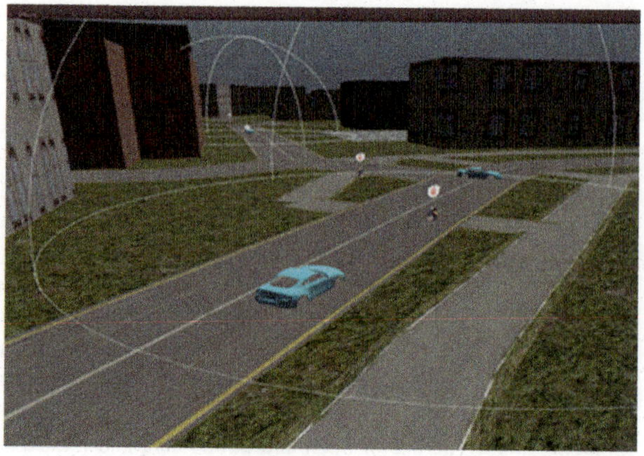

Fig. 4 Scene showing car and pedestrian interaction in Stockholm

Swedish National Transport model. The SUMO clock is controlled through the Traffic Control Interface (TraCI, an interface for online control), and output is streamed from SUMO through a Transmission Control Protocol (TCP) connection.

No COTS pedestrian simulation engine was suitable for this case study; a simple pedestrian multi-agent model was implemented. Pedestrian agents are all students, and have preferences on reaching certain locations (again obtained from the O/D data) by a certain time. They make mode choices by looking up information from publicly available application programming interfaces (APIs) provided by the city of Stockholm. Once they begin moving to their desired location, pedestrians and cars stop for each other as described before and illustrated in Fig. 4, affecting each simulation.

Apart from the default options of playing forward, speeding up, etc., players have the option of making changes in the information that the pedestrians see. They also have the options of changing schedules in public transport to see how the pedestrians would react, experiencing the effects through the animations in the game engine. The range in these options is quite wide, allowing players to settle on optimal designs for a particular simulation configuration.

4.2 Paris

The problem for the city of Paris is to manage crowds and security during large events in two stadiums. The police commanders want to design the physical area around the stadiums, such as placing road barriers, cameras, police personnel, vehicles, etc. around the stadiums to ensure that the crowd behaves accordingly and to ensure appropriate response in case of a crisis.

The 3D geography for the use case is an area around the stadiums, shown in Fig. 5. The crowd is simulated in a proprietary simulation engine developed by a French defense company. The artificial intelligence (AI) model for the pedestrians

Fig. 5 Scene showing crowd simulation in Paris animated in Unity

is very complex, allowing for a wide range of behaviors. There are different types of pedestrian models, such as regular people, very important persons (VIPs), police, terrorists, criminals and so on. Different behaviors are activated by simple tweaking of agent parameters. Traffic is simulated in SUMO, and the same interaction model and interface is used, except that the traffic in this case is minimal since a lot of roads will be blocked. O/D data or other data sources are unavailable for this case. The simulation is controlled through a TCP connection, while data are streamed from the simulation through a User Datagram Protocol (UDP) stream.

The players have the option of completely configuring the physical space around the stadium. They can block/unblock roads, move or introduce cameras, add objects such as ticket barriers, change the operational status of different objects such as automated teller machines (ATMs), change the schedules of VIP arrivals, deploy additional forces, change pedestrian parameters and so on. Once again, there is a wide design space available for the players. Different cameras provide different views, and the views available to each player can even be filtered based on model subtype; for example, some players can see only pedestrians, some can see only VIPs, and so on.

5 Process Design

While the technical framework serves as the tool to provide realism and participation, the context of use of the framework in a decision-making process still needs to be thought of by the game designer. While we have focused on the product design (the design of the physical-technical environment,or the design of schedules for example), research has consistently shown that it is not possible to ignore the implications of social interactions among actors or the institutional space, which structure the actors roles, institutional routines, governance structures and so on (Klabbers 2003).

Klabbers presents a meta-framework that tries to span a theoretical space that embodies all types of gaming. Klabbers distinguishes between two types of design. 'Design-in-the-large' refers to interventions and devising courses of action that turn situations in social systems into more preferable ones. Such situations provide the frame of reference and specifications for simulation games, which are referred to as 'design-in-the-small'. The design-in-the-small simulation game should mirror the design-in-the-large situation, in terms of roles of the actors, resources, communities of practice, social organization and other facets. In a 'design-in-the-small' game, solutions, future situations or problems can be enacted and analyzed, leading to a hypothesized possible solution for the design-in-the-large (Klabbers 2003, 2008).

In other words, game designers should be aware of the social and institutional conditions under which the actors are operating, even if the actors themselves are not. Knowledge of these conditions should be made use of by the designer, shaping the game in terms of roles of the actors, the rules of the game and resources.

The game artifacts in the chosen form of representation mirror the structures of social systems, and will probably belong to different systems; for example, the schedules of classes in the university and the schedule of public transport. The meaning of these artifacts is mediated through the tacit knowledge embedded in multiple communities of practice from which relevant actors for the problem are identified. The evolution of these artifacts leads to new solutions and forms of knowledge. The form of representation is also crucial, especially in our case since we use computer interfaces between the actors and artifacts. The interface also plays a role in structuring the interaction between the player and situation, and between players. These artifacts, otherwise also called as *actants* serve as boundary objects (Fleischmann 2006; Lo et al. 2013; Mutch 2002; Star and Griesemer 1989).

In games for the Dutch railways, some design requirements and functions were observed. Many of these games started with the request to test (formally) the effects of a new artifact, being either a track layout or a new timetable (Lo et al. 2013; Meijer 2012). While setting up these sessions in the frame of gaming for research, it increasingly appeared that the artifact was not complete. The design lacked proper development on the effects on roles and institutional rules, like the responsibilities of train traffic controllers and network controllers, and the balance between the public and private actors on the railways, being infrastructure and train operators. In the running of the sessions, the games changed to assessment of artifacts, delivering worthwhile (but also complex to interpret and manage) lists of potential bottlenecks arising from social interactions or institutional structures the artifact would be put into operations. Since the game participants in these sessions were stakeholders, gaming in such a way gave influence to the design process of a layer in the organization that is usually not present in the design activity itself. The games for design were sometimes the consequence of games for policy making, and sometimes lead to games for true theory testing (Meijer et al. 2012, 2014).

6 Conclusions

In demonstrations with stakeholders, the requirements are reinforced. Stakeholders are concerned mostly with the realism of the simulation. The realism of the data, the models used, the scales of the geography, etc. should be easily identifiable and recognizable. Another crucial requirement is the ability to simulate social aspects of the system under study.

With respect to the development of the technological framework, several design decisions become crucial, like decisions on which component should carry which piece of logic; for example, the logic of city generation or the interaction and synchronization logic of simulations and so on. Placing it in the web service as opposed to the game engine has implications on future development and more importantly on performance. The ability to animate and visualize large numbers of objects at the scale of cities is necessary, and special attention needs to be paid to computational performance.

Some of this is achieved by distributing all the components, so they can run on different machines. All the components, the simulations, web service, and the game engine can run on different machines. Other optimizations include tuning the update of game objects in the game engine, rendering only objects in the field of view and so on. While most optimizations are under the control of the developer, the biggest impact on performance comes from the simulation engines. Simulation engines are engineered to be run as fast as possible, and most engines do not inter operate with other software easily. They are meant to run the scenario, and then data can be analyzed. The clock cannot be controlled; scenario or parameters cannot be changed run time. Most importantly, data output from the simulations is in formats that are cumbersome to parse. Output is usually in time steps, expensive to parse. Developing such frameworks or games for design using realistic simulations and models will require some changes in the software development practices in simulations; the software should be open, provide fast binary streaming output and should allow embedding in other tools.

This relates to another requirement from the stakeholders, as well as the design of the game. Stakeholders are sensitive to what is hidden in the model, hidden biases and limitations, and so on. Modeling or designing a city necessarily involves multiple communities of practice, and while the artifacts in the game can create open dialogue and cross boundaries, they are constrained to a large extent by what the underlying model provides. Enhancing dialogue, providing open access to data and information requires that these models not be black boxes, but should be open reflecting the requirements of the design space.

There have been some initial steps in using gaming for design. While this is still tentative, evidence exists to support the fact that games can create open spaces for dialogue, participation, experimentation and the creation of new artifacts. However, these processes also threaten existing structures, and the game cannot be designed without taking cognizance of these threats. These methods are a new form of planning; one that bridges the traditional paradigms of urban planning, from where it derives its strength. Bridging multiple disciplines, the methods do not pursue a reductionist optimal solution but aim to transform an existing situation into a preferred one in a scientific and participatory manner.

Acknowledgments The research described in this paper was conducted while the authors were affiliated with the Department of Transport Science, School of Architecture and Built Environment, KTH, Sweden. The authors would like to acknowledge EIT ICT Labs and TRENoP who jointly funded this research effort. The authors would also like to acknowledge Mohammed Azhari, Miguel Ramos Carretero and Michael van den Berg for the programming effort that went into this. The authors also acknowledge the reviewers comments, which improved the article.

References

Alfasi, N. (2003). Is public participation making urban planning more democratic? The Israeli experience. *Planning Theory and Practice, 4*(2), 185–202.

Allen, P. M. (2012). *Cities and regions as self-organizing systems: models of complexity*. London: Routledge.

Batty, M. (2007). *Cities and complexity: Understanding cities with cellular automata, agent-based models, and fractals*. Cambridge: MIT Press.

Batty, M., Dodge, M., Jiang, B., & Smith, A. (1999). Geographical Information Systems and Urban Design. In D. J. Stillwell, P. S. Geertman, & D. S. Openshaw (Eds.), *Geographical information and planning* (pp. 43–65). Berlin Heidelberg: Springer.

Behrisch, M., Bieker, L., Erdmann, J., & Krajzewicz, D. (2011). SUMO—Simulation of urban mobility—An overview. Presented at the SIMUL 2011, The Third International Conference on Advances in System Simulation, (pp. 55–60).

Brail, R. K. (2006). Planning support systems evolving: When the rubber hits the road. In J. Portugali (Ed.), *Complex artificial environments* (pp. 307–317). Berlin Heidelberg: Springer.

Burke, E. M. (1979). *A participatory approach to urban planning*. New York: Human Sciences Press.

Cecchini, A., & Rizzi, P. (2001). Is urban gaming simulation useful? *Simulation and Gaming, 32* (4), 507–521.

Checkland, P. B. (1978). The origins and nature of "hard" systems thinking. *Journal of Applied Systems Analysis, 5*(2), 99–110.

Davidoff, P. (1965). Advocacy and pluralism in planning. *Journal of the American Institute of Planners, 31*(4), 331–338.

de Roo, G., & Rauws, W. S. (2012). Positioning planning in the world of order, chaos and complexity: On perspectives, behaviour and interventions in a non-linear environment. In J. Portugali, H. Meyer, E. Stolk, & E. Tan (Eds.), *Complexity theories of cities have come of age* (pp. 207–220). Berlin Heidelberg: Springer.

Dearden, J., & Wilson, A. (2011). Using participatory computer simulation to explore the process of urban evolution. *Transactions in GIS, 15*(3), 273–289.

Devisch, O. (2012). *The metaverse as lab to experiment with problems of organized complexity*. Farnham: Ashgate Publishing.

Duke, R. D., & Geurts, J. (2004). *Policy games for strategic management*. Amsterdam: Rozenberg Publishers.

Dykes, J. (2000). An approach to virtual environments for visualization using linked geo-referenced panoramic imagery. *Computers, Environment and Urban Systems, 24*(2), 127–152.

Fleischmann, K. R. (2006). Boundary objects with agency: A method for studying the design-use interface. *The Information Society, 22*(2), 77–87.

Gaber, J. (2007). Simulating planning simcity as a pedagogical tool. *Journal of Planning Education and Research, 27*(2), 113–121.

Gordon, E., Schirra, S., & Hollander, J. (2011). Immersive planning: a conceptual model for designing public participation with new technologies. *Environment and Planning-Part B, 38* (3), 505.

Greenblat, C. S., & Duke, R. D. (1981). *Principles and practices of gaming-simulation*. CA: Sage Publications Beverly Hills.

Hajer, M. A. (2003). *Deliberative policy analysis: Understanding governance in the network society*. Cambridge: Cambridge University Press.

Hanzl, M. (2007). Information technology as a tool for public participation in urban planning: A review of experiments and potentials. *Design Studies, 28*(3), 289–307.

Harvey, D. (1973). *Social justice and the city*. Baltimore: Johns Hopkins University Press.

Healey, P. (1997). *Collaborative planning: Shaping places in fragmented societies*. UBC Press.

Hughes, J., & Sadler, S. (2013). *Non-plan: Essays on freedom, participation and change in modern architecture and urbanism*. London: Routledge.

Innes, J. E. (1995). Planning theory's emerging paradigm: Communicative action and interactive practice. *Journal of Planning Education and Research, 14*(3), 183–189.

Innes, J. E. (1996). Planning through consensus building: A new view of the comprehensive planning ideal. *Journal of the American Planning Association, 62*(4), 460–472.

Jacobs, J. (1961). *The death and life of great American cities*. New York: Random House LLC.

Klabbers, J. H. G. (2003). Gaming and simulation: Principles of a science of design. *Simulation and Gaming, 34*(4), 569–591.

Klabbers, J. H. G. (2006). Guest editorial. Artifact assessment vs. theory testing. *Simulation and Gaming, 37*(2), 148–154.

Klabbers, J. H. G. (2008). *The magic circle: Principles of gaming and simulation*. Rotterdam/Taipei: Sense Publishers.

Lee, D. B. (1973). Requiem for large-scale models. *Journal of the American Institute of Planners, 39*(3), 163–178.

Lindblom, C. E. (1959). The science of "muddling through". *Public Administration Review, 19*(2), 79–88.

Lo, J., van den Hoogen, J., & Meijer, S. (2013). *Using gaming simulation experiments to test railway innovations: Implications for validity* (pp. 1766–1777). IEEE.

Mayer, I. S. (2010). The gaming of policy and the politics of gaming: A review. *Simulation and Gaming, 40*(6), 825–862.

Meier, R. L., & Duke, R. D. (1966). Gaming simulation for urban planning. *Journal of the American Institute of Planners, 32*(1), 3–17.

Meijer, S. A. (2009). *The organisation of transactions: Studying supply networks using gaming simulation*. Wageningen Academic: Wageningen.

Meijer, S. (2012). Gaming simulations for railways: Lessons learned from modeling six games for the Dutch infrastructure management. In X. Perpinya (Ed.) *Infrastructure design, signalling and security in railway*. doi:10.5772/35864.

Meijer, S. A., Mayer, I. S., van Luipen, J., & Weitenberg, N. (2012). Gaming rail cargo management: Exploring and validating alternative modes of organization. *Simulation and Gaming, 43*(1), 85–101.

Meijer, S., Reich, Y., & Subrahmanian, E. (2014). The future of gaming for design of complex systems. In R. D. Duke., & W. Kriz. (Eds.), *Back to the future of gaming*. Gutersloh: W. Bertelsmann.

Minnery, J., & Searle, G. (2014). Toying with the city? Using the computer game simcity^{TM}4 in planning education. *Planning Practice and Research, 29*(1), 41–55.

Mutch, A. (2002). Actors and networks or agents and structures: Towards a realist view of information systems. *Organization, 9*(3), 477–496.

Poplin, A. (2011). Games and serious games in urban planning: Study cases. In B. Murgante, O. Gervasi, A. Iglesias, D. Taniar, & B. O. Apduhan (Eds.), *Computational science and its applications—ICCSA 2011* (pp. 1–14). Berlin Heidelberg: Springer.

Poplin, A. (2012). Playful public participation in urban planning: A case study for online serious games. *Computers, Environment and Urban Systems, 36*(3), 195–206.
Portugali, J. (1999). *Self-organization and the city*. Berlin: Springer Science & Business Media.
Portugali, J. (2006). Complexity theory as a link between space and place. *Environment and Planning A, 38*(4), 647–664.
Portugali, J. (2012). Complexity theories of cities: Implications to urban planning. In J. Portugali, H. Meyer, E. Stolk, & E. Tan (Eds.), *Complexity theories of cities have come of age* (pp. 221–244). Berlin Heidelberg: Springer.
Raghothama, J., & Meijer, S. (2014). A review of gaming simulation in transportation. In S. Meijer., & R. Smeds (Eds.), *Frontiers in gaming simulation* (Vol. 8264, pp. 237–244). Berlin: Springer International Publishing.
Schoenwandt, W. (2012). *Planning in crisis? Theoretical orientations for architecture and planning*. Farnham: Ashgate Publishing Ltd,
Star, S. L., & Griesemer, J. R. (1989). Institutional ecology, `translations' and boundary objects: Amateurs and professionals in Berkeley's museum of vertebrate zoology, 1907-39. *Social Studies of Science, 19*(3), 387–420.
Tan, E., & Portugali, J. (2012). The responsive city design game. In J. Portugali, H. Meyer, E. Stolk, & E. Tan (Eds.), *Complexity theories of cities have come of age* (pp. 369–390). Berlin Heidelberg: Springer.
Van Bueren, E. (2009). *Greening governance: an evolutionary approach to policy making for a sustainable built environment*. Amsterdam: IOS Press.
Ware, C. (2000). *Information visualization: Perception for design*. San Francisco, CA: Morgan Kaufmann Publishers.

Chapter 17
The Everyone City: How ICT-Based Participation Shapes Urban Form

Sara Levy, Karel Martens and Rob van der Heijden

Abstract Citizen participation is a cornerstone of urban planning. One common criticism is that the process can be cumbersome and slow. However, in the face of recent advances in information and communication technologies (ICT), those problems can be easily overcome, making it possible to extend public participation to a wider sphere of urban planning matters. But what do we know of how ICT-based public participation affects urban form? What does a city shaped by social networks and other ICT-tools look like? We develop an agent-based model of urban growth to improve our understanding of these issues. Our model consists of a spatially disaggregated, micro-economic-based, real estate market model coupled with an ICT-based planning process. In the model, public participation is based on social network affiliation and preferences over the height of buildings.

1 Introduction

1.1 ICT, Citizen Participation and Urban Planning

That citizens should participate in the creation and management of their own cities is a much revered idea, but one that has been proven difficult to implement in practice. The most common criticisms are that participation processes can be cumbersome, slow, frustrating and fail to give a real voice to the citizens, especially the underprivileged classes and minorities. Urban planners continually struggle to

S. Levy (✉) · K. Martens · R. van der Heijden
Institute for Management Research, Radboud University Nijmegen,
P.O. Box 9108, 6500 HK Nijmegen, The Netherlands
e-mail: sarocaslevy@gmail.com

K. Martens
e-mail: k.martens@fm.ru.nl

R. van der Heijden
e-mail: r.vanderheijden@fm.ru.nl

S. Geertman et al. (eds.), *Planning Support Systems and Smart Cities*,
Lecture Notes in Geoinformation and Cartography,
DOI 10.1007/978-3-319-18368-8_17

enhance the participation process, to maximize and diversify stakeholder input in the design of solutions for urban problems (Brabham 2009).

Many authors have welcomed the recent advances in information and communication technologies (ICT), which, by making transactions of information much faster, cheaper and easier, promise to help overcome some of the problems with citizen participation in urban planning. To some extent this is already happening. Mobile apps and webpages are being used to help citizens navigate zoning and urban plan information (Desouza and Bhagwatwar 2012); blogs and social networks are being successfully used as platforms for expressing opinions about planned developments, and for gathering support for those opinions (Evans-Cowley and Hollander 2010); and online petitions against new developments are now a common presence on the web.

While new methods and arenas for citizen participation keep emerging, it is still too soon to know for sure what final forms ICT-based participation will take. Formal public participation is usually reserved for large or complex developments, especially those expected to have wide environmental, social or economic implications. However, nowadays, citizens are no longer dependent on the official arenas for providing feedback over planning matters. They can express their views about a wider sphere of urban planning matters. Opinions expressed through unofficial channels do not have to be considered by planners. However, if support spreads, it might become hard to ignore them and, as a minimum, they are likely to delay approval processes.

Not a lot is known about how public participation affects urban form. But the question takes on new and greater relevance if ICT-based participation fulfils its promise and is able to successfully involve a significant number of citizens in commenting on a wider number of projects. While technology may seem to democratize access to information, it may also introduce undesirable side effects in the outcomes. For example, blogs and social networks may give full-blown visibility to some opinions, even if they do not represent the norm. In addition, ICT tools are not neutral in terms of whom they empower, since the wealthy have privileged access to technology. Furthermore, social networks tend to be composed of similar individuals (except for a few well connected individuals that tend to have very diverse networks), and therefore something that spreads through a social network does not necessarily reach a diverse pool of people.

1.2 Urban Density, Planning Process and Externalities

The urban environment is characterized by a number of dimensions, of which the height of buildings and the amount of green areas are two of the most visible and controversial. On the one hand, the high densities achieved by tall buildings are important for the reduction of negative environmental externalities of human occupation. On the other hand, green and open areas lower residential densities, but they are important for keeping citizens happy and healthy. These are matters that people care deeply about but have little control over.

Urban economics' models tend to assume that demand for housing is seamlessly translated into supply of housing, and therefore, to treat urban form as a direct product of households' preferences. But decisions over building height and amount of open areas are ultimately taken by developers, in dialogue with architects and engineers, in the creative process, and then with planners, in the approval process. Glaeser and Gyourko (2002) argue that, more than a housing affordability crisis, it is the difficulty in obtaining permits for new buildings, compounded with strict zoning laws, that has affected the supply of new housing in some cities. Schaeffer and Hopkins (1987) describe how projects can be transformed in the process of obtaining approval from the planning authorities. According to the authors, if the project stays within the current constraints set by zoning regulations and building codes, the approval should be obtained easily. If not, a negotiation process may ensue, regarding zoning changes and other regulations, or the rights and obligations attached to the land. It is in this process of negotiation that the final shape of the project is determined.

These insights suggest that the relationship between households' preferences and urban form is not as simple as supply and demand. Approval time, for example, is an important variable. Long permit approval times affect developers' decisions because they imply a delay to realize profits, during which the money invested in the project is unproductive. Furthermore, the final form of the development (the density or building height) is affected by negotiation during the permit application period.

On the demand side, things are equally complicated. Households may prefer low-density neighbourhoods, even if higher densities would be more beneficial in terms of environmental impact and economic efficiency. Cinyabuguma and McConnell (2013, p. 332) frame this problem as an externality of urban growth: "growth often creates benefits for an entire urban area, but the costs of that growth are born primarily by residents of the neighborhoods where new development occurs. An externality problem arises because existing residents perceive the local costs associated with admitting new residents, but not the full benefits which accrue to the city as a whole". The same argument can be made for compact (tall) developments. While it allows increasing urban densities and thereby limiting sprawl and its associated environmental costs, the costs of compact development are born by the residents in the neighbourhoods where it occurs. The result is that existing residents have an incentive to block new residents to their neighbourhoods, resulting in cities that are less dense than is optimal, or too spread out (Cinyabuguma and McConnell 2013).

The full-blown impact of this effect, however, depends on the ability of incumbent residents to effectively halt or transform planned developments. In most cases, their ability is small because most developments do not require an official public hearing. But the last decades have delivered the tools to boost the visibility of any one opinion, by spreading it cheaply and quickly to virtually everyone.

What happens to urban form when citizens are able to influence the approval or rejection of residential projects? How does such a city compare to one where citizens only get to shape it as consumers? These are relevant questions if urban planning is serious about the goal of expanding public participation.

1.3 Agent-Based Models of Urban Growth and Urban Planning

Despite being a relatively recent methodology, agent-based models (ABMs) already have a long tradition of modelling urban growth, land-use change and land and real estate markets. Most ABMs of land use are based on von Thünen's (1826) agricultural land rent theory, and its application to urban land markets by Alonso (1964). Alonso's monocentric city model has since been refined by several authors (such as Muth and Mills), and remains a staple of urban economics. Its basic premises are that employment is located in the central business district (CBD), and households face a trade-off between housing lot consumption and the costs of commuting to the CBD. The Alonso-Muth-Mills models have also been extensively criticized on the grounds of their simplistic assumptions, especially: the adoption of the utility maximizing framework and the assumption of perfect rationality; the reliance on commuting costs as the only determinant of urban spatial structure; the neglect of the impact of hedonic value of housing characteristics and neighbourhood characteristics on residential choice; the neglect of various forms of externalities; the assumption of monocentricity; the static nature of the models; and the assumption that location is continuously variable (for a summary, see Briassoulis, undated). Many subsequent models have addressed these limitations. Here, we report only on ABMs.

One important line of ABMs focuses on modelling land markets and land-use change, with an emphasis on the externalities of land-use decisions. Parker and Meretsky (2004), Brown et al. (2004, 2008), Caruso et al. (2007, 2009) and Zellner et al. (2010) explore how both negative externalities of urban land use and positive externalities of agriculture or green land cover affect other agents' location and land-use decisions. Caruso et al. (2007, 2009) present, in great detail, a framework for modelling households residential location decisions subject to two neighbourhood externalities: local residential density and open space. We adapt this framework in our model, but add accessibility as a further location criterion, and adapt the shape of the utility functions to reflect a more complex attitude of households towards open space and residential density (building height, in the case of our model).

Iovanna and Vance (2013) and Filatova et al. (2009) model the bilateral interactions between buyers (urban dwellers) and sellers (farmers) of land. Iovanna and Vance (2013) introduce landscape heterogeneity in terms of different agricultural productivities, and focus on predicting the place of future conversion of land to urban uses in real landscapes, while Filatova et al. (2009) use an abstract landscape to explore how differences in sellers' and buyers' market power affect urban form. Sun et al. (2014) extend the Filatova et al. (2009) model to show how the level of detail of market representation affects the outcomes of the model. Following these authors' recommendations, our model includes representation of essential market features such as budget constraints and competitive bidding to ensure a more realistic representation of the real estate market. CHALMS (Magliocca et al. 2011) is a comprehensive ABM of both land and real estate markets. It includes developers that face the decision of where to develop and at what density. However, in CHALMS, developers and households choice set is composed of fixed combinations of house size and lot size that are prevalent in exurban developments. No neighbourhood attributes are considered in the residential choice. Our model, on the other hand, focuses on how households' choices (both as consumers and as citizens) shape and are shaped by the neighbouring environment. Therefore, the focus is on the most relevant aspects of a residential location: building height, the amount of green space and accessibility.

Another line of research focuses on modelling real estate markets. Devisch et al. (2009) and Ettema (2011) use ABMs to model housing markets as the outcome of households' relocation decisions and bargaining processes with sellers. These models emphasize the decision process as one that incorporates learning from previous market interactions, the effects of uncertainty inherent in decisions about the future, and life cycle effects, leading to changed preferences over time.

Only a handful of ABMs explicitly include planning agents (Monticino et al. 2007; Li and Liu 2007, 2008; Zellner et al. 2009; Robinson and Brown 2009; Ligmann-Zielinska and Jankowski 2010), and of these, in only a few is the focus on approving new developments. These models include some ad hoc features of the planning process and of land markets, but no solid micro-economic foundation. In Li and Liu (2007), government agents decide whether to approve applications for land development, based on a number of factors, including existing land use, existing land-use plans, and the number of times the location has been targeted for development by developers or residents. In Monticino et al. (2007), government agents can approve, modify or reject a development proposal. The decision is based on the government agent type, proposal type, and environmental information provided by the natural system model. In addition, homeowner agents can decide whether to protest the proposed development or not. The protest decision is based on the homeowner agent type, the development proposal, and the type of residential development in which the homeowner agent resides and the weights of the government agent's decision to approve the project. Our model is thus far, to the best of our knowledge, the only ABM to represent the interaction between citizen participation and the real-estate market.

2 Model Description

2.1 Purpose of the Model

This chapter presents an ABM of a real estate market coupled with a participatory planning process, developed to explore the impact of broad ICT-based citizen participation on urban form. The model represents a growing urban area, where developers search for developable land, households search for houses, and a planner handles permit requests. The model extends existing microeconomic models, such as Czamanski and Roth (2011), which provides a solid foundation for addressing the developer's decision of where and how tall a building to build, while at the same time establishing a link with both institutional variables and consumer demand for residential locations. The model used a very simple stylized landscape but can easily be improved to make use of spatial data from real cities.

2.2 Agents

The ABM comprises a landscape consisting of cells, three 'human' agent types—households, developers and planners—and three instrumental agents—alternatives, projects and buildings.

2.3 Process and Scheduling

There are four main processes (A, B, C, D) that occur sequentially in every step of the model.

A. **Developers: Surveying and Selecting**
A number of developers operate in the city. Their goal is to build houses and sell them at the highest profit, and their program is to: (i) find potential locations; (ii) assess the potential revenue from each alternative location and building height; (iii) select a project and submit it for approval; (iv) build; and (v) sell the houses to the highest bidders.

A.1. **Finding potential locations**: Developers search for developable land, probing a high (though limited) number of randomly chosen locations, from which they select the three most promising for further evaluation. Different locations have different values of 'Accessibility' and 'OpenSpace' (undeveloped cells around them).

A.2. **Estimating revenue for different locations and development types**: For each of the three selected locations, developers estimate the potential revenue and costs for each possible building height (1–10). The potential profit from each alternative depends on the cost of land, construction costs, the duration of the approval process, and the estimated demand for each location and height. They select the option with highest potential profit, purchase the land, and submit it for approval to the planning authorities.

B. **Permitting and Participation Process**
In the baseline scenario, the approval process consists only of an interval of time between selection of an alternative and its construction where no action is taken. In the participation scenarios, on the other hand, during this time the planner has to inform citizens of the projects submitted for approval. After he/she informs the public, the following processes can take place:

1. Households: platform creation and support.
2. Planner: gauging strength of public opposition to a project.
3. All agent types: search for consensus solutions.

These processes are detailed in the Design Concepts section below.

C. **Construction**
The construction process consists of transforming the selected and approved developments into buildings, whose houses are put on the market with an ask price set by the developer.

D. **Bidding and Trading**
Households from the 'rest-of-the-world' migrate into the city at a fixed rate in search of a place to live. Households search for available residential locations within their housing budget. They select the option with the highest utility, based on their preferences for three aspects of a location: Accessibility, OpenSpace and Height of buildings. Households place a bid on the selected house that equals their willingness-to-pay (wtp) for the house. Households' wtp is calculated based on housing budget, estimated commuting costs and utility. For each building, developers sell the houses to the highest bidders.

D.1. **Market adjustments**
The bidding and trade process is repeated 20 times or until there are no more available houses or no more households searching for a house. Unsuccessful households will be able to search and bid again in the next step. Developers who haven't been able to sell at least half of the houses they built are not allowed to engage in new projects. Every four steps of the model, the amount of unsatisfied demand is queried, and more developers can be created in response to high levels of unsatisfied demand.

2.4 Design Concepts

2.4.1 Households' Residential Location Choice

The theoretical framework for the households' residential location choice is utility theory. Households calculate the utility of available residential locations based on their preferences regarding the type of neighbourhood environment they would like to live in. Note that households consider only attributes of the location (neighbourhood), and not the actual estate. This corresponds to assuming that, for each location, there is a wide enough array of housing options so that the choice of a house can be nested within the choice of location, but considered independent from it.

Household utility for a particular location depends on three neighbourhood attributes. These are: Accessibility (A), representing the proximity to employment, services and shops in the CBD; OpenSpace (O), representing the amount of undeveloped land in the building's immediate surroundings; and Height (H), representing the height of the surrounding buildings.

Households use a (de)compositional approach (Louviere and Timmermans 1990), in that they evaluate residential opportunities by first valuing the attribute levels and then combining these attribute level values into a value (utility, U) for the entire bundle according to some simple algebraic rule—in this case, a Cobb-Douglas utility function:

$$U = A^{\alpha} O^{\beta} H^{\gamma} \tag{1}$$

Households are heterogeneous in what concerns their preference profile for the location's attributes. Some households attribute more importance to accessibility while others find that the presence of nature or height of development weighs more in their decision. The households' preference profile is the vector of weights [α, β, γ]. These weights reflect the strength of households' preferences concerning, respectively, Accessibility, OpenSpace and Height attributes. The preferences of the households for each factor are assumed to be independent of each other and of income.

Accessibility of a location to the CBD (A_{ij}) is computed using a negative exponential formula, which is one of the standard forms of measuring accessibility:

$$A_{ij} = c \times \exp(-\theta \times \text{distanceCBD}_{ij}) \tag{2}$$

for which ij is the cell for which accessibility is being computed; c is a constant; θ is a distance decay parameter and distanceCBD_{ij} is the topographical distance of the cell to the CBD. For chosen parameters, Eq. (2) renders values that normalize to a 100–0 scale (Fig. 1).

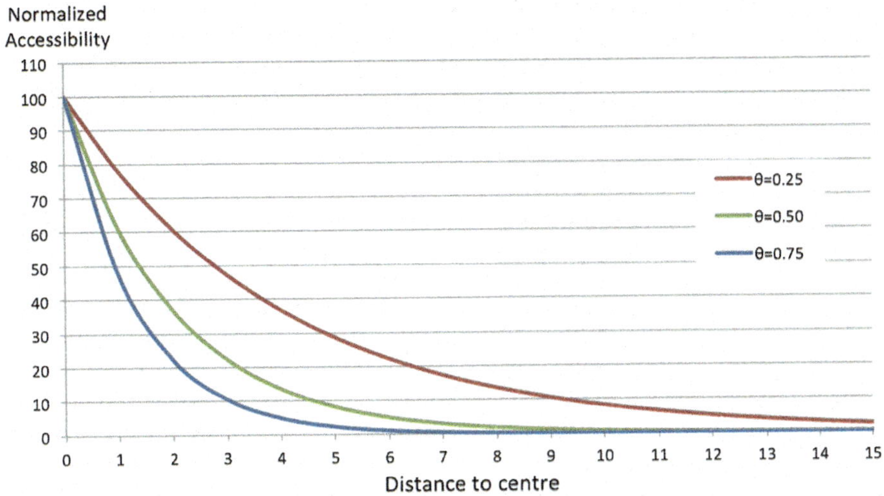

Fig. 1 Normalized accessibility values versus distance to the CBD for different values of the exponent **θ**

The amount of OpenSpace (O) is calculated based on the number of undeveloped cells in the 8-cell neighbourhood of the building. Like other authors [see, for example, Caruso et al. (2007)], we assume O to take a negative exponential form. However, we do not assume that individuals' utility always increases with OpenSpace (more Accessibility is always better, but more OpenSpace is not always better). We assume that each individual *n* has an ideal value of OpenSpace (openspace^*_n) that gives them maximum utility, and they evaluate each location in relation to the ideal value using:

$$O_{n_{ij}} = c \times \exp\left(-\eta \times \left|\text{openspace}^*_n - \text{openspace}_{ij}\right|\right) \tag{3}$$

Figure 2 shows the values taken by O_{nij}, for chosen parameters, as calculated by Eq. (3).

Building Height (H) refers to the height of the tallest building present in the 8-cell neighbourhood of each location. We assume that individuals have a certain height they tolerate, below which they are indifferent and above which their utility decreases quickly with increases in height. Overmax represents how many floors the height is above the maximum tolerated height, h_n*. Figure 3 shows the values taken by H_{nij}, for chosen parameters, as calculated by Eq. (4):

$$H_{n_{ij}} = \exp(-\kappa \times \text{overmax}_{n_{ij}}) \\ \text{overmax}_{a_{ij}} \begin{cases} = 0 & \text{if } h \le h^*_a \\ = h - h^*_a & \text{if } h > h^*_a \end{cases} \tag{4}$$

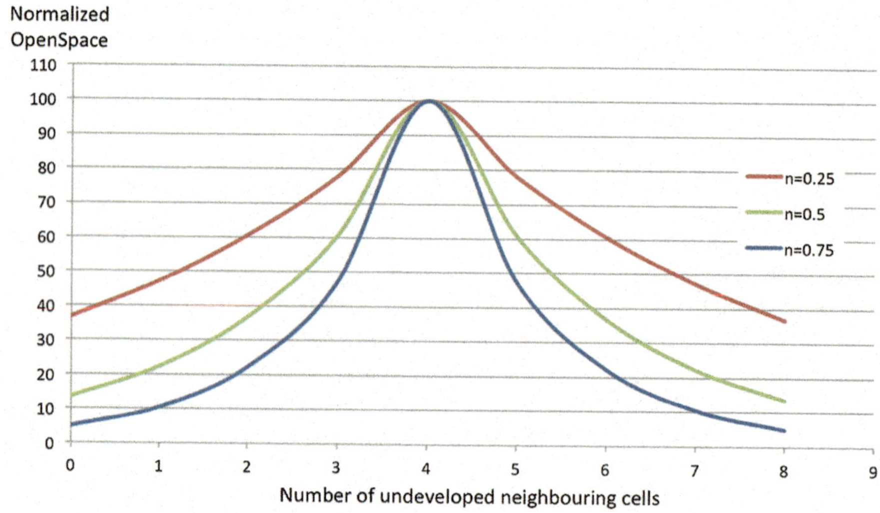

Fig. 2 Normalized OpenSpace values versus number of undeveloped cells in the location immediate neighbourhood, for different values of the exponent η, and for a household with ideal openspace* = 4

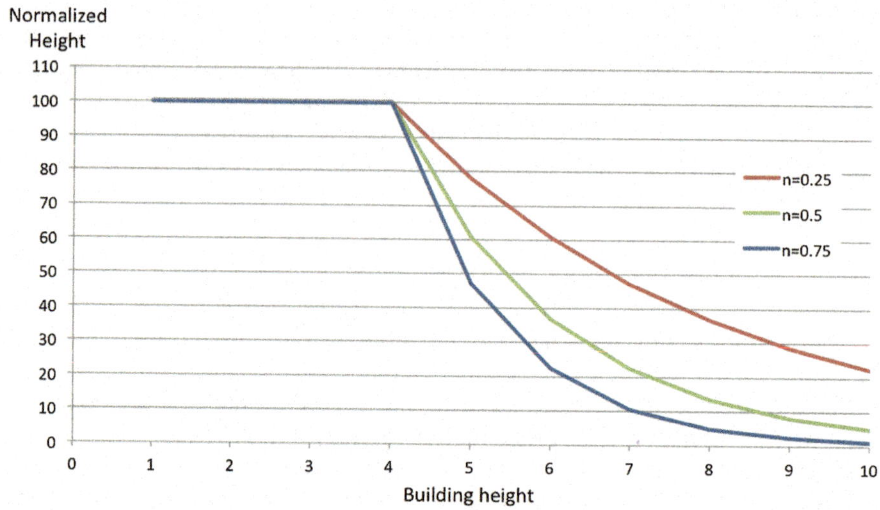

Fig. 3 Normalized height values versus building height, for different values of the exponent η, and for a household with maximum tolerated height, h* = 4

When selecting a house, households choose, out of the available housing opportunities, the one that maximizes their utility U and is affordable given their budget, net of transport costs. The exponents θ, η and κ are chosen after a sensitivity analysis. Alpha, beta and gamma are random parameters controlling for taste heterogeneity and are taken out of a normal distribution, with mean chosen by the user. Ideal OpenSpace (openspace*) and maximum tolerated height (h*) are also taken out of a normal distribution, with means and standard deviation chosen by the user.

2.4.2 Developers' Decision Making

Micro-economic foundations We adapt the model of developer decision-making developed by Czamanski and Roth (2011) and Czamanski and Broitman (2012). This model was chosen because it presents a sound micro-economic foundation for the developers' simultaneous choice of location and building height, and its links to institutional variables such as the duration of the development process. In Czamanski and Roth's model, developers search for developable land while simultaneously deciding on the type of development density to build (building height). Each developer chooses the location and height of development that maximizes the future value of their cash-flow (Czamanski and Roth 2011). Equation (5) assumes the developer has two costs: the initial investment (I), which is essentially the cost of purchasing the land, and 'overnight costs' (c) that represent all the other multi-period costs, such as building costs. Because the investment on land has to be made in a time well before the realization of income from the project, it is penalized by interest rate r for the number of years that separate investment from revenue realization. This time is the Characteristic time τ, and includes construction time plus approval time. Approval time varies according to location and height:

$$\max_{x,h} FV(t = \tau) = -I_x \times (1 + r)^{\tau} - C_d(h) + P(A, G, H) \times h \tag{5}$$

Developers' heterogeneity and rationality Developers are rational profit-maximizers, but their rationality is bounded due to limited access to information about consumers' preferences and the duration of the planning process, and because the costs of searching for information are high. In our model, this translates in a number of ways. Developers do not evaluate all possible locations for building, but rather probe a limited (though high) number of randomly chosen locations, from which they select the three most promising for further evaluation. Also, developers' estimates of consumer preferences are based on random samples of households and not on their future buyers. Developers are heterogeneous in regard to their technology, meaning the overnight costs are developer specific. The cost function for a building of height h is:

$$C_d(h) = cte_d + slope_d \times h \tag{6}$$

where cte_d and $slope_d$ for each developer d are taken from a normal distribution, and h is the height of the building.

Two-step decision-making Developers follow a two-step protocol that allows them to consider a broad range of locations and alternative developments. First, they carry out a preliminary estimation of demand for a high number of possible development locations. This first estimate only considers the characteristics of the location in terms of Accessibility and OpenSpace. The first estimate allows developers to narrow down location choices for their developments. Once choice is narrowed down to three locations, they carry out a second estimation exercise, which also takes into account all possible building heights (1–10) in each of the three selected locations.

Land market We do not explicitly model a land market. Usually, the price of a housing good unit is set in relation to a reservation bid rent, set to be equal to the price of undeveloped land, which is assumed to be at least no less than the average price of agricultural land. As the argument goes, the plots that are not converted to urban use only provide the agricultural rent to the landowners, or at least, average agricultural land price serves as the opportunity costs for conversion into urban land (Filatova et al. 2011). But as these authors point out, in the real world, agricultural productivity, cost of production and prices for agricultural goods, affecting the price for agricultural land are heterogeneous (Filatova et al. 2011), and thus do not provide a solid base for price formation. In our model, land is homogenous except for distance to the CBD. Developers' bids for land are directly proportional to its accessibility, and always above agricultural productivity (farmers always sell).

Duration of development process The duration of the development process depends on the duration of the permitting process, which is endogenous to the participation process.

Estimating demand/sale price of real estate We assume that the developer is able to access stated preference data about residents' preferences directly, and estimate the wtp of households for the concrete alternatives being considered using Eq. (7), adapted from Filatova et al. (2009), where U is the household's Utility, $Y_{housing}$ is the budget for housing, C_{comm} are commuting costs, and b is a proxy for the affordability of all other goods to reflect their relative influence on the WTP for housing (Filatova et al. 2009):

$$WTP = \frac{\left(Y_{housing} - C_{comm}\right) \times U^2}{\left(b^2 + U^2\right)} \quad (7)$$

For each alternative, the developer can use stated preference data for a number of random households. The developer can then consider the top 20 wtp (average of the 20 highest values of wtp), commit to the alternative that maximizes the future value estimated with Eq. (5), and turn the alternative into a project that is submitted for approval. The wtp estimated will serve as ask price when the house enters the market.

2.4.3 Participation Process

Social network formation On arrival, each household starts to form its social network. It adds the first n agents at random, or based on (budget) similarity if the parameter 'status homophily' is on. Agents then add any other agent that has befriended them. Social networks are of varied length.

Platform creation and support When developers submit a project for approval, planners inform the public of the characteristics of the new project submitted. Households examine the impact of the project on their utility. If any household suffers a big enough loss of utility, it can create a platform (website, blog, etc.) to express its disapproval and gather support from other households. The opinions expressed in the platform travel through the social network via the platform-owner's personal network, and the networks of any supporters. The degree of support for the opinion of others depends on many factors. According to research on homophily reported by McPherson et al. (2001) both status and value similarity between two people tends to increase the likelihood of issue-related interpersonal communication, and in turn, leads them to have more influence over one another. In the case of this model, since status similarity may already determine the composition of agents' networks, support from each member of the platform-owner's social network depends on:

- the degree of value similarity (s) between the two households, as expressed by differences in preferences for building height;
- the difference between the height of the proposed new development and the height tolerance of the household (t); and
- a random variable (ξ).

These values are transformed into a probability of support, which is then translated into support if it is higher than 0.5.

$$P_{support} = \frac{1}{1 + e^{-st\xi}} \tag{8}$$

Gauging public opposition Planners have access to the platform and continuously gauge public opposition for a project, only approving projects for which opposition is down to acceptable levels.

Search for consensus When the planner becomes aware of public opposition to the project, he/she can facilitate a negotiation process between citizens and the developer about the height of the development. In the negotiation, the first proposal on the negotiation table is the original project. If there is minimal public opposition, it can be approved. But if public opposition is high, the developer re-examines the profit estimates while lowering the building height by one floor. If this is still a profitable project, the developer introduces this project alteration. If not, the project is withdrawn. The negotiation progresses in this manner from one round to the next.

It ends when public opposition reaches a minimum, or when the developer withdraws the project. Only one negotiation round takes place per time step. Therefore, the negotiation process can span through several time steps, as many as necessary until the loop reaches an exit condition. The longer the negotiation takes, the greater the difference between the original and the consensus solution. This process does not only affect the duration of the development process, but also transforms the project.

Duration of permitting process The duration of development process depends essentially on the duration of the permitting process, which is constant for the baseline scenario and depends on the number of negotiation rounds in the search for consensus in the participation scenario.

2.5 Implementation

The model is implemented in Netlogo v5.0.4. (Wilensky 1999).

2.6 Limitations

Our modelling approach presents a few limitations. With respect to the households' residential location choice, several factors are not accounted for, such as the location of employment, school quality, taxation, and others. With respect to the real estate model, several processes are not accounted for, such as the heterogeneity of housing products, cycles of decay and gentrification, and the cyclical nature of real estate markets (Wheaton and Simonton 2007). In what concerns the participation process, citizens only oppose protests. If a project actually improves their utility, they remain passive. Also, their only two possible motivations are utility and homophily. We do not include other motivations, such as caring for the environment, caring for other parts of the city that are not their own neighbourhood. In addition, negotiations are complex processes that usually involve multiple dimensions of a project. In our model, the only dimension open to negotiation is height.

3 Results

The model was run on a 41 × 41 gridded landscape, with the CBD in the middle at coordinates (0, 0). Incoming households at each step of the model are characterized by a set of preferences and a budget for housing (which also serves as a proxy for wealth). Preferences are drawn randomly from a normal distribution with variable mean and variance, and budget is drawn from a normal distribution with average $200,000 and standard deviation $40,000.

The model was run 10 times for each group of settings. The statistics presented below are averages of the 10 runs, and the snapshots are of one typical run, at step 150.

3.1 Baseline Scenario

In the baseline scenario, only market processes are at work; there is no participation. The results are consistent with urban economic theory: urban density tends to decrease from the centre to the periphery, along with land prices. In the model, arriving households are assigned a set of preferences that are drawn from normal distributions, defined by a mean and a variance. Figure 4 shows the model's state at time step 150 for different runs of the model, that differ in the mean of the random distribution from which the households' preferences for OpenSpace and Height are drawn. For populations with high average ideal OpenSpace, the urban core is surrounded by a mixed use (urban-rural) fringe, confirming previous results by Caruso et al. (2007). For populations with low average ideal open space, the mixed used fringe is less noticeable or disappears completely. For populations with a high average tolerance for building height, a lot of buildings with heights 10 and 9 floors are built, reflecting not only household preferences, but also the developers' scale economies in building high-rise buildings. The model runs in Fig. 4 were conducted with very low variance (homogeneous preferences).

3.2 Participation

3.2.1 Urban Form in Participated Scenarios

Participation has an impact on urban form, as can be seen in Figs. 5 and 7. As seen in Fig. 6, participation it tends to lead to a greater stock of low-rise buildings, as renegotiation of heights pushes the height of some buildings down. For homogenous populations, participation tends to lead to more construction activity and lower unsatisfied demand than any of the market scenarios, and thus a considerably larger city. Note that this model does not produce a measure of sprawl, since households that cannot find a place in the model might still contribute to sprawl outside the bounds of model.

3.2.2 Agent Heterogeneity

Population taste heterogeneity, as controlled by the standard deviation (sd) of the normal distribution of values of height tolerance (h*) and ideal open space (openspace*) also has noticeable effects on the landscape, as seen in Figs. 8 and 9.

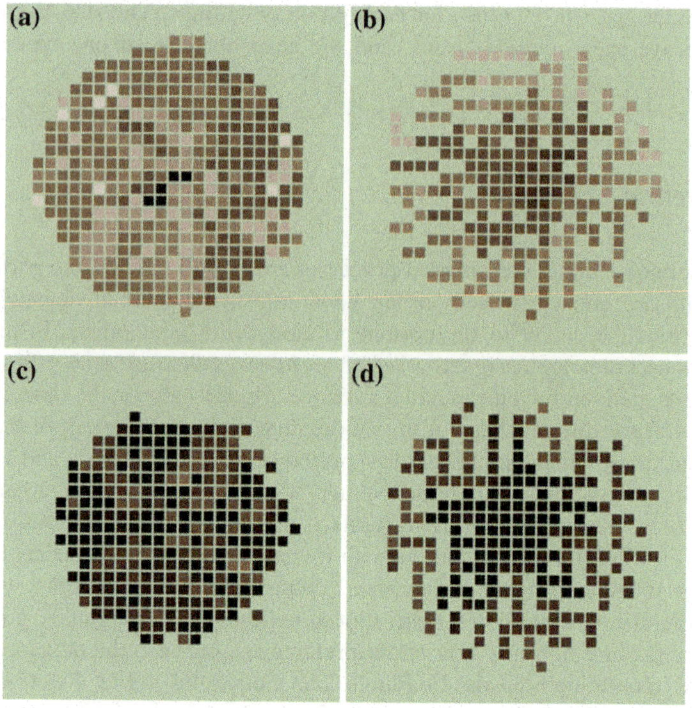

Fig. 4 Urban form for different populations in the market scenario. Different households' preferences in terms of average (and standard deviation, in parenthesis) ideal open space, openspace*, and height tolerance, h*, lead to different urban forms. Green is undeveloped land, lighter shades of brown are lower rise buildings, and darker shade browns are taller buildings (black is height 10). **a** openspace* = 2 (0.5); h* = 2 (0.5), **b** openspace* = 7 (0.5); h* = 2 (0.5), **c** openspace* = 2 (0.5); h* = 7 (0.5), **d** openspace* = 7 (0.5); h* = 7 (0.5)

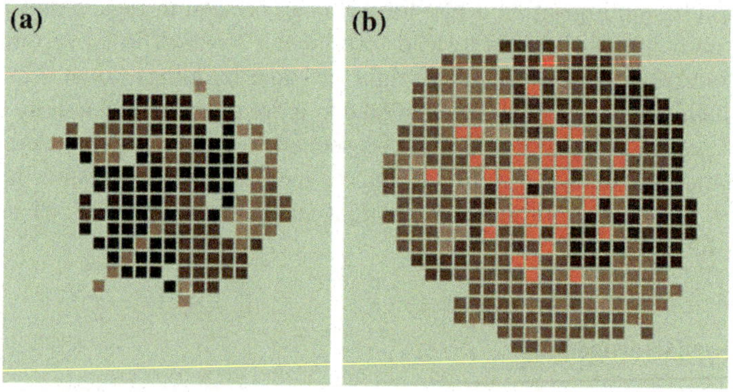

Fig. 5 Urban form in the market scenario (**a**) and participated scenario (**b**), for a population with openspace = 4 (0.5) and h* = 5 (0.5). Colours have the same meaning as above, with the addition of *red*, marking buildings that have undergone negotiation

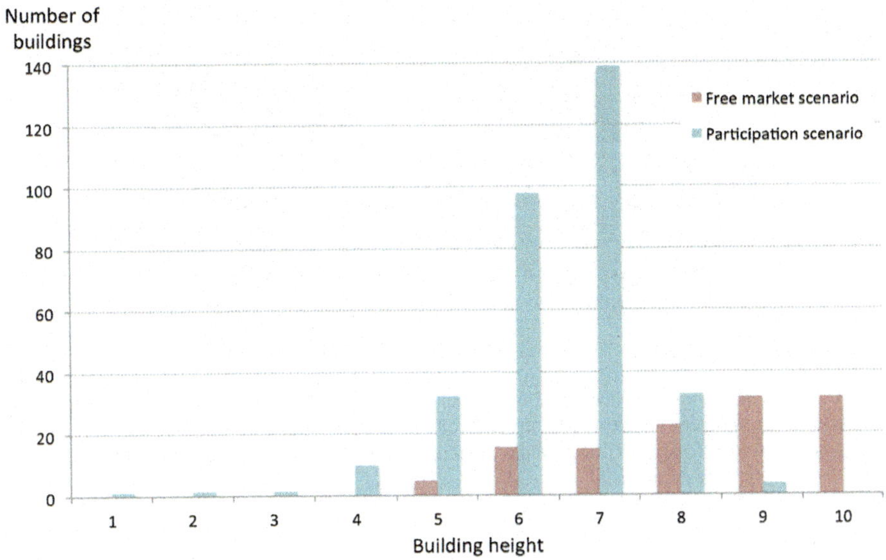

Fig. 6 Number of buildings per height

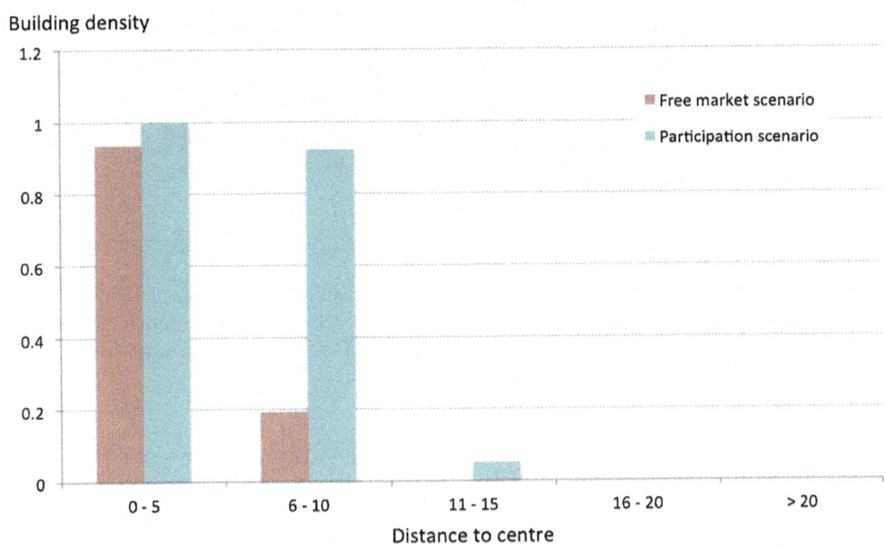

Fig. 7 Distribution of building density (average number of buildings per cell) in space. Each category represents a concentric ring of cells at equal intervals away from the centre cell

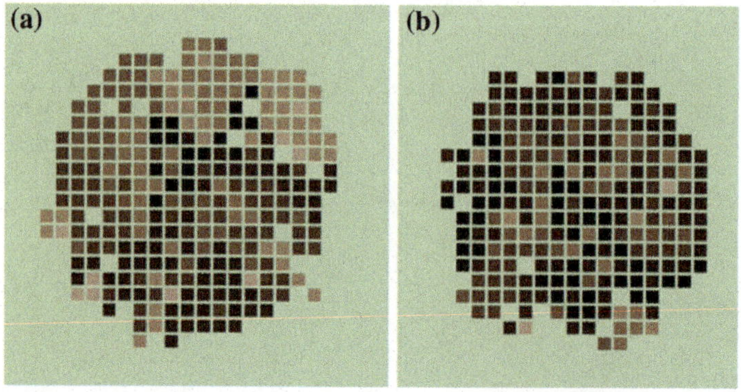

Fig. 8 Urban form, market scenario, for populations with different degrees of heterogeneity. Households' preferences are expressed in terms of average (and standard deviation, in parenthesis) ideal open space (openspace*) and height tolerance (h*). Colours as above. **a** openspace* = 4 (.05); h* = 5 (0.5), **b** openspace* = 4 (0.5); h* = 5 (2)

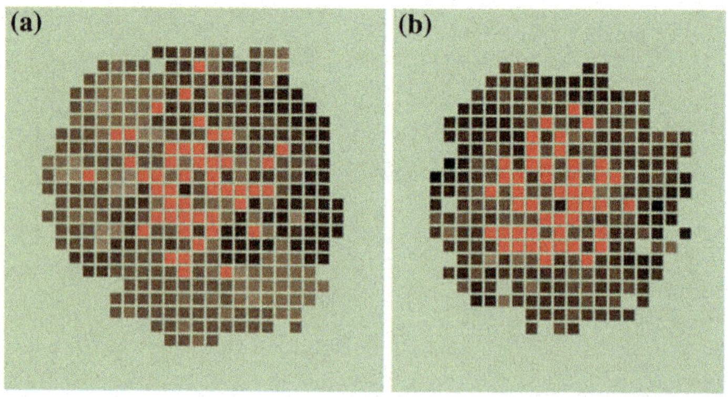

Fig. 9 Urban form, participated scenario, for populations with different degrees of heterogeneity. Households' preferences are expressed in terms of average (and standard deviation, in parenthesis) ideal open space (openspace*) and height tolerance (h*). Colours as above. **a** openspace* = 4 (0.5); h* = 5 (0.5), **b** openspace* = 4 (0.5); h* = 5 (2)

In the market scenario, when there are populations with heterogeneous taste with respect to building height, this will lead to a larger amount of tall buildings than when there are more homogenous populations with the same average height tolerance.

For heterogeneous populations, participation results in lower supply of buildings but perhaps a more adequate supply, since unsatisfied demand is lower than the market scenario.

3.2.3 Participation and Planner's Sensitivity

Participation is initiated from the bottom-up, when a household that opposes a planned development creates a platform to gather the support of others. If there is substantial opposition, the project is re-negotiated. The negotiation involves an alteration of the number of floors for the building proposed, and it can end in agreement (a low-rise building is built) or the developer might have to abandon the project (no building is built).

To evaluate the extent of the participation process, we report the number of projects that are negotiated, both successfully and unsuccessfully, out of all projects proposed by the developers. The number of projects that invoke the public's reaction is a significant proportion of the total number of projects proposed by the developer—from around 13 % to just under 25 %. Out of these projects, about half are successfully renegotiated, meaning that each project is ultimately approved although with substantial alterations (different height than that originally proposed) and built. These numbers vary according to the type of population present and the sensitivity of the planner to the protests. The number of total projects negotiated is higher in more diverse populations, but diversity does not seem to affect the chances of a successful negotiation. The planner's sensitivity threshold is the percentage of the population that needs to be protesting on social media in order for the planner not to approve the project. For homogeneous populations, the more sensitive planner, the higher the number of projects negotiated. However, for heterogeneous populations, less sensitive planners can lead to more project negotiation (Fig. 10).

The total number of projects that are presented by the developer to the planner is considerably higher in the scenarios in which the public was allowed to participate than in the market scenario. This is due to the fact that in the former scenario many projects are abandoned (Fig. 11).

3.2.4 Utility and Distribution of Utility

Finally, we evaluate the utility of households in relation to their residential location. In the case of a homogenous population, the free market seems to lead to a better distribution of utility as shown by the lower Gini coefficient in Table 1. However, in the case of a heterogeneous population, participation leads to slightly higher average utility of households and a slightly more equal distribution, as measured by the lower Gini coefficient in Table 2.

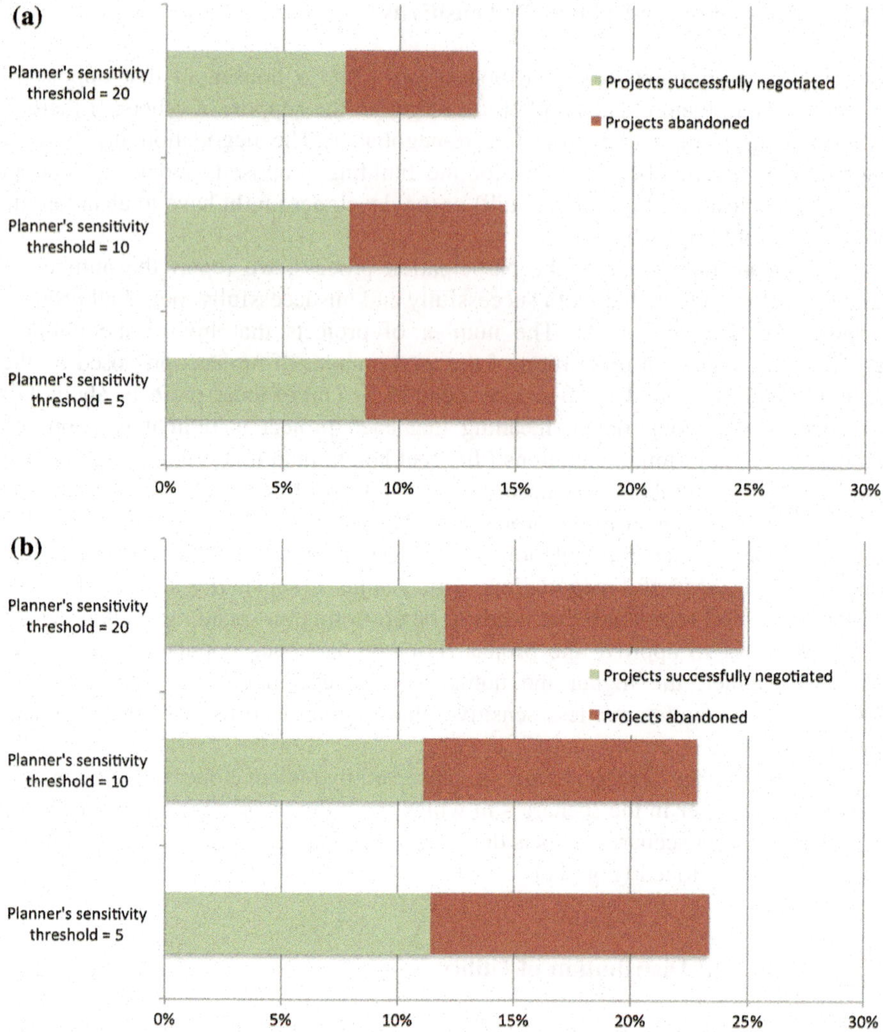

Fig. 10 Percentage of all projects that undergo a participation process (whole bar), and of those, how many are successfully renegotiated and how many are abandoned, for different levels of planner sensitivity (from 5: very sensitive to 20: relatively insensitive); **a** homogenous population; **b** heterogeneous population

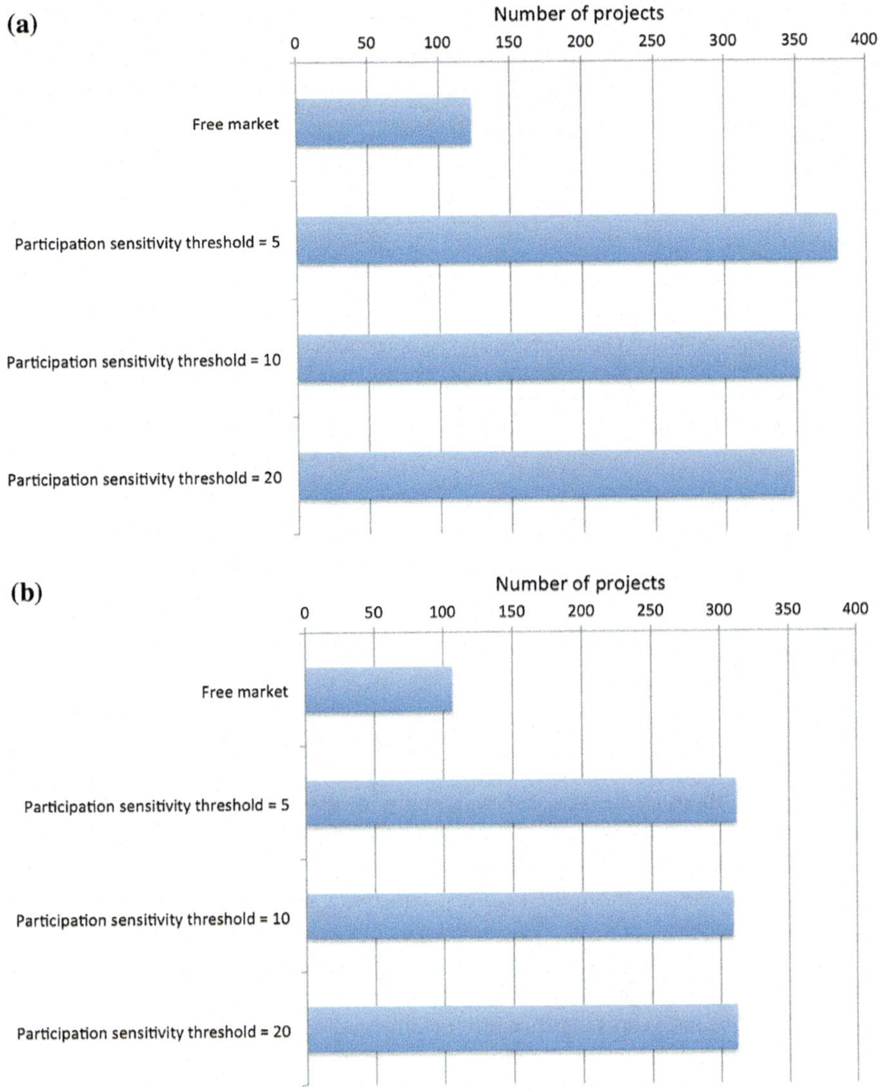

Fig. 11 Total number of projects per run, for different levels of planner's sensitivity threshold. **a** homogenous population; **b** heterogeneous population

Table 1 Average utility and Gini coefficient of the distribution of utility of households in relation to their residential location for a homogenous population and several combinations of status homophily and planners' sensitivity

	Planner's threshold	Average utility	Gini
Free market		59.81	0.328
Participation	5	50.53	0.334
	10	50.88	0.349
	20	50.27	0.365

Table 2 Average utility and Gini coefficient of the distribution of utility of households in relation to their residential location for a heterogeneous population and several combinations of status homophily and planners' sensitivity

	Planner's threshold	Average utility	Gini
Free market		65.08	0.470
Participation	5	55.18	0.423
	10	55.24	0.419
	20	54.20	0.415

4 Conclusions

People shape the cities they live in as consumers of land and real estate, by making choices about the types of neighbourhoods they would like to live in. However, they also shape them as citizens, by engaging with the decisions of developers and planning agencies regarding new developments in their neighbourhoods.

The market tends to produce cities that are not the reflection of consumer preferences, but the compromise between consumer preferences, production constraints and economies of scale. The developer has an incentive to build vertically, especially in the centre, where land is expensive. However, the developer is also bound by the approval process. If the permitting process pays due importance to citizens' opinions, as expressed formally or informally on social networks, blogs or petitions, it can delay approval of unpopular projects, or the developer may be forced to negotiate the height of the buildings.

This chapter starts with the premise that, if up until now citizens' ability to express their opinions about planned developments has been limited, in the future it will be increasingly easy to do so, thanks to new technology that enables any person to make their opinion visible. ICT will no doubt be a part of how people interact with their cities, how they share information, express opinions and gather support for those opinions. Although it is too early to tell what new forms of participation will exist, it seems plausible that social networks will be a key tool for spreading information and enticing the participation of others. The model presented in this chapter was developed to explore the implications of ICT-based participation on urban form.

The results generated by the model suggest that the impact of citizen participation on urban form can be sizeable. Depending on the preferences of the population, and how homogenous those preferences are, participation will tend to bring heights down and increase the number of buildings. In general, participation seems to work towards pushing developers choices in the direction of consumer preferences. However, this might not be true if the incoming population becomes very different from the established population, as could be the case with migration waves or tipping points for taste shifts.

References

Alonso, W. (1964). *Location and land use. Toward a general theory of land rent.* Cambridge: Harvard University Press.

Brabham, D. C. (2009). Crowdsourcing the public participation process for planning projects. *Planning Theory, 8*(3), 242–262.

Briassoulis, H. (n.d.). *Analysis of land use change: Theoretical and modeling approaches. The web book of regional science.* http://www.rri.wvu.edu/Webbook/Briassoulis/contents.htm. Accessed January 22, 2015.

Brown, D. G., Page, S. E., Riolo, R., et al. (2004). Agent-based and analytical modeling to evaluate the effectiveness of greenbelts. *Environmental Modelling and Software, 19*(12), 1097–1109.

Brown, D. G., Robinson, D. T., An, L., et al. (2008). Exurbia from the bottom-up: Confronting empirical challenges to characterizing a complex system. *Geoforum, 39*(2), 805–818.

Caruso, G., Peeters, D., Cavailhès, J., et al. (2007). Spatial configurations in a periurban city. A cellular automata-based microeconomic model. *Regional Science and Urban Economics, 37*(5), 542–567.

Caruso, G., Peeters, D., Cavailhès, J., et al. (2009). Space—time patterns of urban sprawl, a 1D cellular automata and microeconomic approach. *Environment and Planning B: Planning and Design, 36*(6), 968–988.

Cinyabuguma, M., & McConnell, V. (2013). Urban growth externalities and neighborhood incentives: Another cause of Urban Sprawl? *Journal of Regional Science, 53*(2), 332–348.

Czamanski, D., & Broitman, D. (2012). Developers' choices under varying characteristic time and competition among municipalities. *The Annals of Regional Science, 49*(3), 733–743.

Czamanski, D., & Roth, R. (2011). Characteristic time, developers' behavior and leapfrogging dynamics of high-rise buildings. *The Annals of Regional Science, 46*(1), 101–118.

Desouza, K. C., & Bhagwatwar, A. (2012). Citizen apps to solve complex urban problems. *Journal of Urban Technology, 19*(3), 107–136.

Devisch, O. T. J., Timmermans, H. J. P., Arentze, T. A., et al. (2009). An agent-based model of residential choice dynamics in nonstationary housing markets. *Environment and Planning A, 41*(8), 1997–2013.

Ettema, D. (2011). A multi-agent model of urban processes: Modelling relocation processes and price setting in housing markets. *Computers, Environment and Urban Systems, 35*(1), 1–11.

Evans-Cowley, J., & Hollander, J. (2010). The new generation of public participation: Internet-based participation tools. *Planning Practice and Research, 25*(3), 397–408.

Filatova, T., Parker, D., & Van der Veen, A. (2009). Agent-based urban land markets: Agent's pricing behavior, land prices and urban land use change. *Journal of Artificial Societies and Social Simulation, 12*(13). http://jasss.soc.surrey.ac.uk/12/1/3.html. Accessed October 29, 2013.

Filatova, T., Voinov, A., van der Veen, A. (2011). Land market mechanisms for preservation of space for coastal ecosystems: An agent-based analysis. *Environmental Modelling and Software, 26*(2), 179–190.

Glaeser, E.L., & Gyourko, J. (2002). *The impact of zoning on housing affordability*. Working paper National Bureau of Economic Research. http://www.nber.org/papers/w8835. Accessed January 21, 2015.

Iovanna, R., & Vance, C. (2013). Land conversion and market equilibrium: Insights from a simulated landscape. *The Annals of Regional Science, 50*(1), 169–184.

Li, X., & Liu, X. (2007). Defining agents' behaviors to simulate complex residential development using multicriteria evaluation. *Journal of Environmental Management, 85*(4), 1063–1075.

Li, X., & Liu, X. (2008). Embedding sustainable development strategies in agent-based models for use as a planning tool. *International Journal of Geographical Information Science, 22*(1), 21–45.

Ligmann-Zielinska, A., & Jankowski, P. (2010). Exploring normative scenarios of land use development decisions with an agent-based simulation laboratory. *Computers, Environment and Urban Systems, 34*(5), 409–423.

Louviere, J., & Timmermans, H. (1990). Hierarchical information integration applied to residential choice behavior. *Geographical Analysis, 22*(2), 127–144.

Magliocca, N., Safirova, E., McConnell, V., et al. (2011). An economic agent-based model of coupled housing and land markets (CHALMS). *Computers, Environment and Urban Systems, 35*(3), 183–191.

McPherson, M., Smith-Lovin, L., & Cook, J. M. (2001). Birds of a feather: Homophily in social networks. *Annual Review of Sociology, 27*(1), 415–444.

Monticino, M., Acevedo, M., Callicott, B., et al. (2007). Coupled human and natural systems: A multi-agent-based approach. *Environmental Modelling and Software the Implications of Complexity for Integrated Resources the Second Biannual Meeting of the International Environmental Modelling and Software Society: Complexity and Integrated Resources Management, 22*(5), 656–663.

Parker, D., & Meretsky, V. (2004). Measuring pattern outcomes in an agent-based model of edge-effect externalities using spatial metrics. *Agriculture, Ecosystems and Environment, 101*(2–3), 233–250.

Robinson, D. T., & Brown, D. G. (2009). Evaluating the effects of land-use development policies on ex-urban forest cover: An integrated agent-based GIS approach. *International Journal of Geographical Information Science, 23*(9), 1211–1232.

Schaeffer, P., & Hopkins, L. (1987). Behavior of land developers: planning and the economics of information. *Environment and Planning A, 19*(9), 1221–1232.

Sun, S., Parker, D. C., Huang, Q., et al. (2014). Market impacts on land-use change: An agent-based experiment. *Annals of the Association of American Geographers, 104*(3), 460–484.

Von Thünen, J.R. (1826). *The isolated state. English edition* 1966, Oxford: Pergamon.

Wheaton, W. C., & Simonton, W. E. (2007). The secular and cyclic behavior of 'true' construction costs. *The Journal of Real Estate Research, 29*(1), 1.

Wilensky, U. (1999). *NetLogo: Center for connected learning and computer-based modeling*. Evanston, IL: Northwestern University.

Zellner, M., Page, S., Rand, W., et al. (2009). The emergence of zoning policy games in exurban jurisdictions: Informing collective action theory. *Land Use Policy, 26*(2), 356–367.

Zellner, M., Riolo, R., Rand, W., et al. (2010). The problem with zoning: Nonlinear effects of interactions between location preferences and externalities on land use and utility. *Environment and Planning B: Planning and Design, 37*(3), 408–428.

Chapter 18
Usability of Planning Support Systems: An Evaluation Framework

Patrizia Russo, Maria Francesca Costabile, Rosa Lanzilotti and Christopher J. Pettit

Abstract Previous research on Planning Support Systems (PSS) showed that low usability of these computer-based tools is one of the reasons why they are not widely used by planning professionals. Few studies for evaluating PSS usability are performed, possibly because developers do not regard it as their task, do not have enough skills to conduct them, and have not been stimulated so far to appreciate their value. In this chapter, a framework is described that aims at guiding usability evaluation of PSS; it is developed on the basis of a more general usability evaluation framework. The current version of the framework has been applied to evaluate three PSS by performing a test with a small group of land use planners. Results of this user test are discussed, also providing some recommendations for the design of PSS, specifically those addressing Land Suitability Analysis (LSA), capable to generate a positive user experience.

1 Introduction and Motivation

Planning Support Systems (PSS) are computer-based tools designed to support the activities of planning professionals. Both literature and field experts indicate that their adoption is very limited so far. In particular, low usability of PSS has been

P. Russo (✉) · C.J. Pettit
Faculty of Architecture Building and Planning, University of Melbourne,
Parkville, VIC 3010, Australia
e-mail: p.russo@student.unimelb.edu.au

C.J. Pettit
e-mail: cpettit@unimelb.edu.au

M.F. Costabile · R. Lanzilotti
Dipartimento di Informatica, Università degli Studi di Bari Aldo Moro, Bari 70125, Italy
e-mail: maria.costabile@uniba.it

R. Lanzilotti
e-mail: rosa.lanzilotti@uniba.it

S. Geertman et al. (eds.), *Planning Support Systems and Smart Cities*,
Lecture Notes in Geoinformation and Cartography,
DOI 10.1007/978-3-319-18368-8_18

identified as one of the most important factors for this (Te Brömmelstroet 2010; Vonk et al. 2005). Based on this finding, it is time to carry out further research on PSS usability.

Geertman and Stillwell (2003) reported that most PSS are not subjected to well-considered design processes and evaluations. Our experience indicates that PSS design does not sufficiently take into account the user interface, which is the most important component from the point of view of users. User-centred design and usability evaluation of PSS are rather rare (e.g. Vonk and Ligtenberg 2010; Arciniegas et al. 2013). During design and development of PSS, very little attention has been devoted to user-oriented aspects that might improve the overall experience of planning professionals, such as the effectiveness and efficiency of the interaction with such systems and the user engagement and satisfaction. Instead, PSS development has been very much technology-oriented (Vonk and Geertman 2008). Few studies for evaluating PSS usability have been performed, possibly because developers do not regard it as their task and have not so far been stimulated to conduct them. Allen (2008) suggested to rigorously conduct evaluations of PSS in order to encourage developers to increasingly consider user-oriented aspects and to facilitate the selection of appropriate PSS by potential users. It is worth remarking that conducting usability evaluation requires specific skills that PSS designers usually do not possess.

A framework that might serve as a guide for evaluating the usability of PSS is the focus of this research. It has been inspired by DECIDE, a more general usability evaluation framework provided in Rogers et al. (2011). The presented framework represents a first attempt towards more systematic PSS usability evaluation. It is expected to increase and improve PSS evaluation, by providing appropriate guidance to those people that should evaluate a PSS but have not yet enough expertise on usability. In fact, this framework aims to be used by different stakeholders, namely: (i) designers to conceive innovative and more effective PSS, (ii) developers to evaluate new PSS during their development and before their release, (iii) potential users to identify the PSS to adopt in their planning organisations, and (iv) teachers to select PSS suitable for didactic purposes. Thus, this evaluation framework could constitute an important step towards advancing the state-of-the-art in PSS usability.

After an introduction and definition of the concepts of usability and user experience in Sects. 2 and 3 describes the methodological approach and the current status of this framework. As an example of its application, a user test was conducted with a small group of land use planners; they engaged with three PSS that address Land Suitability Analysis (LSA) tasks, namely CommunityViz (http://placeways.com/communityviz), Envision (Newton and Glackin 2013) and Online What if? (Pettit et al. 2013). The participants were not familiar with the three PSS and tested them in a within-subjects design (Graziano and Raulin 2012). The user test was performed with a thinking-aloud approach, complemented with questionnaires. In addition, other methods of predominantly qualitative nature, including video and screen recording were triangulated. Section 4 briefly reports the user test and discusses some results. This section also provides recommendations for the design of PSS, specifically those addressing LSA, capable of supporting a good user experience. Section 5 provides conclusions and outlines next steps in the research.

2 Usability and User Experience

There are different quality factors which interactive systems can be evaluated against. This research focuses on usability evaluation of the user interface, to determine how well the functionality can be used for specific circumstances and environments, also depending on the tasks to be performed and users' characteristics. Usability is a complex concept, which has been defined in different ways in the literature, emphasising specific attributes. The definitions provided by ISO 9241-210 (ISO 9241 2010) and ISO 9126-1 (ISO 9126 1998) are widely acknowledged by the Human–Computer Interaction (HCI) and the Software Engineering communities; they focus on attributes such as effectiveness, efficiency, understandability, learnability and satisfaction. For more on usability definitions and attributes, refer to e.g., Rogers et al. (2011) and Ardito et al. (2014).

Nowadays, the importance of people's overall experience when interacting with software has been recognised by the HCI community. This increasing interest in user experience (UX) has occurred along with some developments in the field of Information and Communication Technology (ICT) over the years (Costabile and Buono 2013; Rogers et al. 2011). In the past, ICT aimed at providing useful and usable functionality, today it attempts to involve users in positive and engaging experiences. The occurrence of mobile systems and internet, which accompany people everywhere, have generated great attention on UX, going beyond the common usability attributes. Depending on the software, positive experiences can be determined, for example, by the level of pleasure and fun (e.g. video games) or satisfaction (e.g. business software that allows accomplishing tasks). Designers of ICT put increasingly emphasis on aesthetics and attractiveness that, according to the literature, generally help improving users' experiences. With respect to the user interface, a good layout with appropriate fonts and colours and engaging interaction techniques are likely to contribute to a better user experience. The research presented in this chapter primarily addresses usability; however, it also focuses on more subjective attributes typical of UX, such as pleasure and user satisfaction, which are also important aspects of usability.

3 A Usability Evaluation Framework for PSS

Evaluation is a central activity in the software design process that aims to assist in developing a product that satisfies users' requirements. Understanding user requirements is fundamental: it is a process of negotiation between designers and users over a certain period of time. Evaluation facilitates the understanding between designers and users, because involving users in evaluating a system prototype is instrumental in highlighting to designers what are the users' needs and expectations. Deciding when and how to evaluate a prototype (or a complete system) requires some expertise and may differ among products. While there are many books and websites that describe evaluation methods, novice evaluators actually need more guidance on how to plan and perform the overall evaluation. The DECIDE framework proposed by Rogers et al. (2011) addresses this need by providing a structure for performing evaluation studies, focusing on the issues that have to be taken into account. Its name derives from the initial of the name of each of the six main activities proposed by the framework (Determine, Explore, Choose, Identify, Decide, Evaluate). The main contribution of the research presented in this chapter is the definition of a framework that, even if it is inspired by DECIDE, is specific for PSS evaluation. It aims at supporting various stakeholders, e.g. researchers and PSS developers, in performing usability and UX evaluation of such systems, without being expert evaluators. In the PSS evaluation framework, the six activities indicated by DECIDE are specialised for the evaluation of PSS, as described in the following subsections.

3.1 Determine the Evaluation Goals

The first activity when planning a usability evaluation is to determine what the goals of the evaluation are. This helps to clarify what is the scope of the evaluation and what should be achieved once the evaluation is carried out. For example: are we going to evaluate the ease of learning of a PSS because we want to use it for a planning course at University? Are we interested in the ease of use of a PSS because it should be adopted by a planning organisation? Are we comparing the usability of similar functionality of two or more PSS? Are we evaluating a PSS to inform the design of its next version? These goals are some of the typical goals of PSS evaluations provided by the framework.

3.2 Explore the Questions

This activity includes the formulation of more specific questions that underpin the goals and should be answered through the evaluation. Indeed, the evaluation is performed just because there are questions about the system usability which should be answered. The creation of a hierarchy of questions and sub-questions allows focussing on issues that need to be addressed, in order to plan the evaluation more in detail. For example, the question "is the PSS usable?" can be decomposed in sub-questions like: Is the user interface easy to navigate? Is the terminology confusing because it is inconsistent? Is the feedback provided to users sufficient? Is the response time too slow?

In relation to the goals presented in 3.1, possible questions are: which steps are difficult to understand by students of planning courses? Can students efficiently use the PSS within a time period of two weeks? What are the skills of the planning professionals who should adopt the PSS? Which PSS provides the most efficient functionality? What could potentially be improved in the current version? Should the next version be web-based or implemented as a plugin of a Geographic Information System (GIS)?

3.3 Choose the Evaluation and Data Collection Methods

Methods for achieving the goals and answering the questions have to be chosen. Many evaluation methods are proposed in the literature (see for example, Rogers et al. 2011; Lanzilotti et al. 2011). The most commonly adopted are inspection methods and user-based methods. Inspection methods involve experts who analyse a system and provide judgments based on their expertise on usability and UX. Their great advantage is that they are cost-effective, since they do not require users nor special equipment or lab facilities, and experts can detect a wide range of problems of complex systems in a limited amount of time. A well-known inspection method is heuristic evaluation (Nielsen 1993). It involves a small set of experts who inspect the system and evaluate the interface against a list of heuristics. The main drawback of such a technique is the dependency on the inspectors' skills and experience, as heuristics are often generic and underspecified (Lanzilotti et al. 2011).

Among user-based methods, user tests allow the analysis of users' performance on the tasks which the system is designed for. They are generally considered the most complete form of evaluation, because usability is assessed through samples of real users. However, reproducing realistic situations of usage in a laboratory is difficult, for example, the selection of adequate user samples and the reproduction of ecological settings for usage in a limited amount of time (Rogers et al. 2011). A cost-effective technique in usability testing is thinking-aloud, which requires users to speak out loud their thoughts while performing tasks. This technique offers a window over the users' mental models, allowing evaluators to detect possible misconceptions about the system and the interface elements which cause them. It provides useful results even with a small number of users (Nielsen 1993).

Studies have outlined that inspection methods and user tests are complementary and can be effectively coupled to obtain a reliable evaluation process (Lanzilotti et al. 2011). Some evaluation methodologies have been proposed that rely on these methods. For example, the e-Learning Systematic Evaluation (eLSE) methodology prescribes a structured flow of activities for evaluating e-learning systems (Lanzilotti et al. 2006). Specifically, eLSE suggests coupling inspection activities and user testing, and precisely indicates how to combine them to make evaluation more reliable and still cost-effective. The PSS evaluation framework proposes a similar approach, suggesting to perform inspection first, while user testing will be performed only in specific cases, when the evaluator feels the need of a more objective evaluation that can be obtained through user involvement. In particular, the PSS framework suggests to use the thinking-aloud technique; it can be complemented with other methods involving users, namely interviews and questionnaires, which are especially valuable to assess user satisfaction and other hedonic qualities of UX.

When performing a user test, methods for collecting data have to be chosen in advance. Screen recording and measures of user performance are often adopted. Multiple data collection methods are recommended as (i) they provide rich datasets that are most valuable in rather unexplored research fields such as PSS usability evaluation, and (ii) the convergence of information from different sources can be investigated and reliability examined (Hilbert and Redmiles 2001).

3.4 Identify the Practical Issues

Many practical issues have to be considered when conducting an evaluation. The expertise of evaluators to conduct the study and the needed resources are very important. Indeed, one of the reasons why most developers neglect usability evaluation is because they feel that they do not have the right expertise and/or that too many resources are required (Ardito et al. 2014). As stated above, one of the goals of the PSS evaluation framework is to provide enough guidance to PSS stakeholders, suggesting methods appropriate for specific situations and indicating the required resources. For example, usability inspection should be performed by at least four usability experts (Nielsen 1993). If only one expert is available, more novice evaluators should be involved.

In user-based methods, an important issue is the choice of the participants, who are ideally representatives of the target user group. At least 4–5 participants should engage with the software (Nielsen 1993). PSS is considered specialised software to be used by planning professionals, with limited application to other user groups. As such, planning professionals are recommended as participants of user-based PSS evaluations. In order to find appropriate participants, a set of requirements that planning professionals have to meet should be defined. Examples of requirements might be planning professionals' level of experience in planning practice, their specialisation (for example, statutory or strategic planning), their familiarity with computer applications.

If two or more products are evaluated with user testing, a decision has to be made whether a between- or within-subjects design is adopted. In a between-subjects design, each participant is randomly assigned to each of the various experimental conditions, i.e. he/she is going to use only one product; in the within-subjects design, each participant is tested under each experimental conditions, i.e. he/she is using all products (Graziano and Raulin 2012). Other decisions about the selection of tasks that participants have to perform during the test, about facilities and equipment to be used, etc., have to be made. The PSS framework provides a description of such issues.

3.5 Decide How to Deal with the Ethical Issues

When people are involved in evaluations, ethical issues have to be addressed, in order to protect participants and their privacy. Participants have to be informed about the data that is gathered during the evaluation and how this is used. This is often done through a plain language statement. Participants are commonly also provided with a consent form that they have to sign, in order to state that they know about the procedure and to agree to participate. An example of a plain language statement and consent form is provided in the PSS framework. To warrant that the evaluation is carried out in alignment with ethical codes, the evaluation study has usually to be presented to the responsible authority where the evaluation is carried out and which has to approve it. However, how ethical issues are treated differs among institutions. As the evaluation cannot be conducted until the study has been approved, enough time should be allowed for getting ethics clearance.

3.6 Evaluate, Analyse, Interpret and Present the Data

Before actually running the evaluation, decisions have to be made about what data are needed, how they are collected, how they are analysed, how they are presented. Several books address those issues. For more details, the reader may refer to Graziano and Raulin (2012). Another delicate choice is whether qualitative data versus quantitative data have to be considered and the identification of proper methods/metrics to evaluate them. Several examples for measuring UX attributes are in Albert and Tullis (2013). It is part of future work on the framework to add specific examples of methods for collecting, analysing and presenting data.

Most importantly, it has to be defined how the quality of the collected data can be demonstrated. To this aim, important factors, such as reliability and validity, have to be considered. *Reliability* is an index of the consistency of the measures applied to the data. Good measures give consistent results, regardless of who does the measuring. A measure is not wholly reliable or unreliable, but varies in its degree of reliability. A correlation index can be used to quantify the degree of reliability. An example of correlation index is the coefficient *alpha*, although much more complicated ones exist (Graziano and Raulin 2012). *Validity* of the measure is its effectiveness in reflecting the characteristic measured. Validity, like reliability, varies in its degree. Once again, a correlation index is typically used to quantify the degree of validity.

The data collected during the evaluation have to be analysed, interpreted and properly presented. Statistics are powerful tools for analysing and understanding data. Specifically, descriptive statistics summarise, simplify and describe a large number of data. Examples of descriptive statistics are: mode, median, mean, average deviation, variance, standard deviation. They are the basis for later analyses in which inferential statistics could be used. Inferential statistics help researchers to interpret what the data mean. Examples of inferential statistics are: T-Test (it applies on independent groups, i.e., when the samples are different, as in the between-subjects design), Correlated T-Test (used in within-subjects design), Analysis of Variance or ANOVA (when more than two groups are involved), Repeated measures ANOVA (used in within-subjects design involving more than two groups).

Besides video recording, which provides further qualitative data, it is suggested to take pictures during the evaluation (e.g. of participants during an interview or a user test), because they are also useful for documentation purposes. Data are better presented if organised in tables, histograms, pie diagrams, etc., whenever it is possible.

4 Applying the PSS Evaluation Framework: A User Test

A user test was conducted with land use planners working in the metropolitan area of Melbourne (Australia) to evaluate the usability of three PSS. The PSS framework guided this usability evaluation: as a first step, it helped to make decisions about the goals of the user test, which was conducted with the thinking-aloud protocol. The goals were: (i) the analysis of PSS characteristics, in order to identify whether they are appropriate for the planners and for LSA tasks, and (ii) the identification of possible usability problems that the planners encountered while using each PSS.

4.1 PSS Evaluated

Three PSS were selected after a comprehensive review of available PSS which has resulted in the formulation of an online resource of PSS (http://docs.aurin.org.au/projects/planning-support-systems/). The following criteria were used to select the PSS:

- perform LSA through *Spatial Multi-Criteria Evaluation*;
- are transferrable to other case studies;
- offer a reasonable level of guidance, in order to not require step-by-step instruction during the user test;
- are either open-source or not too expensive.

LSA determines the suitability of each land unit (usually land parcel) for a specific purpose, based on a set of parameters that are taken into account for calculating the output. The PSS modules devoted to LSA were evaluated: they allow users to give different importance to parameters by assigning weights. This method, also known as *Spatial Multi-Criteria Evaluation*, is well-supported in decision-making processes (Arciniegas et al. 2013). The outcome includes a map that displays the suitability score for each land parcel through colour-coding. The map is also referred to as land suitability layer. The three PSS implement LSA differently, providing different weighting scales or parameter processing, as illustrated in the following.

CommunityViz is a commercial and well-known PSS (http://placeways.com/communityviz, Vonk et al. 2005). It is implemented as a desktop extension in ArcGIS (GIS software). The *suitability wizard* of Scenario 360™—one of its two main components—was evaluated. Datasets for performing LSA are added as layers (shapefile format) to ArcMap (ArcGIS component). The user has to establish the LSA parameters that he/she wants to consider. In the so-called *assumptions* window, the user assigns to each parameter a weight in the range of 0–10 through *sliderbars* (see Fig. 1). The outcome (suitability layer) is provided in ArcMap. Spatial analysis functionality available in ArcMap can be used for analysing the obtained suitability layer. Detailed help documentation is provided by CommunityViz through hyperlinks in the user interface.

Envision is a PSS implemented currently as a plugin in QuantumGIS (GIS software), primarily developed to support sustainable redevelopment of precincts (Newton and Glackin 2013). The module that performs LSA was evaluated. Datasets and parameters for effectuating LSA are stored in a database. 34 parameters, related to property, demographics and location, which the LSA can be based on, are provided. Users select the desired parameters and assign weights in the range of 0–20 to each parameter through *sliderbars* (see Fig. 2). Very few actions

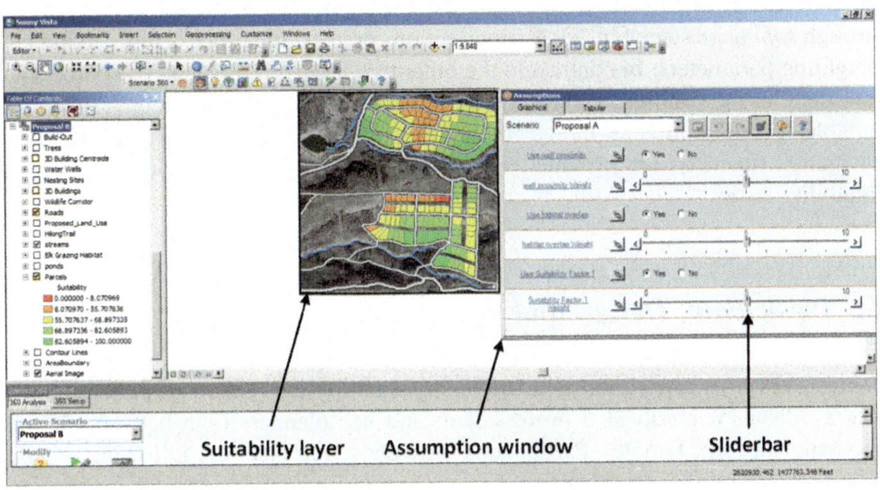

Fig. 1 CommunityViz: the assumption window including sliderbars and the suitability layer in ArcMap

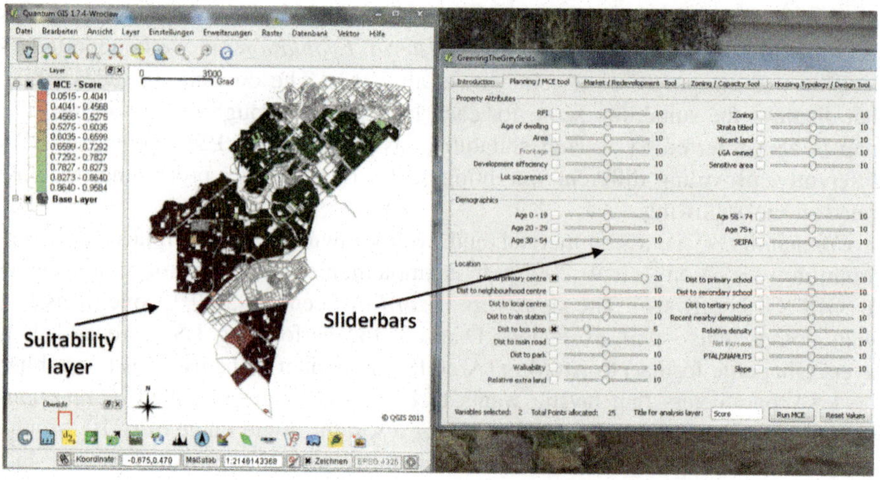

Fig. 2 Envision: the window to input parameters through sliderbars (*right*) and the suitability layer in QuantumGIS (*left*)

precede the process of selecting and weighting parameters, thus facilitating user interaction. The outcome (suitability layer) is provided in QuantumGIS. Similar to CommunityViz, spatial analysis functionality is available in QuantumGIS and can be used to analyse the suitability layer.

The *Online What if?* (OWI) (Pettit et al. 2013) is a web-based PSS version of the well-known What if? (Klosterman 1999; Vonk et al. 2005), slightly modified. The suitability analysis module was evaluated. The datasets and parameters for effectuating LSA are stored in an attribute table (shapefile format). The user has to select the desired parameters and assign weights in the range of 0–100 to each parameter through *spin boxes* (see Fig. 3). Various actions precede the process of selecting and weighting parameters. In contrast to the other two PSS, in OWI land use parcels are classified, in the resulting suitability layer, as not developable, not convertible, not suitable, or suitability as indicated in a 5-point scale from low to high. Basic navigation functions are provided such as panning and zooming for exploring the suitability layer.

4.2 Participant, Design and Procedure

The user test was carried out in a University usability lab, where proper equipment was available. It involved 6 professional land use planners (age between 25 and 45 years old, three female). Participants were chosen to meet the following criteria: (i) are strategic land use planners, (ii) are familiar with LSA and GIS operations such as layer (de)activation, map zooming and panning, and (iii) have never used the three PSS before.

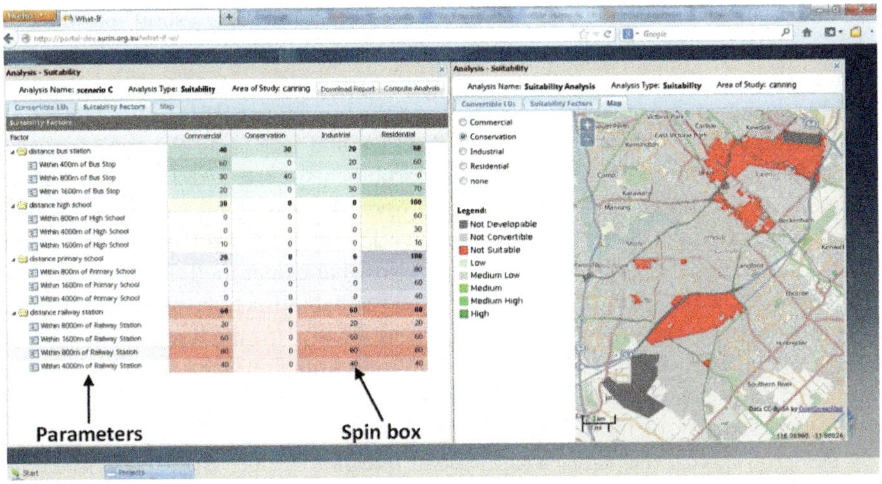

Fig. 3 Online what if?: parameters and spin boxes to assign their values (*left*) and the suitability layer (*right*)

Fig. 4 A land use planner interacting with a PSS during the user test (*left*) and the experimenter (*right*)

During the test, each participant was assigned the task of identifying land most suitable for residential redevelopment by using the PSS on a desktop (see Fig. 4). For providing the same conditions, all three PSS used datasets of a case study area, namely Canning in Western Australia. The participants engaged with the three PSS in sequence, but in different order to avoid potential learning effects (Graziano and

Raulin 2012). As the participants had never used the three PSS, a short introduction was given to each tool before using it, providing information required for performing the task.

Table 1 shows the sequence of steps for each participant P1, …, P6, indicating the order in which the participant used the PSS and providing the time for each step. After engaging with each PSS, the participants were asked to complete an intermediate questionnaire, composed of 5 open questions, in which participants had to write down: (1) their experiences while engaging with the PSS; (2) the context in which these experiences occur; (3) to what extent the PSS's features and functions were satisfactory for identifying land suitable for redevelopment; (4) the three most positive aspects of the PSS used and, finally, (5) the three most negative aspects of the PSS used. At the end, a final questionnaire was completed. It was composed of 2 questions: the former was a close question that required participants to rate the diversity of the three PSS on a 5-point scale (1 = strongly disagree—5 = strongly agree); the latter was an open question which asked participants to indicate to what extent learning how to use the PSS become easier through engaging with additional PSS in the testing.

A technical problem prevented recording the first part of the interaction of P5 and made it not possible to begin with CommunityViz and go on with Envision and OWI, which was the planned design that considered all permutations of the three PSS.

4.3 Some Results and Recommendations

As suggested by the PSS evaluation framework, the user test was carried out with thinking-aloud protocol. This method was triangulated with video recording of user interaction, screen capturing including measurements of some specific actions performed by the participants, and questionnaires. The detailed results of this study are currently described in a technical report (Lanzilotti and Russo 2014) and will be published in a forthcoming paper. In the following, some findings are discussed with respect to the two evaluation goals, which are the analysis of the appropriateness of PSS characteristics for LSA tasks and the identification of usability problems encountered by planners during the test. Based on these findings, some recommendations for the design of PSS, specifically for those addressing LSA, are provided.

User performance was analysed by measuring the time spent to perform the task and by counting those actions, performed by the user, which revealed usability problems. The scheme used to code user's actions followed the one developed by (Vermeeren et al. 2002) to detect usability problems in task-based products for adults. Specifically, the actions were operationalized with three values: (1) *random actions*, referring to those actions not belonging to the correct sequence of actions to accomplish the task; (2) *puzzled actions*, concerning actions for which the user indicated either (i) not knowing how to perform the task or what action is needed

Table 1 User test procedure, indicating for each participant P1, …, P6 the order in which each PSS (CommunityViz (CViz), Envision (Env), OWI) was used and reporting the time in minutes (m) for each step

	General intro & intro to 1st PSS	1st PSS	Questionnaire on 1st PSS & intro to 2nd PSS	2nd PSS	Questionnaire on 2nd PSS & intro to 3rd PSS	3rd PSS	Questionnaire on 3rd PSS & final questionnaire	Total time
P1	(3 m)	Env (13 m)	(9 m)	CViz (13 m)	(6 m)	OWI (17 m)	(11 m)	(1 h 12 m)
P2	(4 m)	CViz (26 m)	(14 m)	OWI (23 m)	(9 m)	Env (20 m)	(16 m)	(1 h 52 m)
P3	(4 m)	OWI (27 m)	(6 m)	Env (19 m)	(6 m)	CViz (14 m)	(7 m)	(1 h 23 m)
P4	(3 m)	Env (19 m)	(9 m)	OWI (23 m)	(7 m)	CViz (15 m)	(10 m)	(1 h 26 m)
P5	n/a	Env (14 m)	(10 m)	CViz (13 m)	(7 m)	OWI (17 m)	(10 m)	(1 h 11 m)
P6	(4 m)	OWI (34 m)	(8 m)	CViz (13 m)	(8 m)	Env (14 m)	(8 m)	(1 h 29 m)

The last column shows the test total time for each Pi. For P5, the time of the first step is not available (n/a) due to a technical problem

Table 2 Performance in accomplishing the task with CommunityViz, Envision and OWI

	CommunityViz	Envision	OWI
Random actions (*N*)	3	1	14
Puzzled actions (*N*)	4	0	15
Uncomfortable actions (*N*)	2	3	9
Time (*min*)	M = 15 SD = 5	M = 16 SD = 3	M = 23 SD = 6

for it, or (ii) not being sure whether a specific action is needed or not; (3) *uncomfortable actions*, referring to actions for which the user indicated that executing the action was difficult or uncomfortable. Table 2 reports the number of the identified users' actions alongside the mean (M) and the standard deviation (SD) of task execution time for the three PSS.

During the interaction with OWI, participants carried out a total 14 random actions: they randomly clicked on interface objects and moved among different windows. Three random actions were observed during the interaction with CommunityViz and only one with Envision. Regarding to the analysis of puzzled actions, during the interaction with OWI participants appeared to be more confused than with the other two PSS, i.e. 15 puzzled actions were observed. With CommunityViz and Envision, 4 and 0 puzzled actions were identified, respectively. Examples of utterances highlighting puzzled actions were: *"I don't really understand!"* or *"How do you actually run it?"*. In many situations in which puzzled actions arose, users could not proceed without help and either asked for it or the experimenter had to intervene. Finally, during the interaction with OWI participants indicated that they are not able to execute an action, as highlighted by the 9 uncomfortable actions coded. A less number of such actions were registered for the other two PSS, i.e. 2 for CommunityViz and 3 for Envision.

The following reasons might explain these results. First of all, the three investigated PSS are different in the way they allow users to specify parameters for LSA and to assign weights to them, which also represent the most significant activities for accomplishing the task. While Envision provides in a window a predefined set of parameters among which the user can select, CommunityViz and OWI require the user to perform more steps, in order to indicate the parameters of interest. Both solutions have strengths and weaknesses. The interface should facilitate user interaction by limiting the number of actions the user has to perform. Thus, the solution proposed by Envision is easier to learn and use, because the user has only to tick off the desired parameters from the provided set. Indeed, the total number of actions highlighting usability problems in Envision is the lowest (see Table 2). However, some participants complained that this set was lacking some parameters, one participant explicitly mentioned the lack of parameters related to heritage, ecological and site contamination. The other two PSS give more freedom to the user, at the expense of more work to do, in order to indicate the parameters of interest. The user does not get proper guidance from the interface, as indicated by the higher number of random, puzzled and uncomfortable actions.

It also emerged that the layout of the assumption window in CommunityViz was not clear. Participants had difficulties in understanding the meaning of the different widgets due to their misleading position, as well as the used fonts and colours.

It was observed that the sliderbars provided by both CommunityViz and Envision was an easier mechanism for selecting the weight to be assigned to a parameter than the spin boxes used in OWI. The latter also requires more time for selecting the weight. Furthermore, users appreciated a smaller weighting range, indicating 0–10 as most preferable.

The calculation of the outcome (suitability layer and scores) is much shorter in Envision than in CommunityViz and OWI. Nonetheless, even for Envision a participant said: “it’s a bit slow” and explained that, in order to better support planning practice, efficient software is desired.

Because the participants were familiar with GIS software, they expected similar functionality when analysing the suitability layer. Indeed, three participants were surprised that, even if Envision is a plugin of QuantumGIS, this functionality was not available. Another participant appreciated the web-based PSS (i.e. OWI) because it is ready to be used and does not require the installation of any software.

Some participants used the help documentation in CommunityViz accessible through hyperlinks; however, they did not appear to be able to assimilate information in that short period of time during the user test. The help button was not selected in Envision (it was actually not working), while OWI does not currently provide any.

For all three PSS, participants remarked the need for more explanation on what the weighting actually mean, how the computation is performed and how to interpret the results. For example, participant P4 said in relation to the suitability scores in the legend: “this doesn’t mean anything to me, to be honest”, “I don’t know what the numbers mean” and “it’s not very useful, not helpful, without knowing more information”.

From the above discussion, it is evident that much more attention has to be devoted to PSS user interfaces and to the whole efficiency of the systems. In particular, the following recommendations can be derived for the design of PSS, specifically the modules addressing LSA, in order to create systems than can satisfy their users, generating a positive user experience.

- Layout, colours and fonts for the presentation of the parameters have to be carefully chosen to make it easier to understand for the user.
- Explanations about the weighting system, the calculation and interpretation of suitability scores should be provided.
- Help documentation, including examples and short demo of system use, should be available.
- The speed of PSS operations, through the selection of appropriate methods and technology optimisation, should be maximised.

5 Conclusions

This chapter presents a framework that serves as a guide for performing usability evaluation of PSS. Several stakeholders should evaluate PSS for various reasons, but they do not have the required expertise. The PSS evaluation framework has been defined to support such stakeholders, so it is expected to increase and improve PSS evaluation.

As a first application of this framework, a user test was performed for evaluating the usability of three PSS modules devoted to LSA. Even with a small number of participants (6 professional land use planners), the test highlighted some strengths and weaknesses of the three PSS with respect to user needs and expectations. Further studies with more participants and/or other methods have to be conducted.

The test results allowed the derivation of some recommendations for PSS designers. The more general recommendation is to devote more attention to the design of the system-user interaction, involving target users, i.e. planning professionals, according to user-centred approaches.

Currently, the framework is rather concerned with having PSS evaluated by participants individually. However, some PSS are supposed to be used in group settings. Future works could, therefore, involve an extension of the framework in order to adapt, where necessary, the various steps to group evaluation. As another future development, the framework could take into account the pattern-based inspection proposed in Lanzilotti et al. (2011). This technique uses evaluation patterns that indicate which are the critical aspects of the application to look at, and which actions to perform during the inspection in order to analyse such aspects. Evaluation patterns are defined to provide support primarily to novice and not professional evaluators; the rationale is that they capture the expertise of skilled evaluators (i.e., their behaviour in conducting an inspection), and express it in a precise and understandable form, so that this expertise can be reproduced, communicated, and exploited. In order to apply the pattern-based inspection to the evaluation of specific systems, a proper set of evaluation patterns has to be provided. Once patterns specific for PSS will be defined, the pattern-based inspection will be the inspection method to be used.

Acknowledgments The support of CRC for Spatial Information is acknowledged. The authors are grateful to the land use planners who participated in the user test.

References

Albert, W., & Tullis, T. (2013). *Measuring the user experience: Collecting, analyzing, and presenting usability metrics*. Newnes: Morgan Kaufmann.

Allen, E. (2008). Clicking toward better outcomes: Experience with INDEX. In R. K. Brail (Ed.), *Planning support systems for cities and regions* (pp. 139–166). Cambridge, Massachusetts: Lincoln Institute of Land Policy.

Arciniegas, G., Janssen, R., & Rietveld, P. (2013). Effectiveness of collaborative map-based decision support tools: Results of an experiment. *Environmental Modelling and Software, 39*, 159–175.

Ardito, C., Buono, P., Caivano, D., Costabile, M. F., & Lanzilotti, R. (2014). Investigating and promoting UX practice in industry: An experimental study. *International Journal of Human-Computer Studies, 72*(6), 542–551.

Costabile, M. F., & Buono, P. (2013). Principles for human-centred design of IR interfaces. In A. M. Ferro, N. Former, P. Mueller, & G. Santucci (Eds.), *Information retrieval meets information visualization* (pp. 28–47). Berlin: Springer.

Geertman, S., & Stillwell, J. (2003). *Planning support systems in practice*. Berlin: Springer.

Graziano, A. M., & Raulin, M. L. (2012). *Research methods, a process of inquiry* (8th ed.). New York, USA: Pearson.

Hilbert, D. M., & Redmiles, D. F. (2001). Extracting usability information from user interface events. *ACM Computing Surveys, 32*(4), 384–421.

ISO 9126. (1998). ISO/IEC 9126-1: Information technology—Software product quality.

ISO 9241. (2010). ISO/IEC 9241-210: Ergonomics of human-system interaction—Part 210: Human-centred design for interactive systems.

Klosterman, R. E. (1999). The what if? Collaborative planning support system. *Environment and Planning B: Planning and Design, 1*(26), 393–408.

Lanzilotti, R., Costabile, M. F., Ardito, C., & De Angeli, A. (2006). eLSE methodology: A systematic approach to the e-learning systems evaluation. *Educational Technology and Society, 9*(4), 42–53.

Lanzilotti, R., De Angeli, A., Ardito, C., & Costabile, M. F. (2011). Do patterns help novice evaluators? A comparative study. *International Journal of Human-Computer Studies, 69*(1–2), 52–69.

Lanzilotti, R., & Russo, P. (2014). Applying a usability framework to evaluate planning support systems. IVU Technical Report nr. 4.2014. Dipartimento di Informatica, Universita' degli Studi di Bari Aldo Moro, Bari, Italy.

Newton, P., & Glackin, S. (2013). Using geo-spatial technologies as stakeholder engagement tools in urban planning and development. *Built Environment, 39*(4), 473–501.

Nielsen, J. (1993). *Usability engineering*. Boston: Academic Press.

Pettit, C. J., Klosterman, R. E., Nino-ruiz, M., Widjaja, I., Russo, P., Tomko, M., & Sinnott, R. (2013). The online what if? Planning support system. In S. Geertman, F. Toppen, & J. Stillwell (Eds.), *Planning support systems for sustainable urban development* (Vol. 195, pp. 349–362). Berlin: Springer.

Rogers, Y., Sharp, H., & Preece, J. (2011). *Interaction design: Beyond human-computer*. West Sussex: Wiley.

Te Brömmelstroet, M. (2010). Equip the warrior instead of manning the equipment: Land use and transport planning support in the Netherlands. *The Journal of Transport and Land Use, 3*(1), 1–17.

Vermeeren, A., den Bouwmeester, K., Aasman, J., & de Ridder, H. (2002). DEVAN: A tool for detailed video analysis of user test data. *Behaviour and Information Technology, 21*(6), 403–423.

Vonk, G., & Geertman, S. (2008). Improving the adoption and use of planning support systems in practice. *Applied Spatial Analysis and Policy, 1*(3), 153–173.

Vonk, G., Geertman, S., & Schot, P. (2005). Bottlenecks blocking widespread usage of planning support systems. *Environment and Planning A, 37*(5), 909–924.

Vonk, G., & Ligtenberg, A. (2010). Socio-technical PSS development to improve functionality and usability—Sketch planning using a Maptable. *Landscape and Urban Planning, 94*, 166–174.

Chapter 19
Facilitating PSS Workshops: A Conceptual Framework and Findings from Interviews with Facilitators

Peter Pelzer, Robert Goodspeed and Marco te Brömmelstroet

Abstract Recent research has emphasized the importance of workshops as a venue where planning support systems (PSS) are used in planning processes. Empirical studies of these workshops have previously largely overlooked facilitation, in particular the role of a moderator (steering the discussion) and/or a chauffeur (steering the PSS). Drawing on existing facilitation research, we identify four main categories of facilitation interventions: substantive, procedural, relational, and tool-related. We use these categories to develop a novel conceptual framework for facilitation at PSS workshops. We test and develop this framework through semi-structured interviews with eight experienced facilitators of PSS workshops in the US and the Netherlands. The interviews confirm the validity of the intervention categories, but also revealed a wider range of PSS-specific workshop outcomes. We conclude that successful facilitation of PSS workshops requires two different types of facilitation interventions: some to encourage PSS use, and others to prevent PSS domination of the group discussion. Facilitating PSS workshops is mainly about finding the delicate and context-dependent balance between these two extremes.

P. Pelzer
Faculty of Geosciences, Utrecht University, Heidelberglaan 2,
P.O. 80.115, 3508 TC Utrecht, The Netherlands
e-mail: P.Pelzer@uu.nl

R. Goodspeed (✉)
Taubman College of Architecture and Urban Planning,
University of Michigan, 2000 Bonisteel Blvd., Ann Arbor 48109, MI, USA
e-mail: rgoodspe@umich.edu

M. te Brömmelstroet
Faculty of Social and Behavioural Sciences, University of Amsterdam,
Nieuwe Achtergracht, 166, 1018 WV Amsterdam, The Netherlands
e-mail: M.C.G.teBrommelstroet@uva.nl

S. Geertman et al. (eds.), *Planning Support Systems and Smart Cities*,
Lecture Notes in Geoinformation and Cartography,
DOI 10.1007/978-3-319-18368-8_19

1 Introduction

It has almost become a cliché to emphasize the importance of collaboration in planning. A vast literature of empirical and theoretical contributions describes and prescribes the characteristics of collaborative planning (e.g. Forester 1989; Healey 1992; Innes and Booher 1999). Simultaneously, planning support systems (PSS) have been developed to support such collaborative planning processes with geo-information, modelling, visualization, or other functionality (e.g. Geertman 2006; Klosterman 1997).

We define PSS as "geoinformation technology-based instruments that incorporate a suite of components that collectively support some specific parts of a unique professional planning task" (Geertman 2008, p. 217). PSS, especially when they aim to support the more strategic phases of planning, are often applied in the context of a collaborative workshop setting, where they are used by groups of professionals or stakeholders engaged in a specific, delineated planning task (examples of this can be found in; Arciniegas and Janssen 2012; Goodspeed 2013; Pelzer et al. 2013; Te Brömmelstroet and Bertolini 2008; Shiffer 1992).

Researchers' focus at these workshops is typically on the technical characteristics of the PSS (Geertman and Stillwell 2003; Geertman 2008). However, it has become clear that PSS technology is only one ingredient to successful PSS applications. Largely outside the view of the PSS literature, practitioners have developed techniques which utilize facilitation to skillfully integrate PSS into planning workshops (Huntsman 2004). Recent academic contributions emphasize the need to expand the research focus to a wider definition of PSS performance (Geertman and Stillwell 2009; Te Brömmelstroet 2013), or by analyzing the influence of process characteristics on project outcomes, such as the structure of the workshop and characteristics of workshop participants (e.g., Goodspeed 2013; Nyerges et al. 2006; Salter et al. 2006). These studies reinforce the importance of *knowledge exchange* (Nonaka and Takeuchi 1995) as a fundamental prerequisite for PSS usefulness. This perspective follows Shiffer (1992), who argues that collaborative planning requires a synthesis of information drawn from analytical tools, diverse media, and collective cognition found in group discussion. A central, but often overlooked, issue here is the fundamental importance of a moderator (steering the discussion) and/or a chauffeur (steering the PSS) in supporting this knowledge exchange among planning actors and the PSS during a workshop.

This chapter reports on an empirical study into the role of facilitation in PSS workshops. Its central research question is: what interventions lead to effective facilitation of PSS workshops? In Sect. 2, we draw on the scholarly literature on facilitation and PSS to develop a novel conceptual framework for facilitation of PSS workshops. The following sections describe our effort to validate and develop this framework through semi-structured interviews with experienced planning facilitators in the United States and the Netherlands. Section 3 describes our research methods, Sect. 4 presents the findings, and Sect. 5 concludes and discusses directions for future research.

2 Conceptual Framework

2.1 Intended Outcomes of PSS Workshops

Analysing effective facilitation at PSS workshops requires defining the intended workshop outcomes. Recent studies show that the specific intended outcomes of PSS workshops vary significantly from case to case (Brail 2008; Geertman et al. 2013; Geertman and Stillwell 2003, 2009; McElvaney 2012; Steinitz 2012). The desired outcomes depend on elements like cultural context, planning style, backgrounds of involved participants, the planning issue at hand, the phase of the planning process and the scale of the planning issue (Geertman 2006). For instance, a workshop with local residents to create a vision for their community has a very different aim than a workshop with professionals to refine a detailed proposal. Drawing on our own recent work (Goodspeed 2013; Pelzer 2015; Te Brömmelstroet 2013) we identify three important outcomes for PSS workshops: social learning among participants, a more integrated analysis, and a more informed planning product (be it a plan, project design, strategy, or vision). These three generic goals are operationalized below.

First, planning workshops can be seen as a venue for *social learning* by individuals, organizations, or society (Wenger 1998; Argyris and Schön 1996; Friedmann 1989). This concept has been operationalized in planning in different ways. For example, shifts in individual views and knowledge have been documented as one outcome of a collaborative planning process (Deyle and Slotterback 2009). In addition to this, Goodspeed (2013) measured self-reported individual learning, as well as dimensions of organizational learning drawn from Argyris and Schön's (1996) double-loop learning model: evidence seeking, valid information, free and informed choice, and internal commitment to choice. However, Gudmundsson (2011) reminds us that a PSS may have uses beyond learning. He argues that knowledge technologies like PSS can not only be used instrumentally (to provide direct insight into a problem) but can also be used in symbolic ways (to make policymaking seem scientific and thorough), or to provide broad conceptual enlightenment (provide broader insights).

Second, PSS can help achieve a *more integrated analysis* of the planning issue at hand. We define an integrated analysis as one which systematically considers all relevant dimensions of a particular planning problem. While the shift towards more communicative approaches to planning has reduced the emphasis on the quality and nature of analysis (Geertman 2006; Klosterman 1997), PSS can arguably serve as a link between different types of information that are brought to the workshop table by planning participants. Many PSS can also offer the potential for integrating design and analysis, by facilitating the rapid analysis of proposals. For example, planning for sustainability requires analysing a complex variety of environmental, social and economic factors (e.g. Pelzer et al. 2013).

Third, the last potential outcome of PSS workshops is *a more informed product*. Klosterman remarks that "the development of PSS can be seen as part of a larger effort to return the planning profession to its traditional concern with using information and analysis to more effectively engage the future" (2009, p. iv; see also Couclelis 2005; Hopkins and Zapata 2007). This resonates with Hopkins's (2001; see Pelzer et al. 2014) work on the effectiveness of plans. He argues that, among other things, the internal consistency of plans should be scrutinized. In addition to plans, PSS workshops may lead to other types of products, such as a project design, strategy, or vision.

We acknowledge that organizers and participants of PSS workshops may pursue a variety of additional process and outcome goals. These might include the development of consensus and commitment on the part of participants, as well as the production of plans with novelty and specificity. The three generic aims outlined above, however, have arguably emerged in the PSS literature as particularly important.

2.2 Facilitation Interventions

Group facilitation is both the topic of practical guides (e.g., Schwarz 1994; Kolb 2011) and scholarly literature (e.g., Frey 1995; Kolfschoten and Rouwette 2006). Facilitation as a field of practice and research emerged from earlier research on group dynamics (e.g., Fisher and Ellis 1980; Poole and Hollingshead 2005). Early studies showed that group performance depends not only on how well tasks are completed by group members, but also on the quantity and quality of relationships within the group (Bales 1950). The facilitation literature identifies "group performance" (Hirokawa and Gouran 1989) or "effectiveness" (Schwarz 2002) as key group outcomes, and are measured as brainstorming creativity, group problem solving ability, and group decision quality. The communication and facilitation groups require for high performance varies considerably by task. For example, Hirokawa and Gouran (1989) observed that simple tasks can be facilitated with centralized and restricted communication patterns where group member contributions are channeled through a strong facilitator who serves to coordinate group communication. Conversely, this study showed that more complex tasks benefited when communication took the form of a decentralized network, with greater face-to-face communication among group members and less active involvement of a facilitator.

In the management literature, the role of facilitation is to maximize the relevant goals of a specific meeting. Good facilitation involves general *behaviours*, specific *interventions*, and *roles* that together seek to maximize process gains and minimize process losses (Hayne 1999). An example of a process gain is additional insight produced by group feedback, and a process loss might be good ideas which are ignored by the group or not recorded. Hirokawa and Gouran (1989) divide activities of facilitators into three areas: *substantive*, or related to topics discussed, *procedural*, or drawing attention to process elements like the agenda, and *relational*, or attending to social or emotional issues.

Although there is no scholarly literature specifically on the role of facilitation in PSS workshops, researchers have examined the role of facilitation in two similar settings: group support systems (GSS) and group use of systems dynamics computer models. These literatures provide additional issues that may apply to PSS workshops too. In a study based on participants' perceptions, de Vreede et al. (2002) propose several unique roles for facilitators at meetings using GSS: technical GSS knowledge, provide an introduction and explanation of the GSS technology, and provide a warm-up to introduce the technology. Similarly, interviews with GSS facilitators found that knowledge of the tool was a critical factor for meeting success, and that several issues related to technology could be barriers to successful meetings: participant anxiety, systems inflexibility, low reliability, technology distraction, and the technology learning curve (Niederman et al. 1996). Building on the existing facilitation literature, Khalifa et al. (2002) found mixed results between facilitation choices and GSS-supported learning outcomes. This study considered the role of two types of facilitation interventions: *procedural* interventions that impose process restrictiveness, i.e., ensure a group is following an agenda, and *substantive* interventions that encourage content restrictiveness, or ensuring focus remains on the assigned topics. In this experiment, although the relationship between content restrictiveness and learning was unclear, they found some evidence that high process restrictiveness reduced student knowledge acquisition.

Researchers working in the area of system dynamics modelling have arrived at similar conclusions. Vennix (1999) identified several types of facilitation skills as important for workshops: process structuring, conflict handling, and communication skills. He argues that system dynamics model building must be "skilfully combined with adequate facilitation" in order for the projects to result in group learning (Vennix 1999, p. 392). Since PSS workshops are similar sociotechnical contexts to these, we anticipate the need for facilitators to engage in PSS-specific interventions. We focus particularly on these tool-related interventions as they are a very relevant topic for both academics and practitioners involved in PSS applications that has thus far received little scholarly attention.

2.3 Conceptual Framework

For the purpose of this chapter, we have simplified the abovementioned debate in Fig. 19.1. In this figure, we define a category of dependent variables (intended outcomes of PSS workshop), which is influenced by an important confounding variable (social interaction during the PSS workshop), the latter is influenced by both context and facilitation interventions. The structure is also informed by earlier interviews of professionals about PSS by one of the authors in the Netherlands, and observations from intensive study of PSS workshops in the United States by another author. We capture as context a range of explanatory variables for a successful PSS workshop, including usability of the instrument, the physical space and the audience. Next, the facilitation interventions aim to influence social interaction

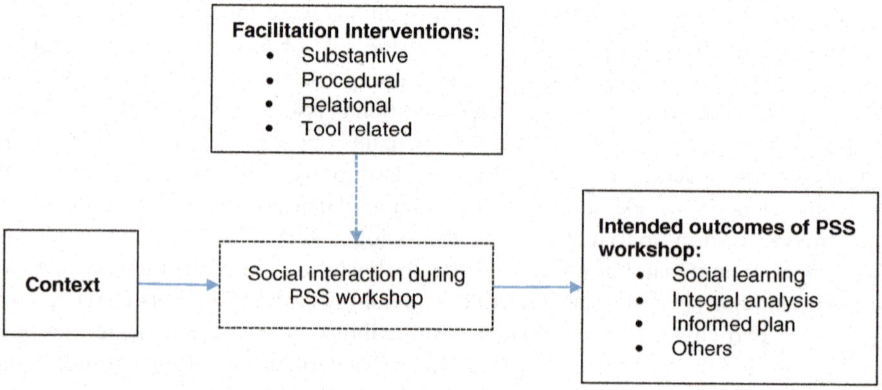

Fig. 19.1 Conceptual framework of chapter

during a workshop, and contribute to certain intended workshop outcomes. The remainder of this chapter draws on our empirical investigation to further explore, refine and illustrate this conceptual framework.

3 Methodology

Our aim is to investigate the nature and effectiveness of facilitation interventions at PSS workshops. Since this is a research topic that has to our knowledge not been explored before and requires in-depth practitioner insights, we adopted a research design featuring key informant interviews among a purposeful selection of practitioners with diverse PSS workshop experience. We relied on the relatively easy access of our own national contexts, which gave the additional advantage that we could compare the two contexts and could draw on our knowledge of the planning cultures and popular PSS in each country. We interviewed eight experienced facilitators of PSS workshops, five from the United States and three from the Netherlands.[1] All eight had participated in multiple PSS workshops, ranging from four workshops within one planning project, to over 50 workshops across many diverse projects.

The workshops described by informants took different forms. In addition to the portions dedicated to PSS use, the events typically combined multiple elements within one meeting, such as slide presentations, display boards, small group discussions, and brainstorming exercises. The specific PSS mentioned were Urban Strategy, Phoenix, CommunityViz, Envision Tomorrow+, as well as non-domain specific tools like Google Earth. Figures 19.2 and 19.3 show pictures taken during

[1]We do not mention the names of our interviewees to allow them to reflect freely on their experiences.

Fig. 19.2 A PSS-supported workshop in Lockhart, Texas (*Source* Robert Goodspeed)

Fig. 19.3 A workshop supported with the PSS Urban Strategy in Utrecht, the Netherlands (*Source* Stan Geertman)

two workshops where the informants served as facilitators. At these events, the PSS were used in diverse ways:

- Two informants had experience at PSS workshops where participants entered land use plans on a paper map, which were digitized on laptops by PSS operators in order to calculate indicators;
- One informant is a specialist in using MapTable, a large touch-sensitive computer monitor (touch table), which allows a small group to simultaneously view and manipulate a computer model;
- One informant described meetings where a PSS is projected from above onto a table, and participants interact with the model through a light pen;
- One informant is a specialist in using dedicated user-friendly drawing software, which is often applied in combination with a touch table;
- Finally, three informants had experience with workshops where groups viewed PSS outputs on computer monitors.

Our semi-structured interviews followed the main variable categories in the conceptual framework. In order to minimize bias, although two authors conducted almost all of the interviews together, a Dutch author asked questions during the interviews of Americans, and an American author asked the questions with the Dutch professionals. Below, we briefly discuss differences we observed between the two countries. The interviews were conducted by phone or online videoconferencing. One interview was conducted in person by only one interviewer. The transcripts were divided among two authors for coding. In the coding process, the four types of intervention were identified. Relevant quotes were assigned to one of the four categories and then given an inductive detailed code. If quotes overlapped with existing codes, they were attributed to the similar code, which was then sometimes slightly modified to encompass multiple quotes.

4 Findings

4.1 Overview of Interventions

A full overview of the resulting codes and corresponding quotes is reported in Appendix 1. Unlike the substantive interventions, most procedural and relational interventions that were mentioned by the facilitators are arguably not specific to planning or PSS workshops, but relevant to any group meeting. With regards to procedural interventions, the most common was time keeping, an unsurprising but necessary dimension of facilitation. Moreover, we found the substantive interventions fell into a varied set of codes. For instance, ensuring the ideas that are proposed are feasible was described this way:

> If somebody puts a park in an area where that park couldn't go for whatever reason, […] (the facilitators) might say: 'That place isn't eligible for this and they would explain why it is not eligible. And then they could parlay that into kind of a discussion, why might that not be there?

On the other hand, substantive interventions also involve ensuring the workshop stays focused in a broader sense: "*You have to draw out commonalities (…) make sure people don't jump ahead.*"

A key set of relational interventions refer to handling difficult or dominant people. A starting point for almost all facilitators is that difficult or dominant people should be listened to and treated with respect, but not allowed to take over the workshop, prevent others from speaking, or prevent the group from working. To deal with these people, a facilitator has various interventions at his or her disposal, for instance, one strategy is to document the complaints: "*If somebody is angry, you try to write down or to make notes on the screen, [so] that he sees that you take him seriously. If he is aggressive, let him draw or type in himself what are the remarks he has, and you even print it out. They are all tools to create confidence and to create trust.*" Another way to address this issue is to proactively create a structure for the workshop that ensures all participants are heard: "*[we apply] Round-Robin techniques [that] prevent one person to dominate the conversation. There are two pens we would we be using, and we would change these pens on a regular interval, so people won't feel we're picking on them.*" In addition, we coded several interventions that are particularly aimed at ensuring a positive atmosphere. As one facilitator notes: "*I use humor purposely as a device to kind of get the room loosened up. You gotta get folks relaxed.*"

Among the tool-related interventions, there is a clear consensus among the facilitators that PSS should not take over the workshops. Several facilitators described their goal was to ensure the PSS is used only in certain ways and times. One facilitator states that: "*You don't want the tool to become the center of attention. (…) I tell people over and over again, this is one arrow in the quiver.*" Another facilitator adds that: "*(we decided) to keep the digitizer role a little more discrete (…) we wanted to prioritize the conversation that people were having around the table [a MapTable] with one another (…) we didn't want the software aspect to have a negative impact on the conversation.*" However, in other instances, facilitators emphasize interventions aimed at increasing the use of the PSS. For instance, a facilitator remarks: "*The facilitator might push the group in a direction the facilitator wants to work in. If I want a group to work with [a PSS], I ask them questions which can only be answered by working with [the PSS].*" Another facilitator remarks that the best way for effectively using the tool is to involve the participants in defining the assumptions used by the PSS. "*So I say, folks, you got to help me get this thing [the PSS] ready. (…) We call that user-defined parameters, UDPs.*" Tool involvement can also take a different form, by giving people time to get accustomed to the PSS: "*The first ten minutes, you have to get people involved. For instance, they want to see their house on the map, and generally we tend to allow that, because it gets people interested and involved, puts them in the right mood, and puts the map central (...) They get more convinced in working around the table [a MapTable].*"

We also inductively coded reactive interventions, which were used if workshop participants question a PSS. For instance, when someone is critical about whether the PSS actually contributes to the planning process: "*The only way to deal with them is to have a quite flawless workshop. (...) I would counter them not by explaining the instrument, but trying to get their attention away from the instrument, try to get their attention to the actual problem they have. Trying to get a dialogue about their problem and their challenges. Letting the instrument be for a while.*"

Finally, some of the tool-related interventions are closely related to the substantive interventions. For instance, by drawing attention to information contained in the PSS: "*They [facilitators] might need to remind people about a data point that was brought up. (...) Facilitators sometimes need to say 'Hold on one second: is that the data that was just presented.' Continue to ground the conversation in the data that was presented.*"

Our binational research design also allows us to tentatively explore differences in PSS facilitation between two planning cultures. Although the variation among the cases is arguably larger than between the two countries, in general, Dutch facilitators described projects with greater participation by government officials and with larger budgets for more technically complex PSS than the American ones. American facilitators described the challenges of using PSS at workshops attended by activists concerned about property rights, which reflects more skeptical attitudes of government authority in U.S. political culture. The hardware also differed between the two countries—MapTables were only used in the Netherlands, and projectors and light pens were used only in the United States. However we did not find evidence these technical differences affected the facilitation or overall workshop dynamics, since both allow for very similar group interactions with the PSS. Despite these differences, the distribution of interventions across our four categories for each country was roughly similar. The most commonly raised interventions, *handling dominant or difficult people* and *prevent tool domination* were top concerns by facilitators in both countries.

4.2 Interventions and Outcomes

Finally, we asked our informants about the relationship between the interventions and what they perceived to be the main goals of the workshops. First, we are very aware that we need to be sensitive to the diverse nature and aims of workshops. Several informants explicitly identified different types of workshops with various outcomes, discussing how facilitation would differ for each workshop. For example, one facilitator mentioned three types: workshops for communicating a design or proposal (for a new highway, for example), workshops focusing on brainstorming and collecting opinions, and workshops for "drawing and calculating" where proposals are sketched and analyzed. Another facilitator described how learning was a necessary stage for all planning projects before group problem-solving could take place.

A second reflection is that several of the substantive and relational interventions were related to planning-specific tasks, such as selecting feasible options, thinking spatially, and drawing on local knowledge. Although all informants stressed the importance of involving all participants in the workshop, they did advocate substantive interventions to move discussions forward. The interventions *feasibility* , *strategic guidance*, and *synthesize* involve providing selected expert opinions on the topics being discussed. Furthermore, *aid with spatial thinking* and *thinking big* are interventions that involve assistance on two types of thinking that participants may not be as well versed in, but are required for planning. One informant stressed the importance of substantive knowledge in establishing credibility among participants: "*The first thing I do when I fly in is to buy a local newspaper. I try to work it into the content of the workshop. (…) If I'm able to say: does that proposal for that new shopping center south of town makes sense in this respect? (They think) God, he knows what's going on.*" While the informants reported a range of substantive interventions, they also reported interventions aimed at focusing attention on the PSS, when it was appropriate for the discussion.

The tool-related interventions that were mentioned contributed in various ways to the objectives we identified: learning, integrated analysis, and informed products. Many of these were aimed at ensuring participants understood and used the PSS when appropriate: *steer towards tool, prevent tool domination, connect to data*, and *guide tool use*. One informant stressed the difficulty of participants to digest all information that was presented, and argued facilitators played a key role in reminding participants of the structure of the PSS so that they could understand how to use it: "*(the facilitators) helped them understand what the different categories were that we were playing with. (…) I think that was important for this particular workshop: (remind them) what is the computer actually recording?*"

Several facilitators also mentioned a range of interventions aimed at additional outcomes beyond integrated analysis and informed products. These include both broader learning outcomes, as well as other outcomes not specified in our framework. For example, *avoiding tool domination* was motivated by a desire to ensure sufficient time for other outcomes, such as mutual learning about competing values among participants. One facilitator argued that the PSS helped to make the workshop "fun" and provide a "sense of accomplishment," both of which she felt were important to ensure ongoing engagement in the project, and therefore for the overall project success.

While many PSS interventions were aimed at learning, integrated analysis, and an informed product, we also identified substantive and relational interventions aimed at a range of other outcomes. Perhaps most surprisingly, as described above, achieving workshop objectives often required explaining the PSS, encouraging participants to use it, but also preventing it from dominating the meeting or distracting from the substantive issues.

5 Conclusions and Further Research

The central question in this chapter was: What interventions lead to effective facilitation of PSS workshops? Our preliminary empirical analysis with a relatively small number of experienced facilitators from the Netherlands and the United States, allows us to discern some tentative patterns. First, procedural interventions were mentioned less frequently than other types of interventions, although this could also be explained because these interventions are taken for granted by the facilitators. Several facilitators mentioned the importance of physical conditions for a good PSS workshop, including good Internet connectivity, well-functioning air-conditioning, working shades and lighting controls, etc. We consider these important aspects of the context for successful PSS workshops. Second, we found that the category of tool-related interventions consists of a combination of relational and substantive interventions, resulting, however, in distinct interventions. On the one hand, facilitators point out that the tool should not dominate the workshop and suppress discussion (a relational consideration), whereas on the other hand, participants are encouraged to grasp the underlying ideas of the PSS and often as a consequence the spatial phenomena at hand (a substantive consideration). Combining these two kinds of interventions in one workshop is a fine balancing act. Based on the interviews, it seems, however, critical in order to come to effective facilitation of PSS workshops.

The main added values resulting from our empirical material for the wider PSS debate are two main sets of tool-related facilitation interventions. First, we found interventions aiming to restrict the role of PSS or apply it selectively, indicating that there is a tendency in PSS workshops to put the tool too much on the forefront, and second, we document interventions to improve the usage of the tool, for instance by involving the participants in setting the indicators or by steering questions towards using the tool. These two clusters might be counterintuitive, but strongly relate to each other. Therefore, the act of facilitation is to find a fine balance between using a PSS to little versus using it too much. This requires a delicate mix of interventions. In addition, facilitating is strongly context-dependent. A workshop with local residents requires a very different way of steering PSS use than a setting with planning practitioners with extensive knowledge of the PSS at hand. Which interventions are needed depend not only on workshop goals, but also on the nature of the PSS, the participants, and the particular discussion they have.

The findings in this chapter are related to two issues we will mention that extend beyond our focus here on workshops: participant involvement in designing PSS, and the negative effects of tools like PSS. Previous researchers have argued for the importance of participant involvement in creating or tailoring PSS tools, using terms like mediated modelling, group model building and mediated planning support (e.g. Van den Belt 2004, Te Brömmelstroet 2010; Vennix 1999). Hence, PSS involvement seems to offer a context-sensitive facilitation intervention that, when applied carefully, can have a positive effect in different kinds of PSS workshops. However, this issue extends beyond the boundaries of the workshop to encompass contextual factors related to the planning process. In relation to the second important observation—that PSS should not dominate workshops—recent studies point at the potentially negative effects a technology like PSS can have when dominating planning and policy making. It can for instance hamper communication (Pelzer 2014) or guide and limit the content of a workshop, something referred to as performativity (Smith et al. 2013). Whereas it is widely acknowledged that PSS should connect to the collaborative and communicative nature of planning (e.g. Klosterman 1997, Geertman 2006), this chapter adds a cautionary note: successful collaboration at workshops requires facilitation to ensure effective PSS use.

This chapter follows two trends in PSS research we believe holds much promise for future research: first, learning from professional practice to generate new theoretical insights, and second, a focus on the importance of details of planning workshops in explaining process outcomes. Specifically focusing on facilitation, informants' experiences suggests additional research could probe the relationship between the broader aims of workshops (e.g. learning), the right conditions in a workshop (e.g. open communication), and the type of planning interventions that could facilitate this (e.g. tool involvement).

Our methodological choice to focus on gathering empirical insights through eight semi-structured interviews raises a number of important limitations on the validity of the findings. Future research using a large-N research design could therefore strengthen the external validity of our findings. This would add even more validity by extending it into different planning cultures and workshop types. We also strongly believe that a more rigorous hypothesis-testing research design (i.e. through case studies, observations, or controlled experiments) could strengthen the internal validity of understanding how facilitation interventions influence PSS workshop outcomes.

This chapter attempted to put facilitation more firmly on the agenda of the PSS debate. We believe this discussion is as relevant for practice as it is for academia, since, as one of the interviewees puts it, workshops can be *"spoiled by a lousy facilitator"*.

Acknowledgments The authors would like to thank the eight professionals who made this chapter possible by providing their time and insights, as well as the helpful comments from two anonymous reviewers and the book editors. This research was supported in part by the Netherlands Organization for Scientific Research (NWO) through the CESAR project and the Sustainable Accessibility of the Randstad program.

References

Arciniegas, G. A., & Janssen, R. (2012). Spatial decision support for collaborative land use planning workshops. *Landscape and Urban Planning, 107*, 332–342.

Argyris, C., & Schön, D. A. (1996). *Organizaitonal learning II: Theory, method, and practice* (Vol. 2). Reading, Mass.: Addison-Wesley Pub. Co.

Bales, R. F. (1950). *Interaction process analysis; A method for the study of small groups*. Oxford, England: Addison-Wesley.

Brail, R. (Ed.). (2008). *Planning support systems for cities and regions*. Cambridge, MA: Lincoln Institute for Land Policy.

Couclelis, H. (2005). "Where has the future gone?" Rethinking the role of integrated land-use models in spatial planning. *Environment and Planning A, 37*, 1353–1371.

Deyle, R., & Slotterback, C. S. (2009). Group learning in participatory planning processes an exploratory quasiexperimental analysis of local mitigation planning in Florida. *Journal of Planning Education and Research, 29*, 23–38.

De Vreede, G., Boonstra, J., & Niederman, F. (2002). What is effective GSS facilitation? A qualitative inquiry into participants' perceptions. *Paper presented at the 35th Annual Hawaii International Conference onSystem Sciences*.

Fisher, B. A., & Ellis, D. G. (1980). *Small group decision making: Communication and the group process*. New York: McGraw-Hill.

Forester, J. (1989). *Planning in the face of power*. Berkeley: University of California Press.

Frey, L. R. (1995). *Innovations in group facilitation: Applications in natural settings*. Cresskill, JN: Hampton Press.

Friedmann, J. (1989). *Planning in the public domain*. Princeton, NJ: Princeton University Press.

Geertman, S., & Stillwell, J. C. H. (2003). *Planning support systems in practice*. Berlin, New York: Springer.

Geertman, S. (2006). Potentials for planning support: A planning-conceptual approach. *Environment and Planning B: Planning and Design, 33*, 863–888.

Geertman, S. (2008). Planning support systems: A planner's perspective'. In R. Brail (Ed.), *Planning support systems for cities and regions* (pp. 213–230). Cambridge, MA: Lincoln Institute for Land Policy.

Geertman, S., & Stillwell, J. (Eds.). (2009). *Planning support systems: Best practices and new methods*. Heidelberg: Springer.

Geertman, S., Stillwell, J., & Toppen, F. (Eds.). (2013). *Planning support systems for sustainable urban development*. Heidelberg: Springer.

Goodspeed, R. (2013). Planning support systems for spatial planning through social learning. Ph. D. thesis, Department of Urban Studies and Planning, Massachusetts Institute of Technology, Cambridge, Mass. Available online at http://dspace.mit.edu/handle/1721.1/81739.

Gudmundsson, H. (2011). Analysing models as a knowledge technology in transport planning. *Transport Reviews, 31*, 145–159.

Hayne, S. C. (1999). The facilitators perspective on meetings and implications for group support systems design. *SIGMIS Database, 30*, 72–91.

Healey, P. (1992). Planning through debate. The communicative turn in planning theory. *Town Planning Review, 63*, 143–162.

Hirokawa, R. Y., & Gouran, D. S. (1989). Facilitation of group communication: A critique of prior research and an agenda for future research. *Management Communication Quarterly, 3*, 71–92.

Hopkins, L. (2001). *Urban development: The logic of making plans*. Washington, D.C: Island Press.

Hopkins, L. D., & Zapata, M. (2007). *Engaging the future: forecasts, scenarios, plans, and projects*. Cambridge, Mass.: Lincoln Institute of Land Policy.

Huntsman, S. (2004). *The Revolution needs a good facilitator: Modeling and visualization software in community settings*. Collaborative Edge, Spring/Summer 2004.

Innes, J., & Booher, D. (1999). Consensus building as role playing and bricolage. Toward a theory of collaborative planning. *Journal of the American Planning Association, 65*, 9–26.

Khalifa, M., Kwok, C. W. R., & Davison, R. (2002). The effects of process and content facilitation restrictiveness on GSS-mediated collaborative learning. *Group Decision and Negotiation, 11* (5), 345–361. doi:10.1023/A:1020449317854.

Klosterman, R. (1997). Planning support systems: A new perspective on computer-aided planning. *Journal of Planning Education and Research, 17*, 45–54.

Klosterman, R. (2009). Planning support systems: Retrospect and prospect. In S. Geertman & J. Stillwell (Eds.), *Planning support systems: Best practices and new methods* (pp. iv–vii). Heidelberg: Springer.

Kolb, J. A. (2011). *Small group facilitation: Improving process and performance in groups and teams*. Amherst, MA: HRD Press Inc.
Kolfschoten, G., & Rouwette, E. (2006). Choice criteria for facilitation techniques. In *Proceedings Hawaii international conference for facilitation techniques* (p. 10). Hawaii International Conference on System Sciences.
McElvaney, S. (2012). *Geodesign: Case studies in regional and urban planning*. Redlands: ESRI Press.
Niederman, F., Beise, C. M., & Beranek, P. M. (1996). Issues and concerns about computer-supported meetings: The facilitator's perspective. *MIS Quarterly, 20*, 1–22.
Nonaka, I., & Takeuchi, H. (1995). *The knowledge creating company*. New York: Oxford University Press.
Nyerges, T., Jankowski, P., Tuthill, D., & Ramsey, K. (2006). Collaborative water resource decision support: Results of a field experiment. *Annals of the Association of American Geographers, 96*(4), 699–725.
Pelzer, P. (2014). Understanding the usefulness of planning support systems. Paper presented at the CITTA/COST conference, 23–24 Oct 2014, Porto, Portugal.
Pelzer, P. (2015). Usefulness of planning support systems: Conceptual perspectives and practitioners' experiences. Ph.D. thesis, Department of Human Geography and Spatial Planning, Utrecht University.
Pelzer, P., Arciniegas, G., Geertman, S., & de Kroes, J. (2013). Using MapTable® to learn about sustainable urban development. In S. Geertman, J. Stillwell, & F. Toppen (Eds.), *Planning support systems for sustainable urban development* (pp. 167–186). Heidelberg: Springer.
Pelzer, P., Geertman, S., van der Heijden, R., & Rouwette, E. (2014). The added value of planning support systems: A practitioner's perspective. *Computers Environment and Urban Systems, 48*, 16–27.
Poole, M. S., & Hollingshead, A. B. (2005). *Theories of small groups: Interdisciplinary perspectives*. Thousand Oaks, CA: Sage.
Salter, J., Campbell, C., Journeay, M., & Sheppard, S. (2006). The digital workshop: Exploring the use of interactive and immersive visualization tools in participatory planning. *Journal of Environmental Management, 90*(6), 2090–2101.
Schwarz, R. M. (2002). *The skilled facilitator a comprehensive resource for consultants, facilitators, managers, trainers, and coaches*. San Francisco: Jossey-Bass.
Shiffer, M. (1992). Towards a collaborative planning system, *Environment and Planning B, 19*, 709.
Schwarz, R. M. (1994). *The skilled facilitator: Practical wisdom for developing effective groups*. San Francisco: Jossey-Bass Publishers.
Smith, H., Wall, G., & Blackstock, K. (2013). The role of map-based environmental information in supporting integration between river basin planning and spatial planning. *Environmental Science and Policy, 30*, 81–89.
Steinitz, C. (2012). *A framework for geodesign*. Redlands, CA: ESRI Press.
Te Brömmelstroet, M. (2013). Performance of planning support systems: What is it, and how do we report on it? *Computers, Environment and Urban Systems, 41*, 299–308.
Te Brömmelstroet, M., & Bertolini, L. (2008). Developing land use and transport PSS: Meaningful information through a dialogue between modelers and planners. *Transport Policy, 15*, 251–259.
Te Brömmelstroet, M., & Schrijnen, P. (2010). From planning support systems to mediated planning support: A structured dialogue to overcome the implementation gap. *Environment and Planning B: Planning and Design, 37*(1), 3–20.
Van den Belt, M. (2004). *Mediated modeling : a system dynamics approach to environmental consensus building*. Washington, DC: Island press.
Vennix, J. A. (1999). Group model-building: Tackling messy problems. *System Dynamics Review, 15*, 379–401.
Wenger, E. (1998). *Communities of practice: Learning, meaning, and identity*. New York: Cambridge University Press.

Part IV
Planning Support Systems: New Methods and Tools

Chapter 20
Visualization Methods for Linking Scientific and Local Knowledge of Climate Change Impacts

Scott N. Lieske, Kari Martin, Ben Grant and Claudia Baldwin

Abstract Planning support systems for smart cities and spatial planning more broadly must be able to help communities confront the combined effects of climate change: flooding, sea level rise, storm surge, and severe weather events in coastal areas. The goal of this chapter is to extend ideas about the role of geographic visualization in generating a societal response to top-down inaction on changing climate by testing methods and evaluating the effectiveness of geographic information-based tools for developing 3D scenes within a participatory process. The engagement process took place within an Australian coastal community where residential development and infrastructure are vulnerable to flooding, sea level rise and storm surge. This research employed and assessed multiple visual methods including geographic visualization to illustrate the impacts of climate change on the study area. Participant assessment indicated all methods employed were beneficial but 3D visualization was the most effective method for knowledge exchange.

1 Introduction

The Intergovernmental Panel on Climate Change (IPCC) Fifth Assessment Report projects sea level rise will continue during the 21st century and into the future. Higher emission scenarios and warmer temperatures will result in ice sheet melting

S.N. Lieske (✉) · K. Martin · B. Grant · C. Baldwin
Faculty of Arts and Business, University of the Sunshine Coast, Maroochydore DC, QLD 4558, Australia
e-mail: scottnlieske@gmail.com

K. Martin
e-mail: kmartin3@usc.edu.au

B. Grant
e-mail: bengrant@optusnet.com.au

C. Baldwin
e-mail: CBaldwin@usc.edu.au

S. Geertman et al. (eds.), *Planning Support Systems and Smart Cities*,
Lecture Notes in Geoinformation and Cartography,
DOI 10.1007/978-3-319-18368-8_20

and sea level rise in metres (Church et al. 2013). The Southeast Queensland region of Australia was identified as a climate change "hotspot" in the IPCC Fourth Assessment Report (Hennessy et al. 2007, p. 525) which highlights the pressures and increased climate change risks from rapid population growth in coastal areas, specifically referring to *"high value canal estates"* and intensively developed low lying lands adjacent to estuaries (Hennessy et al. 2007, p. 528). Projections for the Sunshine Coast region in 2100 include temperature increases up to 6.5 °C, sea level rise of approximately 1.1 m and an increase in the number of severe storms. Sea level rise is expected to result in larger storm surges and an increasing threat of coastal erosion (SCRC 2010).

The IPCC (2012) presents the argument that risk associated with climate change, while based on objective and physical conditions, is socially constructed. Disaster and risk are based upon societal choices for action versus inaction. With a societal propensity toward inaction, the idea that there is a clear role for geographic visualization in generating a societal response to top-down inaction on changing climate has been argued by several authors (e.g. Sheppard 2012; Burch et al. 2009, 2010). With the premise that local action may be the best hope for positive change and climate action, Sheppard (2012) addresses the gap, expressed as a *"gaping hole"* (p. 50) between the scientific literature and local climate knowledge. Taking his inspiration from human nature, *"... as visually oriented beings we change our hearts and minds when we see for ourselves what is or what could be"* (p. vi), Sheppard's idea is to make carbon and climate change visible where we care the most, in our local communities and metaphorical and literal backyards. The recognition that visualization offers strong potential for enhancing communication of climate information stems from the linking of visual thinking and visual communication. This idea was presented as an idealized research sequence by DiBiase (1990) who links the effective distribution of ideas, including scientific findings and complex information, to wide audiences through the use of visualization.

The goal of this chapter is to extend Sheppard's ideas about the use of visualization by testing and presenting methods for geographic information systems (GIS) and GIS-based planning support system (PSS) tools for developing visualization products within a climate change focused participatory process. The case study presented here was developed for a coastal community in Australia where residential development and infrastructure are vulnerable to flooding, sea level rise and storm surge. The participatory process was based on two workshops. In the first workshop residents were presented with local government storm surge and flood inundation maps. The goal was to establish baseline information related to the impacts of climate change for the study area. Residents were asked for their personal knowledge of areas they perceived to be vulnerable to the impacts of climate change, such as places where canal walls may be breached in storms, floods or extreme high tides. The second workshop centered on GIS-based visual display of the impacts of the combined effects of climate change for both current and year 2100 conditions. Participants were shown 3D renderings and interactive fly-throughs illustrating how the impacts of climate change might affect the study area.

Assessment of the efficacy of geographic visualization to align scientific with local knowledge was based on facilitated discussion and responses to post-workshop surveys.

2 Background

Differing ideas of visualization–as methods and technology-based tools, as communication, or as product output (pictures)–exist within a larger context of scientific research and application. In the literature there is often a distinction made between visualization for scientific purposes versus visualization for planning and decision making. Scientific visualization provides methods and tools for exploratory data analysis and other scientific inquiry. Visualization for planning and decision making is sometimes referred to as geographical visualization which links methods and tools with communication (Cartwright et al. 2004) and presents opportunities for improvements in rational planning, environmental management and participatory decision making.

The emerging role of geographic visualization in communicating information about a changing climate is built on a considerable body of literature about the benefits of visual technologies for enabling communication and promoting comprehension (Appleton and Lovett 2003). Visual representations enable communication because they are intuitively understandable, convey information rapidly and offer group cognitive benefits. Visualization is an intuitively understandable medium because people recognize and are instantly able to create associations with 3D objects (Andrienko et al. 2007) and because visual representations take advantage of pre-attentive processing, or pre-existing background knowledge about the subject matter (Dickinson 2010). Visualization offers rapid and accessible information transfer due to the large quantity of information presented in a single scene and the intuitively understandable nature of the medium (Kwartler and Longo 2008; Wissen et al. 2008). Visual representations offer group cognitive benefits including promoting group understanding because they are shared objects to talk about, think about, and use as a basis of coordinating actions and perspectives (MacEachren and Brewer 2004). Similarly, Aggett and McColl (2006) mention the use of visualizations as a shared form of reference. They present the idea of visualization as a bridge that allows a group to move from individual mental images or perceptions regarding an issue to a shared perception.

The solution to top-down inaction on climate change advocated by Sheppard (2012) includes making climate change local using a community perspective, making climate change visual to change the way people see and think about it, and making climate change connected by linking big picture ideas and cause and effect relationships with local impacts. This research extends Sheppard's (2012) overview of the potential for geographical visualization to influence local climate knowledge by testing and presenting methods and lessons learned in creating GIS and GIS-based PSS visualization products.

Fig. 1 Study area

3 Study Area

The focus of this case study is the Twin Waters residential community located within the Sunshine Coast Regional Council (SCRC) local government area of Southeast Queensland, Australia (Fig. 1). The Twin Waters suburb was selected for study as it has publicly available flood hazard maps for both the present and future (year 2100) climate. The community also has an established resident group, the Twin Waters Residents Association (TWRA), that served as a point of contact and engagement for the project. The study site is approximately 4 km^2 in size. The majority of residential properties are located on or close to the water's edge.

4 Methods and Methodology

The participatory process was based on two climate scenarios: (1) the proposed impacts from an annual exceedance probability (AEP) of 1 % (or 1 in 100 year) flood event in the current climate and, (2) an AEP of 1 % flood event incorporating projected impacts from climate change in 2100. The future climate scenario is based on an 800 mm increase in mean sea-level and a 20 % increase in rainfall intensity compared with the current climate. This scenario is described by the Australian Government (2014) as the medium scenario which represents the upper end of

IPCC's 2007 4th Assessment Report (AR4) A1FI projections and is representative of recent global emissions and observations of sea-level rise. However, the future climate scenario does not include the impacts of recent warming trends on ice sheet dynamics. This information is captured in a high-end scenario suggesting 1.1 m sea level rise (Australian Government 2014). Both scenarios take into account riverine flooding, storm surge and tides.

The participatory process employed four consultation techniques over two workshops: (1) SCRC flood hazard maps of the two climate scenarios; (2) participatory mapping of vulnerable areas using a 27-inch screen tablet computer; (3) the visual research technique, Photovoice (Baldwin and Chandler 2010) and; (4) 3D scenes representing the two scenarios. Presentation of the flood hazard maps, participatory mapping and an introduction to Photovoice took place in the first workshop. Publicly available SCRC flood hazard maps provide storm surge, flood inundation and sea level rise mapping for the area (SCRC 2014). The maps were presented and discussed in order to ensure workshop participants had baseline information related to the impacts of climate change. The participatory mapping exercise was an opportunity for residents to indicate areas that might be at risk to the impacts of climate change, such as places where canal walls may be breached in storms, floods or extreme high tides. In the Photovoice exercise, residents were asked to take digital pictures of those sites they identified as at risk to the impacts of climate change. At the second workshop, participants shared their photos, linked their photos to locations and explained the reasons they took a photo. The group discussed the main story arising from their photos and organised the photos to tell that story. Results of the large screen tablet computer and Photovoice exercises are described more fully in Grant et al. (2015).

The second workshop centred on 3D scenes representing the two scenarios. Participants were shown 3D renderings and interactive fly-throughs illustrating how the impacts of climate change might affect the study area. 3D scenes were created using GIS and PSS software, ArcScene and the Scenario 3D module of the CommunityViz® PSS, respectively. ArcScene is the interactive 3D scene development environment and 3D viewer incorporated within ArcGIS (ESRI Inc., Redlands CA). The CommunityViz PSS is a modular system built on the ArcGIS platform consisting of two integrated extensions to ArcGIS: Scenario 360 and Scenario 3D. Scenario 360 extends the quantitative capabilities of ArcGIS by allowing formula-based spreadsheet-like calculations to be performed on geographic data. Scenario 3D allows for 3-dimensional display of the built environment and landscape with real-time movement and object manipulation in a semi photo-realistic setting. CommunityViz is a promising tool for geographic visualization as it explicitly links 3D scenes with scenario planning and it also allows sharing scenes via a separate and free Scenario 3D Viewer.

Data used to develop the scenes included a digital elevation model (DEM), LIDAR-based terrain data, a 2013 aerial image of the study site, flood layers, and geographic data layers representing additional features. All elevation data were 1 m resolution. The DEM, LIDAR, aerial imagery, flood height rasters representing the two scenarios, and building footprints were provided by the SCRC. GIS and PSS methods for creating the scenes may be summarized as: (1) create terrain, (2) texture terrain, and (3) add models and detail. The DEM was used as the terrain. In order to texture the terrain the aerial imagery was draped over the digital terrain resulting in an easily identifiable location specific scene.

Feature layers including building footprints, bridges, trees, houses, cars and people were added to the scenes in order to add detail and enhance realism. With the exception of building footprints, feature layers were developed in ArcMap using heads up digitizing to capture realistic locations of features based on the aerial imagery. ArcScene and Scenario 3D allow incorporating models as 3D symbology. 3D models representing cars, people and different types of trees were downloaded from the SketchUp 3D Warehouse. ArcScene allows importing SketchUp, Collada and similar models as 3D marker symbols within the symbol property editor. Scenario 3D allows one to select models within its 3D layer settings. In both programs, characteristics of 3D models may be edited such as model height, base height or offset, and orientation. Both programs allow extruding polygon features such as building footprints to pre-set heights and adjusting colour. Building footprints were extruded to 6 m to represent full building height. A second building footprint layer was extruded to 2.7 m which represents the minimum elevation required for house floor levels (slab) in the study area. LIDAR (.las) files were used to generate a digital surface model (DSM, first return) which was used to generate heights for some 3D features. Bridges were draped over the DSM to gain the true height of the feature. Tree models were assigned different heights based on best estimates and positioned at random orientations. It was thought the details provided by the 3D features would add a sense of reality to the scenes and provided important spatial references for communicating and understanding flood water levels.

Flood hazard data were incorporated in the 3D scenes as raster data. A key visual effect we sought to incorporate in the scenes was water transparency. Semi-transparent flood waters allow the underlying landscape to be seen and serve as a familiar spatial reference. Within ArcScene the transparency of the flood raster layer could not be easily altered. In order to achieve transparency a 2D polygon layer outlining the flood surface area was created and draped over the flood height raster to portray flood elevation. An advantage of ArcScene is it allows several raster layers to be added to a scene. A limitation of Scenario 3D is it allows only one raster layer to be added to a scene, usually the terrain layer. ArcScene represented the flood height data more easily in the 3D setting. The ArcScene renderings were approved by SCRC staff as portraying the correct representation of the data in a pre-workshop meeting. ArcScene was therefore chosen as the program used to create the 3D scenes presented in the workshop.

Assessment of the efficacy of geographic visualization to align scientific with local knowledge was based on both a survey and facilitated discussion. At the end of the second workshop participants were asked to take part in a post-workshop survey. The survey asked questions in a number categories: an assessment of general knowledge about climate change, addressing the impacts of climate change, visualization of climate change, and sources of information about climate change. Most questions solicited a response on a Likert scale where participants could indicate, for example, very low to very high effectiveness of different media types as information sources. Each section of the survey had open ended questions that allowed respondents to step beyond the format of Likert scale responses. The survey also requested basic demographic information such as age, gender, occupation and motivation for participation. The visualization section of the survey focused on establishing which sources of climate change information were utilised by the group and which climate change visualization method was considered to be most effective. Respondents evaluated the four participatory methods (flood hazard maps, participatory mapping, Photovoice, and 3D GIS-based visualization) as well as group discussion by ranking methods on a five point Likert scale from very ineffective to very effective. Facilitated discussion was used to identify climate change mitigation and adaptation options, the use of local action to influence global conditions, and suggestions for future research. Workshop methods including goals for each and necessary preparation steps are summarized in Fig. 2.

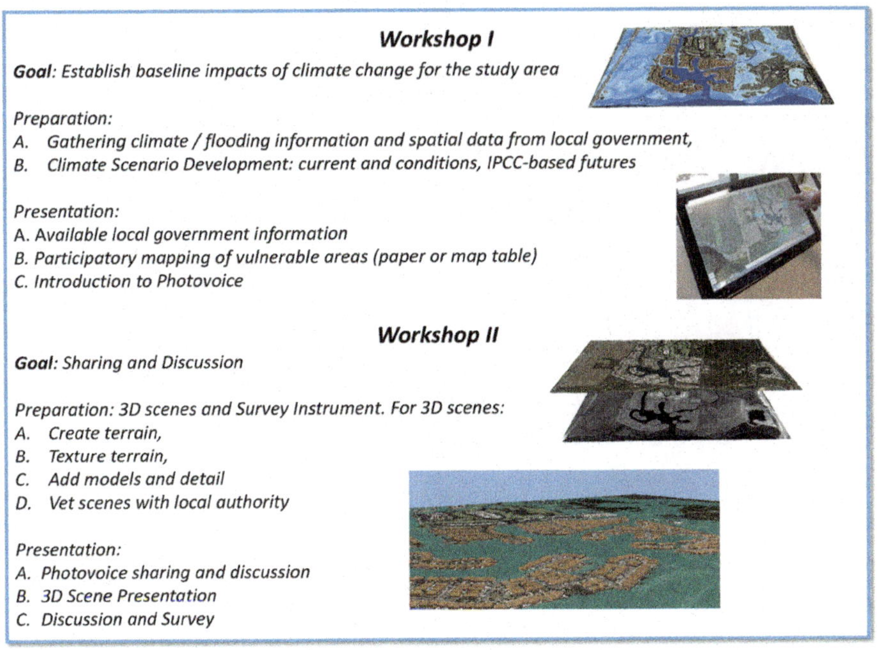

Fig. 2 Methodology for assessing multiple visual methods to influence climate knowledge

5 Results

In the first workshop, participants were shown and discussed the current climate and 2100 flood hazard maps (Fig. 3). Participants found the maps to be both confusing and inaccurate. They pointed out areas such as the vacant farmland on the

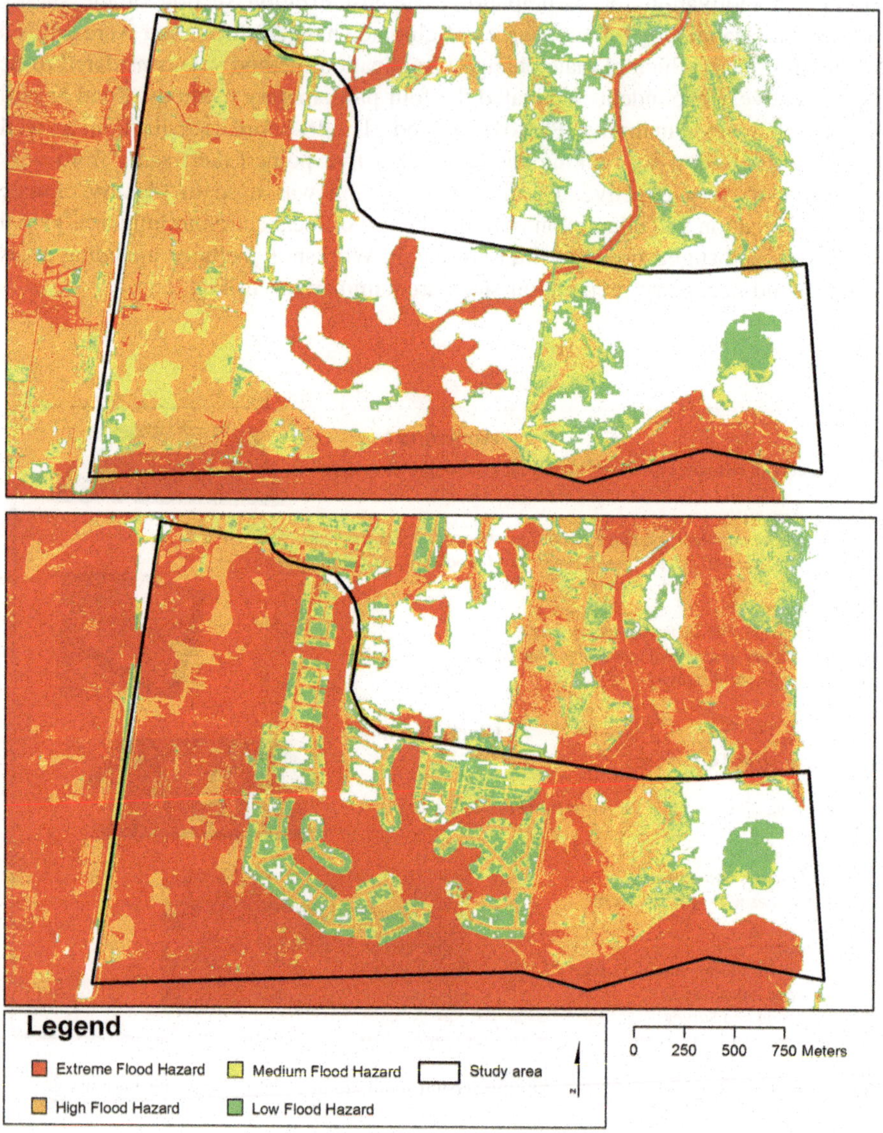

Fig. 3 Sunshine Coast Regional Council flood hazard maps for the current (*upper*) and year 2100 climate (*lower*) scenarios

west of the site adjacent to a highway that had recently flooded, but was not represented on the maps. Facilitators explained to participants the red colour on the flood hazard maps depicted the deepest areas of water during flood, which was more likely the rivers and creek sections.

The second workshop centred on 3D display of the visual impacts of the two scenarios (Figs. 4, 5 and 6). 3D scenes were presented as video fly-throughs on a projector screen. 3D scenes were shown directly from ArcScene rather than as an exported movie file. This allowed for more flexibility in the presentation of information and a much more interactive approach to presenting the scenes. For example, at any stage rendering could be paused and layers added or removed.

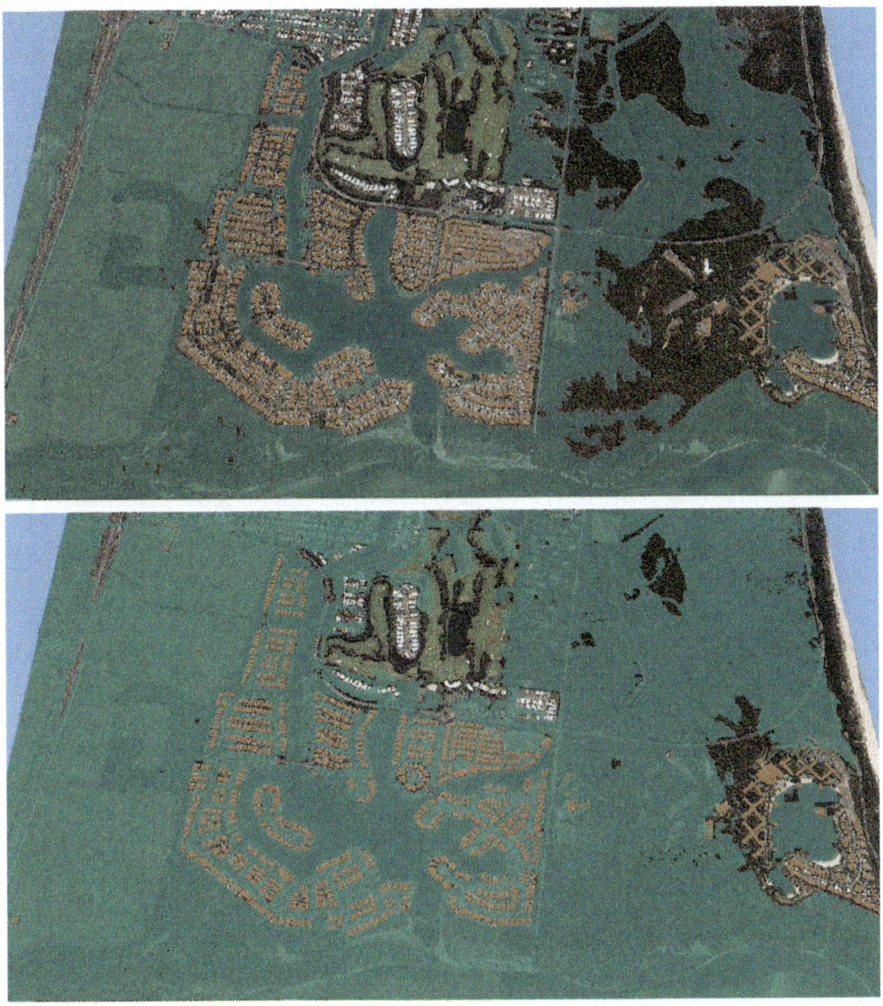

Fig. 4 Study area flood inundation for the current (*upper*) and year 2100 climate (*lower*) scenarios

Fig. 5 Houses (extruded) and road infrastructure for the current (*upper*) and year 2100 climate (*lower*) scenarios

Visual presentation of scenarios demonstrated that in a 1 in 100 year flood event in the current climate virtually no buildings will be impacted by flooding. However, for the 1 in 100 year flood in the year 2100 scenario, the majority of the study site was at a low, medium or high risk of flooding. Approximately 760 residential and commercial properties have been constructed in the subject site, and approximately 658 properties will be affected by flooding in 2100.

Participants sought to see the visual representations of the impacts of both scenarios on their properties, infrastructure such as bridges and revetment walls, and evacuation routes (Fig. 5). Residents were relieved their houses will not be inundated during the short term (current scenario) and noted they would not be alive to see the impacts in 2100. Participants indicated that due to the temporally distant nature of the 2100 flood impacts they were less of a problem. Whilst this does provide some short-term comfort for individuals today, it does not alleviate the

Fig. 6 House floors (*in red*) and flood inundation for the current (*upper*) and year 2100 climate (*lower*) scenarios

impacts or reduce the risks from climate change for future inhabitants of the site. Individuals noted both the presentation of scenarios and the ability to acutely focus on a particular area were beneficial. Participants indicated geographic visualization technology was useful for obtaining low oblique eye level (walk-through perspective) visuals of flood levels on infrastructure such as bridges and their properties.

At participants' request, the building footprints extruded to 6 m were removed and only the minimum elevation required for floor levels was displayed (Fig. 6). This generated significant interest and resulted in the participants requesting to locate their own properties in the scene to determine if they would be inundated or not. Interactive navigation through the scene allowed the participants to visually assess flood impacts on their properties. Another point of discussion raised was access to evacuation routes when roads are flooded. Concern was voiced about the possibility of both the western and eastern exit points from Twin Waters being inundated with water. Participants suggested similar visuals should be developed for all low lying residential areas on the Sunshine Coast to ensure residents have the relevant information relating to their properties and to determine evacuation routes.

Facilitated discussion in the second workshop incorporated in-depth conversations about risk, likelihood, consequences, insurance liability, and infrastructure ownership. Following presentation of the 3D scenes, participants were guided through key objectives, themes and actions from the SCRC *Sunshine Coast Climate Change and Peak Oil Strategy 2010–2020* (SCRC 2010). This generated discussion around adaptation and mitigation actions that can be taken at the local scale. When guided through the Strategy participants noted that they were not previously aware of the majority of the objectives and actions, nor were they aware if actions had commenced or had been completed. One participant indicated that the document was just words on a page and dismissed the potential of the document to reduce the impacts of climate change.

Overall the discussions from the second workshop revealed two key messages. TWRA participants stated the State Government, SCRC and property developers are not actively engaging with the community about the risks posed to the estate by climate change and the information currently available to residents is difficult to understand. Participants also stated they felt overwhelmed by the enormity of the climate change issue noting individuals did not generally undertake mitigation activities as one person cannot change or improve the situation. It was stated by workshop participants that TWRA members are generally retirement age and are more interested in enjoying their lives than taking action on climate change. Participants indicated that members of the TWRA are more inclined to ignore and not worry about the problem. The group stated that there was a delicate balance between fostering economic growth on the Sunshine Coast by promoting the main economic driver, tourism, versus scaring the local and wider communities by over-promoting the risks of climate change on the coast.

Results of the post-workshop survey indicate participants were either highly or very highly concerned about climate impacts overall and at the Twin Waters site. While the number of survey responses was small (n = 3), results were consistent among those surveyed. The Likert scale responses assessing the consultation techniques presented here were rated by all participants as 'effective' and/or 'highly effective'. The 3D scenes of the climate scenarios were considered by all participants to be highly effective as well as the most effective method of portraying climate change information. Second were the SCRC flood hazard maps; Photovoice and participatory mapping followed in third places as equally effective methods for communicating information and generating discussion. Participants indicated the 3D scenes painted a picture of the impacts on Twin Waters and suggested how to react when and if substantial flooding does occur. Participants understood and were concerned about flooding from severe weather events, and especially about evacuation routes.

6 Discussion and Conclusions

The goal of this chapter is to extend ideas about role of geographic visualization in generating a societal response to top-down inaction on changing climate by testing methods and evaluating the effectiveness of GIS and GIS-based PSS tools for developing visualization products within a climate change focused participatory process. The study has implications for a number of areas: development of GIS-based 3D visual tools; use of these tools in a climate change focused participatory process; reducing the gap between scientific and local knowledge; application to low-lying coastal study areas; and future research directions.

Assessment of the effectiveness of geographic visualization to align scientific with local knowledge was based on facilitated discussion and responses to post-workshop surveys. Discussion revealed the combination of consultation techniques: interactive 3D scenes, flood hazard maps, Photovoice, and interactive participatory mapping, were effective in sharing and increasing local knowledge about risk, and reducing the gap between the scientific literature and local knowledge. Survey results indicated that of these consultation techniques, the 3D scenes were the most effective means of communicating information on local climate change impacts. It is our view the 3D scenes were successful because they were based on regional and local scientific data presented in a way that was meaningful to participants and responsive to users' information needs about impacts that might personally affect them and their properties. The process also made clear visualizations do not function as a sole and exclusive means of communication. Rather, they are an important part of a facilitated participatory process. Results demonstrate and affirm the group cognitive benefits of visualization: promoting group understanding, use of visuals as shared forms of reference or objects to talk about and think about, and the use of visualizations as a basis of coordinating actions and perspectives as discussed above.

We encountered one major limit to the software employed. For the purpose of flood visualization with raster data, Scenario 3D was limited to displaying one raster layer per scene. This proved to be a substantial constraint as the scene required a DEM representing the bare earth layer as well as a second raster in order to represent flood levels. This limit was exacerbated by the fact that when using Scenario 3D, flood polygon layers display incorrectly in the complex mix of land and water typical of canal estates. Scenario 3D displays curved surfaces as a number of small triangles (Walker and Daniels 2011) which in our experiments often resulted in inaccurate representation of flooded areas. ArcScene met the needs of the project as it is able to accommodate multiple raster layers and portray flood waters as semi-transparent.

The methods employed and workshop results speak to the issue of necessary level of detail in 3D scenes and the trade-off between the benefits (if any) of increased realism versus time, cost and computing power. Presentation of the 3D scenes that included full detail, the entire suite of building footprints, bridges, trees, houses, cars and people, while they do add realism and provide spatial references, resulted in a staggered and uneven camera perspective and ultimately in degraded scene quality and presentation when recorded as an AVI file. Among the more effective 3D scenes were those that displayed only terrain, imagery, building footprints extruded to minimum floor level height, bridges and flood layers. These key features proved to be enough detail to communicate the climate change risks to the area including the extent of flooding and allowed participants to see if their own properties would potentially be inundated in the two scenarios. This and the discussion of evacuation routes clearly indicate the use of geographic visualization within a participatory process was able to bridge the gap between scientific and local knowledge.

In order to improve the 3D scene for a formal, post-workshop, presentation the focus of the scene was narrowed to a small area and features outside of this area were removed (Fig. 7). This allowed for more detail to be added within the smaller focus area without losing the quality of the fly-through video. Trees, cars, people and house models were added. Three separate flythrough angles were recorded and linked together showing the current climate flood scenario and then repeated showing the future climate scenario. Text screens identifying the context of the scenario were positioned at the beginning of the scene and the scene was exported as a single movie file. In a more formal, less interactive, setting this method of presenting the 3D scenarios worked well.

Sheppard (2012) posits reducing the gap between the scientific literature and local climate knowledge may also lead to local action on climate change. In the post-workshop assessments, TWRA residents indicated they are aware of the 'gap' between their knowledge and actions to reduce climate impacts. Residents were relieved their houses will not be inundated during the current scenario and noted they would not be alive to see the impacts in 2100. Participants also indicated that due to the temporally distant nature of 2100 those flood impacts are less of a problem. The reaction of workshop participants to the severe flooding indicated in the 2100 scenario suggests there are still challenges for using geographic visualization to inspire climate action. These results also suggest the relevance of demographics, in particular age, as a key variable in the assessment of the efficacy of geographical visualization to spur climate action.

The experience of developing the 3D scenes and hosting the workshops resulted in a number of recommendations for future research. Importantly, workshop participants saw the value in the 3D scenes and recommended they be developed for other areas along the Sunshine Coast. The use of 3D scenes was identified in this study as the most effective method of communicating the potential risks of climate change to community members. Participant's lack of knowledge and disregard for SCRC's climate change and peak oil strategy document suggests an opportunity for better and more open community planning and engagement. Communicating climate impact information using 3D scenes has the potential to be an effective planning tool for local governments to bridge the gap between climate science and local climate knowledge. 3D scenes may also be used to bring about more meaningful engagement with communities when developing local planning strategies whether it be for climate change adaptation, emergency response plans or disaster risk management. This

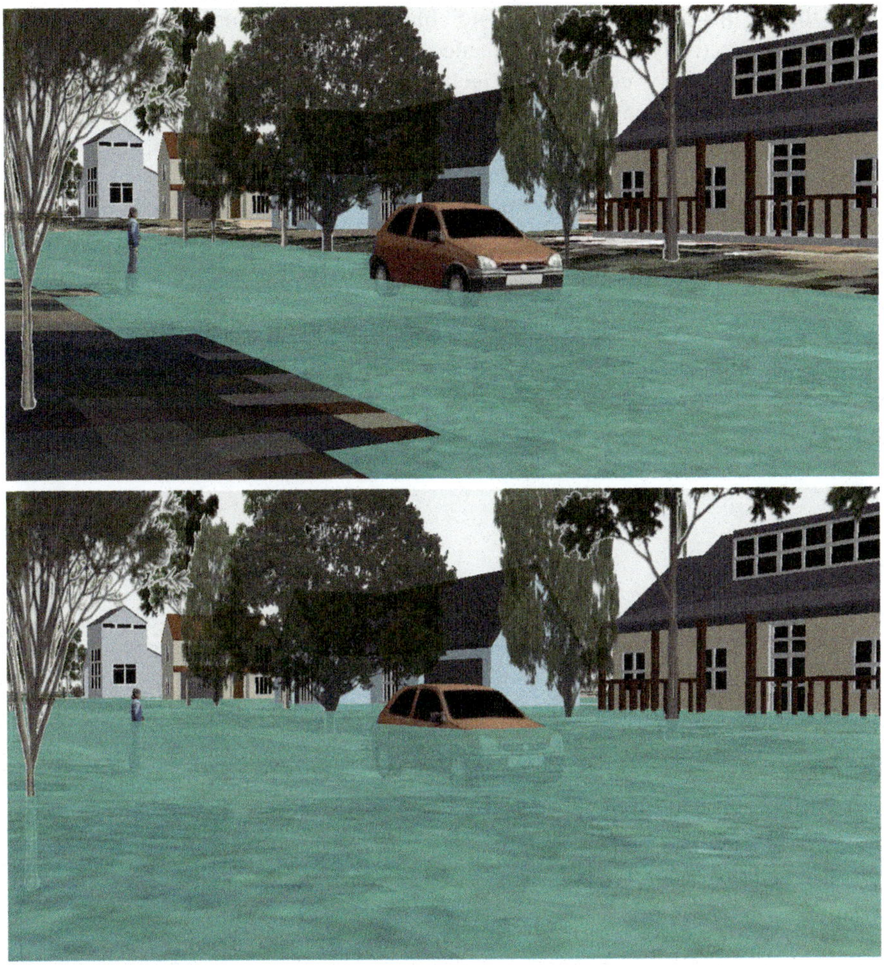

Fig. 7 Localized flood inundation for the current (*upper*) and year 2100 climate (*lower*) scenarios

approach is of benefit to local governments and communities as it assists them in preparing for local impacts of climate change. Further studies should be undertaken in other regions of Queensland, Australia and elsewhere. Burch et al. (2010) augment this argument noting that coastal communities around the world will need to implement plans that incorporate local climate adaptation and mitigation measures. This is an opportunity for PSS and spatial planning to provide needed outputs for a substantial user community well into the future.

Additional areas for future research include larger participating groups as well as evaluation of the efficacy of different participatory processes. Following the methods summarized in Fig. 2 and the rationale articulated above we aim to apply these techniques elsewhere. For evaluation of different participatory processes, it is noted that in this research there was no control group. The use of a control group could more effectively isolate the influence of geographic visualization technologies and other community consultation techniques. There is also an opportunity to integrate landscape change models with future climate models. The year 2100 3D scene portrayed 2100 flood waters on a 2014 built environment. Landscape change in the study area could further exacerbate flooding due to a likely increase in impervious surface. A further limitation of these methods was the lack of consideration of the length of time of inundation. Flood inundation duration has a significant impact on the extent of overall flood damage. Once determined or modelled, inundation time could be presented as auxiliary text within a 3D scene and/or augment the discussion. Finally, future studies could provide opportunities to extend this research to include not only looking at the physical environmental impacts but also the economic and social impacts on local communities by comparing scenarios based on proactive adaptation and mitigation implementation versus the 'business as usual' approach.

References

Aggett, G., & McColl, C. (2006). Evaluating decision support systems for PPGIS applications. *Cartography and Geographic Information Science, 33*(1), 77–92.

Andrienko, G., Andrienko, N., Jankowski, P., Keim, D., Kraak, M. J., MacEachren, A., & Wrobel, S. (2007). Geovisual analytics for spatial decision support: Setting the research agenda. *International Journal of Geographical Information Science, 21*(8), 839–857.

Appleton, K., & Lovett, A. (2003). GIS-based visualisation of rural landscapes: defining "sufficient" realism for environmental decision-making. *Landscape and Urban Planning, 65*, 117–131.

Australian Government Department of Environment. (2014). Mapping sea level rise. http://www.climatechange.gov.au/climate-change/adapting-climate-change/australias-coasts-and-climate-change/mapping-sea-level-rise. Accessed 20 November, 2014.

Baldwin, C., & Chandler, L. (2010). At the water's edge: Community voices on climate change. *Local Environment, 15*(7), 637–649.

Burch, S., Shaw, A., Sheppard, S. R. J., & Flanders, D. (2009). Climate change visualization: Using 3D imagery of local places to build capacity and inform policy. In T. Neset, J. Johansson & B. O. Linnér (Eds.), *State of climate change visualization* (pp. 65–73), CSPR Report No. 09:04, Centre for Climate Science and Policy Research, Norrköping, Sweden: Linköping University Electronic Press.

Burch, S., Sheppard, S. R. J., Shaw, A., & Flanders, D. (2010). Planning for climate change in a flood-prone community: Municipal barriers to policy action and the use of visualizations as decision-support tools. *Journal of Flood Risk Management, 3*, 126–139.

Cartwright, W., Miller, S., & Pettit, C. (2004). Geographical visualization: Past, present and future development. *Journal of Spatial Science, 49*(1), 25–36.

Church, J. A., Clark, P. U., Cazenave, A., Gregory, J. M., Jevrejeva, S., Levermann, A., et al. (2013). Sea level change. In T. F. Stocker, D. Qin, G. K. Plattner, M. Tignor, S. K. Allen, J. Boschung, A. Nauels, Y. Xia, V. Bex & P. M. Midgley (Eds.), *Climate Change 2013: The Physical Science Basis*. Contribution of Working Group I to the Fifth Assessment Report of the Intergovernmental Panel on Climate Change. Cambridge, United Kingdom and New York, NY, USA: Cambridge University Press.

DiBiase, D. (1990). Visualization in the earth sciences. *Earth and Mineral Sciences Bulletin, 52*, 13–18.

Dickinson, B. (2010). *3D Visualizations of cyber security events*. GeoDesign Summit. Redlands, CA. http://www.geodesignsummit.com/videos/day-one.html. Accessed June 14, 2012.

Grant, B., Baldwin, C., Lieske, S. N., & Martin, K. (2015). Using participatory visual methods for information exchange about climate risk in canal estate communities. *Australian Journal of Maritime & Ocean Affairs*, *7*(1), 23–37. doi:10.1080/18366503.2015.1014012.

Hennessy, K., Fitzharris, B., Bates, B. C., Harvey, N., Howden, S. M., Hughes, L., et al. (2007). Australia and New Zealand. In M. L. Parry, O. F. Canziani, J. P. Palutikof, P. J. van der Linden & C. E. Hanson (Eds.), *Climate Change 2007: Impacts, Adaptation and Vulnerability. Contribution of Working Group II to the Fourth Assessment Report of the Intergovernmental Panel on Climate Change* (pp. 507–540). Cambridge: Cambridge University Press.

IPCC. (2012). *Managing the risks of extreme events and disasters to advance climate change adaptation*. A Special Report of Working Groups I and II of the Intergovernmental Panel on Climate Change. In C. B. Field, V. Barros, T. F. Stocker, D. Qin, D. J. Dokken, K. L. Ebi, M. D. Mastrandrea, K. J. Mach, G. K. Plattner, S. K. Allen, M. Tignor & P. M. Midgley (Eds.), Cambridge, UK and New York, NY, USA: Cambridge University Press.

Kwartler, M., & Longo, G. (2008). *Visioning and visualization: People, pixels, and plans*. Cambridge, MA: Lincoln Institute of Land Policy.

MacEachren, A. M., & Brewer, I. (2004). Developing a conceptual framework for visually-enabled geocollaboration. *International Journal of Geographical Information Science, 18*(1), 1–34.

Sheppard, S. (2012). *Visualizing climate change: A guide to visual communication of climate change and developing local solutions*. Washington, DC: Earthscan.

Sunshine Coast Regional Council. (2014). Flood information for a property: Map layers. Sunshine Coast Regional Council. http://www.sunshinecoast.qld.gov.au/sitePage.cfm?code=region-flooding. Accessed November 21, 2014.

Sunshine Coast Regional Council. (2010). Sunshine coast climate change and peak oil strategy 2010–2020. http://www.sunshinecoast.qld.gov.au/sitePage.cfm?code=cc-strategy. Accessed November 12, 2014.

Walker, D., & Daniels, T. L. (2011). *The planners guide to CommunityViz: The essential tool for a new generation of planning*. Chicago: Planners Press, American Planning Association.

Wissen, U., Schroth, O., Lange, E., & Schmid, W. A. (2008). Approaches to integrating indicators into 3D landscape visualisations and their benefits for participative planning situations. *Journal of Environmental Management, 89*, 184–196.

Chapter 21
Virtual Worlds as Support Tools for Public Engagement in Urban Design

Anja Jutraz and Tadeja Zupancic

Abstract The purpose of this chapter is to show how important it is to use visualization techniques for enhancing public participation in creating smart cities. The chapter focuses on exploring people's opinions with regard to urban design, and on exploring new media and digital tools for public engagement. More specifically, it focuses on exploring the potential of using virtual worlds in the process of urban design from the first stages of the design process to the construction and maintenance phases. Quantitative and qualitative research methods have been used for the purposes of this chapter. The Terf virtual world is explored as a laboratory for interdisciplinary collaboration in urban design as well as an interface allowing involvement of different actors within a public participatory process. The final part of this chapter offers guidelines for future development of support tools for public engagement, especially concerning the Terf virtual world, and their potential use by municipalities.

1 Introduction

Urban planning is a complex process (Alexander 2007). It is therefore important to involve as many different experts and people who will eventually use the designed urban space (Hunt 2006) as possible. It is an interdisciplinary activity which brings together various experts offering a wide range of skills. There are several different elements, which should also be taken into account during the planning process: social interaction; the people in a specific urban space; the context of that urban space; the perception of space; and how proposals for redevelopment are associated with the natural, physical and cultural environment (de Roo and Silva 2010). When designing urban spaces, we should not forget the users who breathe life into urban areas and contribute to their final look and feel. More importantly, by educating

A. Jutraz (✉) · T. Zupancic
Faculty of Architecture, University of Ljubljana, Zoisova 12, 1000 Ljubljana, Slovenia
e-mail: anja.jutraz@fa.uni-lj.si

S. Geertman et al. (eds.), *Planning Support Systems and Smart Cities*,
Lecture Notes in Geoinformation and Cartography,
DOI 10.1007/978-3-319-18368-8_21

them through involvement in the planning process and by raising awareness, we can influence their lives. The main aim is to create a space that will be attractive, sustainable and boast high quality, so the people will want to live, work and relax there. With high quality urban design, we can influence the quality of life and the use of urban space, the quality of living conditions and people's opinions (Royal Institute of Chartered Surveyors and Department of the Environment 1996).

Public engagement is vital to the city's aims, as this ensures that the needs of a community are fully understood (Acland 2012). The public participation process in urban design can be defined as co-operative design, allowing designers (professionals) and users (public) to be involved on an equal footing (Jupp 2008; Stiles 2007). Our aim is to build a bridge between the public and the professional sphere, between residents, researchers, stakeholders and municipality officers by promoting virtual collaboration with the help of new media and advanced digital visual tools (mobile applications, virtual worlds, web pages). These tools promote the sharing of the residents' experiences and behaviour concerning their living environment, and it could be seen as a life-long learning process. With the help of digital tools, we should be able to help all those involved in creating a shared urban vision, and increase the satisfaction of users of redesigned areas. Moreover, such techniques could contribute significantly to the transfer of knowledge between all participants involved in the process.

There are several major research problems existing at this point, including a lack of dialogue, difficulties arising from the complexity of dialogue (communication and collaboration between the public and the experts on urban design issues), and a lack of appropriate communication technologies (i.e. that are easy to use and accessible). The chapter further focuses on exploring new media and digital tools for public engagement and interdisciplinary collaboration (Hanzl 2007). More specifically, it deals with exploring the potential of using virtual worlds in urban design from the first stages of the design process (problem definition) to the construction and maintenance phases (Bainbridge 2010; Smith et al. 1998). The chapter explores a 3D visualization software system called Terf to facilitate public engagement and participation in site planning and urban design reviews and workshops. The main research question is whether the proposed virtual world (Terf), along with its characteristics, is suitable for public participation processes and what should the virtual visualization tool look like in order to simplify urban engagement processes. This chapter additionally focuses on exploring people's opinion with regard to urban design, and features they find important within an urban space.

2 Urban Engagement and Smart Cities

The city does not only represent a built environment, but also a community of people who live there. Communities should have the option to express opinions and decide about the future of their city from the early stages of the design process onwards. Different decisions in urban space can affect various individuals or groups of people. They may feel threatened because they usually cannot predict the consequences of such changes. If people know the positive aspects of activities/changes in urban space, they can accept and support them more easily. Public engagement from the first stages onwards, combined with interdisciplinary collaboration, leads to the need for new methods/tools for a participatory planning system. Moreover, in the new era, when smart technology is rapidly changing our environment, urban engagement is also changing a lot. Innovative operational and information technologies are being used, and citizens and city infrastructure generate reliable real-time data. According to Buscher et al. (2014, p. 2) a smart city is *"an efficient city, a liveable city, as well as an economically, socially and environmentally sustainable city"*. We can identify a need for cities to change and become more efficient, attractive, inclusive and competitive. Using visualization techniques and digital tools for enhancing public participation is vital in creating smart cities (Al-Kodmany 2002; Jones 2006). The public engagement from the early stages of urban design process onwards helps create smart communities; and the appropriate digital tools could help by creating a sustainable and smart city.

2.1 Visualization Techniques and Digital Tools

Due to the rapid development of digital media in the last decade, physical participation (face-to-face collaboration) has begun to lose its exclusive role and is being replaced by increased virtual collaboration (Haller 2009; Smith 2003). This has consequently increased the need for an interface between different users. Digital technology is making massive changes to the way we experience and use our cities, as well as to the sustainable development process, and to the way professionals collaborate and involve the public. The main aim of designing smart cities is to connect different systems, to explain how a city as a whole functions as a result of the interplay between its systems where digital technology is used to better integrate these different systems. Digital tools should help the city function even better, and the power of new technologies should serve cities and improve the citizens' lives. Slotterback (2011) examined potential opportunities and constraints related to the development of technology in participation processes, and he states that technology is important as an enhancement to more traditional participation efforts, and not a replacement for them. Moreover, London is in the lead when it comes to smart cities developing and exploring digital technologies and, in 2014, a detailed study was prepared of 'How digital technologies are shaping the city', where different methods and tools are explored (Gann 2014; NLA 2014). Based on the review of literature and current digital tools, some of the categories of and theoretical approaches to visualization can be identified as follows:

(i) *3D visualizations*: 3D city models as digital tools for public involvement and development of smarter cities have already been explored by the Future Cities

Lab and Smarter Better Cities at ETH, Eidgenössische Technische Hochschule Zürich (exploring visualization and decision-making solutions), and by the Urban Simulation Team at UCLA (The University of California, Los Angeles) where they are exploring applications for real-time visual simulation in design, urban planning, emergency response and education (Future Cities ETH 2014; Smarter Better Cities 2014; UCLA—Urban Simulation Team 2014).

(ii) *Space syntax, spatial analysis*: Researchers at the Barlett Centre for Advanced Spatial Analysis (2014) are exploring the science of cities, the latest geospatial methods and ideas in computer-based visualisation and modelling.

(iii) *Immersive virtual labs*: The Collaborative for Advanced Landscape Planning's Landscape Immersion Lab (LIL) at the University of British Columbia (Salter et al. 2009) is designed with a three-screen immersive projection lab, where the combination of the CommunityViz software and LIL's large screen display is used. It allows the participants to interactively view and explore the planning area and the represented scenarios in a shared, immersive setting. Also the SIM Lab in Vienna (an interdisciplinary centre for spatial simulation and modelling) deals with the latest computer-aided technologies for the built environment, focusing on smart cities and 3D virtual reality (VR) environment (SIM Lab Vienna 2014). Both of these have the ability to dynamically explore the visualisations of planning proposals and increase participants' understanding of the plan, but they have both time and location constraints (one exact physical location of the lab, limited time).

(iv) *Virtual worlds*: They present a media established for interdisciplinary collaboration and public participation processes, e.g. Terf, Cloud cities, OULU 3D, etc. (3D Immersive Collaboration Consulting 2014; Cloud Cities 2014; OULO 3D 2014).

Moreover, digital visualization tools are quite commonly combined also with traditional methods/tools, as it is important to keep face-to-face contact in the public participation process. For example, Senbel and Church (2011) have explored different visualization tools in a neighbourhood of Marpole in Vancouver: a community voices film (for capturing and representing the direct voices of community residents); a policy film (for educational purposes on the alternative building and transportation scenarios); a physical model; a digital model (SketchUp model with a fly-through animation showing a variety of perspectives); an energy consumption model (a computerized 3D representation of the physical model using CityCad software); and a digital summary. However, a planning support system (PSS) of tools is still missing that could combine all of them, be used as a medium between different users and different tools/methods used in public participation process in urban design, and be easy to use and available at any time, from any location.

2.2 Virtual Worlds for Urban Engagement

The focus of this chapter is virtual worlds, but when we talk about virtual worlds, we must first define the concept of virtual reality. Strehovec (1992) argues that virtual reality coexists with a given real world and takes place in real time because we are simultaneously present in two worlds: the consciousness is located in the virtual world and time, while the body remains in the real world and time. Kitchinu (1998) claimed that a virtual reality contains the following components: interactivity, interaction and progress in real time. By wanting to build another world inside a computer, we want to have access to another level of existence that is able to offer us the same experience as the real world; we wish to use cyberspace to create a virtual place and thus enable the users to experience the virtual environment as if they were truly present in it (Tuan 1977). The virtual world can also be used as a tool to show the built environment, as it allows us to introduce it in 3D models and through walk-through (Whyte 2010) sessions.

The first 3D models of cities were designed in the 1950s when the first simulations of movement through space (flight simulations) were made possible. Over the years, the level of dedication to detail increased, and methods of presentation and optimizing 3D models improved. Virtual worlds can assist you in experiencing space and gaining experience in the real world; they attempt to emulate physical space, its image, services, and interactions of people. They exist in direct connection to the physical space. The purpose of virtual cities is to emulate physical location, their image, services, and interactions of people, while maintaining a direct link with the physical place (Lenarčič 2002). All these players are planning various support system applications, covering different aspects and methods. Based on the review of different theoretical approaches and different digital tools, we could notice the lack of immersive virtual tools, which could be linked also to other digital tools (other visualization methods and tools e.g. 2D maps, BIM (Building information modeling) programs, 3D city models etc.) and could be accessible from anywhere and anytime.

Virtual worlds must contain a variety of elements: an interface for user enrolment; the time component that shows the phases of development planning at a specific time; a tool for measuring the dimensions (e.g. the building height and the width of roads); the function "teleport", which allows immediate navigation or movement from one point to another; as well as a tool for marking and adding descriptions and a display of several planes among which the user can choose. Most virtual worlds only have a few elements, but our focus is the PSS tool, based on a virtual world, which could be used as a medium between different digital visual tools, especially with 3D models, as they enable full immersion. Users can experience the site by walking through virtual worlds allowing the pedestrian point of view, a view they are used to from their daily lives.

Our research is based on the Terf visualization tool (3D Immersive Collaboration Consulting 2014), which could potentially connect different tools, and, most importantly, it could represent an immersive tool, enabling you to experience 3D models of neighbourhoods from the point of view of pedestrians. Terf is a virtual world, which has already been used for long-distance collaboration in companies that have offices in different locations (mostly in the United States), operating in various fields, such as construction management and medicine. It was our wish to explore the possibility of using it in urban design and in public participation processes. It is virtually based, without time and location constraints, and it can also be combined with traditional ways of communication. It offers different tools for communication and collaboration (Fig. 1). Additionally, it allows importing of large 3D models of urban settlements, which serve as an opportunity to do walk-throughs, thus exploring the 3D models of urban design proposals. It offers its users two aspects of exploration, namely, the experiential design aspect (pedestrian view) and the planning design aspect (bird view). Terf has already been used for meetings, as it also combines the following elements: individual and group chat (you can chat with participants online in real time); whiteboards (for sharing analyses, presentations, 2D maps, etc.); and the educational module (learning takes place through direct interaction with other disciplines). Terf is connected to other programs such as Revit, SketchUp, Microsoft Word and ArcGis.

Terf is a laboratory intended for interdisciplinary collaboration and co-ordination during urban planning and urban design process; it is meant for urban management and it represents an interface allowing involvement of different actors within a public participatory process, ranging from the public to the representatives of

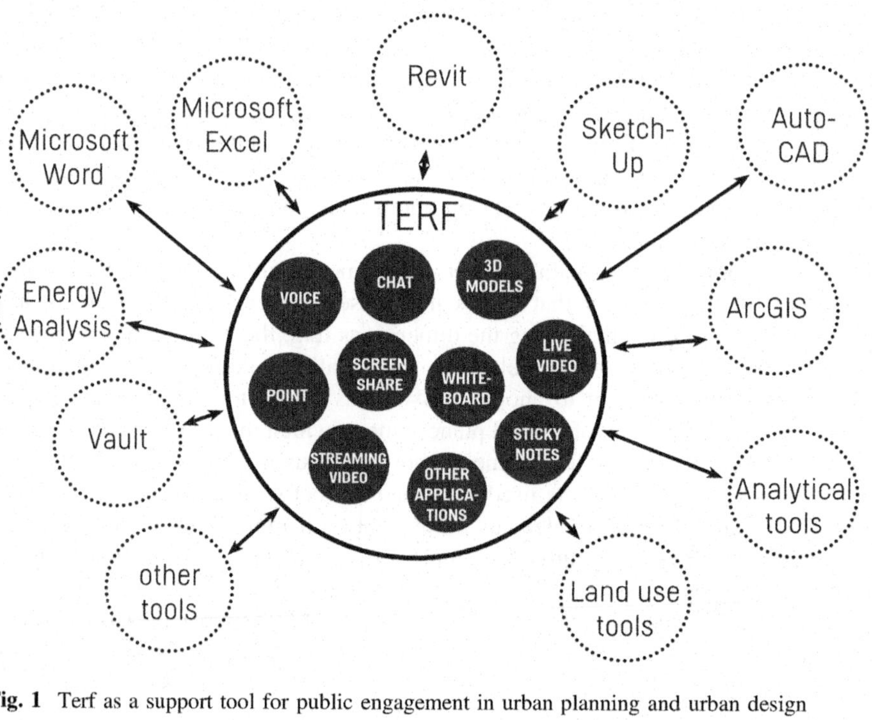

Fig. 1 Terf as a support tool for public engagement in urban planning and urban design

various professions (different disciplines). There are several recommendations concerning the improvement of the Terf virtual world, namely: Terf should be implemented as a hybrid tool; physical and digital spaces should be combined (e.g. physical elements on site, digital on the web); the option of importing larger 3D models should be added, along with measurements in 3D; and better renderings should be prepared. Terf presents a support tool for public engagement in urban planning and urban design as it offers different applications for collaboration and co-ordination, enables 3D model walk-throughs, and, at the same time, it features direct connections to other tools for planning and communication such as Revit, BIM programs, Word and Excel (Fig. 1). For example, you can import 3D models directly from SketchUP or Terf, and you can open Word or Excel files directly in Terf. Thus, Terf is not only a visualization tool, but it is a support system of tools that can help us to come closer to the local community, to their homes, as it is available from anywhere at anytime.

2.3 Evaluating Urban Space

Based on the analysis of previous research and a variety of literature, which lay down the principles and guidelines for the design of urban space, its values, dimensions and categories, we have set up a synthesis scheme consisting of elements and categories to which attention must be drawn in urban space. Some of the principles and guidelines are listed here. These principles of designing urban space are: visibility, composition and volumes of buildings; composition and scale of the openings on the facade; relation to its surroundings; materials used; design of outdoor public spaces (Vandell and Lane 1989); the qualities of urban space such as visual quality, functionality, social use, natural environment and sustainable development (Royal Institute of Chartered Surveyors and Department of the Environment 1996); principles of regulation of urban space: urban design for people (safe, comfortable, attractive), creating and maintaining connections, mixed structure (and form), respect for the natural elements, flexibility, economic vitality (Yeang 2000) and many more (Burton and Mitchell 2006; CABE, DETR 2001; Carmona 2010; MFE/NZ 2005; Stiles 2007). Also Nasar (1990, pp. 41) defined the visual preferences, the *"desirable features"* of urban environments as: *"naturalness, upkeep, openness, order, and historical significance"*.

Based on the theory review and in-depth insight into urban spaces, we have taken a more detailed look at the natural features of urban spaces on the one hand and built features on the other. We were particularly interested in evaluation categories of urban spaces that can be used in the processes of public involvement in urban design, and which of them directly affect the design. We wished to define those categories, which can be seen from photos and evaluated, along with being understandable to the lay public. Based on these criteria, the following categories of evaluation of urban space for the needs of public participation in the urban design were defined: orientation (how well users move around in the designed space, or whether they know where they are); natural elements (greenery, trees, parks, water elements, for example); urban equipment (urban furniture such as benches, bins, lights); connectivity (connection between the ground floor of the building and the public space); flexibility of use (the chosen design offers the possibility of different activities, such as concerts); and variety of activities (the design of urban space can encompass different activities, such as sitting and walking). The evaluation categories for public participation in urban design are fully defined in Jutraž (2013).

3 Methodology

Our research was divided into two stages: (i) exploring user opinions: research among the wider population, focused on exploring evaluation categories for public participation in urban design (quantitative research methods, a survey); and (ii) exploring new methods/tools for public participation in urban design: testing of existing tools, including virtual world Terf and the decision-making tool (qualitative research methods, semi-structured interviews).

A survey was initially conducted between representatives of the public and experts (N = 878) and it focused on respondents' opinions concerning urban spaces. The importance of the evaluation categories of urban space has been studied in order to create a decision-making support tool, which was then tested in the next stage. Semi-structured interviews were conducted among representatives of the public and experts (N = 20), and focused on exploring the usability and appropriateness of the immersive Terf software, a system of support tools for public engagement in urban design. Bratovševa ploščad was chosen as a case study for the interviews. This is a public space in one of Ljubljana's suburban neighbourhoods, which currently lacks a detailed program of events that use this urban space, as well as actual users. The Bratovševa ploščad area lacks life; people do not want to spend time there. The space does not offer enough events and it is not suitably equipped to do so. Since it is located in the heart of residential neighbourhoods, it would be highly advisable to ask the people for their wishes and needs.

At the beginning, four different 3D models were prepared, two of them are presented on the Fig. 2 (left: 3D model of proposal 2—pergola; right: proposal 1—microambients). Figure 2 presents two different views in Terf: bird view (left) and pedestrian view (right). Firstly, a basic 3D model of the urban area (the neighbourhood of Bratovševa ploščad), was followed by four 3D models (the current situation and three alternative renovation proposals). The renovation proposals served as a basis for the qualitative research and decision-making processes. Different proposals were shown to the 11 representatives of the public and 10 representatives from the planning profession. Interviewees had to move around digital 3D models on their own and evaluate the decision-making tool and the Terf virtual world.

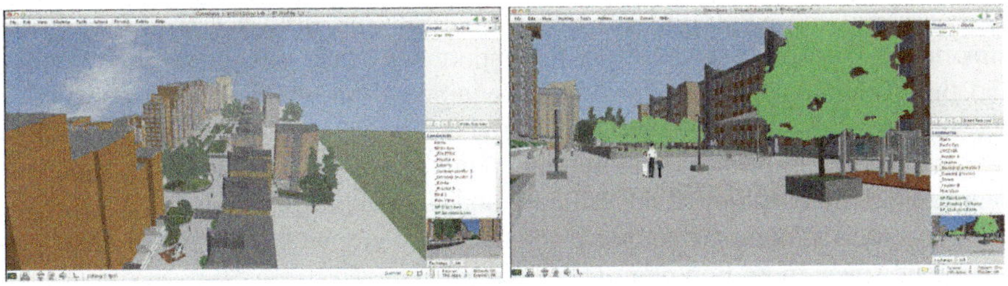

Fig. 2 Terf allows large 3D models of urban settlements to be imported, as well as giving its users the opportunity to do 'walk through' the 3D models of urban design/planning proposals

4 Results

We will present only some of the results of the entire research conducted as part of a Ph.D. (Jutraž 2013). First, we present the results of the quantitative research (survey), and, second, we present the results of the qualitative research (semi-structured interview).

4.1 Evaluation Categories of Urban Space for Public Participation Processes (Survey)

The survey was conducted among 878 people: a total of 674 members of the public whose education, profession or occupation was not related to urban planning and design, as well as 231 representatives of the profession who were professionally, directly or indirectly, involved in designing urban space, and their education, profession or occupation was connected to urban planning or urban design disciplines. The resulting statistical analyses are based on a comparison of the two sample groups, the public on one side and the professionals on the other side.

First, we defined six main evaluation categories of urban space: orientation (how well users move around in the designed space, or whether they know where they are); natural elements (greenery, trees, parks, water elements etc.); urban equipment (urban furniture e.g. benches, bins, lights); connectivity (connection between the ground floor of the building and the public space); flexibility of use (the chosen design offers the possibility of different activities, such as concerts); variety of activities (the design of urban space can encompass different activities, such as sitting or walking); and then the adequacy of the importance of the evaluation categories was tested. Figure 3 presents the importance of the specific categories or elements in an urban space for the public and professionals.

We found that, for both the public and the professionals, orientation was the most important category in the urban space (first place), followed immediately by natural elements (second place). For both, the public and the professionals, the variety of activities (fourth place) and flexibility of use (fifth place) were of lesser importance. There is a difference in the importance of the categories for the public and the professionals when considering equipment and connectivity (connection between the ground floor of the building and the public space): the public found equipment (urban furniture) more important (third place) than the connection between the ground floor of the building and the public space (sixth place), while the professionals found the connection between the ground floor of the building and the public space more important (third place).

Orientation, natural elements and equipment within an urban space were the most important elements for the public, followed by a variety of activities, and the use of flexibility and connectivity of the ground floor of the building to the public space (Fig. 3). It is possible that the last three categories are more abstract, which means the public may find them difficult to understand. The professionals saw orientation as the most important element as well, while there were no significant differences in ranking among other categories. Significant differences between the public and the professionals in the ranking of categories can be observed in relation to the urban equipment, connectivity and the integration of natural elements; equipment such as natural elements seem to be less important to professionals than to the public, while it seems that connectivity is more important to the public when compared to the opinions of professionals. The results of the survey helped us to prepare a tool for decision making, which we tested also in the following semi-structured interviews.

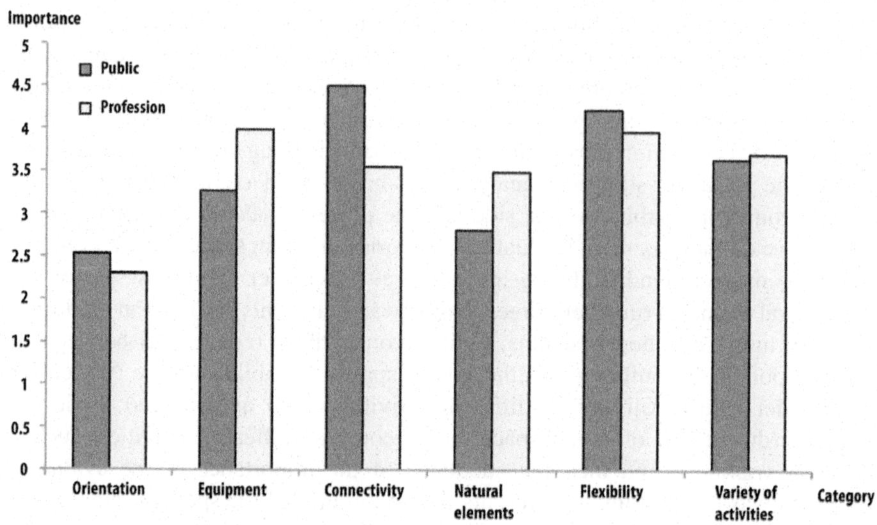

Fig. 3 Comparison of the importance of individual categories between the public and the professionals (1 is very important; 5 is not important)

4.2 Decision-Making Tool and the Terf Virtual World (Semi-Structured Interviews)

The evaluation categories below form the basis of the decision-making tool, which is intended primarily for users of urban space and is used in the process of public involvement in urban space planning/design, and is also directly connected to Terf. We carried out semi-structured interviews with 11 members of the public and 10 interviews with professionals. There was a wide range of individuals from both groups included in this process (according to age, gender, educational level and field of interest). Each interview was based on pre-prepared questions, while also leaving room for additional questions. The interviews lasted approximately 45 min each (minimum 20 min, maximum 67 min). The information about the participants and their ways of thinking was collected through their active participation in semi-structured interviews, direct observation (emotional responses, skills exhibited when using tools), and through testing specific research questions (simplicity of the virtual world Terf and adequacy of tools for decision-making processes on the basis of pre-prepared tasks—respondents were given tasks, printed out on paper, and invited to take some time to solve them).

Interviewees were given a printed table (Fig. 4), and they were invited to perform their own walk through 3D models of the existing situation and the three

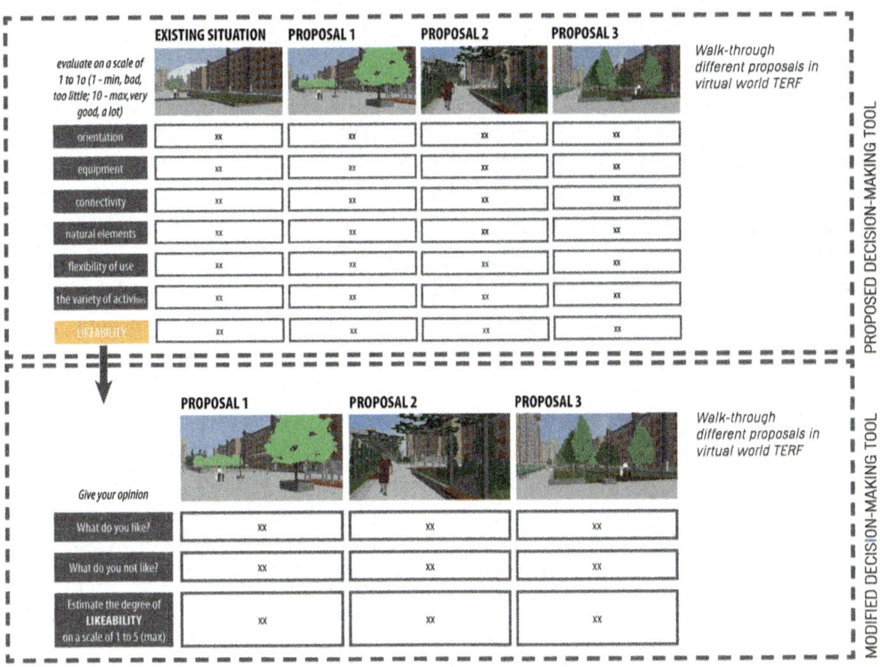

Fig. 4 A decision-making support tool for public involvement directly connected to Terf

proposals for rearrangement of the given urban space (in Terf). With the use of an avatar, it was easier to experience space and show solutions. Interviewees evaluated the 3D models on the basis of evaluation categories of urban space, related to the needs of public involvement in the planning area. The evaluation included a scale from one to ten, with ten representing the maximum value. Table 1 shows the results (mean estimates).

In the survey we were testing the general opinion on evaluation categories of urban space (orientation, equipment, connectivity, natural elements, flexibility, variety of activities; explained above), but in the interview the answers were based on the proposals shown to the interviewees (3D models of different proposals of renovation of urban space: we gave them the option to walk through the 3D models on their own). Individuals in the public and professional categories evaluated the importance of the evaluation categories of urban space on a scale from one to ten, with ten representing the maximum value. The public evaluated orientation much better than the professionals (average estimates of the public are 8.3, 8.8, 8.8 and 8.8, and of the professionals are 6.2, 6.6, 7.4 and 7.5). Significant differences in orientation (similar to the average score) were not observed among the various proposals. Orientation was the best estimated in all 3D models. The second in importance were natural elements in proposals 2 (public 8.1 and professional 7.6) and 3 (public 9.0 and professional 8.1) and the variety of activities in proposal 1 (public 8.5 and professional 6.4).

We were also interested to see whether the results obtained in this part of the study are comparable with the results obtained through the survey. Very comparable results were obtained with regard to natural elements (this category is always among the more important ones, and it always ends up in second or third place), and to the variety of activities (always a medium important category, in fourth place). Interestingly, the professionals provided more comparable results in both stages of the research than the public. The reason for this can be found in the fact that the public is responding specifically in relation to the shown location and not in general (as they were invited to an interview with a specific aim: to provide an opinion on a specific location, Bratovševa ploščad). However, the public and the professionals have managed to come together when faced with solving the many problems related to these tasks: all evaluation categories seem equally important to them, they would rather provide an overall opinion on Bratovševa ploščad than dealing with certain categories. The public found the evaluation categories too abstract, too difficult to understand.

Before testing the decision-making process, Terf was presented to the interviewees; they were also shown some basic functions (move around the room with the arrow keys on the keyboard, the 'landmark' function for pre-set views and walking between the individual 'rooms' with alternative solutions). Interviewees were asked if they considered the experience of space (walk-through of the 3D model space) as a way to improve the presentation of alternative solutions. Except for two participants, they all believed that the experience of space could help improve the presentation of alternative solutions. To create an even better representation of the alternative solution, the public also suggested we add some photographic material.

Table 1 Importance of the evaluation categories of urban space in connection to the shown 3D models of urban space

	Public/profession	Orientation	Equipment	Connectivity	Natural elements	Flexibility	Variety of activities
Current situation	Public	8.3	4.6	5.5	4.6	4.0	4.3
	Profession	6.2	2.8	4.2	4.2	5.8	2.6
Proposal 1–microambients	Public	8.8	8.1	6.9	8.2	7.8	8.5
	Profession	6.6	5.9	5.9	5.9	6.0	6.4
Proposal 2–pergola	Public	8.8	7.6	6.6	8.1	7.3	7.6
	Profession	7.4	6.9	6.3	7.6	6.7	6.7
Proposal 3–park	Public	8.8	8.1	7.0	9.0	8.3	8.2
	Profession	7.5	7.1	6.1	8.1	6.7	7.1

Average 1 is very poor, 10 is very good; we were interested to find out in which proposal a particular category is most present

The profession liked Terf, but they believe it should be used only as one of the tools and as through the whole urban design process. They also believe that it is generally helpful if someone talks the users through the 3D model walk-through session—a combination of visual presentations with sound might be worthy of a discussion. Moreover, professionals have mentioned that it would be sensible to combine Terf with urban physical ambient interventions in the area; the public cannot imagine changes until they are visible in the real world.

5 Discussion and Future Work

Based on qualitative and quantitative analysis, it has so far been observed that a quality urban space must encompass all the categories: equipment, orientation, natural elements, variety of activities, connectivity (connection between the ground floor of the building and the public space) and flexibility. Additionally, safety and tidiness are a must–enabling the creation of a community of people who care for maintenance and regulation of space and its use. Filling the table (Fig. 4 above) was easier for the younger generation, while older participants struggled a bit. The results of the survey can help us understand how the public thinks and how, on the basis of participants' opinions, further designs can be created. The evaluation of urban space in the collaborative decision process also has some positive features: the interviewees start thinking about urban space, and 3D models help them see things and think of things they would not think of otherwise. Expressing opinions suddenly becomes much easier for people because they see the idea in 3D, as opposed to being compelled to consider everything completely on their own and work from memory (Al-Douri et al. 2001, Panagopoulos et al. 2009). Terf as a combination of different tools, therefore, represents a laboratory intended for assistance in developing project requirements through which you can obtain opinions.

In the survey participants were asked to fill in the table shown on Fig. 4 above, where they had to define the importance of each category in different 3D models (current situation and in three alternative proposals). It was found out that this was too complicated for them, all categories had equal importance for them, so it was proposed to simplify the decision-making tool (Fig. 4).

Interviewees were asked whether they believe the experience of space (walk through the 3D model space) can improve the presentation of alternative solutions. Cullen (2007) had already dealt with the elements of visual images and ways to assist him in achieving emotional responses to user space (changing the sequence of images, the structuring of space, urban fabric as a mixture of styles, materials, colours and textures), but we were interested in an experience gained by using the virtual world. Except for two participants who remained unidentified, all others believed that the experience of space could help communicate the features related to alternative solutions better. For a better presentation of ideas, they would upload photos. We were also interested in whether the chosen representatives of the public and the professionals liked the proposed tool for public involvement in decision-making processes (based on the Terf virtual world). We found that the majority of respondents liked the tool very much and that they saw it as an appropriate tool for public participation processes in urban design (nine members of the public and ten representatives of the profession liked the tool, and two interviewees did not wish to identify themselves). Respondents believed that the tool, along with minimal improvements, could be used in the process of public participation in urban design. It should be noted that the general likeability of the tool does not say much about its true usefulness and suitability; its appropriateness still requires a detailed examination.

On the basis of this study, the following advantages of Terf were noticed: it provides a combination of different tools; it is intended for the exchange of information, documents, presentations in virtual environments and for walking through 3D models (experiential mode); it is especially suitable for the younger generation that is usually not involved in the processes of urban design; it is suitable for interdisciplinary collaboration; and it presents a tool excellent for promotion of public participation in the planning area. Moreover, our interviewees really liked the experience of the site provided by the 3D models, landmarks helped them when moving through the 3D models, and they found it simple, efficient and easy to use.

From the current findings, it is clear that Terf presents a tool for public participation in the field of planning of urban spaces; it is an encouraging tool; which helps raise awareness; enables alignment between planning elements and the wishes of the public (people can see what is possible, how they can improve their urban space and what are its alternative uses). There are also other important aspects of the tool worth mentioning: avoidance of general dissatisfaction or disagreement with the final proposal; lifelong learning process (helping educate the public about urban space); and creating a community that will begin to combine and regulate its urban space (actually, usually no major interventions are required, it is only necessary to create a community that is responsible for the regulation of urban space and maintainability). On the one hand, meeting rooms (only virtual or hybrids which combine the physical and the virtual world) enable dislocated meetings (city authorities with the public), while the 3D digital model rooms (with the current situation and alternative solutions) on the other hand provide experience, full immersion and virtual walk-throughs. The unique nature of the tool (engagement anywhere and anytime, full immersion) also affects how you learn about urban space, so it could also be seen as an interface intended for knowledge transfer from the professional sphere to the public, as well as among professionals. The specific results, described above, can only lead us to a general conclusion: virtual worlds could be used as a planning support system for urban engagement.

Future work will involve the development of digital system of tools for public participation in urban design, where Terf will present the core element, and other tools will support it (connection to other existing digital tools, and also to traditional methods/tools). In the next research stage, the PSS tool (Terf) will be tested also in a real-life city project, and it will be used from the first stage of the planning process (problem definition) all the way to the construction and maintenance phase.

Also the connection to smart city technologies will be explored, too. Planning support tools involving virtual worlds for public engagement/collaboration in urban design could potentially be used by municipalities for the purposes of creating smarter cities.

We will also develop Terf for mobile phones and tablets, so it would be accessible from different locations in the urban environment too. Moreover we will develop different packages for different users: locals, AEC (architecture engineering construction) professionals, government workers, visitors. We will test the possibility of using Terf at the physical site by using Google Glasses and Augmented Reality plugins to ensure the option of real immersion at the physical site. Terf will be improved and prepared for importing bigger 3D models and usage by a bigger group of people at the same time (e.g. possibility of having meetings in virtual city hall for 500 people at the same time). It would be used for opinion polling, and improved for adding the ability to simulate nighttime views (lighting), adding the ability to display shadows in urban spaces, and showing the appearance of the discussed urban space at different times of day and in different seasons.

6 Conclusion

To conclude, most visual digital tools (including 3D city models) are suitable only for a specific location, while Terf can be used at any location (easily applicable to other cities around the world). Terf presents a laboratory for interdisciplinary collaboration and co-ordination in urban design.It is an interface allowing involvement of different actors within a public participatory process, ranging from the public to the representatives of various professions (different disciplines). It combines and connects different planning and communication tools, and can be used through the entire process of public participation. An urban mediator is coordinating communication and ensuring the inclusion of the public and members of different professions in order to tackle major problems that a community/city faces. The end goal of such collaboration is to build a more competitive, equitable and sustainable city, and Terf can simplify the mediator's work. Its versatility and technology solutions are adding value to planning support systems (Saad-Sulonen 2005), and it could be considered as one of the tools enabling the creation of smarter cities. Virtual world Terf could be described as a tool for promotion of public participation in urban design. Our findings indicate that, in general, people tend to be better at imagining the content of renovation proposals of urban design if they see 3D models; and through exploring different alternatives, it is easier for them to express their opinions. Different renovation proposals help them remember what they really want to see installed in the urban space. Moreover, if we show them a 3D model of a proposal and let them walk through it, the levels of immersion and consequently of their experience increase. Immersive virtual environment Terf can also help improve communication between the public and the practicing professionals. Therefore, virtual world Terf can be seen as a very useful tool for public participation processes.

Acknowledgments This research project was carried out in co-operation with the company 3D ICC Immersive Collaboration, and the Slovenian Research Agency (ARRS) supported it. The research was part of Ph.D research at the University of Ljubljana, Faculty of Architecture. We would like to thank Julie LeMoine and Ron Teitelbaum from the 3D ICC Immersive Collaboration for their assistance and for the opportunity to explore and work in Terf, an immersive virtual environment.

References

3D Immersive Collaboration Consulting. (2014). *Immersive Terf.* http://3dicc.com/. Accessed December 10, 2014.

Acland, A. (2012). *Dialogue by design: A handbook of public and stakeholder engagement.* London, Dialogue by design. http://designer.dialoguebydesign.net/docs/Dialogue_by_Design_Handbook.pdf. Accessed December 10, 2014.

Al-Douri, F., Clayton, M., & Abrams R. (2001). The impact of 3D digital modeling on 3D design aspects in urban design plans. *eCAADe23, Conference Proceedings*, pp. 331–340.

Al-kodmany, K. (2002). Visualization tools and methods in community planning: From freehand sketches to virtual reality. *Journal of Planning Literature, 17*(2), 189–211.

Alexander, C. (2007). The timeless way. In M. Larice & E. Macdonald (Eds.), *The urban design reader*. London and New York: Routledge.

Bainbridge, W. S. (2010). *Online worlds: Convergence of the real and the virtual.* London: Springer.

Barlett Centre for Advanced Spatial Analysis. (2014). *CASA*. http://www.bartlett.ucl.ac.uk/casa. Accessed December 10, 2014.

Burton, E., & Mitchell, M. (2006). *Inclusive urban design: Streets for life*. Oxford: Architectural Press.

Buscher, V., Doody, L., Webb, M., & Aoun, C. (2014). *Urban mobility in the smart city age.* London: Schneider Electric, ARUP, The Climate Group.

CABE, DETR. (2001). *The value of urban design: A research project commissioned by CABE and DETR to examine the value added by good urban design.* London: Thomas Telford Publishing.

Carmona, M. (2010). *Public places—urban spaces: The dimensions of urban design* (2nd ed.). Amsterdam: Elsevier.

Cloud cities. (2014). https://www.cloudciti.es/#!/. Accessed December 10, 2014.

de Roo, G., & Silva, E. A. (2010). *A planner's encounter with complexity.* Burlington, VA: Ashgate Publishing.

Cullen, G. (2007). *The urban design reader.* London, New Yok, Routledge.

Future Cities ETH. (2014). http://www.futurecities.ethz.ch/. Accessed December 10, 2014.

Gann, D. (2014). *Smart London Plan.* London.

Haller, C. (2009). *Cross-media public participation.* Denver: Place Matters.

Hanzl, M. (2007). Information technology as a tool for public participation in urban planning: A review of experiments and potentials. *Design Studies, 28*(3), 289–307.

Hunt, J. G. (2006). Forms of participation in urban redevelopment projects: The differing roles of public and stakeholder contributions to design decision making processes. In J. P. Leeuwen & H. J. P. Timmermans (Eds.), *Innovations in design & decision support systems in architecture and urban planning* (pp. 375–390). Dordrecht, Netherlands: Springer.

Jones, Q. (2006). Virtual-Communities, virtual settlements and cyber-archaeology: A theoretical outline. *Journal of Computer-Mediated Communication, 3*(3).

Jupp, E. (2008). The feeling of participation: Everyday spaces and urban change. *Geoforum, 39*, 331–343.

Jutraž, A. (2013). *Sistem vizualnih digitalnih orodij za participacijo splošne javnosti pri prostorskem načrtovanju = Visual digital system of tools for public participation in urban design*. Ph.D. Thesis, University of Ljubljana.

Kitchin, R. (1998). *Cyberspace: The world in the wires*. Chichester: Wiley.

Lenarčič, B. (2002). *Fizično vs. virtualno mesto*. Dissertation, University of Ljubljana. http://dk.fdv.uni-lj.si/dela/Lenarcic-Blaz.PDF. Accessed December 10, 2014.

MFE/NZ. (2005). *The value of urban design: The economic, environmental and social benefits of urban design*. New Zealand: Ministry for the Environment.

Nasar, J. L. (1990). The evaluative image of the city. *Journal of the American Planning Association, 56*(1), 41–53.

NLA. (2014). *Smarter London: How digital technologies are shaping the city (NLA Insight Study)*. London: NLA London's Centre for the Built Environment.

OULO 3D. (2014). Retrieved January 01, 2014, from http://www.oulu3d.fi/3d-spaces/. Accessed December 10, 2014.

Panagopoulos, T., Andrade, R. R., & Barreira, A. P. (2009). Citizen participation in city planning and public decision assisted with ontologies and 3D semantics. In A. Rocha, F. Restivo, L. Reis & S. Torrão (Eds.), *4th Iberic conference of systems and information technologies* (pp. 167–172). Portugal: Póvoa de Varzim.

Royal Institute of Chartered Surveyors and Department of the Environment. (1996). *Quality of urban design: A study on the involvement of private property decision-makers in urban design*. London: Royal Institution of Chartered Surveyors and Department of the Environment.

Saad-sulonen, J. (2005). *Mediaattori—Urban Mediator: A hybrid infrastructure for neighborhoods*. Media Lab: University of Art and Design Helsinki.

Salter, J. D., Campbell, C., Journeay, M., & Sheppard, S. R. J. (2009). The digital workshop: Exploring the use of interactive and immersive visualisation tools in participatory planning. *Journal of Environmental Management, 90*(6), 2090–2101.

Senbel, M., & Church, S. P. (2011). Design empowerment: The limits of accessible visualization media in neighborhood densification. *Journal of Planning Education and Research, 31*(4), 423–437.

SIM Lab Vienna. (2014). http://simlab.tuwien.ac.at/. Accessed December 10, 2014.

Slotterback, C. S. (2011). Planners' perspectives on using technology in participatory processes. *Environment and Planning B: Planning and Design, 38*(3), 468–485.

Smarter Better Cities. (2014). http://www.smarterbettercities.ch/about-us/. Accessed December 10, 2014.

Smith, A. H. (2003). *Digitally distributed urban environments: The prospects for online planning*. University College London. www.casa.ucl.ac.uk/andy/. Accessed December 10, 2014.

Smith, A., Dodge, M., & Doyle, S. (1998). *Visual communication in urban planning and urban design*. London: UCL.

Stiles, R. (2007). A guideline for making space: Joint strategy activity 3.3. Dunaj: TU Wien and UrbSpace.

Strehovec, J. (1992). *Umetnost virtualnih strojev* (pp. 150–151). Časopis Za Kritiko Znanosti XX: Virtualna Resničnost.

Tuan, Y. F. (1977). *Space and place: The perspective of experience*. Minneapolis: Minnesota Press.

UCLA. (2014). *Urban simulation team*. http://www.ust.ucla.edu/ustweb/ust.html. Accessed December 10, 2014.

Vandell, K. D., & Lane, J. S. (1989). The economics of architecture findings and urban design: Some preliminary economics. *Journal of Real Estate, 17*(2), 235–260.

Whyte, J. (2010). *Virtual reality and the built environment*. London and New York: Architectural Press.

Yeang, L. D. (2000). *Urban design compendium*. United Kingdom: English Partnerships.

Chapter 22
Recoding Embedded Assumptions: Adaptation of an Open Source Tool to Support Sustainability, Transparency and Participatory Governance

Jennifer S. Minner

Abstract This chapter traces the adoption of Envision Tomorrow, an open source planning support tool, in a large-scale planning effort within the Austin metropolitan region. A regional consortium of public, nonprofit, and private organizations was awarded a U.S. Department of Housing and Urban Development (HUD) Sustainable Communities grant to create and deploy an analytical tool for the assessment of district, community and regional-scale scenarios. Several dimensions of Envision Tomorrow are described in the chapter including its use: as a tool in participatory plan-making; as an analytical process that extends and structures how planners perform analysis; as a PSS that focuses on quantifiable sustainability indicators and thus supports the inscription of particular definitions of sustainability; and as a conduit of exchange between planners and university researchers and between planners and members of the public. The chapter concludes with a reflection on the virtues of transparency and adaptability. It also reveals embedded assumptions that represent both sources of promise and concern in the application of a PSS in planning processes.

1 Introduction

New web-based models for citizen participation (Brabham 2009; Evans-Cowley and Hollander 2010; Ganapati 2010; Koekoek et al. 2009; Seltzer and Mahmoudi 2013; Townsend 2013), advances in geographic information systems (GIS) modeling, and the growing application of scenario planning methods have spurred the production and use of new and improved planning support systems (PSS) (Bartholomew 2007). Scenario planning tools, in particular, seem to hold great promise for the creation, visualization, and analysis of alternative scenarios in

J.S. Minner (✉)
Department of City and Regional Planning, Cornell University, Ithaca, NY 14853, USA
e-mail: j.minner@cornell.edu

S. Geertman et al. (eds.), *Planning Support Systems and Smart Cities*,
Lecture Notes in Geoinformation and Cartography,
DOI 10.1007/978-3-319-18368-8_22

which land use, transportation, the economy, environmental systems and social needs are considered holistically (Goodspeed 2013; Holway et al. 2012).

Scenario planning methods and tools are promoted by nonprofit, governmental and university-based groups. Nonprofit organizations such as the Lincoln Institute of Land Policy, Sonoran Institute and PlaceMatters bring together tool developers, educators and practitioners to advance scenario planning and associated tools (Holway et al. 2012; Open Planning Tools Group, undated; PlaceMatters, undated). Federal government initiatives, such as the Partnership for Sustainable Communities, have directed resources toward the use and development of scenario planning tools and integration of housing and urban development, transportation, and environmental planning through scenario planning (Partnership for Sustainable Communities 2014). While many scenario planning efforts were originally concentrated in the Western United States to address rapid growth, scenario planning tools and processes are increasingly used across the country. This is evidenced not only by the Austin metropolitan region's Sustainable Places Project, but Buffalo Niagara region's One Region Forward, which is aimed at responding to a trend of population loss and abandonment as well as suburban sprawl.

As more planners are drawn to the possibility of deploying PSS in scenario planning, it is necessary to reflect carefully on them. The coding, calculations and mapping functions embedded within PSS have important implications that may not be immediately apparent. What a PSS includes and what it leaves out, the limitations and extensibility of its structure, and the mechanics of its application should be carefully considered. This case study examines the adaptation and use of an open source PSS called Envision Tomorrow (ET) to aid a federally-funded regional planning process in the Austin, Texas, metropolitan region.

The following chapter traces four aspects of ET: (1) its use as a tool in participatory plan-making; (2) its use as an analytical process that extends and structures how planners perform analysis; (3) its use as a PSS that focuses on quantifiable indicators and thus supports the inscription of particular definitions of sustainability; and (4) its use as a conduit of exchange between planners and university researchers and between planners and members of the public. The chapter concludes with a reflection on the virtues of transparency and adaptability in a PSS and its significance for the creation of collaborative networks of planning researchers and practitioners. It also notes areas of critical concern. These include the embedded assumptions that must be carefully considered in the deployment of PSS. Additional concerns include observed limitations in the extent to which the results of PSS are actually shared with citizens in the process and the complications in doing so.

2 Calls for the Use and Production of PSS

As noted above, a convergence of nonprofit and governmental agencies are promoting the use of scenario planning tools and especially GIS-based PSS to deepen planners' analytical capabilities and public participation in planning (Holway et al.

2012; Open Planning Tools Group, undated; PlaceMatters, undated; Partnership for Sustainable Communities 2014; Chakraborty 2011). Consulting firms that regularly aid local and regional governments across the country also promote the application of their own scenario planning PSS (EnvisionTomorrow.org 2014, Calthorpe Associates, undated). Chakraborty notes that with *"a growing literature"* there is a need for *"more examples of innovative participatory methods and quantitative modeling techniques"* (Chakraborty 2011, p. 398).

Despite the promise of new PSS tools, the literature suggests barriers and limitations in their use. For example, Göçmen and Ventura (2010) observed that within a sample of public agencies, GIS is seldom used for modeling or spatial analysis and more often used for simple mapping or to access basic information. Bartholomew (2007) noted in an extensive assessment of 80 scenario planning processes in more than 50 metropolitan areas that: *"In many cases, citizens have not been meaningfully involved, the processes have been agenda driven, the differences between alternative scenarios and trend projections are slight, and effective implementation strategies have not been adopted"* (pp. 397–8).

Drummond and French (2008) have encouraged planners to take a direct role in the production of open source GIS tools that pair participation and analysis. They write that: *"academic planning programs willing to build the necessary expertise for this effort could develop new GIS analysis, design, and participation tools if local and regional planning agencies provided funding and a real world application environment"* (p. 173). Thus, the application of Envision Tomorrow in the Austin metropolitan region provided an important opportunity to respond to Drummond and French's call for the development of open source PSS. It also provided a ripe opportunity to observe the PSS's role in the process.

3 Methodology

Beauregard (2012) writes that in attempts to represent the workings of planning process and society, planning theorists often neglect the 'things' that shape planning processes. In an article that generally applies a Science and Technology Studies (STS) approach and more specifically Actor-Network Theory, Beauregard deconstructs an analysis of a planning meeting (initially recounted by John Forester). At that meeting, a site plan, a three-dimensional model, and photographs are used to augment deliberation over a proposed development. Beauregard elaborates:

"Without these things in the meeting room, the discussion that took place would not have occurred. Another discussion, with a likely different outcome, would have taken its place. And a different relationship would have been established between the planners and the developer's team. Among other consequences, things focus deliberations" (Beauregard 2012, p. 184).

This chapter is a partial response to Beauregard's challenge. The challenge to account for the agency of 'things', and especially technology, is issued from Actor-Network Theory (ANT). ANT is a body of thought, perhaps best described as a methodological approach rather than a theory (Latour 2005; MacMillan 2009) developed by Bruno Latour, John Law and Michael Callon, among others who have taken up the banner of ANT. In *Reassembling the Social*, Latour advises the social scientist to pay close attention to the role of technology in the making of associations, as it is a network of these flexible associations that constitutes society (Latour 2005). ANT is a methodology for tracing these networks of association.

Taking inspiration from Latour, this chapter seeks to share some of the everyday workings of a regional scenario planning project and to detail the ways in which Envision Tomorrow was used to construct alternative futures that encapsulate sustainability within a PSS in order to challenge development 'as usual' and identify and promote alternative sustainability scenarios for the future development of communities in the region.

This research is derived of participant observation in public workshops and charrettes as well as private research team meetings. This chapter also incorporates the assessments of colleagues who participated in and have written about the Sustainable Places Project. It foregrounds observations about the structure and function of Envision Tomorrow and the active and shifting construction of sustainability through the association of various human actors (planners and consultants, representatives of organizations in the consortium, and public participants in charrettes) as well as non-human actors. Non-human actors include the PSS itself, and the buildings, trees, stormwater management systems, and other aspects of the built environment and 'green infrastructure' that are embedded within Envision Tomorrow's calculations.

4 The Sustainable Places Project and Origins of Envision Tomorrow

The story of Envision Tomorrow's use in Central Texas begins with the award of a federal grant, although the development of ET began prior to the Sustainable Places Project. The Capital Area Planning Council of Governments (CAPCOG) and a consortium of other planning agencies, universities and nonprofit organisations applied for and received a U.S. Department of Housing and Urban Development (HUD) Sustainable Communities Regional Implementation grant. The consortium was awarded a 3.7 million dollar grant in 2010 (Capital Area Council of Governments 2012). A competitive selection process invited local governments within the region to participate (U.S. Department of Housing and Urban Development 2010). Participating local governments would serve as pilot areas where organizational partners and a cadre of planning consultants would apply ET in the creation of plans for demonstration sites. The sites were selected for their potential to demonstrate the sustainability of mixed-use and transit-oriented development.

The initial proposal included IBM as a partner to develop a scenario planning analytics tool for the regional process. After a year of discussions with IBM, the university-based research team chose to seek out a PSS that had been deploying an already developed GIS, rather than investing in the creation of a PSS from scratch, as had been their initial intention. The university research team evaluated a number of PSS and decided to expand the capabilities of Envision Tomorrow, an open source tool that had been developed by Fregonese Associates, a well-known consulting firm based in Portland, Oregon.

Fregonese Associate's decision to make their system of linked spreadsheets and ArcGIS open source was a major factor in the decision to use it in Central Texas. While ET is described as open source, it runs as an extension of ArcMap, a desktop application within the proprietary software package ArcGIS and Excel spreadsheets. Although built atop two proprietary systems, the underlying calculations and development types within ET are open, portable and modifiable. The research team perceived ET to be flexible enough to be useful in multiple local jurisdictions, given the near ubiquity of Excel spreadsheet usage among local government planners and the ease at which they could see and modify embedded calculations in the spreadsheets. Another factor in its selection was the market domination of ArcGIS as a GIS platform for local and regional government planning departments.

The research team surmised that the scenario planning tool could be improved and adapted for the Sustainable Places Project and that the process of calibrating ET for use in Central Texas communities could be replicated in the many communities across the country that already use ArcGIS. ET seemed open and much less proprietary than prospects of a new analytical tool developed by IBM, where there had been no reassurance that planners would know how to operate the tool and no clear plan for its relationship to the GIS datasets and software packages planning departments were already using across the region.

5 Deploying Envision Tomorrow: Demonstrations of Its Use as a Participation Tool

Envision Tomorrow was used in five demonstration areas that had been designated 'activity centers' in a regional planning process called Envision Central Texas and in the region's Long-Range Transportation Plan for 2035. The demonstration areas were located in four small suburban communities: Dripping Springs, Elgin, Hutto and Lockhart. The fifth site was in Austin. Called the South Shores Site, it is located immediately across the Colorado River from Austin's downtown. The City of Austin's initial proposal was to focus not only on the South Shores site, but also another potential station stop for urban rail in the Mueller neighborhood, a new urbanist community that had been developed on the former site of an airport (City of Austin 2011).

In the suburban communities, there was a three-part public participation process comprised of a visioning workshop, a planning charrette, and an open house. The visioning workshops were organized by several planning consultant firms with expertise in scenario planning and public engagement as well as CAPCOG and the University of Texas at Austin's research team. In each community, a stakeholder committee was formed that was involved in the organization of an initial community visioning workshop. At the workshop, members of the public and the stakeholder committee selected the most important indicators for assessing scenarios, discussed goals and values, and were invited to write specific development ideas on large base maps.

A second set of charrettes was organized in each of the four communities. At the charrettes, planning consultants provided a brief overview of the project and then participants were invited to break out into smaller groups of 5–10 people. At each table participants met around large aerial maps of the community and a facilitator encouraged participants to look over a menu of 'development types' that represented mixes of buildings and land uses. The menus gave figures for the number of jobs or housing units per acre and images that represented what the development types would look like if built. The facilitator encouraged participants to place 'development chips' on the map in order to create a vision of the community in the future. Each group would then give the map a title to represent their joint vision. Representatives from each table were asked to present their scenario to all attendees toward the end of the charrette.

At each table, a trained digitizer, in most cases a graduate student from the University of Texas at Austin, 'painted' the development types on a laptop with ArcGIS and the ET extension for ArcMap installed. Each digitizer captured citizen-generated scenarios in ArcGIS in real time. The idea was to enable the facilitator and digitizer to share indicators so that the scenario was done in an iterative fashion. Metrics would enable the participants to understand how many jobs and housing units their proposed scenarios would generate.

In reality, there was little time during the workshops for the facilitator and participants to discuss any of the indicators. The time allotted for the entire workshop was two and a half hours. With a presentation at the beginning and time allotted to share scenarios at the end of the evening, there was just enough time to get participants to settle on a joint scenario and to add development chips and other annotations to the base map. Digitizers were occasionally asked to report back to participants as to how the scenarios were performing. The digitizers primarily shared information about basic indicators such as total number of housing units and jobs accommodated or land use mix. It was rare for digitizers to have the time to share other sustainability indicators, such as estimated vehicle miles traveled, building energy or water usage, let alone to have the time to explain to participants what the indicators meant.

Additional factors that may have produced variation in the way the tool was applied at each table were facilitation styles among planners and the skills of digitizers. Direct observation seemed to indicate that the tool was used fairly consistently across the tables, which was likely to be due to digitizer training prior to the meetings and facilitator protocols. The relatively limited amount of time allotted for developing scenarios seemed to keep both facilitators and public participants focused on placing development chips and discussing scenarios, rather than engaging in conversations around sustainability indicators.

At the planning charrette in one community, John Fregonese of Fregonese Associates, compared publicly generated scenarios with a 'business as usual' or 'trend' scenario. This presentation of the indicators was very quick with little time to explain their meaning. At the charrettes in other communities, even less time was allocated to sharing back the results of sustainability indicators. Instead, only a few of the sustainability indicators were discussed at the end of the night. Planners explained that the scenarios provided by participants would be analyzed after the charrettes and that these citizen-generated scenarios would be condensed down into a smaller number of alternative scenarios.

A third meeting in each community consisted of an open house in which members of each community were invited to view a 'base case scenario' or trend scenario alongside two scenarios. The open houses allowed members of the public to come and go in a relatively unstructured way during the time allotted. Members of the public circulated among poster boards illustrating three scenarios and a limited set of about five indicators showing the performance of each scenario. Participants chatted with representatives from the University of Texas at Austin and members of the planning consulting teams, local planners and public officials, among others. Participants were invited to fill out a survey to indicate their preferred scenario.

The planning process within Austin was significantly different than in the smaller communities. Instead of holding three consecutive meetings, planners from the City of Austin worked with Fregonese Associates to analyze existing conditions and to propose scenarios before inviting the public to participate. ET was used to generate redevelopment options that could be built under current market conditions and in accordance with ideas generated from a prior process sponsored by the American Institute of Architects. Students in the University of Texas at Austin's Urban Futures Lab, led by professor Dean Almy, utilized ET to generate an additional scenario for public consideration. A workshop was held in the South Shores District introducing attendees to scenario planning and ET. Instead of a planning charrette, the meeting was aimed at presenting the two scenarios already generated and comparing them to a 'business as usual' scenario.

The role of ET in the South Shores process was to aid local planners and university-based researchers and students in understanding market conditions and the feasibility and outcomes of expert-driven scenarios. The tool was not used to incorporate public participation into the generation of scenarios as it was in the other communities; instead, it was used in preparation for an informational session with the public. Photo simulation and 3D modeling were used in addition to ET to convince the public and city council members that the district should undergo a more extensive planning process. Public participation and finer-grained analyses were promised if next steps were funded by the Austin City Council. The output from ET helped to describe the quantifiable benefits of scenarios. For example, a green infrastructure application developed for ET during the South Shores Project was used to illustrate how stormwater could be prevented from flowing into the adjacent Colorado River under one of the scenarios.

The South Shores planning process demonstrated how ET could be used for persuasive purposes, as a tool for shaping public opinion with or without participatory processes. I perceived no misuse of the tool; faculty and students at the University of Texas at Austin as well as planners at the City of Austin seemed to sincerely have the best intentions of the public in mind as they explored alternatives with the tool. However, it is worth noting that nothing in ET requires public input or participation.

Comments and questions from members of the public at the South Shores meeting seemed to reflect a general acceptance and support for new, mixed use and higher density building typologies. This was not particularly surprising; the City of Austin had been successful in garnering public and city council support for higher density, mixed use development in recent years, as evidenced in the recent adoption of a comprehensive plan (City of Austin 2012) that embraced concepts of smart growth.

More surprising was what appeared to be widespread support for compact, higher density, mixed use development types in the four smaller, suburban communities. Citizen-generated scenarios in these small towns differed significantly from existing development trends that favored single family tract housing and auto-centric big box and commercial strip development. Community participants placed development types in scenarios that included an ample number of multi-family units, compact neighborhoods with a mix of garden apartments and townhomes, as well as mixed use main street and town center typologies. At UT-Austin research team meetings, faculty and students registered both relief and some surprise that there seemed to be broad public support for a wider range of transit-oriented and mixed use development types and that the charrettes had yielded mostly positive results according to sustainability indicators. On the other hand, the research team noted resistance in one community where a group of Agenda 21 protesters distributed leaflets and spoke with great hostility at what they perceived as a United Nations and U.S. federal government conspiracy to force local communities to depopulate rural communities (for more information about Agenda 21 protests, see Wittenmore 2013).

6 How the PSS Structures Planning Analysis and Defines Sustainability

As noted above, the public participation process can be organized in multiple ways and the process for engaging the public is not defined by Envision Tomorrow. In contrast, ET does provide planners with a critical path to performing planning analysis. The structure of the PSS facilitates a specific set of steps that move from modeling individual buildings to aggregating modeled buildings to development types, to sketching these development types as scenarios in ArcMap.

The first step in the process is for planners to define building types that serve as the building blocks of new development. In this first step, planners work with a Return on Investment (ROI) model to model the *"physical and financial feasibility of a proposed development or existing development regulations"* (Envisiontomorrow.org, undated). A planner can download and easily modify a set of building prototypes that range in type, such as mixed use buildings, light or heavy industrial developments, suburban or urban office buildings, and arterial commercial or big box buildings, among other types. The planner can then begin to address questions such as: *"What building types are feasible in the current market? What would X (e.g. rent, land cost, financing) need to be in order to achieve a desired building type? If certain regulatory constraints (e.g. height restrictions, high parking requirements or setbacks) were relaxed or removed, what new building types become possible?"* (envisiontomorrow.org, undated).

In ET training sessions for planners and students, representatives of Fregonese Associates demonstrated how the 'goal seek' function in the ROI spreadsheet can assist planners in finding public subsidy levels or making modifications to a building prototype's physical inputs that will result in a financially feasible development. One of the recommended methodologies for gaining information to be used in the model is through interviews and focus groups with developers and to compare this input to databases such as RSMeans. Thus the financial feasibility of new development is embedded in the first steps of the analytical process and the planner's drive to achieve a sufficient return on investment compels the planner to consider public-private partnerships or strip away requirements that hinder project feasibility.

In the next step, planners aggregate individual building prototypes into development types using ET's Scenario Builder. Development types are defined by percentages of building prototypes. Within development types, planners calibrate the mix of buildings to current development trends and to aspirational goals for development. As scenarios are painted, a set of indicators is available based on area calculations from ArcMap that are linked to the Scenario Builder spreadsheet.

Table 1 shows the indicators that are readily available in the Scenario Builder. While there are extensions of ET with additional functionalities, the following are core indicators readily available for use in evaluating scenarios. Sustainability is coded into the Scenario Builder in two ways. First, sustainability is defined through a set of indicators related to compact, mixed use, transit-oriented development. A narrower set of indicators are specifically defined in Scenario Builder as sustainability indicators and include measures of energy and water consumption and waste in the form of greenhouse gas emissions, waste water and solid waste generation.

An advantage of ET's spreadsheet structure is the transparency of its indicators and the formulae that underlie them. An additional feature is the embedded 'sketch tool' that, as discussed in the context of public workshops, hypothetically allows planners to track citizen-generated scenarios in real time.

Table 1 Indicators embedded in Envision Tomorrow Scenario Builder

Indicator type	Associated indicators	
Basic calculations	Developed acres Population	Net new growth
Compact, mixed use development	Infill development Building square footage mix	Land area mix Density
Housing	Displacement from redevelopment Housing mix Affordability Housing distribution by income	Housing by building type Jobs/housing balance[a] Affordability (Housing + Transportation + Energy costs)
Employment/jobs	Employment mix Jobs/housing balance[a]	Income
Transportation and travel behavior	Walk and transit friendliness Parking Bike trips Walk trips Transit trips Vehicle trips Internal trips	External trips Vehicle miles traveled (VMT) Traffic accidents ULI shared parking savings Transportation affordability Daily emissions
Fiscal impact and infrastructure	Financial Subsidy New infrastructure Tax revenue Tax revenue per new housing unit	Operations and maintenance costs Capital outlays Cost to revenue ratio
Defined as "sustainability" in scenario builder	Energy use (million BTU/yr) Carbon emissions (tons/yr) Landscaping water use (g/day)	Internal water consumption (g/day) Waste water (g/day) Solid waste (lbs/day)

[a]Appears in two indicator type categories

Beyond the core ability of planners to facilitate conversations and make maps, planners must continually decide what constitutes the content and boundaries of applicable planning knowledge. They are continually pressed to assemble best practices and planning knowledge that can be applied to local contexts and combined with citizen input and local knowledge. The flexible spreadsheet of assumptions is a means of collecting planning knowledge about development performance. It serves as a method for calibrating planning research to local conditions. Coded into ET's ROI and Scenario Builder tools are not only formulae, but their sources with references to academic research and common heuristics used by planners and developers from around the country. Certainly in the context of a large-scale federally funded demonstration project that brought together local and regional agencies, multiple planning consultants and university partnerships, the nature of planning knowledge needed to be clearly documented and communicated. The components of ET served that function.

A criticism raised by researchers involved in the project was the omission of equity among ET's core indicators (Oden et al. 2014; Jackson 2013). An attempt to code in equity, as well as other definitions and indicators of sustainability, is described in the following section.

7 Recoding and Extending Envision Tomorrow

Critiques of the limitations of embedded calculations in ET fueled the expansion of sustainability metrics to address equity and green infrastructure, among other indicators. To address the lack of equity indictors, a faculty and student team developed several new applications (Mueller and Torrado 2013). One such application involved the use of ET's Redevelopment Candidate application, which had been created to enable planners to identify parcels that are likely to redevelop. Inverting the purpose of this application, Mueller and Torrado created a process to use the Redevelopment Candidate application to identify parcels where there is redevelopment pressure that threatens to displace low income renters. This modification essentially inserts equity concerns in the equation, without which a planner might otherwise assume that redevelopment is simply a desirable outcome and ignore the impacts to vulnerable populations or the supply of affordable housing in the study area.

In the process of coding in equity, the research team found that ET's indicators were structured into the PSS at the scale of building and district. Equity is by definition a measure of comparison, often at the district, neighborhood, or regional scale and most often beyond the scale of individual buildings (Oden et al. 2014). Therefore the structure of ET made full representation of equity as a sustainability indicator difficult.

Another faculty and student team extended ET to monetize and represent ecosystem services associated with street trees in development scenarios (Hilde and Paterson 2014). This extension of ET built on the U.S. Forest Service's iTree suite. Central to this accomplishment is the union of two disparate practices to extend the definition of sustainability and reinforce the value of urban canopies in land-use planning. The team explained:

"Traditionally ecosystem services and land use planning have existed as separate disciplines and opportunities to consider both in tandem would be unlikely or not feasible due to 'black box' software platforms. Herein lies the benefit of ET's open-source format: utilizing the familiar and accessible MS Excel platform allows underlying tabular data (which drives the scenario measurement and analysis) to be modified and supplemented. Thus while in the past proprietary software tools would be difficult to integrate, and disciplines such as planning and urban forestry

would not interact, the current environment of free tools with transparent coding is allowing cross-disciplinary integration to become more feasible (Hilde and Paterson 2014, *p. 528)".*

The application associates particular development types (assemblages of buildings) with trees. This association can be customized by planners; therefore, it is adaptable and the association between development types and trees is not rigidly specified. Where the potential for rigidity is expressed is in the reduction of the value of trees from object to ecosystem services (which may be qualitative or quantitative, such as aesthetic value or the reduction of urban heat island effect) to an output in terms of economic value. Thus, the embedded assumptions and their ramifications are multiple: (1) the application expands the constellation of non-human actors that are considered in the search for a sustainable future scenario; (2) it reinforces the notion that economic valuation is the best currency for planning analysis; and (3) it holds that valuation of ecosystem services as static and unproblematic. This last point is important, because the notion of static 'valuation' has been challenged by anthropologists who have emphasized that valuation is complex and is continually *"being combined, tinkered with and reinvented"* (Caliskan and Callon 2009, p. 387). This is just the kind continual assembly of associations between human and non-human actors that ANT is preoccupied with revealing.

In Bruno Latour's nomenclature, the additional calculations can be seen as "inscriptions", or the translation of the empirical world through maps, diagrams, calculations, and reports (Latour 1987). Thus a concern for the value of trees was inscribed into the tool, which could then influence considerations in future planning processes. Close study of the iTree application in ET reveals a process of inscription and valuation that can assist the thoughtful planner who is intimately involved in applying the tool. To the planner who just wants to get the analysis done or to the 'consumer' of planning information, the valuation may appear an unproblematic statistic or fact. Upon closer examination, the conversion of ecosystem services into monetary value seems to feed the assumption that the most sustainable solution for urban tree canopy has the highest monetary value, to the potential cost of more nuanced discussion of how best to plan an urban forest and incorporate it into planning.

Planners at the City of Austin and another set of University of Texas faculty and students also participated in the creation of a Green Infrastructure application that extended the ability of planners to incorporate best practices in stormwater management. Figure 1 is a screenshot of this application. In the screenshot, the 'Green Infrastructure Medley' option has been selected which, according to tips on the screen, consists of 'a mix of green roof, rainwater harvesting, and rain gardens'. Although it is not shown in Fig. 1, the model calculates the removal of pollutant loads associated with different options, from 'No Green Infrastructure' to 'Minimum Green Infrastructure' to the 'Green Infrastructure Medley' depicted below.

Interestingly, the financial statistics remain a prominent part of the application. Figure 1 shows the Internal Rate of Return (IRR) on Project Cash as −4.4 % as compared to a Target Return of 12.0 % for an example building prototype. The IRR is the average annual rate of return from a development over a time period for

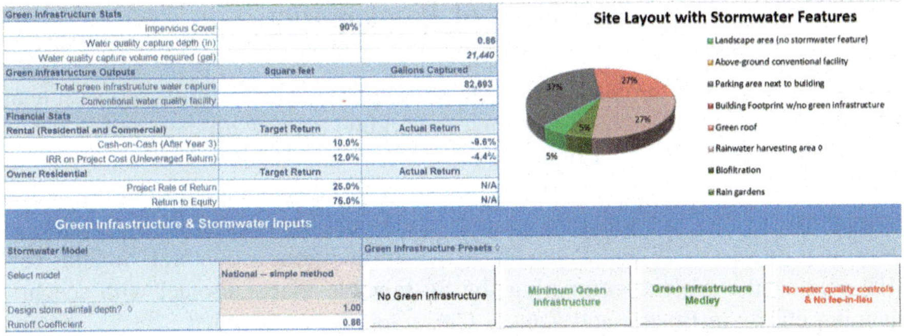

Fig. 1 Screenshot of the Green Infrastructure app embedded in Envision Tomorrow's ROI tool (reproduced from building prototype downloaded from Envision Tomorrow website)

which an investor is involved in a development (Nelson et al. 2012). Therefore, the planner is encouraged to split his/her attention between the potential benefits of green infrastructure and the site level financial feasibility of a development project. As described above, the benefits and costs of an individual building are included within 'development types' that calculate the aggregate values associated with a mix of building prototypes. By including green infrastructure in the ROI tool, planners are encouraged to think about the value of green infrastructure strategies. In a sense the value of trees, plants and soil are encapsulated as agents of positive urban change; in the concepts of ANT, they become assembled as actors within a network that defines sustainable scenarios.

Thus planners are encouraged by what is included in the tool to pursue site-level strategies to reduce pollution. Other benefits of green infrastructure, such as potential monetary value or increase in quality of life indicators are not represented in the quantitative model, although planners were observed in planning processes using attractive renderings and other depictions that spoke to the potential aesthetic and place-making values of vegetation and soil.

In addition to the embedded assumptions of the spreadsheets, the research team noted the inability to analytically explore relationships between development types in ArcMap, pointing out that the GIS sketch tool was primarily used for area calculations. Most of the analytical power of ET is accomplished not in the GIS, but in the Excel spreadsheets. Missing from the GIS extension were the kinds of fragmentation and interaction analyses more common to environmental analyses. This was an area where researchers had limited ability to quickly build in new routines. Modifications to the ArcMap extension required more extensive coordination with ESRI, illustrating the place where the easily adaptable, open source nature of ET stops and the complications of modifying proprietary software begin.

8 Observations on the Limits of Participation

Both during and after the Sustainable Places Project, members of the research team produced documentation in the form of working papers, articles and a thesis. Conclusions ranged from critique of the process and outcomes of the planning process (Oden et al. 2014; Jackson 2013) to optimistic demonstrations of the adaptability of the software to address additional sustainability indicators, as described above (Hilde and Paterson 2014). In addition, a doctoral student from MIT studied the project, comparing the Sustainable Places Project with scenario planning efforts in Boston and Kansas City (Goodspeed 2013).

Oden et al. (2014) identified four problems with scenario planning within the Sustainable Places Project: *"(1) roles and representation of various publics in the planning process; (2) analytic pitfalls; (3) political tensions surrounding development and redevelopment; and, (4) integration of equity concerns"* (Oden et al. 2014, p. 8). The authors emphasized a lack of representation of lower income residents, a segment of the population that, according to HUD criteria, should be a major participant and beneficiary of the planning process. In addition, they noted how the emphasis on individual demonstration sites essentially devolved a regional effort to the scale of four small communities and an urban district. They were disappointed that the process did not result in forging a common regional approach to equity.

In an addition, Jackson interviewed stakeholders from the four small suburban towns (Jackson 2013). Although there was a low response rate and results were mixed, Jackson found sufficient evidence that casts some doubt as to the efficacy of Envision Tomorrow in shaping the development of scenarios in the charrettes:

"Subjects had a general understanding of the indicators used in Envision Tomorrow and what they implied, but they did not appear to use them to guide their scenario development. As a participant at design tables, it appeared that residents found them occasionally useful but for the most part designed scenarios based on the opinions of individuals on certain parcels or areas of town" (Jackson 2013, p. 152).

He noted that *"interviews suggest that the indicator results did serve effectively as a sort of 'check' on plans after the fact, but this is not the same as an iterative process in which residents gain a more organic understanding of the analysis"* (Jackson 2013, p. 158). He observed that a lack of iterative use of the indicators, in which participants understand and discuss trade-offs may indicate *"little true social learning"* (Jackson 2013, p. 159). In fact, he attributed the positive performance of preferred scenarios generated by the public as potentially tied to a social network of workshop participants who were involved in landscape, architecture, urban planning, and urban design. Additionally, economic indicators appeared more of a concern for residents and city officials than the environmental or social indicators.

Jackson also described the development and utility of the aforementioned extensions of ET. These extensions were developed in the planning process, but not fully integrated in the analysis for every community. In fact, many of these indicators were never unveiled or integrated into the public participation process, having been developed too late or deemed too complex for the series of public workshops and charrettes.

Goodspeed (2013) used observational and survey methods to assess the extent to which the Sustainable Places Project achieved social learning in the four suburban communities. He noted that the analytical contributions of ET's indicators were limited to only four indicators of which *"three were summative and one an analysis of the impact of development on tax revenues"* (Goodspeed 2013, p. 173).

These critiques correspond closely with my own observations. Some public participants appeared interested and impressed by the technology used to create scenarios, but actual engagement with ET and its indicators was limited. Much of the material shared with the public, in print or web materials and at community meetings was boiled down to include only a few indicators. The superficiality of the public's engagement with the indicators contrasted with the level of engagement by planners, researchers and students. These participants strove to learn, implement, incorporate and extend the analytical capabilities of ET. The state of planning research related to sustainability and how this could be represented in ET were discussed extensively by the research team and planners. Thus, the tool provided an important conduit of knowledge exchange between practitioners and academics, while the exchange between planners and the public seemed more limited and constrained.

9 Conclusions and Recommendations

Observations on the use of Envision Tomorrow to successfully support sustainability, transparency and participatory governance in the Austin metropolitan region are mixed. There was evidence that ET structured planners' processes and concerns and enhanced their analytical capabilities. Modifications to ET show how the tool can be open and transparent—planners can access the formulae and assumptions behind the sustainability indicator outputs and, in conjunction with software developers or through manual processes, they can change them and add more. On the other hand, the existence of those formulae and the ability to recode them did not ensure that citizens had access to either the indicators or the means of deriving them or that there was an opportunity to discuss them. Nor did it ensure that the indicators were actually used in determining final preferred scenarios.

In examining the use and development of a PSS, there is a danger of assuming a kind of instrumental rationality or a technological determinism in its use. Planners and researchers can too easily and implicitly assume that a PSS is applied in a disinterested, objective fashion in the production of facts and for use in coming to a rational decision about an optimal future scenario. It is evident from observations of the Sustainable Places Project that it is fairly easy to transform a scenario generated either by the public or by planners into a 'black box' of facts that are selectively presented without caveats or context. If open houses, charrettes, and workshops do not reveal the calculations and assumptions behind scenarios or within the PSS, there can be an illusion of rationality and scientificity that transforms heuristics or calculative assumptions based on provisional knowledge into known facts (Jackson 2013) or persuasive materials leading to public consent. Yet, opening embedded calculations to a potentially deeply divided public opens up planners and their knowledge to questions and concerns they may feel ill-equipped to answer.

Perhaps most hopeful are observations that a flexible and transparent PSS can help planning practitioners and researchers collect and transmit best practices, heuristics and the state of planning knowledge. The most substantial learning outcomes seemed to have been among the researchers, practitioners and students who could experience firsthand the complications of both analysis and participation. They were charged with disassembling the technology to understand how it works and reassembling it with new ideas and methods. When properly resourced and coordinated, this case study illustrates how a constellation of smart growth and sustainability concepts can be linked to assist in the generation of new visions of the future and in inscribing concern for more aspects of the built and natural environment. In the collection of these assumptions within a PSS there is hope that planners can continually question them. The PSS and the scenarios generated within them need not be black boxes; if assumptions are available and modifiable, then the technology allows for assembly of new concerns, people, 'things', and concepts into different and potentially more representative models of future scenarios. This case study highlights the importance of inviting professionals and citizens alike to recognize and deconstruct embedded assumptions in the planning process and to recode them anew.

Acknowledgments I would like to thank Tom Hilde, Donald Jackson, Elizabeth Mueller, Michael Oden, Robert Paterson, Marla Torrado, Sarah Wu for sharing their insights and research. I would also like to thank the editors and anonymous reviewers who helped to guide the development of this article.

References

Bartholomew, K. (2007). Land use-transportation scenario planning: Promise and reality. *Transportation, 34*(4), 397–412.

Beauregard, R. A. (2012). Planning with things. *Journal of Planning Education and Research, 32* (2), 182–190.

Brabham, D. C. (2009). Crowdsourcing the public participation process for planning projects. *Planning Theory, 8*(3), 242–263.

Caliskan, K., & Callon, M. (2009). Economization, part 1: Shifting attention from the economy towards processes of economization. *Economy and Society, 38*(3), 369–398.

Calthorpe Associates. (undated). Scenario planning tools: RapidFire and UrbanFootprint. http://www.calthorpe.com/scenario_modeling_tools. Accessed 28 Nov 2014.

Capital Area Council of Governments. (2012). Frequently Asked Questions. http://www.sustainableplacesproject.com/node/2. Accessed 1 Jan 2015.

Chakraborty, Arnab. (2011). Enhancing the role of participatory scenario planning processes: Lessons from reality check exercises. *Futures, 43*(4), 387–399.

City of Austin. (2011). CATS demonstration site application: City of Austin proposal urban rail system. (Central Austin and Mueller Activity Centers) http://austin.sustainableplacesproject.

com/sites/sustainableplacesproject.com/files/files/Austin%20Application.pdf. Accessed 28 Nov 2014.

City of Austin. (2012). *Imagine Austin comprehensive plan*. Austin, Texas: City of Austin.

Drummond, W., & French, S. (2008). The future of GIS in planning: Converging technologies and diverging interests. *Journal of the American Planning Association, 74*(2), 161–174.

Envisiontomorrow.org. (2014). Envision tomorrow—welcome to envision tomorrow. http://www.envisiontomorrow.org/. Accessed 28 Nov 2014.

Evans-Cowley, J., & Hollander, J. (2010). The new generation of public participation: Internet-based participation tools. *Planning Practice and Research, 25*(3), 397–408.

Ganapati, S. (2010). Using geographic information systems to increase citizen engagement. IBM Center for the Business of Government website. http://www.businessofgovernment.org/sites/default/files/GanapatiReport.pdf. Accessed 9 Mar 2012.

Göçmen, Z. A., & Ventura, S. J. (2010). Barriers to GIS use in planning. *Journal of the American Planning Association, 76*(2), 172–183.

Goodspeed, R. (2013). *Planning support systems for spatial planning through social learning*. Dissertation, Massachusetts Institute of Technology, Cambridge, MA.

Hilde, T., & Paterson, R. (2014). Integrating ecosystem services analysis into scenario planning practice: Accounting for street tree benefits with i-Tree valuation in Central Texas. *Journal of Environmental Management, 146*(15), 524–534.

Holway, J., Gabbe, C. J., Hebbert, F., Lally, J., Matthews, R., & Quay, R. (2012). *Opening access to scenario planning tools*. Cambridge, MA: Lincoln Institute of Land Policy.

Jackson, D. E. Jr. (2013). *Fiscal impacts of land use planning decisions: Sprawl, sustainable development, and simulation in public process*. Thesis, University of Texas at Austin, Austin, Texas.

Koekoek, A., van Lammeren, R., & Vonk, G. A. (2009). The potential of integrating e-participation in planning support systems. *URISA Journal, 21*(2), 39–47.

Latour, B. (1987). *Science in action: How to follow scientists and engineers through society*. Cambridge, Mass.: Harvard University Press.

Latour, B. (2005). *Reassembling the social: An introduction to actor-network-theory*. Oxford, New York: Oxford University Press.

Macmillan, B. (2009) *ANT in Westboro: An examination of the applicability of actor-network theory in gentrification studies*. Thesis, Carleton University, Ottawa, Ontario.

Mueller, E., & Torrado, M. (2013). Redevelopment and displacement manual. Document from e-mail correspondence.

Nelson, A., Eskic, D., Kittrell, K. (2012). Return on investment (ROI) model: Envision tomorrow (ET+). Accessed at http://www.arch.utah.edu/cgi-bin/wordpress-etplus/wp-content/uploads/2012/10/Return-on-Investment_7-1-12.pdf. Accessed 28 Feb 2014.

Oden, M., Mueller, E., Jackson, D. E. Jr. (2014). The big picture: Using scenario based planning to achieve environmental and social sustainability. Unpublished Working Paper. Document from e-mail correspondence. 6 Aug 2014.

Open Planning Tools Group. (undated). About. http://www.openplanningtoolsgroup.org/about/, Accessed 28 Nov 2014.

Partnership for Sustainable Communities. (2014). Partnership grants, assistance, and programs. http://www.sustainablecommunities.gov/partnership-resources, Accessed 28 Nov 2014.

PlaceMatters (undated). Place/Matters. http://placematters.org/, Accessed 28 Nov 2014.

Seltzer, E., & Mahmoudi, D. (2013). Citizen participation, open innovation, and crowdsourcing: Challenges and opportunities for planning. *Journal of Planning Literature, 28*(1), 3–18.

Townsend, A. M. (2013). *Smart cities: Big data, civic hackers, and the quest for a new utopia* (1st ed.). New York: W.W. Norton & Company.

U.S. Department of Housing and Urban Development (2010). Notice of funding availability (NOFA) for HUD's fiscal year 2010 sustainable communities regional planning grant program.

Wittenmore, A. (2013). Finding sustainability in conservative contexts: Topics for conversation between American conservative elites, planners and the conservative base. *Urban Studies, 51* (12), 2460–2477.

Chapter 23
Monitoring and Visualising Sub-national Migration Trends in the United Kingdom

John Stillwell, Nik Lomax and Nikola Sander

Abstract Urban policy makers and service providers need to understand the magnitude and dynamics of population migration to and from towns and cities since both the internal and international components are increasingly important in driving urban demographic development. In this chapter, an information system is outlined with a simple interface that allows migration data alongside data for natural change for selected districts or city regions to be tabulated and visualised so that time series trends and spatial patterns can be identified and compared. The data suggest that, during the 2000s, the major cities in the UK collectively experienced significant population growth, a large increase in net international migration and a decline in the relatively longstanding process of counterurbanisation.

1 Introduction

In both the developed and less developed worlds, cities are growing rapidly and smart methods are increasingly required to manage complexity, increase efficiency, reduce expenditure and improve quality of life. Whilst emerging technologies (ultra-low power sensors, wireless networks and web and mobile-based applications) have begun to reshape urban environments and smart cities are fast becoming a reality (Batty 2012; Department for Business Innovation and Skills 2013; Centre for Cities 2014), basic understanding of the demographic evolution of towns and cities is imperative if local authorities are to provide housing, infrastructure and services that accord with demand. This involves understanding the complexity

J. Stillwell (✉) · N. Lomax
School of Geography, University of Leeds, Leeds LS2 9JT, UK
e-mail: j.c.h.stillwell@leeds.ac.uk

N. Sander
Vienna Institute of Demography (Austrian Academy of Sciences),
Wittgenstein Centre for Demography and Global Human Capital,
Wohllebengasse 12-14, 1040 Vienna, Austria

S. Geertman et al. (eds.), *Planning Support Systems and Smart Cities*,
Lecture Notes in Geoinformation and Cartography,
DOI 10.1007/978-3-319-18368-8_23

of flows into and out of metropolitan areas as well as the spatial demographic dynamics taking place within cities and their hinterlands.

Mid-year population estimates for local authority districts in the United Kingdom (UK) are generated by the national statistical agencies. Together with population projections, they are used directly by local authorities for planning and policy making but they also underpin the annual allocation of financial resources from central government to local authorities following the Lyons enquiry (Lyons 2007). Demographic change (in both urban and rural areas of the UK), as in other parts of the world, is driven by the changes in the components of growth: the balance between births and deaths and between in-migration and out-migration. It is important to monitor patterns and trends in each of these components to establish, quantify and understand the processes that have occurred in the relatively recent past (e.g. urbanisation, suburbanisation) and that are currently taking place (e.g. counterurbanisation, reurbanisation) as well as to provide a basis for projecting what might happen in the future.

In this chapter we report on the construction of a system for monitoring population change, natural change, internal migration and international migration over time from the beginning of the twenty-first century for a hierarchical system of spatial units that includes local authority districts, city regions and NUTS1 regions in each of the countries that constitute the UK. Details of the time series data sets estimated for a consistent set of spatial units are outlined in the next section. An information system has been built to support the quick and easy retrieval of estimated data based on user specification of a spatial unit and time period. The system interface is based on the concept of a demographic dashboard for illustrating the time series trends in different demographic components for a selected city by the user, but in this case the data are assembled manually and the user observes time-series rather than real-time indicators. Thereafter, there are two sections which provide analyses of time-series trends and spatial patterns which are, at the moment, not automated as part of the information system. First, key trends occurring in flows between metropolitan and non-metropolitan areas across the UK are presented together with relationships between internal and international migration. Second, London's migration characteristics are examined using conventional mapping methods before directional flows between city regions are visualised using circular plots. Some conclusions are included in the final section with proposals for further improvements and extensions to the system.

2 Demographic Data and Spatial Units

The decennial census of population is a key data source for planning in the UK; not only does it provide the benchmark for mid-year estimates but it also provides a matrix of migration flows taking place in the year prior to the census between Local Authority Districts (LADs). Between censuses, sub-national population estimates for LADs are produced by the Office for National Statistics (ONS), by the

National Records of Scotland (NRS) and by the Northern Ireland Statistics and Research Agency (NISRA) for England and Wales, Scotland and Northern Ireland respectively. These estimates are derived by taking the population measured in the most recent census, ageing this on and then adjusting for births, deaths, international migration and internal migration (see ONS 2014a, for a full explanation of the methods used for producing mid-2013 estimates for England and Wales). Although population estimates from Scotland and Northern Ireland are also collated by ONS to produce UK totals, one of the shortcomings of this approach involving three separate national statistical agencies is that a complete matrix of flows between all LADs in the UK for each mid-year to mid-year period between censuses is not assembled. Each agency estimates flows between LADs within its own jurisdiction using data from administrative databases and surveys, but the cross-border migration flows between local authorities in the three constituent countries are never properly estimated from aggregate totals between the countries that are available. Figure 1 is a schematic diagram of the matrix of migration flows between districts across the UK.

Working in collaboration with the ONS, Lomax (2014) has developed a methodology to estimate complete matrices of internal migration flows between LADs across the UK for a time series commencing in mid-year 2001–02 using data from censuses and administrative and survey sources. Details of the methodology are explained in Lomax et al. (2013) and an initial analysis of the estimated flows is reported in Lomax et al. (2014) for the system of 406 LADs shown in Fig. 1. This system of spatial units, which includes adjustments for boundary changes due to the restructuring of local government in England and Wales in 2009, can be aggregated into more macro regions to facilitate summary and comparative analysis. Two levels of aggregation of LADs include (i) the 12 regions which are equivalent to European Union NUTS1 regions, and (ii) 13 city regions, which in most cases have four component areas: core, rest, near and coast and country and were first used by Stillwell et al. (2000, 2001) for comparison of migration in the UK and Australia (Fig. 2).

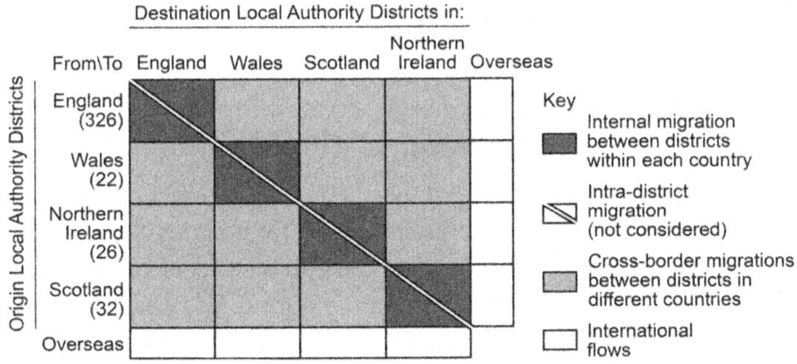

Fig. 1 Schematic diagram of the matrix of estimated migration flows between LADs in the UK

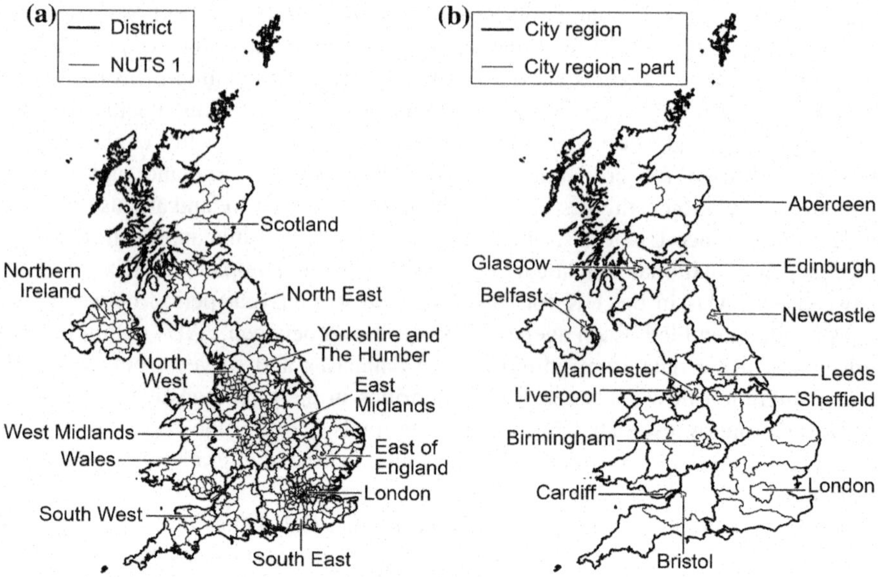

Fig. 2 Boundaries of the spatial units used for monitoring. **a** Local authority districts and NUTS 1 regions. **b** City regions and their component areas

In addition to the estimated inter-censal mid-year to mid-year time-series matrices of internal migration flows, vectors of mid-year populations, annual births and deaths and international immigrants and emigrants have been assembled from official sources and adjusted where necessary, enabling a full accounting of the population change from one mid-year period to the next from 2000–01 to 2010–11 for sets of spatial units with consistent boundaries throughout the period. Considerable time and effort has been invested in building this unique set of data for the UK for monitoring and analysis of changing spatial patterns and processes. The following sections explain the user interface and identify, for the first time, some of the trends and relationships that characterize migration in the UK in the 2000s. Although estimates have been produced for populations by five-year age group and sex, we have confined our analysis to aggregate statistics in this chapter.

3 Structure and Exemplification of the Monitoring Tool

One example of the development of creating, modelling and communicating aspects of the smart city is the 'city dashboard', a service that collects various types of near-live publically available data from different external websites and makes them available through a web interface that enhances public awareness. One example of this is that created for London and other cities in the UK (Roumpani et al. 2013).

The concept of a dashboard underpins the interface of the information system that we have constructed which in our case is based on 'historical' data estimated externally and assembled manually in Microsoft Excel with each of the time-series data sets for every component being assembled in the same workbook. The interface is a spreadsheet with VLOOKUP functions to retrieve values for each of the variables for the spatial unit (LAD) selected by the user from a dropdown list.

In the example of the interface presented in Fig. 3, Leeds LAD has been chosen, an area which is the core component of the wider Leeds city region and which is the regional capital of the Yorkshire and Humber region. We observe from the time-series estimates—presented in tabular and graphic form—that the population increase in Leeds Core over the period has not been linear; there is a steep increase in population between 2003–04 and 2004–05 with more steady growth in previous and subsequent years. Growth over the decade has been less than the national average and that taking place in the city region as a whole. The graph of vital statistics indicates how growth in the natural change component has occurred as births have risen and deaths have declined. Net internal migration, on the other hand, remains relatively marginal over the period with gross inflows, though increasing, being cancelled out by gross outflows. The migration efficiency index measures net migration as a percentage of migration turnover (in-migration plus out-migration) (Stillwell et al. 2000, 2001; Lomax et al. 2014) and gives a clearer picture of how the net migration fluctuates around zero whilst the connectivity index suggests that Leeds is connected by both outflows and inflows involving more than 3 migrants with around 94 % of the other 405 LADs in the UK.

It is apparent from the international migration graph that the time-series trends have been much less stable over time, with a significant increase in immigration taking place between 2002–03 and 2004–05 without a corresponding change in emigration. The demographic trajectory of the Leeds population has therefore been impacted by the net immigration rising from 5,422 to 8,028 to 12,961 over the three years concerned with serious implications for the provision of services such as housing and schools.

Demographic growth in Leeds of 5 % over the 2000s has generated pressures for additional accommodation, some of which has been provided through the construction of private housing in outlying suburbs but much of which has taken place through private sector investment in the development of blocks of flats in areas adjacent to the city centre (particularly along the waterfront) once used for industrial or commercial purposes (Unsworth and Stillwell 2004). The latter phenomenon, involving mostly relatively young single or dual income households without children, and known as city living (Unsworth 2005), occurred in many other cities in the UK as a result of urban regeneration processes and is partly responsible for the high rates of growth that are apparent elsewhere. Rates of population change expressed as an index that takes the value of 100 at the start of the period are shown in Fig. 4a for each of the 13 city region cores. Manchester has experienced the largest growth between 2001 and 2011 at almost three times the national average, with all the city region cores apart from Newcastle, Glasgow and Belfast having experienced relatively more growth than Leeds by the end of the period. In the case

Select Area:	Leeds

Type in area or use dropdown box

Leeds is within the city region part:	Leeds Core
Leeds is within the former GOR of:	YORKSHIRE AND THE HUMBER

Click here for the city region profile

Click here for the Region profile

	2000/01	2001/02	2002/03	2003/04	2004/05	2005/06	2006/07	2007/08	2008/09	2009/10	2010/11	2011/12
Population	715,609	716,380	717,592	723,042	735,054	736,204	738,560	741,665	743,885	747,571	750,683	757,655
Births		7,835	7,995	8,278	8,486	8,930	9,274	9,525	9,913	10,319	10,319	10,388
Deaths		7,052	6,983	6,914	6,708	6,601	6,541	6,643	6,536	6,308	6,613	6,434
Natural Change		783	1,012	1,364	1,778	2,329	2,733	2,882	3,377	4,011	3,706	3,954
Internal Migration In		30,713	31,214	32,772	33,758	34,437	35,264	35,642	36,162	35,175	35,484	39,048
Internal Migration Out		31,371	32,286	32,546	32,704	33,997	35,524	35,296	35,739	36,219	36,596	37,651
Internal Migration Net		-658	-1,072	226	1,054	440	-260	346	423	-1,044	-1,112	1,397
International Migration In		9,593	11,488	11,376	18,316	8,804	10,185	9,113	8,301	8,086	8,406	6,582
International Migration Out		4,807	6,066	3,348	5,355	6,468	6,262	5,101	5,937	3,449	3,818	4,963
International Migration Net		4,786	5,422	8,028	12,961	2,336	3,923	4,012	2,364	4,637	4,588	1,619
Other Changes Net		-4,140	-4,150	-4,168	-3,781	-3,955	-4,040	-4,135	-3,944	-3,918	-4,070	2

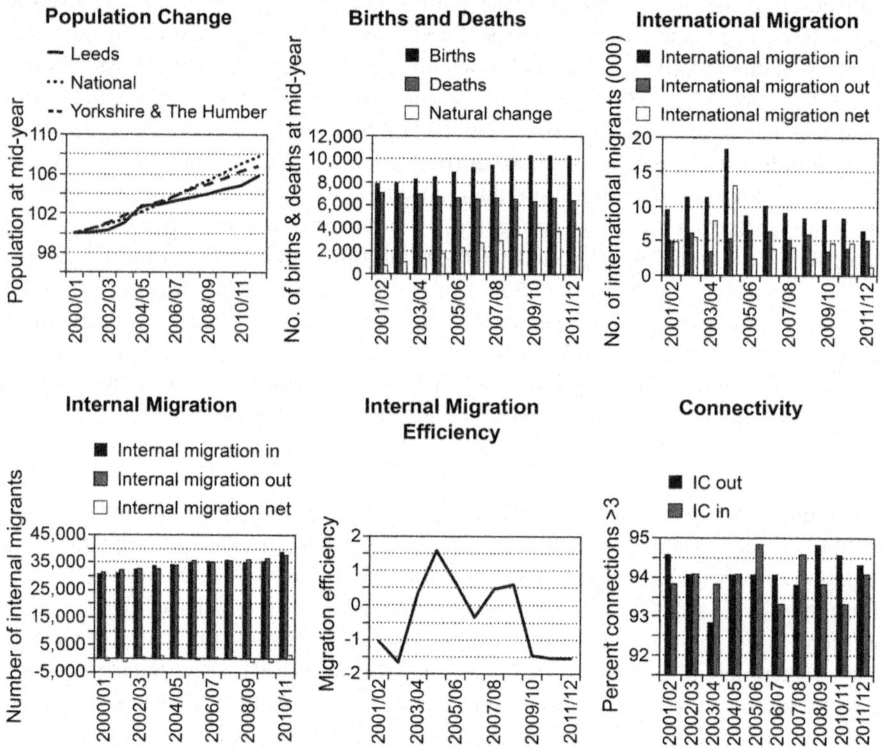

Fig. 3 Example layout of interface with indicators for Leeds Core

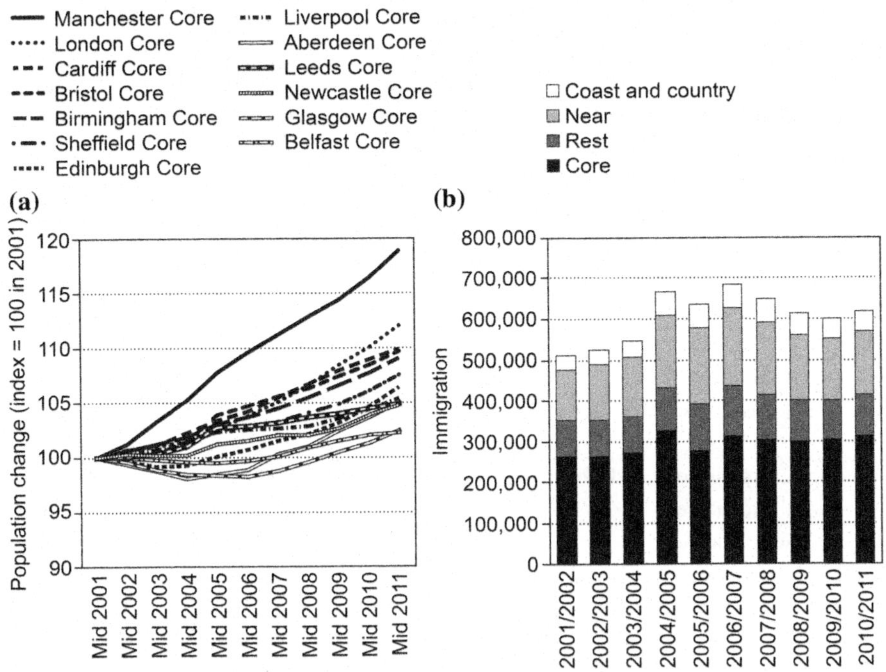

Fig. 4 Population and immigration time series, 2001–2011. **a** Population change. **b** Immigration change

of the large cities in Scotland and Northern Ireland (Edinburgh, Glasgow, Aberdeen and Belfast), the trajectories shown in Fig. 4a indicate population losses in the first half of the decade with overall growth rates that remain below the national level for the decade of 7 %.

One of the key features of demographic development in the UK during the 2000s has been the extent of net immigration from the rest of the world, an issue that has led to much political debate and public discussion, not least in terms of its implications for the labour market and for the provision of services. Figure 4b captures the time series trend in the total number of immigrants into the UK during the decade, rising from around half a million in 2001–02 to 666,000 in 2004–05 and remaining at over 600,000 per year for the rest of the decade. Around one half of this migration has been into the city region cores, with London attracting one third of a total immigration (over 6 million) to the UK throughout the period, whilst contributing only 28 % of the total flow of migrants (3.7 million) leaving the country.

The system allows for the easy generation of league tables for selected variables over user-defined periods of time. Table 1 exemplifies this by showing the top and bottom ten positions in the population change league table for different city region parts over the decade. As indicated previously, Manchester is the UK's 'boom city'

Table 1 Top and bottom positions in the population change league table, 2001–2011

Rank	LAD	2001 Population	Population change	% Change
1	Manchester Core	502,902	79,987	18.9
2	London Core	8,204,407	882,004	12.0
3	Aberdeen Near	347,120	33,180	10.6
4	Sheffield Coast and Country	714,768	67,128	10.4
5	Aberdeen Coast and Country	305,080	28,530	10.3
6	Cardiff Core	472,121	42,756	10.0
7	Bristol Core	428,074	38,025	9.7
8	Belfast Near	726,714	62,050	9.3
9	Birmingham Core	1,074,823	89,641	9.1
10	Belfast Coast and Country	654,245	53,721	8.9
38	Liverpool Coast and Country	191,436	6786	3.7
39	Liverpool Near	496,981	17,068	3.6
40	Newcastle Near	1,492,300	39,026	2.7
41	Glasgow Core	593,060	14,350	2.5
42	Manchester Coast and Country	499,817	12,022	2.5
43	Glasgow Rest	1,112,740	25,390	2.3
44	Belfast Core	433,359	9709	2.3
45	Newcastle Rest	825,049	4474	0.5
46	Glasgow Coast and Country	543,910	170	0.0
47	Liverpool Rest	915,114	−10,383	−1.2

of the 2000s (Crook and Linton 2012) with London appearing in second place. Aberdeen, in Scotland, has benefited from North Sea oil-related development and this is likely to explain the relatively high population growth in its affiliated near and coast and country area. Cardiff, Bristol and Birmingham Cores are also in the top ten, the latter being the UK's second largest city and the only other city with more than a million inhabitants. Whilst Belfast Near appears in the top third of the table with growth of nearly 9 %, Northern Ireland's capital city itself (Belfast Core) experienced a slower rate of growth of 2.3 %. The area immediately adjacent to Newcastle Core (Newcastle Rest) grew only marginally (0.5 %), whilst the equivalent area of Liverpool Rest was the only area at this spatial scale to lose population during the period. These areas contain some of the old industrial LADs in the north of the country including Barrow-in-Furness, Knowsley, Sefton, South Tyneside and Sunderland, all of which experienced population declines of 3 % or more.

4 Trends in and Relationships Between Demographic Components

Whilst the monitoring framework allows the trajectory of each LAD or city region to be profiled against that of its corresponding region or against national figures, and to show its position in the league table *vis a vis* its competitors, it also proves useful to identify trends in the components of change at a more macro level, in order to identify what processes are happening throughout the country as a whole for places of similar type or at different spatial scales. This is illustrated in Fig. 5, which shows two sets of time-series trends. The graph on the left-hand side plots the migration flows between LADs classified as either metropolitan (core and rest areas) or non-metropolitan (near and coast and country areas), indexed to a value of 100 in 2001–02. The graph reveals a decline in flows from metropolitan to non-metropolitan areas suggesting a decline in counterurbanisation (Champion 1989), a process that has characterised internal migration in the UK for several decades. The graph also reveals a decline in movements between rural areas but an increase in movements into metropolitan areas from rural areas and other metropolitan areas. These statistics support the hypothesis of reurbanisation or 'back to the city' movement in the UK that has been occurring over this decade, generating the type of 'city living' environment apparent in Leeds.

The graph in Fig. 5b depicts only migration from the core areas of city regions, again indexed to 2001–02. In this case, we observe that whilst core to core movements have been increasing, regardless of whether London Core is included or excluded in the analysis, it is longer-distance movement from the cores to the coast and country (C & C) areas of the city regions that has declined significantly. Flows out of cores to neighbouring rest or near regions have only shown marginal decline.

The boundary of the core of the London city region coincides exactly with that of the Greater London region which is composed of the City of London and 32

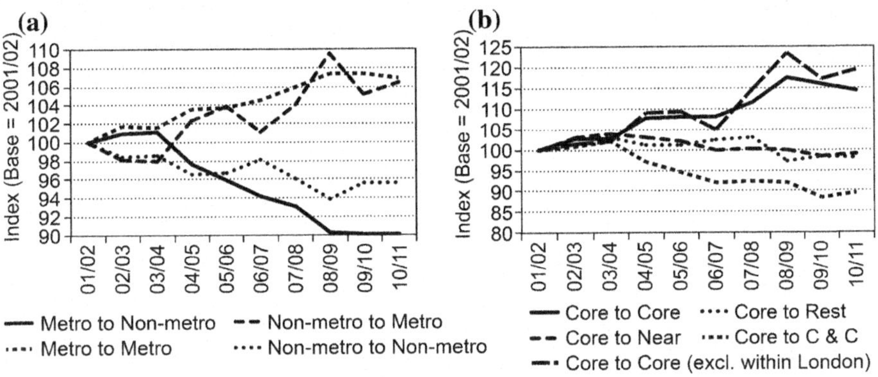

Fig. 5 Time series indices of aggregate flows. **a** flows between metropolitan and non-metropolitan areas. **b** Flows away from city region cores, 2001–2011

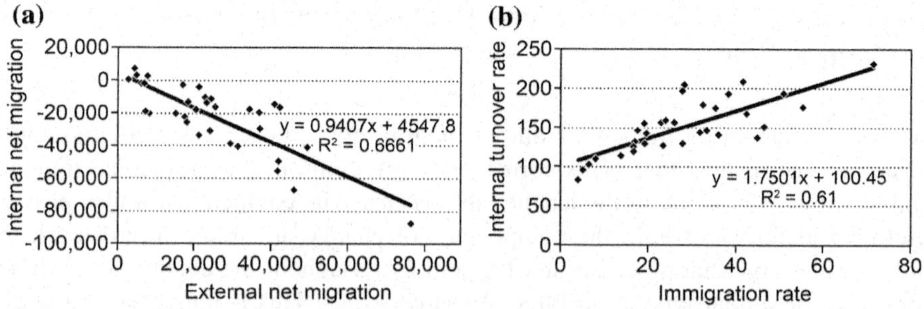

Fig. 6 Linear regression relationships between selected variables for London boroughs, 2001–2011. **a** Internal and external net migration. **b** Internal turnover and immigration rates

London boroughs. It is this (NUTS1) region which is the engine that drives the whole migration system of the country, attracting in-migrants from overseas and from the rest of the UK, but also generating out-migrants elsewhere in large volumes. In net terms, all boroughs bar the City of London, Harrow and Brent lost net migrants over the decade from mid-2001 to mid-2011 to elsewhere in the UK but these losses were more than offset by net migration gains to each borough from the outside the UK. Figure 6a captures this linear relationship whilst Fig. 6b illustrates that whilst London boroughs with relatively low turnover rates, measured as the sum of in-migration and out-migration each year over the decade and divided by a mid-decade population, had lower rates of immigration from overseas, boroughs with high internal turnover had high rates of immigration. In other words, rates of immigration were higher to boroughs where internal migration activity, measured by turnover, was more buoyant.

These estimates of net migration for the decade as a whole provide a useful summary of the extent of internal net losses and external net gains in migration but they conceal fluctuations that may have occurred during the time period from year to year as well as processes of internal migration within London Core. In the next section, we retain our focus in the first instance on London and examine patterns of net and gross migration using conventional mapping techniques but then demonstrate a new method of mapping directional flows using data for city regions.

5 Visualising Internal Migration Flow Patterns

London Core grew by 12 % during the 2000s from an initial population of 7.3 million to 8.3 million by mid-2011. Our estimates suggest that gains from international net migration offset losses from net internal migration in each year of the decade apart from 2002–03 and 2003–04. Overall, London gained around 929,000 from overseas and lost around 724,000 over the ten-year period and Fig. 7 presents the year-on-year breakdown of net migration balances between internal

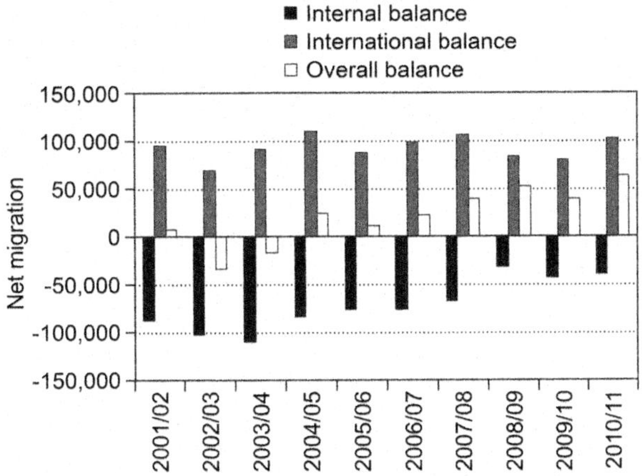

Fig. 7 Net migration balances of London Core, 2001–02 to 2010–11

and international flows, indicating how internal losses reduced during the latter half of the period, resulting in increased overall net gains.

The overall net migration balances for individual London boroughs can be decomposed into three component parts which are shown in Fig. 8 for the first and last years of the time series. Internal net migration is divided into two types with maps a and b illustrating the net result of flows between boroughs within London whilst maps c and d show the pattern of net migration flows between each borough and the rest of the UK. These maps reveal important patterns of net migration that are concealed by the aggregate balances.

Within London, a process of deconcentration is taking place resulting in net losses in inner London and net gains in outer London as migrants reveal their preference for more suburban residential environments in the outer boroughs or fewer opportunities to find suitable accommodation in the inner areas where house prices have risen disproportionately during the decade. However, inner London boroughs are gaining migrants in net terms from the rest of the UK whilst outer London boroughs are losing migrants in significant quantities to the rest of the UK. These patterns align with the concept of London as an 'escalator' region (Fielding 1992). Inner London is an important destination for employment and educational opportunities and therefore, like other cities, attracts large numbers of migrants in higher education or early career life stages, whereas migrants departing London from the outer boroughs tend to be older and motivated to leave for housing and other reasons. Comparison between the two years indicates how more inner London boroughs were experiencing net in-migration from the rest of the UK by the end of the period. The pattern of international balances shown in map e shows net gains across the board in 2001–02 which are particularly high in the north (Haringey, Barnet, Enfield, Waltham Forest and Redbridge) but also in the west (Ealing) and in the east (Newham and Tower Hamlets). A similar pattern is evident for 2010–11 (map f) with the exception of Kensington and Chelsea, whose balance has turned positive.

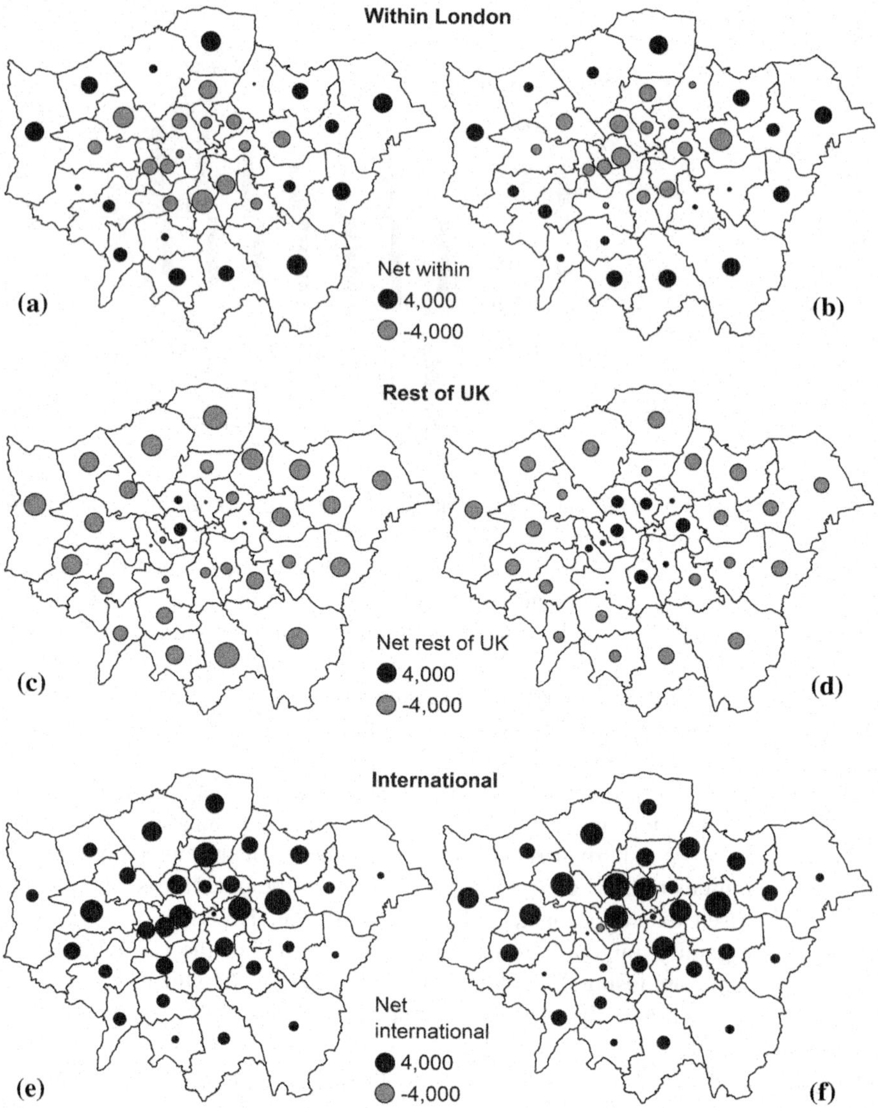

Fig. 8 Net migration balances for boroughs of London Core, **a**, **c**, **e** 2001–02 and **b**, **d**, **f** 2010–11

Whilst the analyses and visualisations shown in the previous sections allow the capture of trends for individual areas or groups of areas in summary form, the net internal migration balances conceal the directional movements that are taking place between origin and destination areas. Conventional migration flow mapping—involving the drawing of lines on a base map—dates back to the early visualisations by Minard (flow map of Napoleon's disastrous Russian campaign in 1812) and Ravenstein (1885) and has been slowly improving since the early experiments with

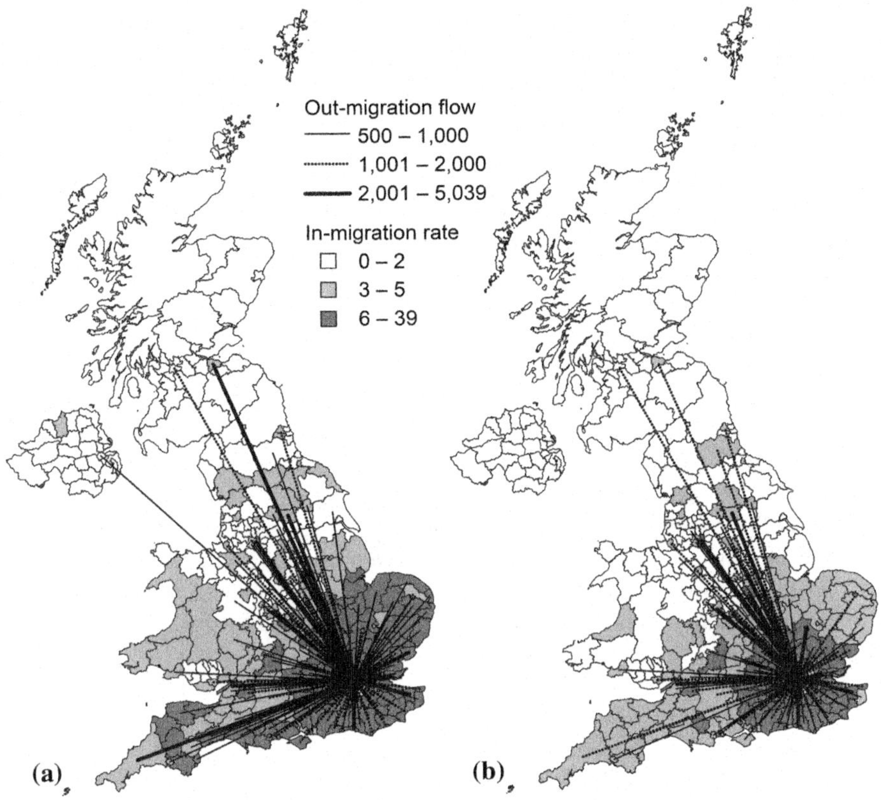

Fig. 9 Flow maps of out-migration from London in **a** 2001–02 and **b** 2010–11

automated cartography (e.g. Tobler 1987). Figure 9 contains two flow maps that illustrate changes in the pattern of flows between London Core and other districts in the UK between the beginning and the end of the period.

Both the maps in Fig. 9 shows gross out-migration flows of over 500 individuals originating from London Core and moving to other LADs in the UK, overlaid on top of choropleth maps showing the rates of in-migration from London into each of the destination LADs. The rates are computed as the number of in-migrants per 1000 resident population in each LAD. At the beginning of the period, outflow destinations are predominant in the south and east of England. By 2010–11, the pattern has changed, with far fewer migrants moving from London to the Midlands and the North of England as well as moving shorter distances to the East, South West and rest of the South East. The number of LADs that receive over 500 migrants from London fell from 155 in 2001–02 to 134 in 2010–11. This decline is driven primarily by a fall in the number of LADs that received between 500 and 1000 migrants (the black lines in Fig. 9), which fell from 69 in 2001–02 to 51 in 2010–11. While London has maintained a substantial influence as a distributor of migrants around the UK, this influence in terms of out-migration rates appears to have declined over the 2000s, and as the maps illustrate, the highest in-migration rates are more spatially confined to LADs close to London in the South East by the end of the time series.

Although flow mapping can be undertaken using proprietary GIS packages and various specialist software packages are now available,[1] the complexity of the patterns being visualised often means that the resulting flow maps are chaotic and indecipherable, particularly if the aim is to consider flows between all origins and destinations simultaneously. In this instance, we use the approach developed by Sander et al. (2014) to visualise migration flow patterns between city regions and their component parts using circular plots. Abel and Sander (2014) have introduced circular migration plots as a way of visualising their estimates of flows of migrants between countries of the world and we have used this approach to produce interactive plots for our time series of migration within the UK as illustrated in Figs. 10 and 11. The basic idea behind the plot in Fig. 10 is to show the relative size of the flows that have been estimated between city regions which are represented by the segments of the circle. Neighbouring city regions are located close to one another in the circle and the size of each migration flow is represented by the width of the links between the regions, with colour coding to identify origins and destinations. The direction of the flow is shown both by the origin colour and by the gap between link and circle segment at the destination.

Figure 10 shows the larger gross flows between each of the 13 city regions and reveals that inter-regional flows are not entirely dominated by London although London does have major links with all the other city regions in England. There are also relatively strong migration flows between those city regions that are adjacent to one another—between Sheffield and Birmingham, Manchester and Leeds, Sheffield and Leeds and Liverpool and Manchester. The numbers around the perimeter of the plot are gross flows in 10,000s and only gross flows over 7,380 are shown on the plot so as not to obscure the image with a lot of smaller links. Consequently, the city regions in Scotland and Northern Ireland appear relatively isolated apart from the link in both directions between Glasgow and Edinburgh.

The second circular plot (Fig. 11) illustrates the flows of net migration in excess of 280 individuals between each of the component parts of the city regions, where the numbers around the perimeter are net flows measured in 1000s. In this instance, the plot is dominated by the net flows from London Core to London Rest although this pattern is replicated for other provincial city regions.

[1]http://www.csiss.org/clearinghouse/FlowMapper/.
http://flowmap.geo.uu.nl/index.php.

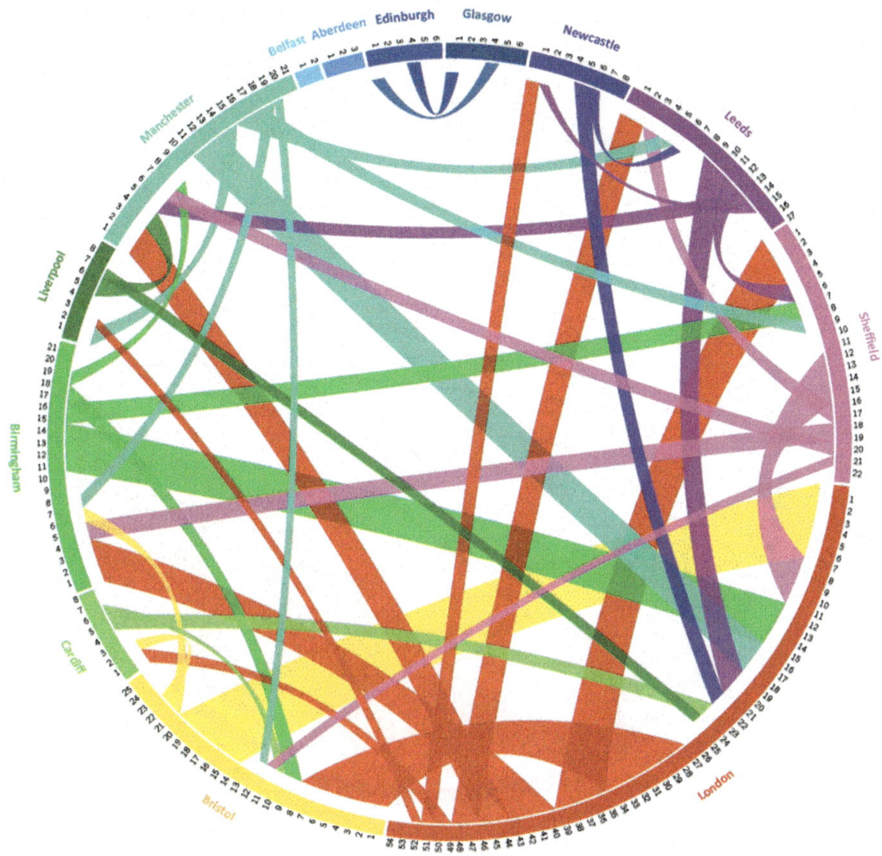

Fig. 10 Circular plot showing gross migration flows between city regions of over 7,380 migrants in 2010–11

If we compare the plots between the beginning and end of the decade,[2] although changes have taken place, there is remarkable stability of directional origin-destination migration patterns over time. The proportion of intra-versus inter-city region migration is fairly stable across the time series, hovering around 63 % of total. Collectively, it is the city region core to core moves that have increased at a faster rate than any other kind of move, while moves from core to coast and country have declined and moves to the core from all other types of area have increased.

[2]The interactive circular plots for city regions for the time series will be available at www.uk.migration.info.

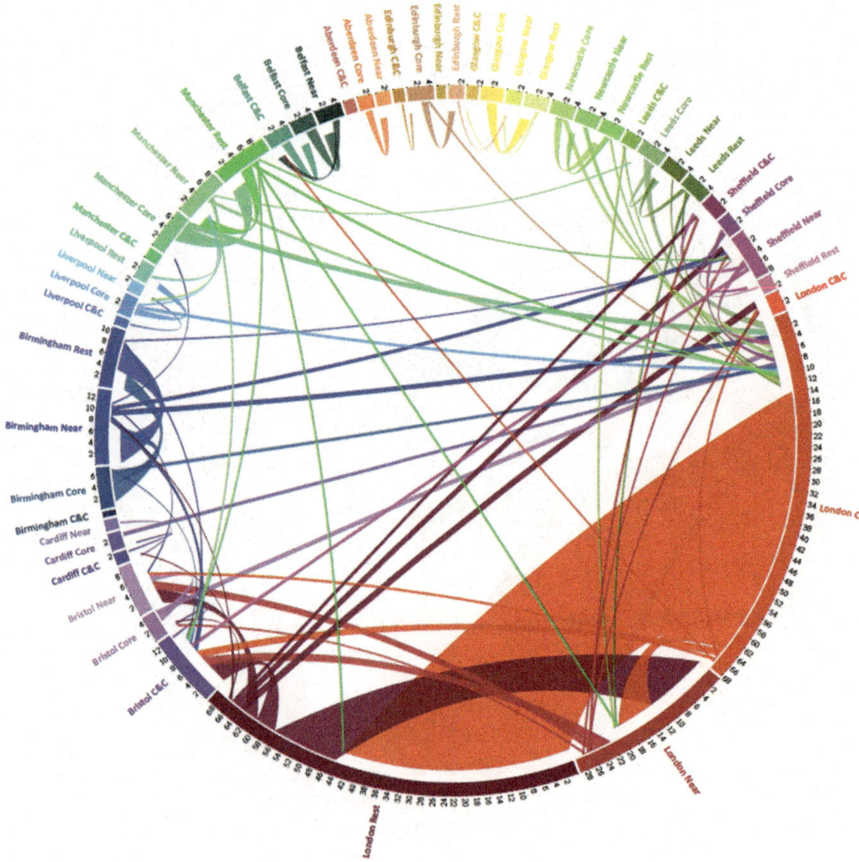

Fig. 11 Circular plot showing net migration flows between city region component areas of over 280 migrants in 2010–11

6 Conclusions

Whilst urban planners must understand the complexities of demographic change within cities, we contend that it is also important that policy makers and service providers have an equally good understanding of the relationship between cities and their migration 'hinterlands', be they within the same country or overseas. This is equally true of other interaction phenomena such as commuting, tourism or trade flows, but migration is a particularly important component of demographic change because of the relative permanence associated with changing place of usual residence and the implications for housing and the labour market as well as the provision of welfare, education and health services. We also believe that grouping like areas together is a beneficial approach for illustrating sub-national patterns and helping to understand the underlying processes.

During the 2000s, the UK as a whole has experienced population growth more rapid (7 %) than for a number of decades and much of this growth has come about in the cities. Collectively, the city region cores have grown by 10 %, due in many cases to the increased levels of immigration rather than internal migration, although the overall magnitude of the latter is far greater. Net international immigration has reached unprecedented levels and captured the attention of politicians and journalists as well as urban planners concerned with the implications for service and infrastructure provision. Concern has also been expressed over the shortcomings associated with the collection and measurement of international migration data, prompting the National Statistician in 2006 to set up an Inter-Departmental Task Force on Migration Statistics (National Statistics 2006), leading to recommendations for an ONS programme on 'Improving Migration and Population Statistics (IMPS)'. In 2008, a Parliamentary Committee reviewed complaints from local authorities about the inadequacy of official population statistics and its report (House of Commons Treasury Committee 2008) resulted in a cross-government programme to deliver the Task Force recommendations by 2012. In other words, the primary consideration given by central government to improving the measurement of migration is indicative of the importance that is attached to this dimension of urban and regional development.

The information system that we have constructed provides access to a valuable new collection of both internal and international migration estimates and natural change components over time which allows users to explore and compare trends and patterns for selected districts or their wider city regions or NUTS1 regions. At the moment, we make use of the lookup functions of Excel to perform the queries that generate the dashboard of indicators for selected areas but the creation of graphics, maps and circular plots presented as part of the subsequent analyses of the data are not automated or linked to the spreadsheet. Although this platform satisfies the basic demands for handling the data, it would be beneficial to move to a database system with a web interface for access and analysis, akin to the Web-based Interface to Census Interaction Data (WICID)[3] used by the UK Data Service-Census Support to provide users with access to migration and commuting flow data from the last five successive censuses (Stillwell and Duke-Williams 2003; Dennett et al. 2010). This transition would seem imperative because estimates have also been made of inter-district migration flows for each year for five-year age groups for males and females and a spreadsheet is less suitable for handling a much larger number of cell counts. It would also facilitate linkage with the software for creating circular plots and necessitate the adoption of web mapping tools. We envisage using a bundle of open source software (postgres, R and QGIS) with some scripts to automate the SQL queries and support a series of options that take the multi-scale origin-destination flow matrices, plus the boundary files and populations at risk, and create the charts and maps required for analysis from pick lists.

[3]http://census.ukdataservice.ac.uk/get-data/flow-data.aspx.

Aside from these technical or computational developments, the next steps in terms of data acquisition and estimation include: (i) adjustment of the time-series estimates for each of the inter-censal years so that they are consistent with the flows due to be released from the 2011 Census; (ii) updating the time-series estimates for the years following the 2011 Census so that more contemporary dynamics can be monitored and analysed; (iii) extending the time series beyond the immediate past by estimating future trends in different components or incorporating future projections generated by the national statistical agencies; and (iv) the addition of some key socio-economic time-series indicators such as employment, house prices, GDP and deprivation so that relationships between demographic and socio-economic development can be investigated.

Finally, there is the challenge of estimating the inter-censal time series of annual migration flows taking place within districts. In London this has been partly achieved since data are available on flows between boroughs but, both in London and elsewhere, planners and policy makers would find it extremely useful to know something about annual flows taking place between smaller spatial units on an annual basis. Although the current census provides flows of migrants between small sub-district areas such as wards and super output areas, annual estimates would rely on using data collected from administrative sources which, as yet, are not available on a national level across the UK. It is to be hoped that current and future work by the national statistical agencies on the use of administrative data as an alternative to the traditional census (ONS 2014b), will be fruitful in this respect, particularly if the traditional population census is replaced in due course by an alternative approach in which small area migration statistics are no longer collected.

Acknowledgments The first author is grateful for funding to support his involvement in this research from the Economic and Social Research Council under project ES/J02337X/1 (UK Data Service Census Support Service). We are grateful for the editorial comments provided by Joe Ferreira.

References

Abel, G., & Sander, N. (2014). Quantifying global international migration flows. *Science, 343* (6178), 1520–1522.

Batty, M. (2012). Smart cities, big data. *Environment and Planning B, 39*(2), 191–193.

Centre for Cities. (2014). *Smart cities*. Briefing Paper. London: Centre for Cities. http://www.centreforcities.org/wp-content/uploads/2014/08/14-05-29-Smart-Cities-briefing.pdf. Accessed February 20, 2015.

Champion, A. G. (1989). *Counterurbanization: The changing pace and nature of population deconcentration*. New York: Routledge.

Crook, A., & Linton, D. (2012, July 17). Its official: Manchester is Britain's boom city. *Manchester Evening News*. http://www.manchestereveningnews.co.uk/news/greater-manchester-news/its-official-manchester-is-britains-boom-691227. Accessed February 20, 2015.

Dennett, A., Stillwell, J., & Duke-Williams, O. (2010). Access to census interaction data. In J. Stillwell, O. Duke-Williams & A. Dennett (Eds.), *Technologies for migration and commuting analysis. Spatial interaction data applications* (pp. 31–50). Hershey: IGI Global.

Department of Business, Innovation, & Skills. (2013). *Smart cities*. Background Paper. London: BIS. https://www.gov.uk/government/uploads/system/uploads/attachment_data/file/246019/bis-13-1209-smart-cities-background-paper-digital.pdf. Accessed February 20, 2015.

Fielding, A. J. (1992). Migration and social mobility: South East England as an escalator region. *Regional Studies, 26*(1), 1–15.

House of Commons Treasury Committee. (2008). *Counting the population*. Eleventh Report of Session 2007–08, House of Commons. http://www.publications.parliament.uk/pa/cm200708/cmselect/cmtreasy/183/183.pdf. Accessed February 20, 2015.

Lomax, N. (2014). *Internal and cross-border migration in the United Kingdom: harmonising, estimating and analysing a decade of flow data*. Unpublished Ph.D. Thesis, University of Leeds, Leeds.

Lomax, N., Norman, P., Rees, P., & Stillwell, J. (2013). Sub-national migration in the United Kingdom: producing a consistent time series using a combination of available data and estimates. *Journal of Population Research, 30*(3), 265–288.

Lomax, N., Stillwell, J., Norman, P., & Rees, P. (2014). Internal migration in the United Kingdom: analysis of an estimated inter-district time series, 2001–2011. *Applied Spatial Analysis and Policy, 7*(1), 25–45.

Lyons, M. (2007). *Lyons inquiry into local government place-shaping: A shared ambition for the future of local government*. London: The Stationery Office. http://www.lyonsinquiry.org.uk/. Accessed February 20, 2015.

National Statistics. (2006). *Report of the inter-departmental task force on migration statistics*. London: National Statistics. http://collections.europarchive.org/tna/20080729015235/http://statistics.gov.uk/about/data/methodology/specific/population/future/imps/updates/downloads/TaskForceReport151206.pdf. Accessed February 20, 2015.

Office for National Statistics (ONS). (2014a). *Methodology guide for mid-2013 UK population estimates (England and Wales)*. Titchfield: ONS. http://www.ons.gov.uk/ons/guide-method/method-quality/specific/population-and-migration/pop-ests/population-estimates-for-las/index.html. Accessed February 20, 2015.

Office for National Statistics (ONS). (2014b). *Beyond 2011, final options report (O4)*, May. Titchfield: ONS. http://www.ons.gov.uk/ons/about-ons/who-ons-are/programmes-and-projects/beyond-2011/reports-and-publications/index.html. Accessed February 20, 2015.

Ravenstein, E. (1885). The laws of migration. *Journal of the Statistical Society of London, 48*(2), 167–235.

Roumpani, F., O'Brien, O., & Hudson-Smith, A. (2013). Creating, visualizing and modelling the real-time city. In *Proceedings of Hybrid City II 'Subtle rEvolutions' Conference*, May 2013. http://casa.oobrien.com/misc/presentations/roumpani2012a.pdf. Accessed February 20, 2015.

Sander, N., Abel, G., Bauer, R., & Schmidt, J. (2014). *Visualising migration flow data with circular plots*. Working Paper 02/2014. Vienna: Vienna Institute of Demography.

Stillwell, J., & Duke-Williams, O. (2003). A new web-based interface to British census of population origin-destination statistics. *Environment and Planning A, 35*(1), 113–123.

Stillwell, J., Bell, M., Blake, M., Duke-Williams, O., & Rees, P. (2000). A comparison of net migration flows and migration effectiveness in Australia and Britain: Part 1, Total migration patterns. *Journal of Population Research, 17*(1), 17–41.

Stillwell, J., Bell, M., Blake, M., Duke-Williams, O., & Rees, P. (2001). A comparison of net migration flows and migration effectiveness in Australia and Britain: Part 2, Age-related migration patterns. *Journal of Population Research, 18*(1), 19–39.

Tobler, W. (1987). Experiments in migration mapping by computer. *The American Cartographer, 14*(2), 155–163.

Unsworth, R. (2005). *City living in Leeds—2005*. Leeds: KW Linfoot Plc. and University of Leeds. http://www.leeds.gov.uk/docs/CD5-3%20KW%20Linfoot,%20Rachel%20Unsworth,%20City%20Living%20in%20Leeds%202005.pdf. Accessed February 20, 2015.

Unsworth, R., & Stillwell, J. (Eds.). (2004). *Twenty-first century Leeds: Geographies of a regional city*. Leeds: Leeds University Press.

Chapter 24
Urban Data and Building Energy Modeling: A GIS-Based Urban Building Energy Modeling System Using the Urban-EPC Engine

Steven Jige Quan, Qi Li, Godfried Augenbroe, Jason Brown and Perry Pei-Ju Yang

Abstract There is a lack of building energy modeling in current planning support systems (PSS) while building energy efficiency is getting greater attention. This is due to the current limitations of energy modeling at the urban scale and the inconsistency between the available urban data and that required for modeling. The chapter seeks to fill this gap by developing a GIS-based urban building energy modeling system, using the Urban-EPC simulation engine, a modified Energy Performance Calculator engine. This modeling system is compatible with other planning tools, enhanced by the combination of physical and statistical modeling, and adjustable in its resolution, speed and accuracy. Through processing the Data Preparation, Pre-Simulation, Main Simulation and Visualization and Analysis models in this energy modeling system, the urban data related to the basic building information, mutual shading, microclimate and occupant behavior are collected, modified, and synthesized in the GIS platform and then used as the input of the Urban-EPC engine to get energy use of every building in a city, which could be further visualized and analyzed. The method is applied in Manhattan to show its potential as an important component in PSS to inform urban energy policy making.

S.J. Quan · Q. Li · G. Augenbroe · J. Brown · P.P.-J. Yang (✉)
College of Architecture, Georgia Institute of Technology, Atlanta, GA 30308, USA
e-mail: perry.yang@coa.gatech.edu

S.J. Quan
e-mail: stevenchuan6656@gmail.com

Q. Li
e-mail: liqithu@gmail.com

G. Augenbroe
e-mail: fried.augenbroe@coa.gatech.edu

J. Brown
e-mail: jason.brown@coa.gatech.edu

S. Geertman et al. (eds.), *Planning Support Systems and Smart Cities*,
Lecture Notes in Geoinformation and Cartography,
DOI 10.1007/978-3-319-18368-8_24

1 Introduction

The energy used in buildings is a large share of overall energy use, e.g. 20–40 % in developed countries, and the potential for reduction is drawing the attention of planners and policy makers (Perez-Lombard et al. 2008). Although it was stated from a long time ago that energy use is an important aspect in planning support systems (PSS) for the sustainable urban development (Harris and Batty 1993; Mohammadi et al. 2013; Snyder 2003), the investigation of building energy use is still rare in the PSS field. Only a few researchers have incorporated a "Performance/Evaluation Model" that measures building energy use in PSS and Geodesign and used the results as the criteria to evaluate different planning scenarios (Quan et al. 2013; Yeo et al. 2013).

The discrepancy between the lack of building energy assessments in PSS and Geodesign models and the increasing need to assess the energy performance of different urban forms is due to two reasons: modeling limitations and data inconsistencies. First, there are significant modeling limitations in measuring large-scale building energy use. Traditional engineering-based building energy modeling tools, including the US standard program DOE-2 and its successor EnergyPlus (Crawley et al. 2001), IES-VE (Virtual Environment by Integrated Environmental Solution, a commercial building simulation tool) (Integrated Environmental Solutions Limited 2012), analyze only an individual building as a single system for simplification (Al-Homoud 2001). There are four major groups trying to scale this up to the city level, but none have provided a sufficient solution to account for the influence of urban contexts at different spatial scales. The first group scales energy assessment from single buildings to urban areas directly by using simple building stock approaches. However despite the discussion of spatial variations in the "*second order uncertainty*" (Booth et al. 2012), little concern was placed on the influence of building locations and their urban contexts. The second group is considerably aware of the urban context in their modeling (Pisello et al. 2012; Wong et al. 2011). However their approaches tend to be confined by specific urban settings and are hard to apply to other places. The third group has developed fully fledged energy modeling methods for the urban environment, including CitySim and UMI (Reinhart et al. 2013; Robinson et al. 2009). But as stand-alone software, they require tedious data transferal and rebuilding from ArcGIS data as widely used in urban studies. The last group led by Steemers and Ratti has developed the LT model to measure building energy based on raster data in GIS (Ratti and Richens 2004). But their specific assumptions of occupant behavior and the resolutions of the raster limit its use at the city scale.

Besides the unawareness of urban context, the engineering-based methods in the four groups require basic knowledge of building physics, HVAC (heating, ventilation and air conditioning) systems, etc., which are unfamiliar to most planners. This calls for a simplified, cross-scale, context-sensitive and GIS-based modeling method to measure urban building energy use in PSS.

Second, data inconsistency is another obstacle in developing and applying the building energy modeling in PSS. On the one hand, there are plenty of urban data available such as building intensity and population. On the other hand, building energy modeling methods require very detailed building component data, such as building shapes, materials, fenestrations, occupant schedules and HVAC systems, which are lacking at the urban scale (Al-Homoud 2001; Flaxman 2010). Due to such discrepancies between the available urban data that could provide general building information and the missing specific building-level data for detailed building information required by the building energy modeling, researchers and planners have only been able to estimate building energy use in small areas using surveys of detailed building data (Reinhart et al. 2013; Robinson et al. 2009), or to roughly estimate the building energy performance in large areas with simplified assumptions (Quan et al. 2013). Developing urban building energy modeling requires finding a way to use related information in urban data to get the detailed information as building scale data, which connects data at urban and building scale.

In order to fill those research gaps, this chapter aims to develop a GIS-based urban building energy modeling system using what we call the Urban-EPC engine, a modified version of a reduced-order building energy model called the Energy Performance Coefficient Calculator, or EPC for short. The EPC is an implementation of the ISO 13790:2008 standard, which lays out a calculation recipe for normatively estimating building energy performance using basic physics-based equations involving a comparatively small set of parameters and normative statements about the assumed usage scenario, system efficiency, etc. per functional type of building (ISO 2008). The underlying model of the EPC is much smaller than tools such as EnergyPlus, resulting in not only faster computational speed, but also an input parameter set that is much smaller and simultaneously aggregated to a level more commensurate with urban GIS data. Through its simplicity and unified modeling assumptions, this approach forms the basis for assessing building energy performance in a standardized and transparent way (Hogeling and Van Dijk 2008). Because of this, the EPC is well-suited for rating the energy performance of both new and existing buildings. In addition, the normative assumptions were calibrated on a large collection of different buildings, making the calculator well suited for the macro-level calculations as reported in this chapter, e.g. where the information about individual buildings is thin.

The EPC recipe is based on the hourly heat balance of the whole building using inputs such as wall and window areas, shading coefficients, material properties, net functional floor area, lighting density, internal heat production from appliances, plug loads, temperature set points and occupancy schedules. The calculation goes through three steps. Based on hourly calculations in the local weather conditions, the heating and cooling loads are calculated. This thermal demand is then translated into the delivered energy (electric and gas) used by the building systems. The translation is driven by macro system efficiency factors, normatively defined per system type. Finally, with the addition of data on other electricity usage in the building, the total consumption can be calculated and translated into primary energy units, i.e. the summation of the primary energy (gas, coal, oil) that is consumed by the generation sites. Comparative analyses have shown that the calculator is accurate for the purpose described in this chapter (Kim et al. 2013; Lee et al. 2013).

This energy modeling method has five advantages over previous methods:

Urban context sensitive: the modeling takes the influence of urban context into account and is able to estimate building energy performance in different urban environments.
Urban data driven: it utilizes abundant urban data to inform building energy modeling, providing building details and urban contexts, using DOE (Department of Energy) reference buildings (Deru et al. 2011) as the complement.
GIS based: it is based on the ArcGIS platform, widely used software in PSS, and therefore it is relatively easy for planners to run the modeling and visualize the results.
Planning oriented: as a geo-based modeling method, it allows planners to easily map the simulated energy use and overlay them with other planning mappings for further analysis.
Resolution controlled: the temporal resolution of the modeling could be changed to provide hourly, daily, weekly, monthly or annual building energy use results. Similarly, the accuracy resolution can also be changed to the high, medium and acceptable levels. It allows users to adjust the trade-offs between accuracy and speed with purposes of analysis.

This modeling system requires ArcGIS 10.x and Microsoft Excel to be installed on the PC being used.

2 Methodology

The methodology of the modeling system incorporates three aspects: the influence of the urban context on building energy use; the role of urban data in building energy simulation; and the integration of data processing and energy simulation as one modeling system. Urban-EPC enhances EPC to account for these first two aspects; a larger software architecture then coordinates the Urban-EPC to realize the third.

2.1 The Influence of Urban Context on Building Energy Use

It is well discussed that building design, system efficiency and occupant behavior have considerable impacts on building energy consumption (Al-Homoud 2001). Besides the three factors, some scholars are aware of the importance of urban context in building energy (Golany 1996; Mitchell 2005; Ratti et al. 2005; Steadman 1979). However, few comprehensive studies have been conducted to date and the ways in which urban context influences building energy use are still unclear. Such influences can be explored using a systems perspective. Although a single building is already a complex system, in a larger urban system it is seen as only one component. In such a "system of systems" or "network of networks" (Ackoff 1971; Batty 2013; Maier 1998), the interactions among different components can significantly affect the individual performance and the overall system performance. The influence of urban context on building energy captures such system interactions including the following three types:

Interactions between a building and other buildings and obstructions: As solar radiation on the building facades influences building energy use significantly, the obstruction of sunlight by surrounding buildings, trees and other obstructions plays an important role in building energy use. Such interactions among geometries are known as external shading effects or mutual shading effects (Littlefair 1998; McPherson and Simpson 2003; Ok 1992; Quan et al. 2014; Ratti et al. 2005; Rode et al. 2013; Yezioro and Shaviv 1994; Yi and Malkawi 2009). The effects generally increase building energy use during winter and reduce it during summer because of less solar gain.

Interactions between a building and the microclimates around it: Aspects of the local climate—including air temperature and wind patterns—can be modified by urban form-related factors, e.g. reduced radiative heat loss and turbulent heat transfer in urban canyons, increased thermal storage within buildings and impervious surfaces, anthropogenic heat release in the urban context, etc. (Eliasson 2000; Hassid et al. 2000; Oke et al. 1991; Steemers 2003; Wong et al. 2011). Often known as the 'urban heat island effect', modified urban microclimates can reduce building energy in winter and increase it in summer (Kolokotroni et al. 2006; Santamouris et al. 2001).

Interactions between buildings and occupants: The occupancy pattern, including the density and behavior, could lead to variations of building energy use in the same building. Although the influence of occupant behavior is still unclear (Branco et al. 2004; Guerra Santin et al. 2009), the impact of occupant density is straightforward.

The influences of these interactions are measured in this modeling system as mutual shadings, zonal microclimates and occupant densities by different tools and engines to inform the building energy simulation. To assess those influences, urban scale data are needed.

2.2 The Role of Urban Data in Building Energy Simulation

There is an ever-increasing supply of urban data. Enormous amounts of information are produced and collected through traditional commercial and administrative censuses and surveys, and more recently from mobile and social media data such as real-time geo-labeled tweet data, traffic data, etc. (Döllner and Hagedorn 2007; Reades et al. 2007). Urban data are of diverse types (e.g. population, economics, transportation), available at various scales (e.g. census tracts, neighborhoods, cities), in different formats (spatial and aspatial data), all collected for different points in time or time periods. Such rich resources could greatly inform urban building energy modeling after a careful selection of what is required directly and indirectly by the simulation engine.

The core simulation engine of the Urban-EPC consists of the EPC, reference building models from the US Department of Energy (DOE), a shading engine, a microclimate engine, and an occupancy engine. In the EPC, general building component information such as floor areas and room volumes are taken from building footprint data, parcel data and land use data in the urban dataset. However, some detailed building data such as materials, heating, ventilation and air conditioning (HVAC) systems, and window-wall ratios, are also required by the EPC which cannot be found in the urban data. A set of commercial reference building models, developed by the US Department of Energy (DOE) and representative of the national building stock, are used to provide missing building model inputs. These contain three categories of building vintage (based on the construction year), each of which includes 16 building types representing most of the commercial buildings across 16 US climate zones. Model inputs, including geometry, envelope, material properties, building usage and operational schedules were developed from several building databases such as F.W. Dodge building stock and forecast data (Dodge Data and Analytics 2005), engineering studies, design standards and guidelines such as ASHRAE (1989, 2004), and statistics such as the Commercial Building Energy Consumption Survey (CBECS) (U.S. Energy Information Administration 2005). Detailed building information can be determined by linking reference building types with the information of the building function and construction year provided by the urban data. Using the reference building, the Urban-EPC model manages to get the detailed information of each building based on its related urban data, and thus reduces data inconsistency between available urban data and required building data.

In the Urban-EPC core engine, three sub-engines capture the influence of urban context on building energy. The shading engine uses the building footprint, tree canopy, topography and parcel data to calculate external/mutual shading effects on the windows. The microclimate engine takes weather information, building footprint, land cover, vegetation, street, and block data as input to estimate local air temperature and wind patterns. The occupancy engine utilizes population and job data to generate occupant density and use schedules of residential, commercial and public buildings.

This Urban-EPC model requires the parameters shown in Table 1 as inputs, all of which could be found in the urban data complemented by the reference building database.

However, the availability of related urban data varies from city to city. In some cities, open-source urban data may be greatly limited.

Table 1 Input requirements of the Urban-EPC model and related urban data

Model components	Required input parameters	Data source	Related urban data	Specific urban data sources
Detailed building information	Building shapes (e.g. total floor areas, volumes, façade areas, rooftop areas)	Urban data	Building footprint data, parcel data, topography data	City Department of Planning
	Window to wall ratios	Reference building and urban data	Building footprint data, parcel data, land use data	City Department of Planning
	Building materials	Reference building and urban data	Building footprint data, parcel data, land use data	City Department of Planning
	HVAC system	Reference building and urban data	Building footprint data, parcel data, land use data	City Department of Planning
Shading engine	Building geometries	Urban data	Building footprint data, parcel data, topography data	City Department of Planning
	Other obstruction geometries	Urban data	Tree canopy data, topography data	City Department of Planning, City GIS Portal
Microclimate engine	Urban canyon parameters	Urban data	Urban block shape data, building footprint data, parcel data, street network data	City Department of Planning
	Percentage of pervious surfaces and building rooftops	Urban data	Land cover data, building footprint data, vegetation data, tree canopy data	City Department of Planning, City GIS Portal
	Weather data	Imbedded in EPC and urban data	Weather station data	NOAA (National Oceanic and Atmospheric Administration)
Occupant behavior modification	Occupant density	Reference building and urban data	Population distribution data, job distribution data	U.S. Census Bureau
	Occupant behavior	Reference building and urban data	Detailed population distribution data, detailed job distribution data	U.S. Census Bureau

2.3 The Integration of Data Processing and Energy Simulation as One Modeling System

Based on an understanding of the influence of the urban context, the core Urban-EPC engine and its related urban data, a GIS-based urban building energy modeling system is developed.

2.3.1 Structure of the Modeling System

The modeling system developed in this chapter contains four major models: the Data Preparation Model, the Pre-Simulation Model, the Main Simulation Model and the Visualization and Analysis Model. This modeling system uses urban data from various sources as its input, integrates and refines them into a new set of data required by the pre-simulation engines, provides the resulting data to the main simulation model, and visualizes and analyzes the final results with other urban data. The modeling system structure and its data flow are shown in Fig. 1.

The whole system runs on the GIS platform except for the main simulation engine, which runs in Excel but is organized and linked to GIS by ArcPy, "*a site package that builds on the successful arcgisscripting module*" (Esri 2012). The whole modeling system is developed in two GIS tool forms based on a trade-off between usability and speed: the ModelBuilder toolboxes and the ArcPy codes. The ModelBuilder approach is easy to use for planners, but is also relatively slow. The ArcPy codes run much faster but require basic knowledge of ArcPy. Generally the ModelBuilder approach is more useful for energy simulation of small areas while the ArcPy approach is the better choice for large urban areas.

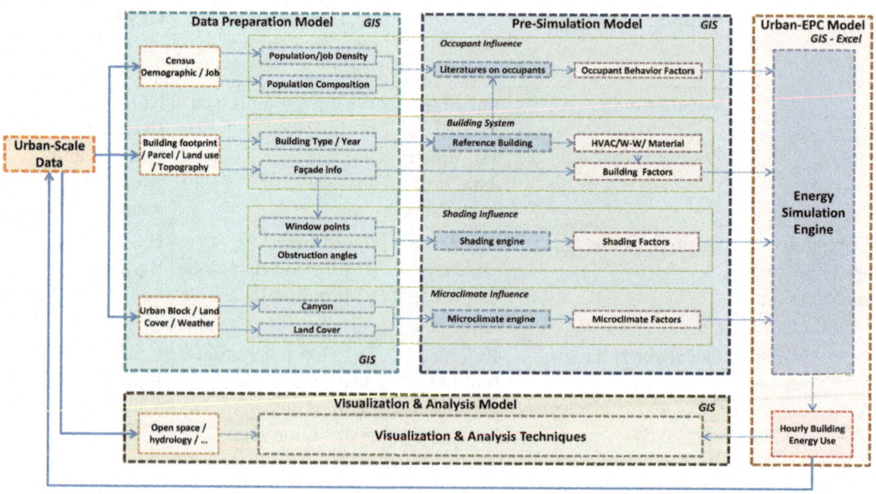

Fig. 1 Structure and the data flow of the urban building energy system

2.3.2 Data Preparation Model

The Data Preparation Model organizes, modifies and integrates the urban data from different sources into a new dataset to inform pre-simulation engines. Within this model, there are four streams of data flow. The first data stream starts with the building, parcel, land use, and topography data, sorting and transforming them into inputs, encoding the building function, construction year, building shape and detailed façade dimensions in eight orientation categories (as required by the EPC). From the façade information, the second data stream provides sample window point matrices and calculates their obstruction angles for the shading engine; this engine measures where solar direction radiation is blocked due to surrounding obstructions within a certain distance, as shown in Fig. 2. The obstruction angle then equals arctan(H/D). The third data stream extracts population density, job density and population composition information such as age, gender and education information from the demographic and job data. The fourth data stream uses urban block, street, land cover, and weather data to calculate canyon height, canyon ratio (the canyon height/the canyon width) and impervious surface ratio, which are important input parameters in the microclimate engine.

2.3.3 Pre-simulation Model and Main Simulation Model

The pre-simulation model includes the reference building dataset and the three urban context engines and provides important inputs to the Urban-EPC engine. The reference buildings are used to determine more detailed building information based on building functions and construction years. The shading simulation engine takes the obstruction angles of each window from all possible directions, and compares them with the zenith angle of the sun every hour throughout a year to find whether a window is shaded and then calculates the shading factor as the percentage of shaded windows on each façade. The microclimate simulation engine parameterizes urban characteristics from the urban block, street and land cover data into four urban parameters in each microclimate zone: canyon height, canyon ratio, pervious road

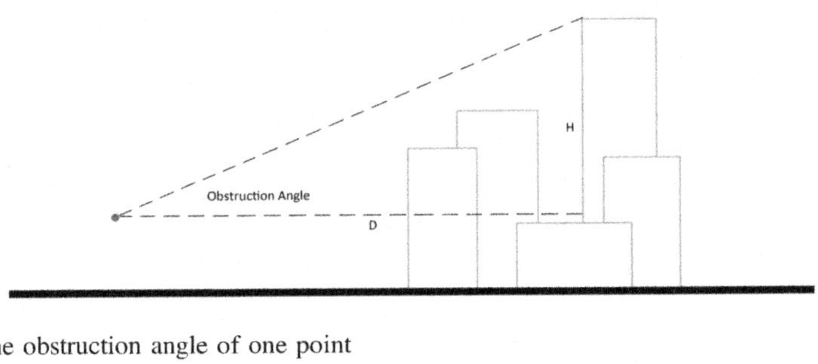

Fig. 2 The obstruction angle of one point

fraction and building roof fraction. It relates microclimate changes to these urban parameters through explicit statistical formulations such that buildings with similar urban parameter values share a similar microclimate, i.e. hourly ambient temperature and wind data. Then, by dividing all of the buildings into several climate zones and assuming that buildings in the same zone share a set of typical urban parameter values, the number of microclimate simulations can be reduced to only the number of zones. This dividing process is performed using a K-means clustering algorithm (MacQueen 1967), which selects 50 climate zones and the typical values for each zone such that the average difference between values of a single building and the typical values of its microclimate zone is minimized. The modification of occupant density is based on the population density and job density from urban data while the occupant behavior change is referred to the related literature.

Taking as input the detailed building information from the reference building approach, the shading factors of each building from the shading engine, the ambient temperature and wind data of each microclimate zone from the microclimate engine, the modified occupant behavior data from the occupancy engine, as well as the general building information extracted directly from the urban data, the core engine generates the hourly energy consumption of each building throughout a year in the Main Simulation Model. The results are then aggregated into monthly and annual building energy use data.

2.3.4 Visualization and Analysis Model

In the Visualization and Analysis Model, simulated building energy use is mapped to building GIS data for visualization. The simulated energy data are generated in this modeling system from the urban data and now they become a new part of the urban data in GIS format. They can be easily overlaid with other urban data for further analysis such as the density-energy relations. Also, since the models and engines are loosely coupled in the whole system, new engines and modules could be added to this system to extend its analysis capacity.

3 Case Study: Energy Performance of Buildings in Manhattan

The Manhattan borough in New York City is taken as a test case to demonstrate how the proposed urban building energy modeling system works in an actual urban area. The simulation results are then compared with measured data from 2012.

3.1 Case Area

According to the borough boundary and building footprint data of Manhattan (NYC Department of City Planning 2014; NYC Department of Information Technology and Telecommunications 2014), there were 45,920 buildings with the total floor area of 43,743,004 sq. ft. in 2013.

3.2 Data Preparation

In the Data Preparation Model, related urban data were organized to provide the inputs for the simulation. Although data production dates range from 2010 to 2013, this study makes the assumption that urban changes during these four years are minimal and that all the data represent Manhattan in 2012.

3.2.1 General Building Information

The building footprint data only contain the buildings' geometric information. In order to get other information such as building types, built years, etc., the parcel-level PLUTO (Primary Land Use Tax Lot Output) data was joined to the building footprints (NYC Department of City Planning 2014), as shown in Fig. 3a. However, the geometric data in both dataset does not quite match and so some corrections were made to estimate the building heights and number of stories, with references to Google Earth 3D buildings.

Based on the orientations, the facades were categorized into eight groups, as shown in Fig. 3b. Because the Manhattan grid has its particular orientation of 29°, the eight default orientations in the EPC calculator were modified in this case study by adding 29 degrees to each, which became 29°, 74°, 119°, 164°, 209°, 254°, 299° and 344°. The areas of those categorized facades, as well as the building floor areas and volumes were calculated as the input of the energy simulation model.

3.2.2 Mutual Shading Data

In this case study, only buildings are considered as obstructions in the shading effect. Since the GIS building footprint data already contains elevation data, the facades can be readily located without additional topography data. Then, point matrices were generated on facades wider than 5 ft. and higher than one storey to represent samples of windows, assuming that windows are evenly distributed at each floor on the facades. The vertical spacing of the point matrices is the storey

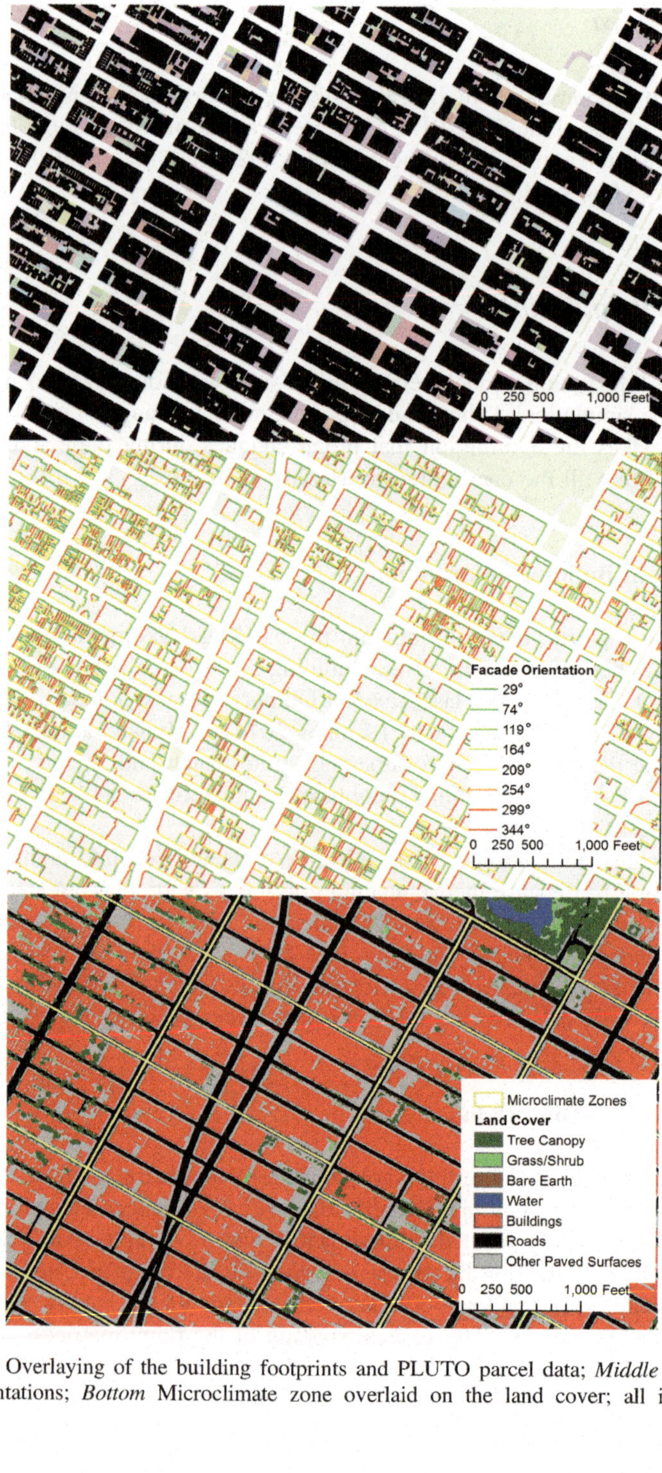

Fig. 3 *Top* Overlaying of the building footprints and PLUTO parcel data; *Middle* Categorized facade orientations; *Bottom* Microclimate zone overlaid on the land cover; all in Midtown, Manhattan

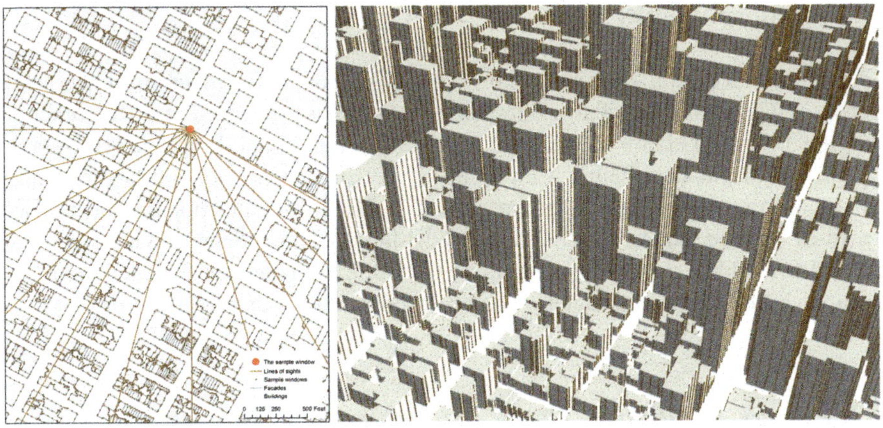

Fig. 4 Point matrices on the façades and the lines of sight from a sample point

height, while the horizontal spacing is set at 40 ft. for simplicity. Overall there are 397,404 points generated on the facades of 45,920 buildings in Manhattan.

From each point representing a sample window position, lines of sight were generated in GIS with the angle interval of 15°, which is the average angle that the sun's position changes by every hour throughout a year in New York City (NYC) (SunEarthTools.com 2014). The maximum obstruction angle along each line was calculated by intersecting the buildings with the line of sight. The length of the line of sights was set to be 3281 ft. (1 km) to intersect buildings lower than 500 and 6562 ft. (2 km) to intersect buildings taller than 500 ft. Considering the solar path throughout 2012 in Manhattan, the possible angles of the lines of sight are limited. Therefore a 397,404 × 17 matrix was generated with the rows as the points and the columns as the maximum obstruction angles along lines of sight. The point matrices and the lines of sight are shown in Fig. 4.

3.2.3 Microclimate Data

In this study, census tracts were chosen as the spatial units to divide Manhattan into 288 parts and further aggregated to 257 microclimate zones to match the spatial extents of the PLUTO data. In each part, average street width and average building heights were calculated to get the urban canyon widths and heights based on the building footprints, PLUTO, and street data. Pervious road fractions and building roof fractions were measured by extracting information from the land cover data overlaid by census tracts in Manhattan, as shown in Fig. 3c.

3.2.4 Occupant Density Data

To estimate occupant densities, the block level population data from the 2010 Census TIGER (Topologically Integrated Geographic Encoding and Referencing) and the block level job data from LODES (Longitudinal Employer-Household Dynamics Origin-Destination Employment Statistics) were joined to the census blocks and aggregated to the microclimate zones to get the population and job densities (United States Census Bureau 2014a, b). Those density data were then applied to buildings within the zones as the occupant densities for residential and commercial buildings. Occupant behavior modification was not tested in this case.

3.3 Energy Simulation Using the Urban-EPC Engine

The Pre-Simulation Model uses the shading, microclimate, and occupancy engines to determine hourly shading factors for building facades, hourly ambient temperatures for each microclimate zone, and occupant density data. Detailed building information were obtained from the reference building dataset for NYC. These data, together with the building information, were used as the input of the simulation to estimate the hourly total energy use, electricity use and gas use of each building in Manhattan throughout the year of 2012.

3.4 Visualization and Analysis

The resulting estimates of building energy use were joined back to the building data and visualized in ArcGIS to show the distribution of annual building energy use in Manhattan in 2012, as in Fig. 5. It is clear that the buildings consuming the most energy are located in downtown and midtown. However, the mapping of building energy use intensity tells another story. The comparison of building energy use and its intensity shown in Fig. 6 suggests that although the skyscrapers in the downtown and midtown areas consume the most energy, their energy intensities measured by energy use per floor areas are moderate compared to the mid-rise and low-rise buildings on the island. How building energy efficiency varies with building form and density could be further analyzed based on the results of this energy modeling system.

There is also a temporal dimension in the output of the modeling system. The hourly result data can show the fluctuation of building energy over 24 h in a typical day in Manhattan, or can be aggregated to show the monthly variation of average building energy use intensity to better understand the dynamics of the building energy use, as in Fig. 7.

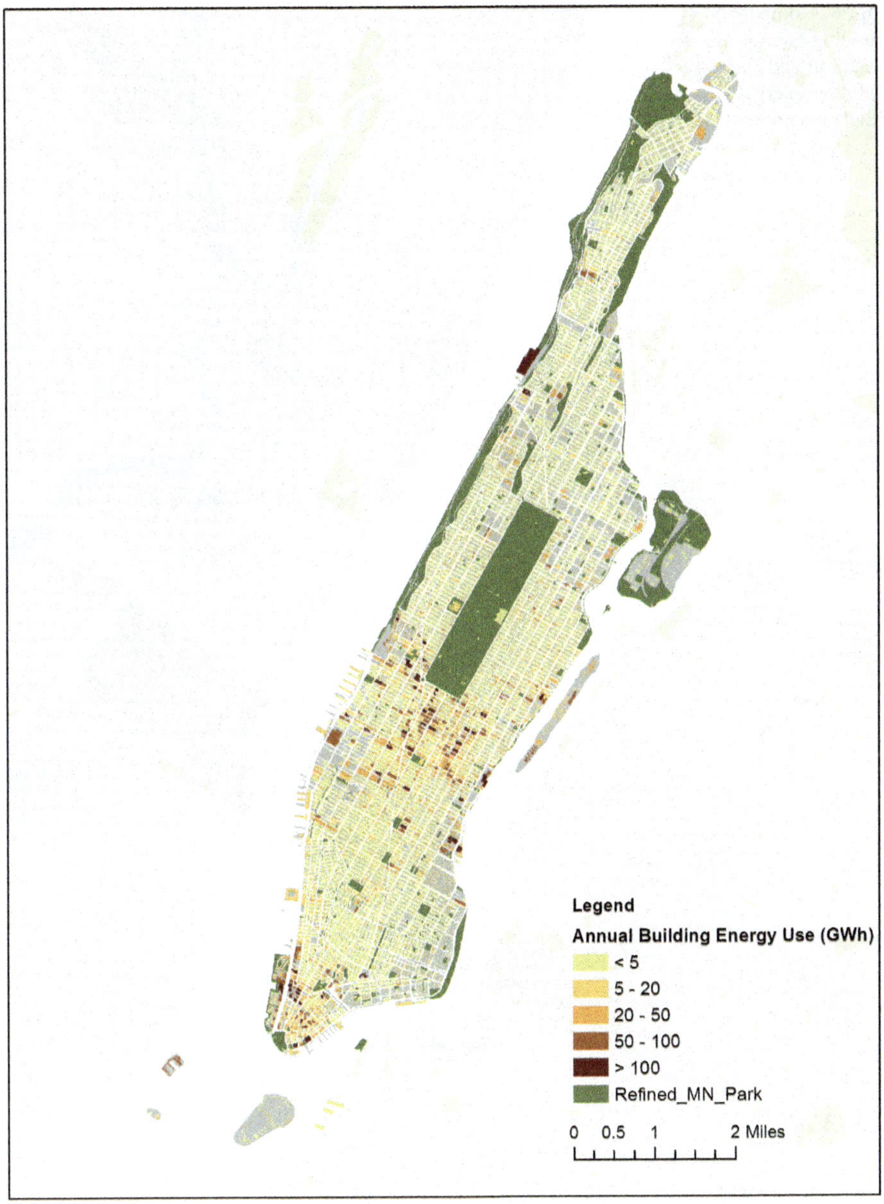

Fig. 5 Building energy use mapping in Manhattan in 2012 (kWh)

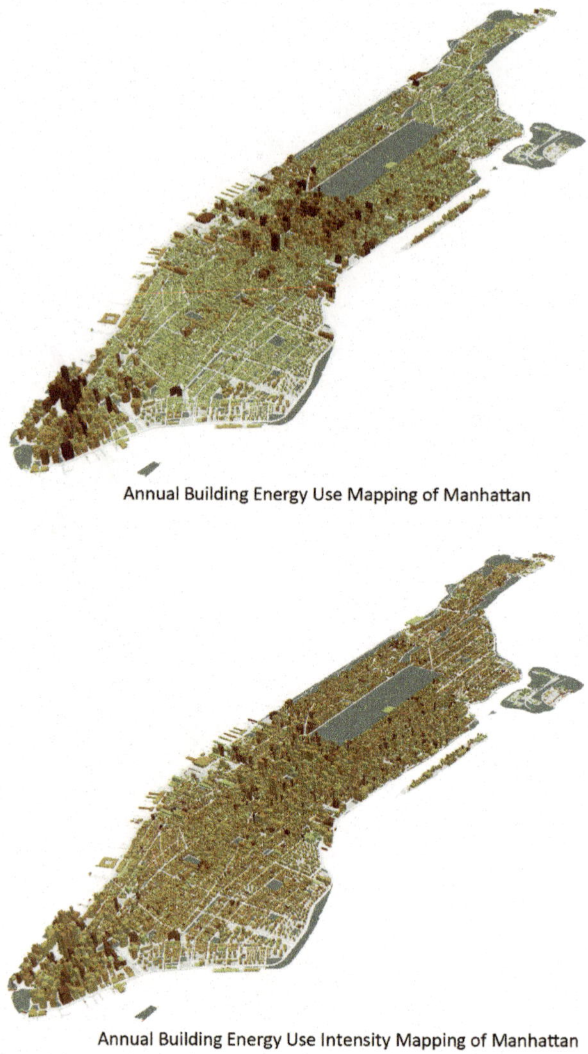

Fig. 6 Comparison of (*top*) building energy use (EU) mapping and (*bottom*) building energy use intensity (EUI) mapping of Manhattan

3.5 Validation

The reliability and accuracy of the urban building energy modeling system is critical when applied to support planning practice and policy making. However, so far, few building energy modeling studies at the urban level has been rigorously validated based on a large dataset of measure data. In this case study, the validation used a building energy use dataset in 2012 provided under the Local Law 84 (LL84) published by The New York City Department of Buildings (DOB), which requires "*annual benchmarking data to be submitted by owners of buildings with*

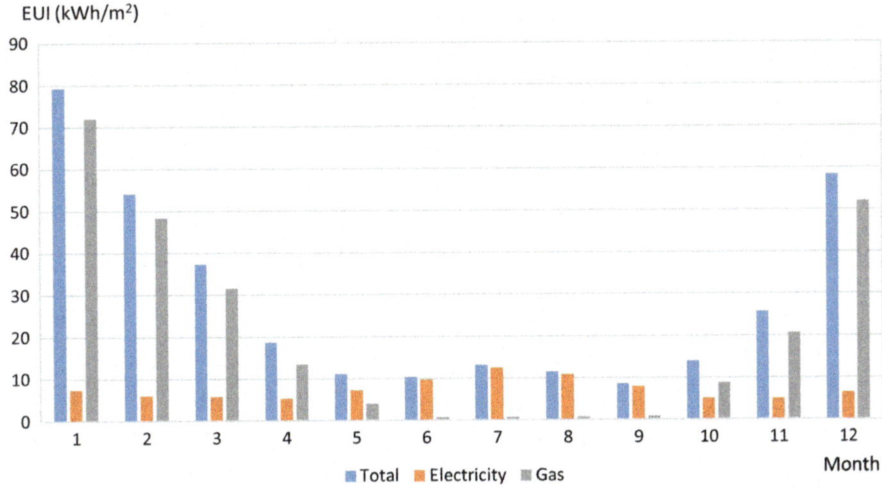

Fig. 7 Average monthly building energy use intensity (EUI) of Manhattan

more than 50,000 sq. ft. for public disclosure" (The City of New York 2014). The 2012 LL84 dataset contains the annual energy use data of 1680 buildings in Manhattan, and after cleaning up missing and outlier data, 1118 buildings were left as the dataset for validation.

The validation shows that in 80.1 % of the buildings, the estimated energy use is within the range of 0.5–2 times of the measured energy use, suggesting an overall good fit (Fig. 8). The NMBE (net mean bias error) is 0.28, which suggests that the total estimation of all EUI is larger than the reported by around 28 %. The CVRMSE (coefficient of variance root mean square error) is 0.69 at the same time, indicating that for estimation of a single building the average error is around 69 %. Comparing the two indices to the ASHRAE (American Society of Heating, Refrigerating Air-Conditioning Engineers) standard of 0.05 for NMBE and 0.15 for CVRMSE for monthly energy consumption of a single building, the accuracy is sufficient for the urban level building energy simulation given so many uncertainties in the assumed data and modeling parameters.

To understand to what extent the urban context engines improve the modeling system, results from five modeling method scenarios were compared, including modeling with no urban context engines, modeling with the shading engine, modeling with the microclimate engine, modeling with the occupancy engine and modeling with all three engines (i.e. the full urban context). The results indicate that the urban context engines improve the modeling considerably, and that there is the trade-off between the influences of shadings which tends to increase heating loads which are the major loads in NYC and other factors which are likely to reduce heating loads, as shown in Table 2 and Fig. 9.

The total time used for the above simulation of the hourly energy use of 45,920 buildings in Manhattan was 80 h using a desktop computer with an Intel i7 CPU

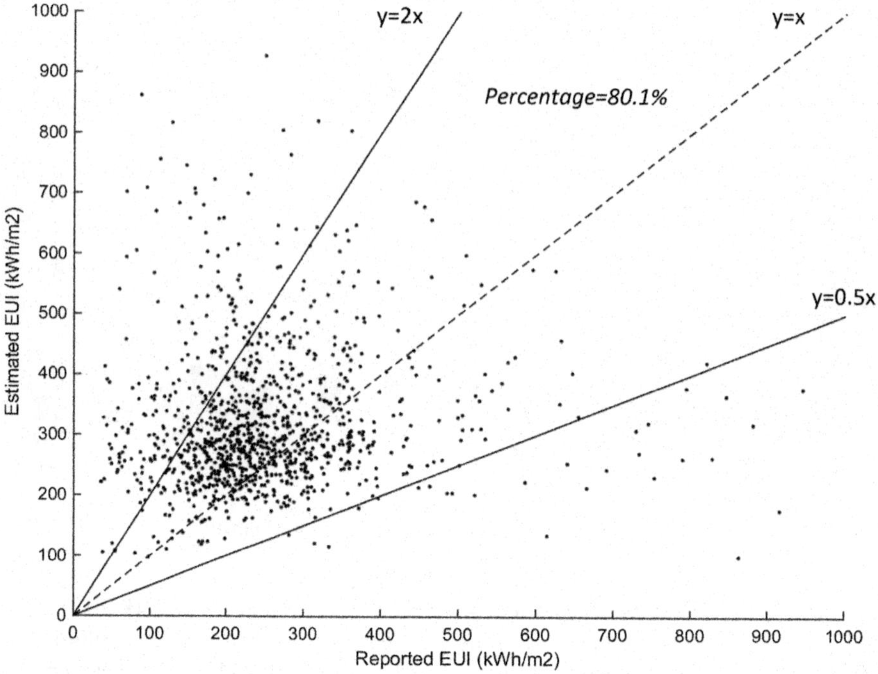

Fig. 8 Scatter point chart of the estimated and reported EUI (*upper line* estimated data = 2* reported data; *lower line* estimated data = 0.5* reported data)

Table 2 Validation results of the urban-EPC modeling and the traditional EPC modeling methods

Modeling method	NMBE	CVRMSE
EPC with urban context (all 3 engines)	0.28	0.69
EPC with shading	0.52	0.85
EPC with microclimate	0.33	0.70
EPC with occupant behavior	0.43	0.81
EPC	0.50	0.83

and 32G RAM, which is quite good for such large and dense urban area. The use of more computers with lower modeling resolutions can reduce the simulation time to one day or less.

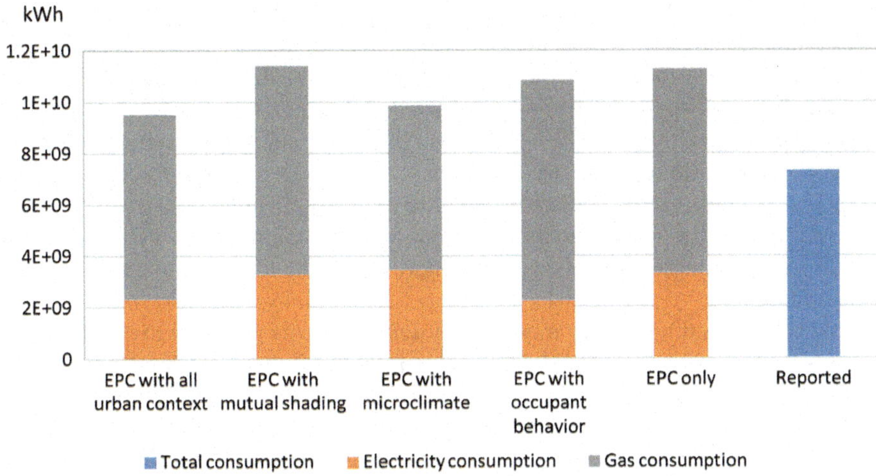

Fig. 9 Comparison between different modeling scenarios

4 Conclusions

There is a lack of an urban-context-aware building energy modeling method in PSS despite the fact that the issue of building energy efficiency is receiving greater attention. This is due to the inability of building energy modeling to account for the urban context and the inconsistency between the available urban data and the required building data in traditional building energy modeling tools. Although planners have access to abundant urban data, applications using these data for energy performance assessments of different urban design scenarios rarely occur.

This chapter has tried to explain a methodology that aims to fill this gap. The GIS-based urban building energy modeling system that has been outlined can be applied to other planning studies, enhanced by the combination of the building physical modeling and statistical dataset, and adjustable in its resolution, speed and accuracy. The modeling system as a process of using urban data to inform urban building energy use was demonstrated using the case study of Manhattan. The results show an acceptable level of accuracy for modeling such a large and dense urban area based on a relatively simple method.

The modeling method also reveals some problems with data management and tool platforms. On the data side, data inconsistency, low-level detail, and missing information are common in urban data for many cities, especially for the information required by building energy modeling. For example, building data and PLUTO data have huge inconsistencies in building heights. Better management is required to collect, examine, organize and share urban data in each city. In some cities where open-source urban data are limited, planners need to use the BAD (Best Available Data) to inform energy assessment and support policy making, as long as reasonable assumptions are made or appropriate substitute data are chosen (Klosterman 2008).

On the tool platform side, although GIS is a powerful analytical platform, it becomes less capable when dealing with detailed data at a large scale. Data volume increases considerably when the spatial level of urban studies goes down from cities and districts to buildings, facades and even windows. As a consequence, GIS tools become slow and often show errors because of memory limitations. ArcPy codes run faster than the tools but are still much slower than the previous VBA language. Therefore, GIS computation needs to be improved for handling big data in building energy modeling.

This urban building energy modeling system shows its potential to contribute to PSS. Its inclusion in PSS could help planners better understand how urban form performs in terms of building energy use. It can also evaluate the environmental, economic and social impact of large-scale energy-related renovation proposals, e.g. implementation of white roofs or low-e glass, so as to support policy making at the urban level. More importantly, it provides estimates of the spatial distribution of the building energy use in a city, which allows planners and policy makers to adjust energy supply to optimize the whole energy system. When applied to design proposals, this modeling system could assist designers to reconfigure the land use patterns and building layouts for better building energy performance.

The development of this urban building energy modeling system exemplifies how to link building-scale engineering modeling with meso-scale urban data to inform planning practice. It allows planners and policy makers to look at urban data through the lens of energy performance, and to reconsider where related urban data are, how urban data can be managed, and most importantly, what urban data can inform urban energy policies.

References

Ackoff, R. L. (1971). Towards a system of systems concepts. *Management Science, 17*(11), 661–671.

Al-Homoud, M. S. (2001). Computer-aided building energy analysis techniques. *Building and Environment, 36*(4), 421–433. doi:10.1016/S0360-1323(00)00026-3.

American Society of Heating, Refrigerating Air-Conditioning Engineers, & Illuminating Engineering Society of North America. (1989). *Energy efficient design of new buildings except low-rise residential buildings: ASHRAE/IESNA Standard 90.1-2004*. American Society of Heating, Refrigerating Air-Conditioning Engineers. Atlanta, GA.

American Society of Heating, Refrigerating Air-Conditioning Engineers, & Illuminating Engineering Society of North America. (2004). *Energy efficient design of new buildings except low-rise residential buildings: ASHRAE/IESNA Standard 90.1-2004*. American Society of Heating, Refrigerating Air-Conditioning Engineers. Atlanta, GA.

Batty, M. (2013). *The new science of cities*. Cambridge, MA: The Mit Press.

Booth, A. T., Choudhary, R., & Spiegelhalter, D. J. (2012). Handling uncertainty in housing stock models. *Building and Environment, 48*, 35–47. doi:10.1016/j.buildenv.2011.08.016.

Branco, G., Lachal, B., Gallinelli, P., & Weber, W. (2004). Predicted versus observed heat consumption of a low energy multifamily complex in Switzerland based on long-term experimental data. *Energy and Buildings, 36*(6), 543–555.

Crawley, D. B., Lawrie, L. K., Winkelmann, F. C., Buhl, W. F., Huang, Y. J., Pedersen, C. O., et al. (2001). EnergyPlus: creating a new-generation building energy simulation program. *Energy and Buildings, 33*(4), 319–331.

Deru, M., Field, K., Studer, D., Benne, K., Griffith, B., Torcellini, P., et. al. (2011). *US Department of Energy commercial reference building models of the national building stock.* National Renewable Energy Laboratory. Retrieved from http://www.nrel.gov/docs/fy11osti/46861.pdf.

Dodge Data and Analytics. (2005). *McGraw Hill construction.* Retrieved from: http://construction.com/dodge/.

Döllner, J., & Hagedorn, B. (2007). Integrating urban GIS, CAD, and BIM data by service based virtual 3D city models. In M. Rumor, V. Coors, E. M. Fendel, & S. Zlatanova (Eds.), *Urban and regional data management-annual* (pp. 157–160). London: Taylor & Francis Group.

Eliasson, I. (2000). The use of climate knowledge in urban planning. *Landscape and Urban Planning, 48*(1), 31–44.

Esri. (2012). *What is ArcPy? ArcGIS Help 10.1.* Retrieved from http://resources.arcgis.com/en/help/main/10.1/index.html#//000v000000v7000000. Accessed December 1 2014.

Flaxman, M. (2010). Fundamentals of Geodesign. In E. Buhmann, M. Pietsch, & E. Kretzler (Eds.), *Peer reviewed proceedings of digital landscape architecture 2010 at Anhalt university of applied sciences* (pp. 28–41). Heidelberg, Germany: Wichmann Verlag.

Golany, G. S. (1996). Urban design morphology and thermal performance. *Atmospheric Environment, 30*(3), 455–465.

Guerra Santin, O., Itard, L., & Visscher, H. (2009). The effect of occupancy and building characteristics on energy use for space and water heating in Dutch residential stock. *Energy and Buildings, 41*(11), 1223–1232.

Harris, B., & Batty, M. (1993). Locational models, geographic information and planning support systems. *Journal of Planning Education and Research, 12*(3), 184–198.

Hassid, S., Santamouris, M., Papanikolaou, N., Linardi, A., Klitsikas, N., Georgakis, C., et al. (2000). The effect of the Athens heat island on air conditioning load. *Energy and Buildings, 32* (2), 131–141.

Hogeling, J., & Van Dijk, D. (2008). *P60 More information on the set of CEN standards for the EPBD.* European Communities. Retrieved from http://www.buildup.eu/sites/default/files/P060_EN_EPBD_CEN_March2008_p3031.pdf.

Integrated Environmental Solutions Limited. (2012). VE-Ware. Retrieved from http://www.iesve.com/software/ve-ware.

International Organization for Standardization. (2008). *CEN-ISO Standard 13790-2008: Energy performance of buildings—Calculation of energy use for space heating and cooling.* International Organization for Standardization. Retrieved from http://www.iso.org/iso/catalogue_detail.htm%3Fcsnumber=41974.

Kim, J.-H., Augenbroe, G., & Suh, H.-S. (2013). Comparative study of the leed and ISO-CEN building energy performance rating methods. In E. Wurtz (Ed.), *Building simulation 2013: Proceedings of BS2013: 13th conference of IBPSA (International Building Performance Association)* (pp. 3104–3111). France: International Building Performance Simulation Association.

Klosterman, R. E. (2008). A New Tool for a New Planning: The What if?TM Planning Support System. In R. K. Brail (Ed.), *Planning support systems for cities and regions.* Hampshire: Puritan Press Incorporated.

Kolokotroni, M., Giannitsaris, I., & Watkins, R. (2006). The effect of the London urban heat island on building summer cooling demand and night ventilation strategies. *Solar Energy, 80*(4), 383–392.

Lee, S. H., Zhao, F., & Augenbroe, G. (2013). The use of normative energy calculation beyond building performance rating. *Journal of Building Performance Simulation, 6*(4), 282–292.

Littlefair, P. (1998). Passive solar urban design: ensuring the penetration of solar energy into the city. *Renewable and Sustainable Energy Reviews, 2*(3), 303–326.

MacQueen, J. (1967). *Some methods for classification and analysis of multivariate observations* (Vol. 1: Statistics, pp. 281–297). Berkeley, CA: University of California Press.

Maier, M. W. (1998). Architecting principles for systems-of-systems. *Systems Engineering, 1*(4), 267–284.

McPherson, E. G., & Simpson, J. R. (2003). Potential energy savings in buildings by an urban tree planting programme in California. *Urban Forestry & Urban Greening, 2*(2), 73–86.

Mitchell, G. (2005). *Urban development, form and energy use in buildings: A review for the solutions project*. EPSRC SUE SOLUTIONS Consortium. School of Geography and Institute for Transport Studies, University of Leeds. Retrieved from http://web.mit.edu/cron/Backup/project/urban_metabolism/TGOFF/readings%20and%20websites/Urban%2520development,%2520form%2520and%2520energy%2520use%2520in%2520buildings.pdf.

Mohammadi, S., de Vries, B., & Schaefer, W. (2013). A comprehensive review of existing urban energy models in the built environment. In S. Geertman, F. Toppen, & J. Stillwell (Eds.), *Planning support systems for sustainable urban development* (pp. 249–265). Heidelberg: Springer.

NYC Department of City Planning. (2014). *BYTES of the BIG APPLE*. Available from NYC Department of City Planning, http://www.nyc.gov/html/dcp/html/bytes/applbyte.shtml.

NYC Department of Information Technology & Telecommunications. (2014). *Building footprints GIS file*. Available from NYC Department of Information Technology & Telecommunications, https://nycopendata.socrata.com/.

Ok, V. (1992). A procedure for calculating cooling load due to solar radiation: The shading effects from adjacent or nearby buildings. *Energy and Buildings, 19*(1), 11–20.

Oke, T., Johnson, G., Steyn, D., & Watson, I. (1991). Simulation of surface urban heat islands under 'ideal' conditions at night part 2: Diagnosis of causation. *Boundary-Layer Meteorology, 56*(4), 339–358.

Perez-Lombard, L., Ortiz, J., & Pout, C. (2008). A review on buildings energy consumption information. *Energy and Buildings, 40*(3), 394–398. doi:10.1016/j.enbuild.2007.03.007.

Pisello, A. L., Taylor, J. E., Xu, X. Q., & Cotana, F. (2012). Inter-building effect: Simulating the impact of a network of buildings on the accuracy of building energy performance predictions. *Building and Environment, 58*, 37–45. doi:10.1016/j.buildenv.2012.06.017.

Quan, S. J., Economou, A., Grasl, T., & Yang, P. P.-J. (2014). Computing energy performance of building density, shape and typology in urban context. *Energy Procedia, 61*, 1602–1605.

Quan, S. J., Minter, J. D., & Yang, P. P.-J. (2013). A GIS-based performance metrics for designing a low energy urban agriculture system. In S. Geertman, F. Toppen, & J. Stillwell (Eds.), *Planning support systems for sustainable urban development* (pp. 225–247). Heidelberg: Springer.

Ratti, C., Baker, N., & Steemers, K. (2005). Energy consumption and urban texture. *Energy and Buildings, 37*(7), 762–776. doi:10.1016/j.enbuild.2004.10.010.

Ratti, C., & Richens, P. (2004). Raster analysis of urban form. *Environment and Planning B-Planning and Design, 31*(2), 297–309.

Reades, J., Calabrese, F., Sevtsuk, A., & Ratti, C. (2007). Cellular census: Explorations in urban data collection. *Pervasive Computing, IEEE, 6*(3), 30–38.

Reinhart, C., Dogan, T., Jakubiec, J. A., Rakha, T., & Sang, A. (2013). Umi-an urban simulation environment for building energy use, daylighting and walkability. In E. Wurtz (Ed.), *Building simulation 2013: Proceedings of BS2013: 13th conference of IBPSA (International Building Performance Association)* (pp. 476–483). France: International Building Performance Simulation Association.

Robinson, D., Haldi, F., Kämpf, J., Leroux, P., Perez, D., Rasheed, A., & Wilke, U. (2009). CitySim: Comprehensive micro-simulation of resource flows for sustainable urban planning. In E. Wurtz (Ed.), *Building Simulation 2009: Proceedings of BS2013: 11th Conference of IBPSA (International Building Performance Association)* (pp. 1083–1090). Scotland: International Building Performance Simulation Association.

Rode, P., Keim, C., Robazza, G., Viejo, P., & Schofield, J. (2013). Cities and energy: urban morphology and residential heat-energy demand. *Environment and Planning B: Planning and Design, 40*(2013).

Santamouris, M., Papanikolaou, N., Livada, I., Koronakis, I., Georgakis, C., Argiriou, A., & Assimakopoulos, D. N. (2001). On the impact of urban climate on the energy consumption of buildings. *Solar Energy, 70*(3), 201–216. doi:10.1016/S0038-092x(00)00095-5.

Snyder, K. (2003). Tools for community design and decision-making. In S. Geertman & J. Stillwell (Eds.), *Planning support systems in practice* (pp. 99–120). Berlin: Springer.

Steadman, P. (1979). Energy and patterns of land use. In D. Watson (Ed.), *Energy conservation through building design* (pp. 246–260). New York: McGraw-Hill.

Steemers, K. (2003). Energy and the city: Density, buildings and transport. *Energy and Buildings, 35*(1), 3–14. doi:10.1016/S0378-7788(02)00075-0.

SunEarthTools.com. (2014). Annual sun path. Retrieved from http://www.sunearthtools.com/dp/tools/pos_sun.php, Accessed December 1 2014.

The City of New York. (2014). About LL84. Retrieved from http://www.nyc.gov/html/gbee/html/plan/ll84_about.shtml, Accessed December 1 2014.

U.S. Energy Information Administration. (2005). *2003 CBECS (commercial buildings energy consumption survey) survey data*. Retrieved from http://www.eia.gov/consumption/commercial/data/2003/index.cfm?view=consumption.

United States Census Bureau. (2014a). *LEHD origin-destination employment statistics (LODES)*. Retrieved from: http://lehd.ces.census.gov/data/.

United States Census Bureau. (2014b). *TIGER Products*. Retrieved from: https://www.census.gov/geo/maps-data/data/tiger.html.

Wong, N. H., Jusuf, S. K., Syafii, N. I., Chen, Y. X., Hajadi, N., Sathyanarayanan, H., et al. (2011). Evaluation of the impact of the surrounding urban morphology on building energy consumption. *Solar Energy, 85*(1), 57–71. doi:10.1016/j.solener.2010.11.002.

Yeo, I., Yoon, S.-H., & Yee, J.-J. (2013). Development of an Environment and energy Geographical Information System (E-GIS) construction model to support environmentally friendly urban planning. *Applied Energy, 104*, 723–739.

Yezioro, A., & Shaviv, E. (1994). Shading: a design tool for analyzing mutual shading between buildings. *Solar Energy, 52*(1), 27–37.

Yi, Y. K., & Malkawi, A. M. (2009). Optimizing building form for energy performance based on hierarchical geometry relation. *Automation in Construction, 18*(6), 825–833. doi:10.1016/j.autcon.2009.03.006.

Chapter 25
MassDOT Real Time Traffic Management System

Russell Bond and Ammar Kanaan

Abstract MassDOT is pursuing an innovative business model that provides real time travel time information to the public using dedicated highway signs covering over 700 miles of state highway and encompassing the entire metropolitan Boston area. This investment in infrastructure means that MassDOT owns and archives these data, maintains and controls their quality, and freely provides them to third party developers in real time. This business model differs from all other state Departments of Transportation (DOT) around the United States that currently purchase travel time data from private companies. The development and operation of this Real Time Traffic Management (RTTM) system has resulted in a shift towards new measures of system performance and has triggered the production of new strategies for multi-modal transportation system management—all in a manner directly supportive of emerging trends: open public data, big data and the development of smart cities.

1 Introduction

The emergence of low cost, wireless communications technology and the rapid adoption of this technology by the public have presented public transport managers with new opportunities and new challenges for the management and operation of their transportation systems. Transportation agencies are increasingly taking advantage of wireless technology, such as Bluetooth, to improve their ability to monitor traffic flow in real time to deploy new, dynamic management strategies.

R. Bond (✉)
Massachusetts Department of Transportation—Highway Division, 10 Park Plaza, Suite 7410, Boston, MA 02116, USA
e-mail: russ.bond@state.ma.us

A. Kanaan
Kanaan Consulting US, Inc., 7 North Market Building, Suite 5125, Boston MA 02109, USA
e-mail: akanaan@portmanagers.org

S. Geertman et al. (eds.), *Planning Support Systems and Smart Cities*, Lecture Notes in Geoinformation and Cartography,
DOI 10.1007/978-3-319-18368-8_25

The purpose of this chapter is to describe an innovative public sector approach to transportation system management using new Bluetooth sensor technology, along with the current impact that this approach has had on the public, on transport managers, and on the business community.

2 MassDOT Real Time Traffic Management (RTTM) System

The Massachusetts Department of Transportation (MassDOT) is responsible for the operation and maintenance of over 3000 miles of state highway, over 5000 bridges, and seven tunnel systems that are used by over 4.7 million licensed Massachusetts drivers. MassDOT oversees four divisions: Highway, Mass Transit, Aeronautics and the Registry of Motor Vehicles (RMV), in addition to an Office of Planning and Programming. As evidence of the new paradigm, MassDOT has deployed a Real Time Traffic Management (RTTM) system, the purpose of which is to dynamically calculate the travel times on major Interstate and State highways between key interchanges or decision points. Travel times are calculated between Bluetooth readers typically installed every 5–10 miles on highly-traveled highways.

The first RTTM system was first installed as an operational test in May of 2012 on 43 miles of Interstate 93 (I-93) between New Hampshire and Canton, Massachusetts. The system consisted of 22 portable variable message signs (PVMS) that dynamically displayed travel times to passing motorists. For the technical convenience of sharing power and communications, a Bluetooth reader was co-located with each PVMS. Based on a very positive public response, MassDOT made the decision to expand the RTTM to Interstate 90 (I-90) and Route 3 in July of 2013. At present, there are 64 co-located PVMS and Bluetooth readers installed on these three highway corridors. The significance of this operational test was threefold:

- the use of Bluetooth was validated as a feasible, easily deployable and accurate means of estimating travel time;
- providing up-to-the-minute travel time information to the traveling public using PVMS was validated as a safe (not cellphone/smartphone based) and reliable way to inform the public of travel delays; and
- the choice to own the source of travel time data rather than to buy travel time data from private service providers created new opportunities to manage the transportation system.

Based on the lessons learned above, the RTTM is being expanded statewide within Massachusetts and is evolving from a system that utilizes PVMS (Fig. 1) to display travel times to one that utilizes permanent ground mounted signs with dynamic travel time light-emitting diode (LED) inserts. These signs are designed in a manner consistent with the Manual of Uniform Traffic Control Devices

Fig. 1 A portable variable message sign displaying travel times on Interstate 93

(MUTCD) and consist of standard green guide signs with up to three destination displays (Fig. 2). Travel times are shown in minutes and can be followed with a plus or minus sign to indicate if the travel time is increasing or decreasing.

The new travel time system design was launched in April 2014 on Cape Cod and consisted of nine travel time signs and 23 Bluetooth readers. At that time, Governor Deval Patrick branded the RTTM with the name and logo "Go Time" and announced the expansion of this system statewide. The statewide expansion involved re-advertising the project for an additional 137 travel time signs (including the replacement of existing PVMS) with co-located Bluetooth readers and another 129 standalone Bluetooth readers—a total of 266 locations. The low-bid contract for the physical construction of the RTTM was awarded to an electrical contractor in late 2014 for approximately $13.5 million dollars. The system integrator/system

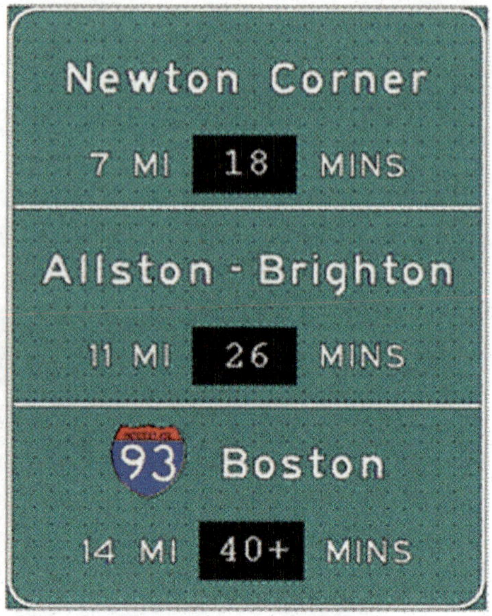

Fig. 2 Permanent ground mounted sign with a dynamic travel time LED insert

operator contract was awarded to the current RTTM system operator, Kanaan Consulting US, Inc. (KCUS), through a best value contract worth \$1.8 million dollars. The latter contract will provide all necessary software development, system integration, data warehouse, public data feed, 24/7 operations, and maintenance for three years after system acceptance from the construction contractor.

When the deployment of the statewide RTTM system is completed at the end of 2015, it will be the largest single deployment of a Bluetooth-based travel time system. The system will cover over 700 miles of state highway and will encompass the entire metropolitan Boston area. It is estimated by MassDOT that over 2.2 million motorists will view the travel time signs on a daily basis. Figure 3 delineates the anticipated coverage area of the statewide RTTM system.

3 Bluetooth Technology

As part of the technological transition to wireless technology, Bluetooth-based technology has emerged as an efficient way of estimating traffic flow by tracking the movement of vehicles that carry Bluetooth-enabled devices. This method of estimating travel time and speed is superior to other methods (radar detectors, loop detectors, global positioning systems and cellular phones) of travel time and speed flow estimation using conventional vehicle detection systems (Cambridge Systematics 2012). This superiority results from the capability of the Bluetooth system to measure actual traversal time between two sensor points as opposed to a single-point measurement process that is less accurate. A detector that measures

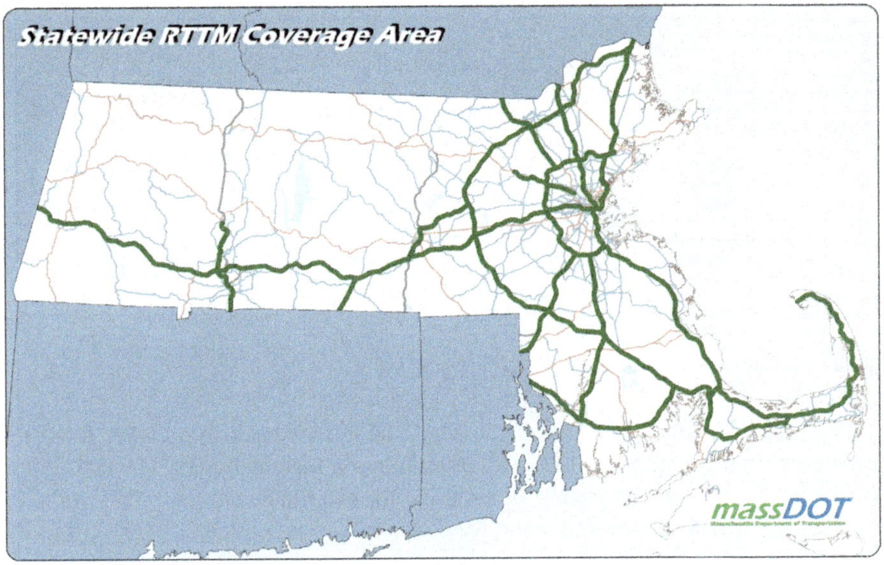

Fig. 3 Massachusetts RTTM highway coverage area

speed from a single point along the highway assumes this speed is constant over the given highway segment and cannot incorporate delays caused by incidents, intermittent backups, or other factors such as toll booths. This is not the case with using the Bluetooth method.

Bluetooth is a wireless communications specification, similar to Wi-Fi, that is used to connect two Bluetooth-enabled devices over short distances. When a cellphone or any Bluetooth-enabled electronic device is turned on, has Bluetooth transmission enabled, and is set to be discoverable, it will continuously broadcast a Media Access Control (MAC) address. This allows the device to communicate with another Bluetooth-enabled device. Typical Bluetooth-enabled devices include smart cellphones, tablets, headsets and vehicles with hands-free calling capability. A MAC address is a unique identifier assigned to all wireless electronic devices as part of the manufacturer's registered identification number. As such, there is no personal identification involved in the Bluetooth based RTTM and it does not identify the person using the device. As a result, all the MAC data collected by the RTTM is anonymous.

Figure 4 illustrates a typical highway segment with two Bluetooth readers that are 2 miles apart. The two readers create a segment that can be measured for travel time and speed. Within a detection radius of approximately 300 ft, each Bluetooth reader identifies, timestamps, and stores the MAC address of every vehicle that has a Bluetooth-enabled device that is turned on and broadcasting its MAC address. Each reader uses an omni-directional antenna to increase signal strength sufficiently to capture MAC broadcasts from both travel directions of a highway, unless blocked by trees or obstacles in the median.

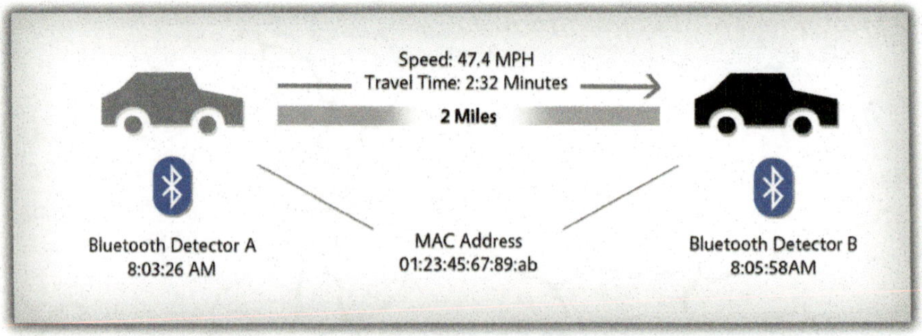

Fig. 4 Typical bluetooth detector pairing configuration

The readers wirelessly transmit the MAC addresses back to a server that uses software to match identical MAC addresses creating matched pairs. The travel time is calculated using the registered timestamps for each matched pair. This software uses sliding time windows and recursive estimation to process data every 1 min. At each process time period, the most recent N observations before the current time that are within a specified time window are selected. The number of points to process and the duration of the time window are chosen based on the distance between sensor pairs to optimize filtering performance. The distribution of these data points is used to remove both high-speed and low-speed outliers. The software filters the matched pair travel times to produce an average for the given time period. Commonly-detected outliers are vehicles that stop or leave the road for gas or rest and return to the road at a later time.

4 RTTM Travel Time Accuracy

The RTTM system does not need to track every vehicle on the highway in order to accurately estimate travel time. Experience has shown that as little as 4 % of the vehicles on medium- to high-volume roadways are sufficient to reliably estimate travel times. A study conducted for PennDOT favorably compared Bluetooth travel times on I-76 that had a match rate of 4 % with E-ZPass Toll Reader-based travel times that had a match rate between 10 and 37 % (KMJ Consulting Inc. 2010). Data collected by MassDOT indicates an average match between 4 and 7 % on I-90, I-93 and Route 3.

Another factor in travel time accuracy is the spacing of the Bluetooth readers. Readers that are spaced too far apart will have lag times between when the data are collected and when they are displayed on the sign. Although spacing may vary depending on highway characteristics, Bluetooth readers are generally spaced 5 miles apart in order to minimize the lag time effect.

Data accuracy and reliability are extremely important in order to ensure continued public confidence in the RTTM. Two control points built into the design of

Fig. 5 Pole-mounted bluetooth reader

the RTTM are the ability to see how many matched pairs were used in the travel time estimation process and the capability to measure the difference (lag) between the travel time display and the actual average travel time experienced by the motorists who just traversed a certain segment. It should be recognized that the RTTM does not predict travel time, it only calculates the average travel time of the motorists that journeyed ahead of the driver. The RTTM software will have an automatically-updated "lag time" performance indicator that allows for the real time survey of the accuracy of the data being produced. The lag time performance indicator will allow users of the system to compare the travel time displayed on a travel time sign to the average travel time that drivers passing by the sign are experiencing. Lag time will be displayed as a duration (in minutes) or as a percentage. Users will be able to output lag time reports for any specified interval, including minutes, hours, calendar day, week or month for any segment or aggregation of segments. The lag time performance indicator reporting will be developed as a customization.

The RTTM system is powered by batteries and is constantly recharged using solar power. The Bluetooth reader is pole-mounted and takes up very little space, as shown in Fig. 5.

5 MassDOT Traveler Information Experience

MassDOT has a long history of providing traveler information to the public. The 'SmarTraveler' public/private partnership between the former MassHighway (now MassDOT), and SmartRoute Systems, Inc. was launched in 1993. This partnership was the first operational test of a telephone-based traveler information system in the United States. The system provided route-specific traffic and public transportation information to travelers in eastern Massachusetts. Initially funded by the Federal Highway Administration (FHWA), the SmarTraveler Program became a cornerstone of MassDOT's Intelligent Transportation System (ITS) deployment program over the years at a cost averaging approximately $1.5 million dollars a year. This cost provided for a separate SmartRoute System operations center in Cambridge and full-time staff that proactively sought incident information and then manually voice recorded updates to travelers. The cost of this service also included a per-minute telephone charge to MassDOT. At its peak, the system averaged 500,000 calls a month.

The SmarTraveler business model was based on initially providing a free service to the public; however, it was envisioned that eventually this service would become valued by customers who would be willing to pay a small fee per transaction or for a subscription service. This business model was analogous, at that time, to the automatic teller machine (ATM) that was initially free to users. Eventual attempts to make this service self-sustaining through revenue sharing derived from advertising were never realized. While SmartRoute Systems did generate certain revenue from radio and television, there were never any significant revenues generated that reduced the cost of public sector financial support. By the year 2010, it was apparent that the market for traveler information services was evolving beyond a dial-in telephone service and that users were increasingly getting their travel information from the Internet and mobile devices. In an attempt to reduce costs and update the service, a private company, Sendza, was selected to operate a 511 phone based traveler information service at no cost to MassDOT. Sendza's business model was based on a software-as-a-service message platform that provided Mass511 message services to residents in Massachusetts and also provided a free 511 phone call to a computerized voice recognition system for traffic updates. Travel time information was provided by the company Inrix. In 2013, the number of 511 phone calls logged to Sendza was approximately 100,000 calls a month. This number continues to decline and is projected to be approximately 71,000 calls per month in 2014 if current trends hold.

6 Statewide RTTM Deployment

In 2012, MassDOT was looking for ways to improve public traveler information services. As the 511 model declined in efficiency and popularity, MassDOT decided to invest in a RTTM system that would post travel times on PVMS located at strategic points along the highway and that would give this information to the public and third party developers as a real time XML data feed. These data could be combined with other types of traveler information such as accidents, delays, congestion levels and weather.

This strategy was based on the success that the Massachusetts Bay Transit Authority (MBTA) had with providing their data through open data portals in order to improve traveler information for transit users. The MBTA began posting transit delays and next bus arrival data to the developer community as a real time XML data feed beginning in 2010. The success of this strategy was based on the awareness that customers were increasingly turning to mobile communications and social networks to get real time status information on the transit system. Upon opening their data to the public, the first cell phone app that used that data was created in 2010 by a student at no cost to the MBTA (http://dailyfreepress.com/2010/09/07/mbta-theres-an-app-for-that/). Based on the success of this strategy, MassDOT moved forward with plans to implement the RTTM.

The RTTM was first bid for in 2012 as a best-value Request for Response (RFR) procurement. This RFR required a system operator to provide a turn-key travel time system for two years on I-93 using PVMS. There were four bidders that responded to the RFR. However, only one bidder (KCUS Inc.) proposed a Bluetooth based system that included unrestricted data ownership of the travel time data. The success of the I-93 operational test has led to its adoption as a central element of MassDOT's Intelligent Transportation System (ITS) Program and the lessons learned from this experience formed the basis for a new business model for delivering real time information to the public. This model consists of providing transportation-related data freely to the public and private sectors in order to promote private sector innovation that was in the best interest of the public. This business model is based on the following precepts:

- Public users of the multi-modal transportation system no longer rely on government funded programs such as 511 to obtain real time travel information. The rise of phone apps and social networking has changed how customers want to receive travel updates. The Government cannot keep up with private sector innovation and the rapid changes in technology, and, therefore, should exchange the current business practice of being a traveler information service provider for the role of being an open data provider.
- It is in the best interest of MassDOT to install their own RTTM sensor system used to measure travel time on major highways. The primary purpose of the RTTM is to improve the highway monitoring and management capabilities of MassDOT and is no different from the need to use other field sensors such as cameras, traffic volume detectors, vehicle classification/weight detectors, and weather sensors.
- It is in the best interest of MassDOT to own the data generated from the RTTM rather than to buy these data from third party service providers. It is important to have the capability to ensure data quality by having unrestricted access to and control over the raw data generated by the RTTM, not just the travel time meta data typically provided by third party service providers. This capability will also make it more efficient for MassDOT to develop internal performance measures as part of a data warehouse.
- It is in the best interest of MassDOT to provide real time data to third party developers as part of the Massachusetts Open Data Initiative that promotes new and innovative transportation smartphone applications at no additional cost to

MassDOT. http://www.massbigdata.org/initiative/—the Mass Big Data Initiative was announced on May 31, 2012 at the Massachusetts Institute of Technology's Department for Computer Science and Artificial Intelligence Laboratory in Cambridge.

- It is in the best interest of MassDOT to meet the Federal Highway Administration (FHWA) 1201-mandated mobility information requirements, including MAP 21 performance monitoring and measurement requirements, using the RTTM system.

This approach or business model leverages MassDOT-owned infrastructure, encourages private sector innovation, takes advantage of emerging trends in technology, and seeks to meet the increasingly sophisticated demands of the public for real time information that increases their ability to make informed travel choices. These activities are consistent with the roles of the private sector and government sector and takes advantage of emerging trends in technology and how customers want to get and use travel information.

7 RTTM System Architecture

As can be seen from the system architecture diagram (Fig. 6), the RTTM system consists of field devices (Bluetooth reader, cellular communications modems, LED displays, and power supply) that are integrated into a functionally-integrated system. Field devices connect through cellular communications to the host server that processes the Bluetooth-derived MAC addresses and calculates segment travel times. The processed travel time information is transmitted back to the appropriate LED sign displays and is concurrently transmitted to the public as an XML data feed. These data are also stored in the system's data warehouse. Because the system components utilize open Application Programming Interfaces (APIs) that any third party can write to for integration purposes, any vendor hardware or software module can replaced in the future if it becomes necessary or desirable.

As the system integrator for the statewide expansion of the RTTM, KCUS Inc. will use vendor software to build a fully-functional RTTM back office. As part of the statewide expansion, KCUS is upgrading the functionality of the current software. KCUS will use Traffax Inc.'s BluFAX suite, which will serve is the central travel time data processing and filtering hub and National Transportation Communications for Intelligent Transportation System Protocol (NTCIP) communications interface to the Bluetooth readers. This interface will allow for the automated remote retrieval of collected MAC addresses at configurable intervals or in an ad hoc manner. The interface will also have the ability to receive operational status, detected fault information, and real time solar power and battery monitoring information.

The interface will also allow for the development, configuration, and implementation of multiple types of Bluetooth reader device drivers. A sample of the type of data set collected by the central software from the field devices is listed below:

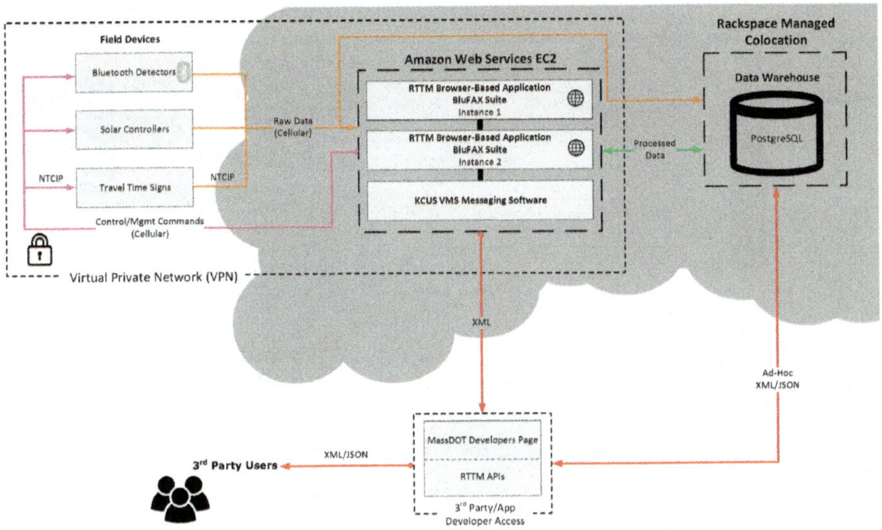

Fig. 6 RTTM system architecture

- captured MAC addresses, through single or multiple read requests;
- timestamp of MAC address capture;
- captured location information (whether defined by device ID, segment ID, or geographic coordinates);
- Bluetooth device status (on/off, system status, etc. as made available by Bluetooth reader device);
- solar power and battery monitoring subsystem, as made available by the device; and
- other detected faults or anomalies, as made available by the field device.

The RTTM software suite will include a GIS-based interactive map page that will provide the end user with an overall view of the system and devices (Fig. 7). The user will also able to view detailed information by clicking on a segment; this will produce a popup that lists origin/destination, current speed, and speed limit information.

8 RTTM Database and Open Data Warehouse

All RTTM information will be archived indefinitely on a cloud-accessible secure database hosted at a remote data center (Rackspace). Rackspace data centers are designed to host mission-critical servers and computer systems and feature advanced physical security and fully-redundant subsystems (cooling, power, network links, storage, etc.). In order to ensure that the RTTM data warehouse serves

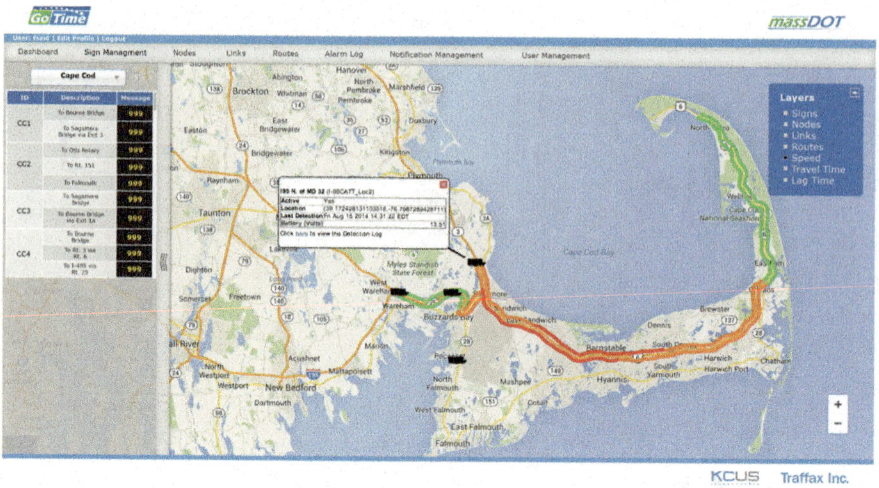

Fig. 7 RTTM GIS-based interactive map

the interest of the State's Open Data initiative, MassDOT has formed an RTTM Governance Committee consisting of representatives from state transportation agencies and other public and private stakeholders, such as representatives from the "Hacker" community. The RTTM Governance Committee will serve as the principal stakeholder group that will guide the development, design, and implementation of the Statewide RTTM Database and Data Warehouse. Specific deliverables resulting from the Committee's initial work will include the RTTM Database/Data Warehouse Concept of Operations, System Needs, and System Requirements documentation which will be submitted to the MassDOT RTTM Governance Committee. These documents will serve as the roadmap for the design of the RTTM database/data warehouse and will ensure the development of a system that will best meet the needs of all stakeholder and users.

The proposed software solution will utilize a powerful, extensible, scalable and open PostgreSQL-based data warehouse to store RTTM-related data. All raw and processed RTTM data will be assimilated, organized, fused, and stored within the data warehouse. A set of Representational State Transfer RESTful APIs will allow third party developers to programmatically query the data warehouse to remotely retrieve general and specific real time and historical data. Responses will be formatted in industry-standard XML or JSON formats. The data warehouse will be fully extensible and scalable to allow it to grow organically with the RTTM system and to house new data and data types for mining and reporting purposes. New data sources that will be evaluated for inclusion into the RTTM data warehouse and xml feed, including the following available transportation data: traffic volumes; incidents; pre-planned construction or events; and weather.

The data warehouse will meet the following requirements:

- MassDOT will retain ownership of all data housed within the Warehouse, including the database design and schema. The database schema will be provided to the MassDOT Governance Committee during the design phase.
- All personal, non-public data transmitted to the data warehouse will be encrypted and will not be accessible to the public. The KCUS Team will work with the MassDOT Governance Committee to define access levels for different data types.
- Upon termination of the contract, KCUS, Inc. will provide an orderly return of all assets to MassDOT via a physical data backup format and will securely dispose of all assets housed within the cloud-hosted data warehouse.

The RTTM software will have built-in measures to ensure that no personal or potentially identifying information will be accessible or stored by the system. The RTTM software will only retrieve MAC addresses in an anonymized, hashed format to protect the identity of detected devices.

9 New Measures of Transportation System Performance

The RTTM is designed as a multi-purpose system that can be used for planning, real time operations, and for measuring system performance. For example, the RTTM system can produce travel time graphs (Fig. 8) that can be used to deduce what day and time it would be best to depart for Cape Cod before the Labor Day holiday weekend. RTTM data were graphed for a segment of the Southeast Expressway on the three days leading up to the weekend holiday in 2013. Assuming that traffic conditions in 2013 are a good indication of those in 2014, the data would indicate that, in order to minimize delay, a traveler should leave by 2:00 p.m. if they are to make that trip—regardless of the day they choose to start their holiday.

The proposed RTTM Software system will be deployed using two platforms: Amazon's state-of-the-art Elastic Compute Cloud (EC2) hosting platform for live operations and a "Colo" managed hosting solution at Rackspace for the data warehouse. The Amazon EC2 platform allows the configuration of additional capacities and instances with minimal friction and within minutes, which allows the application to be scaled dynamically as storage and performance needs change. Rackspace's Colo solution is also extremely scalable to accommodate an organically-growing data warehouse.

The system will be designed in a manner that ensures that it maintains real time operations and remains sufficiently responsive to user input. Expected response times, not including web response/transfer/rendering times, to standard operations are identified as follows:

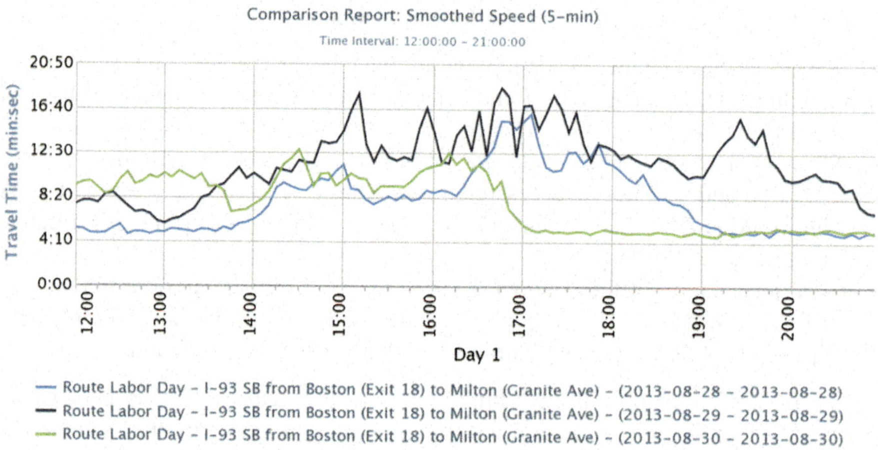

Fig. 8 RTTM travel time graph (Kanaan Consulting US, Inc.)

- screen changes (navigation): <0.5 s;
- real time data updates [travel time posting to Travel Time Sign (TTS)]: <0.5 s;
- graphical map updates (system-to-map transition): <1.5 s;
- playback queries: <5 s for most queries;
- report generation (varies based on multiple factors): 5–30 s for most reports;
- dashboard updates: <5 s for most updates.

The EC2 hosting platform guarantees an uptime of more than 99.95 %, measured monthly. Rackspace guarantees a 100 % uptime for its Colo platform, not including scheduled outages. The RTTM data warehouse will be able to serve content to 50 external users simultaneously.

10 Lessons Learned

The Statewide RTTM has been procured using two different contracting methods:

- a low-bid construction procurement for the installation of all field equipment: concrete foundations, guide signs, solar panels, Bluetooth readers, modems and associated electronics; and
- a best value procurement for a service contract that includes all software development and integration, back-office management of the RTTM system, and 24/7 operation and maintenance of the entire system for three years after the system has been built and accepted by MassDOT.

After a one-year contractor warranty period, the system operator, KCUS, will assume full responsibility for the statewide RTTM. Because the system operator is also the system integrator of the RTTM hardware and software design, they are in the best position to manage and operate the live system. There are many points of failure between the field devices and central processing software and the system demands a high degree of monitoring to ensure problems are quickly identified and corrected.

This combined procurement approach was deemed the least risk to MassDOT compared with using a single contractor to build and operate the system. The combined approach would ensure that the expertise of the construction contractor and the system integrator/system operator were fully utilized. However, despite this approach, the system cost has exceeded estimates. There were four construction bids from qualified contractors and the low bid was 37 % higher than the inhouse design cost estimate. A partial explanation for the high bids was the perceived risk that construction contractors apparently factored into their bids. For example, one of the requirements for their contract was a 4-h response time to remediate any system errors resulting from the field equipment. This may require the contractor to conduct a field visit to test and replace an actual part at any one of 268 locations.

Because the RTTM uses two separate contracts, one to install the field equipment and one to integrate and operate the system, close coordination between the operator and the construction contractor is required. After the system has been operational and fully accepted by MassDOT, the system operator will have complete responsibility for all operations and maintenance of the Statewide RTTM. However, until the project is accepted by MassDOT, the contractor is responsible for all field installations and communications to the system software. The following describes the different responsibilities of each.

- The contractor shall be responsible for installation, setup, configuration, testing and maintenance of all field equipment in addition to operational performance of the Bluetooth Readers, Sign Controllers, LED display modules, solar power systems and batteries and as well as the 4G/LTE cellular data links.
- The contractor shall be responsible for device restart in the event of failure, device firmware upgrades to be coordinated with system operator, device performance adjustment and tuning (i.e. radio signal strength) as well as coordinating and working with the system operator.
- The contractor shall respond to travel time application automated faults via email.
- The contractor shall log all maintenance activities via an online log and reporting system.

System operator responsibilities are as follows:

- The system operator shall be responsible for monitoring all of the field equipment including Bluetooth Readers, Sign Controllers, LED display modules and solar power controllers.
- The system operator shall be responsible for the travel time application uptime and maintenance, as well as the accuracy of the travel time algorithm. The system operator shall be responsible for collection of the field device MAC addresses and matching pairs for travel time calculations.

- The system operator shall be responsible for posting travel times on LED display modules, for software and database maintenance, for network security from a hosted location to the field devices (to be coordinated with the contractor), for hosting a device alert system and log files as well as delivering the XML feed to application developers.
- The system operator shall be responsible for device performance issue identification; this will be coordinated with the contractor to make field corrections.

The RTTM system is designed to run in a fully-autonomous fashion. However, the system requires full-time status monitoring in order to identify and respond to any system failure. Because the system is designed as a network of sensors, one site failure can affect the ability to accurately estimate travel times at upstream or downstream locations. To compensate for a single sign or Bluetooth location failure, the software is being modified by KCUS Inc. to automatically create a new segment from the adjacent upstream or downstream Bluetooth sensor until the problem location is addressed and the downed device is back online.

The software is designed to detect any faults or errors reported by the travel time signs and/or the Bluetooth reader devices in the field. These errors are reported on the RTTM website's alarm page, including automatic notifications to appropriate personnel. Typical device faults are communication errors between the modem and cellular service provider, low power issues, and stale data caused by too few MAC address matches. The latter occurs due to periods of low traffic volumes such as off-peak times, weekends, or winter on holiday travel-heavy Cape Cod.

At present, the weakest link in the RTTM system on Cape Cod appears to be the power supply that feeds locations that include collocated Bluetooth readers and LED sign displays. There are nine collocated sites on Cape Cod and 13 standalone Bluetooth readers. Travel time signs with three destinations (three LED inserts) are supported by eight 24-V batteries, two destination signs by six-24 V batteries, and the standalone reader uses only one-24 V battery. All batteries are warranted for a 30 continuous operation without solar recharge. If a co-located site gets below 19 V, it is likely to cause a low voltage disconnect (LVD) if the batteries are not immediately recharged by the solar panels or a portable generator. The winter months on Cape Cod have witnessed a number of LVD failures due to the low angle of the sun and the proximity of signs to trees that further block out available sunlight. The Cape Cod system was launched in April of 2014, but such failures did not occur until the month of December. In response to the above problem, KCUS plans to modify their software to automate the voltage monitoring of co-located sites utilizing the solar charge controller. The solar controller is a network-enabled device and allows battery levels to be monitored remotely; the controller also provides a 60/90 day history of this information. Monitoring the voltage will allow lead time that is sufficient to recharge the batteries using a portable generator to prevent device failure.

11 Conclusions and Next Steps

The RTTM represents a major commitment by the Commonwealth of Massachusetts and MassDOT to use mobile, wireless communications technology to enhance travelers' driving experience, alleviate congestion, increase mobility options, and improve overall highway safety. Furthermore, this investment in Bluetooth-based roadside infrastructure means that MassDOT owns and archives this data, maintains quality control of the data, and freely provides this data to third party developers in real time via standard data communications protocols.

The development and implementation of the RTTM is an outgrowth of new and emerging technology. At the same time, it represents a commitment from the MassDOT to expand how it interacts with the customers of the transportation system by using mobile technology and social media. While MassDOT will not invest in the actual phone apps themselves, it will work with stakeholders to facilitate access to open public data that can be used for the public good.

The vision for this effort is one where the customer can make real time decisions based on incidents, traffic delays, weather or special events. The future traveler in the smart city of tomorrow will use mobile phone apps to make choices that include route changes, mode changes and departure time adjustments. While it is not envisioned that the public sector will invest in predicting travel times, university researchers and the private sector have been investigating predictive travel time estimation through dynamic traffic assignment models for over a decade (Milkovits et al. 2010). These models are now ready for application as continuously on-line prediction models—as opposed to off-line models. Historical information on traffic flow by time of day, day of week, month of the year can be used to develop trip reliability or buffer indexes for individual trip makers. This would allow a user to customize his/her trip between an origin and destination with varying degrees of certainty and could be used to inform the traveler of the best mode and most reliable (statistically) departure time. The private sector will add value to MassDOT's open data and integrate this data with private sector data such as parking availability and parking reservation systems. Most of the major corridors into Boston are paralleled by light rail or commuter rail lines; open data systems, such as the RTTM, will create the connections between all modes of travel, increase customer choice, and ultimately allow travelers to make the best use of limited highway and rail capacity.

Acknowledgments The authors would like to acknowledge the contributions and support of the following individuals without whom the success of this project would not be possible:

Jerry Allen, Former MassDOT Chief of Highway Operations; Leonard Walsh, Current MassDOT Chief of Highway Operations; Richard Davey, Former Secretary of Transportation; Frank DePaola, Former MassDOT Division Administrator and Current General Manager of MBTA; Rachel Bain, MassDOT Assistant Secretary for Performance Management & Innovation; Marjorie Weinberger, MassDOT Legal Counsel; Promise Otaluka, ITS Engineer, FHWA Division Office Cambridge; Firas Said, Lead Systems Engineer, KCUS; Rebecca Morgan, Formerly of KCUS; Chris Costello, Jacobs Engineering; Michelle Boucher, former MassDOT Director of ITS; Neil Boudreau, MassDOT State Traffic Engineer; Steve Timmins, MassDOT Highway Signing

Engineer; Jared Kadich, ITS Research Analyst, MassDOT; Ken Charlton, MassDOT Resident Engineer for Construction; Marco Pereira, Senior ITS Engineer, MassDOT; Cassandra Asvestas, MassDOT Assistant Resident Engineer.

References

Cambridge Systematics. (2012). *Travel time white paper*. Prepared for Florida Department of Transportation, September.

KMJ Consulting Inc. (2010). *Bluetooth travel time technology evaluation using the BlueTOAD*. Prepared for the Pennsylvania Department of Transportation, January.

Milkovits, M., et al. (2010). DynaMIT 2.0: The next generation realtime dynamic traffic assignment system. In *2010 Second International Conference on Advances in System Simulation, Nice, France*, 2010.

Chapter 26
Pragmatic Incremental or Courageous Leapfrog [Re]Development of a Land-use and Transport Modelling System for Perth, Australia

Sharon Biermann, Doina Olaru, John H.E. Taplin and Michael A.P. Taylor

Abstract Responding to land-use and transport modelling requirements, identified through a rigorous stakeholder engagement process, current land-use and transport modelling practices in Perth, Western Australia were examined and benchmarked against world-wide best practice. Three alternative model systems were proposed and evaluated. The preferred option, PLATINUM (Perth LAnd and Transport INtegrated Urban Model), is the more radical option, avoiding duplication and other resource inefficiencies, yet not discarding specialised and advanced work already undertaken. The unique contextual design challenges relate to the current modelling situation in Perth. It is concluded that designing model systems should explicitly acknowledge the current system in use and solutions should specify the pathway from the existing situation to the new model system. In addition, the two-edged sword of experience should be recognised as both a positive influence in terms of innovation awareness but carefully handled in relation to potential negative influences of path-dependent, 'incrementality' at the expense of embracing more radical innovations.

S. Biermann (✉)
School of Earth and Environment, Planning and Transport Research Centre,
The University of Western Australia, Perth, WA 6009, Australia
e-mail: sharon.biermann@uwa.edu.au

D. Olaru
Business School, Planning and Transport Research Centre,
The University of Western Australia, Perth, WA 6009, Australia
e-mail: doina.olaru@uwa.edu.au

J.H.E. Taplin
Business School, The University of Western Australia, Perth, WA 6009, Australia
e-mail: john.taplin@uwa.edu.au

M.A.P. Taylor
School of Natural and Built Environments, University of South Australia,
Adelaide, SA 5001, Australia
e-mail: map.taylor@unisa.edu.au

S. Geertman et al. (eds.), *Planning Support Systems and Smart Cities*,
Lecture Notes in Geoinformation and Cartography,
DOI 10.1007/978-3-319-18368-8_26

1 Introduction

New generation strategic urban land-use and transport models, increasingly linked with GIS and visualisation tools, can be considered as Planning Support Systems (PSS), developed to "*enlighten*" (Geertman 2006, p. 864) certain tasks involved in the process of integrated land-use and transport planning. Predominantly supporting the tasks of projecting, simulating and evaluating alternative scenarios, land-use and transport models can be classified as "*analysing PSS*" (Vonk 2006, p. 79), specialist tools that are technically complex and difficult to implement, but with the potential to improve strategy-making phases of planning (Te Brommelstroet 2013). Costs and risks of successfully implementing analytical PSS are high and usually tend to outweigh the less tangible benefits of improved scenario modelling and analysis (Vonk 2006).

More traditional land-use and transport models formed the core of the so-called large-scale system models developed and used in the 1960s and 1970s for metropolitan planning. Largely discredited by Lee (1973) on the basis of, amongst other reasons, being too complex and data-hungry, planning support instruments reverted to simple tools supporting routine planning tasks, with more sophisticated analysis, simulation and modelling tools largely neglected in the subsequent two decades (Geertman 2006). According to Waddell (2011), during the 1970s and 1980s, only two integrated modelling systems prevailed in practice: Putman's (1983) spatial interaction model system (ITLUP) and the spatial econometrics model systems of TRANUS (de la Barra 1995) and MEPLAN (Echenique et al. 1990). In 1995, the prognosis was that many of the challenges originally identified by Lee (1973) and subsequently updated by him in 1994 (Lee 1994), still applied (Waddell 2011). Since then, at least in the US, in response to legislation enforcing the coordinated approach to land-use, transportation and air quality planning, there has been a resurgence in urban modelling. Consistent with the intent of contemporary PSS in general, this urban modelling revival has been accompanied by a shift to responsiveness to accommodating a more diverse range and increasingly complex set of policy questions relating to multi-modal and demand-side approaches, incorporating behaviour change and becoming more open and accessible to a wider range of participants involved in the planning process (Waddell and Ulfarsson 2004). This has resulted in model development within the GIS environment, agent-based approaches and increased visualisation (de Ortúzar et al. 2011; Hunt et al. 2005; Iacono et al. 2008 and Miller 2003).

Despite the renewed interest in urban modelling, there are still considerable challenges in transferring the advances made from academic research environments into policy and practice, including aspects of transparency, flexibility and ease of use, behavioural and empirical validity, uncertainty, computational performance and data availability and quality (Waddell 2011). These modelling-specific challenges are echoed in relation to PSS more generally with a significant body of literature devoted to understanding barriers to uptake (Batty 2003; Geertman 2006; Geertman and Vonk 2006, Te Broemelstroet 2013; Vonk 2006; Vonk et al. 2005,

2006, 2007a, b; Waddell 2011). There is agreement that the overriding reason for disappointing levels of diffusion of PSS into practice, despite rapid advances in computer hardware and software, increasing availability of data and information and the growth and diffusion of the Internet, can largely be attributed to the mismatch between supply and demand (Geertman 2006; Harris and Batty 1993; Vonk 2006; Vonk and Geertman 2008).

To narrow the gap between demand and supply, thus increasing uptake, Vonk and Geertman (2008) deem it essential that lessons are learnt and captured by studying the design and implementation of PSS in practice. On this basis, a number of conceptual frameworks, incorporating both supply-side and demand-side criteria, have been proposed to be used in the design and implementation of new PSS to increase the probability of uptake. These frameworks identify sets of development context, internal instrument quality and diffusion factors as being critical in the design and implementation of a successful PSS (Biermann 2011; Geertman 2006; Geertman and Stillwell 2004; Vonk and Geertman 2008). Waddell and Ulfarsson (2004) similarly consider context, policy objectives, policy options to be tested and model requirements in the process they propose and use in the design of a model system for the Puget Sound region in the State of Washington in the US.

The purpose of this chapter is to describe the application of current theoretical insights into critical success factors for PSS to the design of an analytical PSS (Vonk 2006), an integrated land-use and transport model system for the Perth metropolitan area (Perth) in Western Australia (WA). Lessons learnt from other jurisdictions, as well as new lessons generated through application in the specific Perth development, planning and policy context, will in turn be captured to further inform the successful design of subsequent PSS. This will contribute to breaking the "*vicious cycle*", identified by Vonk and Geertman (2008, p. 153), of lack of uptake leading to lack of applications to be studied resulting in few lessons learnt, yielding more, poor quality PSS, in turn resulting in lower levels of uptake.

In order to harness recent advances in addressing the unique contextual and growing challenges of transport modelling and improve modelling operational efficiency in Perth, a transport modelling review process was initiated in response to an agreement between the Director General of Transport and the Chairman of the Western Australian Planning Commission of the Government of Western Australia, to move towards the development of a new strategic transport model and a single modelling team. As part of this broader review process, an independent review of transport modelling was undertaken with a view to providing design options for a new and responsive model system. This chapter describes the process, outcomes and unique design challenges of this review and design initiative with a particular focus on path dependency and the critical choices in moving from current modelling practices to a new model system.

After summarising the methodological approach, the results of the model design process are described for the case of Perth. A discussion of the lessons learnt for the future design of analytical PSS follows concluding with recommendations for future model design processes and conceptual frameworks to enhance the uptake of PSS.

2 Methodological Approach

Waddell and Ulfarsson (2004) incorporate a set of essential criteria identified in the literature for successful PSS as part of the process they propose and use in the design of a model system for the Puget Sound region in the State of Washington in the US. The first four steps in the process are to assess: (1) the development context (institutional, political and technical); (2) the planning and policy objectives; (3) the policy options to be tested; and (4) the model requirements. The outcome of these steps informs the subsequent process steps of (5) making preliminary design choices and (6) selection of modelling approach.

Broadly following the approach by Waddell and Ulfarsson (2004), but with adaptations to suit the particular local situation, the approach to the model system design for Perth is summarised in the following steps (Fig. 1):

1. Assess the development context.
2. Assess the current modelling situation.
3. Assess stakeholder-identified needs, capabilities and gaps identified through a stakeholder engagement process followed in order to obtain an understanding of requirements for the model, current capability to meet requirements and to inform the evaluation framework.
4. Review current international best practice advances in land-use and transport model development from the literature.

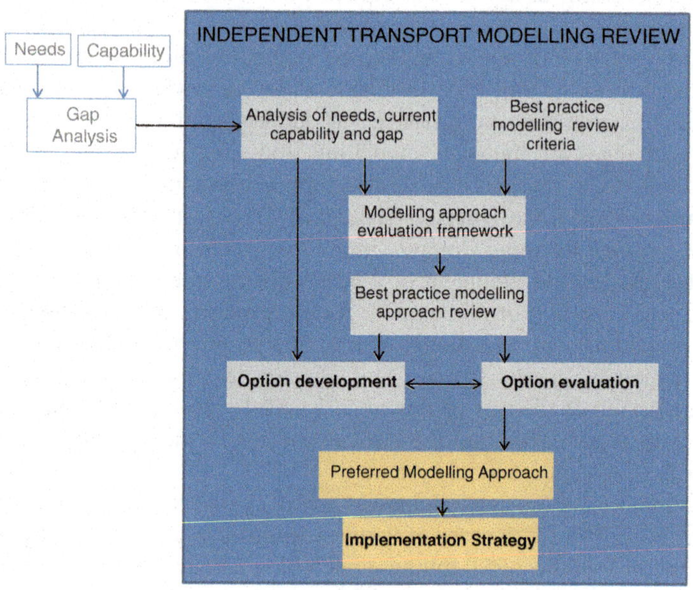

Fig. 1 Methodological approach to the design of a model system for Perth

5. Develop options on the basis of best practice approaches feasible to meet identified needs.
6. Evaluate the options.
7. Select the preferred modelling system approach.
8. Propose an implementation strategy.

3 Results

3.1 Assessment of Development Context

With a 32 % population growth from 1.52 million (2004) to 1.97 million (2013) and an average annual GDP growth of 5.7 % (from 2002/3 to 2012/3) in the Perth metropolitan area, demand for transport and other infrastructure is high, with commensurate urgent demands for infrastructure planning and policy development to meet the growth, supported by modelling. The state government is responsible for planning and transport across Perth and Peel (essentially the metropolitan region of Perth). In WA, as is the case in Australia generally, there is a high degree of centralisation in planning with state governments being the authorities for city planning as well as for remote and regional areas of the state. Local governments exist institutionally and legally, only as an arm of state government with certain designated local planning functions, but ultimately with decision-making power held at the state level. Stakeholder engagement in planning processes is largely limited to formal consultation.

There is a strong tradition of transport modelling in Perth and a high degree of confidence placed in modelling results even to the extent that concern has been expressed by modellers that modelling output is used and interpreted by planners without due recognition of modelling assumptions and limitations. An artefact of a previous dispensation when planning and transport were both part of a single mega-department, subsequently split into a number of different departments, transport modelling is undertaken by two separate, experienced and capacitated modelling teams in two departments: Planning and Main Roads. While there is co-operation between the two teams, there is increasing competition and duplication of effort. Transport modelling in Perth is used primarily for operational purposes and as input to cost-benefit evaluation of discrete land-use and transport projects, precinct and local (structure) plans. There is little evidence of land-use and transport models being used for longer term, scenarios-based, integrated strategic planning. Land-use modelling is limited to highly inflexible and bespoke population and employment forecasting with, trend- and capacity-based allocation to small areas.

Directions 2031 is the current high level spatial framework and strategic plan that establishes a vision for future growth of the metropolitan Perth and Peel region to guide the planning and delivery of housing, infrastructure and services necessary to accommodate a range of growth scenarios (Government of Western Australia 2010). The strategic vision is that "*By 2031, Perth and Peel people will have created a world class liveable city; green, vibrant, more compact and accessible with a unique sense of place*" (p. 2). The Directions 2031 strategic framework is based on the five key themes for a liveable, prosperous, accessible, sustainable and responsible city. The objective for the "*accessible*" theme is that "*all people should be able to easily meet their education, employment, recreation, service and consumer needs within a reasonable distance of their home*" (p. 23) with strategies to achieve that objective identified as being to:

- *Connect communities with jobs and services.* This strategy places emphasis on activity centres and transit-oriented development to achieve better integration of land-use and transport services.
- *Improve the efficiency and effectiveness of public transport.* The focus here is on linking key activity centres and nodes, and significantly improving integration, efficiency and patronage, featuring rail and other rapid transit and high frequency bus services. Achieving minimum residential densities, significantly higher than present densities, around key transport nodes is a key part of this strategy.
- *Encourage a shift to more sustainable transport modes.* The promotion of walking and cycling through the application of transit-oriented development and liveable neighbourhoods design principles, TravelSmart programs and cycling network improvements.
- *Maximise the efficiency of road infrastructure.* Transport demand cannot be satisfied only by building more roads and thus improved efficiency of the road network, congestion management including real time road user information, incident management and ramp metering and intelligent transport systems are part of the strategy.
- *Manage and reduce congestion.* In addition to road use efficiency, an effective public transport network with superior service in terms of travel time and reliability, together with adequate levels of passenger security and comfort and a transport system, land-use pattern and urban design that are conducive to walking and cycling, are part of the strategy.
- *Protect freight networks and the movement economy.* Protecting major transport corridors and freight operations from incompatible urban encroachment, particularly those leading to major ports, is a key element.
- *Consider parking in the overall transport picture.* Parking supply, demand and rights of parking allocation must be carefully managed to support broader accessibility objectives.
- *Plan and develop urban corridors to accommodate medium-rise higher density housing development.* Planning for existing and potential activity centres into the future with an increased focus on transport integration and transit-oriented development, agglomeration of economic activities and mixed-use development including higher density housing.

The following list, which is not exhaustive, draws attention to the policy-relevant modelling outputs and measures relevant to the types of policies under consideration in Perth:

- passenger demands, by time period during the day;
- trips by mode—the basic mode choices giving share of trips between modes;
- travel times and link speeds;
- trip forecasts:
  - metropolitan region flows in future years;
  - flows resulting from particular land-use developments;
  - flows resulting from particular highway or railway extensions;
- road traffic flows by link or at screen lines;
- road vehicle queue lengths;
- public transport patronage: system wide and at particular locations;
- impacts of specific policies:
  - traffic management;
  - price changes (including road pricing and tolls, congestion charges, fare changes and differentials, operating cost changes due to rising fuel prices, mobility charging);
  - parking;
  - reduced crowding through the provision of more rail cars or buses;
  - reduced variability in freeway travel times though, e.g., ramp metering;
- cost-benefit ratios; and
- measures of accessibility.

3.2 Assessment of Current Modelling Practice in Perth

There are essentially two strategic transport models in operation in Perth, each operated by separate modelling teams in different state government departments. The Strategic Transport Evaluation Model (STEM), at the Department of Planning, is used to assess the impacts of alternative land-use scenarios on Perth's multimodal metropolitan transport systems. STEM is a model used for land-use and transport policy assessment. There are currently two versions of STEM in existence—one using the established strategic network travel demand analysis package EMME/2 software platform and the other, the CUBE VOYAGER platform. It uses a multimodal transport network and land-use (zoning system) database for the metropolitan area that is suitable for broadbrush studies of differences in travel patterns on that network for different land-use scenarios and transport policy options. The outputs of the model are flows of vehicles and travellers at the cordon or screenline level, and measures of the performance of the metropolitan transport system in terms of economic efficiency, social impact and broad environmental impact. STEM also produces several accessibility indicators.

The Regional Operations Model (ROM24), hosted at Main Roads, is a strategic network model, suitable for more specific studies of traffic impacts of road infrastructure projects, land-use developments and metropolitan-wide area traffic management measures. It provides traffic volume data for use in the planning and design of elements of the road traffic system, such as interchanges and intersections. It is further used to study regional traffic impacts of land-use development projects in the metropolitan area. It is built using an established strategic network travel demand analysis package (CUBE/TRIPS) and a main road transport network and land-use (zoning system) database. The ROM24 outputs are flows of vehicles at the network link level, and measures of metropolitan road network performance in terms of economic efficiency. Outputs may also be used in evaluating social and environmental impact.

A key input to the STEM and ROM24 is the land-use forecasts provided by the Department of Planning's Metropolitan Land Use Forecasting System (MLUFS). MLUFS currently provides estimates of population, dwellings and employment at five yearly intervals to 2031. MLUFS is used to forecast dwellings and population and employment by industry for small areas within the Perth Metropolitan Region. Small areas are Census Collection Districts (CCD) but the MLUFS outputs can be aggregated to traffic and other zones (Government of Western Australia, Department of Transport 2013). The dwellings/population module estimates dwellings by type and population by age and sex. The employment module estimates employment by industry. Each module is controlled by reference to global dwelling, population and employment projections made independently of MLUFS. MLUFS allocates total population, dwellings and employment to spatial areas (CCD) based on a combination of trend analysis and analytical procedures which iteratively adjust allocations to balance capacities and growth trends in each area (Government of Western Australia, Department of Transport 2013).

3.3 Assessment of Stakeholder Requirements

A consultative process to assess current and anticipated future transport modelling needs of government stakeholders was undertaken (Government of Western Australia 2013, p. i). Key stakeholder groups comprising Main Roads, the Department of Transport, the Public Transport Authority, the Department of Planning, local government and the private sector, participated in a needs assessment survey and a series of stakeholder workshops held for each stakeholder group. The purpose of the survey questionnaire and workshops was to determine: current use of modelling outputs, whether current modelling outputs are considered fit for purpose, anticipated future modelling needs and suggested areas for improvement.

The key findings were:

- There was a broad level of support shown for the development of one integrated strategic level transport model for the Perth region.
- The current suite of transport modelling outputs from current strategic models was generally considered to be fit-for-purpose.

- Land-use inputs and feedback were an issue of concern and were considered a significant risk in the successful implementation of a new strategic level transport model.
- The need for mesoscopic traffic modelling was expressed, especially for the modelling of congested areas.
- State government should take a lead role in developing mesoscopic modelling and microsimulation modelling expertise through a Centre of Excellence approach.

Stakeholder requirements for the strategic model included:

- capability to provide fit-for-purpose transport modelling outputs for policy analysis, strategic and operational uses;
- multimodal capability, with the ability to calculate accurate mode split shifts;
- incorporating longer-term (2041 and 2051) land-use and transport scenarios;
- time of day modelling capability—peak period modelling, peak spreading, departure choice;
- improving quality, timeliness and consistency of land-use inputs, particularly in relation to the land-use forecasts;
- interactive land-use and transport modelling;
- appropriate application of macroscopic, mesoscopic and microscopic models;
- improved freight modelling capacity;
- achievable transport modelling data requirements;
- enhanced level of detail—disaggregation and segmentation—scale, zonal, socio-economic and modal attributes;
- predict behavioural responses to policies and demand management measures; and
- incorporating dynamics (treatment of time).

3.4 Review of International Best Practice Advances

In order to find best practice solutions to meet the stakeholder-identified requirements, a review of international and in particular Australasian developments in strategic transport modelling practice was undertaken. In terms of best practice, the focus of the review was on production as opposed to R&D models. The production model is defined as that available for use by or on behalf of users outside the model development-management group, currently being employed in applications on real world plans and projects. The review identified a number of recent advances:

- Replacement of the trip-based modelling approach by a tour-based schema, while remaining in the overall four-step modelling system.

- Introduction of a time of day modelling capability, certainly for production of 'n-hour' origin-destination matrices to cover the hours of the day, but not necessarily incorporating peak spreading.
- Extensive segmentation of demand by household type, travel purpose and other input variables, to give discrete choice modelling formulations for destination, mode and departure time choice.
- Dynamic traffic assignment, including a number of alternative formulations including dynamic equilibrium, non-equilibrium and quasi-dynamic assignment, which enable modelling of delays, queuing and congestion dynamics.
- Mesoscopic traffic assignment models for large networks, covering entire metropolitan areas, and including a hybrid modelling capability so that small parts of the entire network can be modelled microscopically in the mesoscopic model, on a case by case or project specific basis.
- Land market economic theory-based land-use models, connected to transport models (as opposed to being fully integrated) with feedback to land-use choices through accessibility/composite utility values.
- [Re]Develop models within a GIS environment for data handling capabilities, enhanced visualisation and user-interfacing and spatial aggregation and disaggregation capabilities.
- Freight model development continues to be hampered by the lack of suitable and reliable freight-specific data.

The review concluded that no one city or one model combines all of the features collectively forming current best practice. It did, however, indicate that different cities and regions have included best practice features in their production models. Extracting and combining best practice features from different cities into a general purpose production model of metropolitan travel demand yields an 'ideal' tour-based strategic model which also:

- includes departure time choice, and possibly extending into peak spreading;
- has extensive segmentation for modelling of destination, mode and trip timing choices;
- has the opportunity to provide feedback to land-use;
- is supported by a freight modelling capability which provides separate vehicle origin-destination matrices;
- is possibly (but not necessarily) using quasi-dynamic macro assignment;
- most certainly works in conjunction with a mesoscopic traffic network model for the entire study region;
- is connected to a land market economic theory-based land-use model with feedback to land-use choices through accessibility/composite utility values;
- is developed within a GIS environment with enhanced visualisation and user-interfacing; and
- is part of an integrated modelling suite.

3.5 Development of Options

On the basis of the literature review of current state-of-the-art international advances in transport modelling, stakeholder requirements and importantly, accounting for the unique current modelling practice situation in Perth, three feasible options for model development were identified and evaluated:

- *Option 1*—continued (separate, parallel) development of both STEM and ROM24 models, providing specialised services to current clients, but with improved integration.
- *Option 2*—gradual development into a single modelling system, which incorporates the top features of the current two models, plus additional best practice developments.
- *Option 3*—development of a new best practice modelling system, with strong feedbacks from the transport model to the land-use model and from the dynamic traffic assignment to the previous stages of the strategic model, also integrating a freight component.

3.6 Evaluation of Options

The analysis of best practice of production models, highlighted a number of commonalities, summarised in the form of assessment criteria:

- land-use and transport models directly interacting;
- tour-based trip modelling;
- simultaneous mode and destination choice giving logsum benefits;
- time of day modelling taking account of peak spreading;
- static traffic assignment as a base case;
- mesoscopic traffic assignment;
- hybrid mesoscopic and microscopic modelling; and
- increased focus on detailed modelling of traffic streams.

Each of the options was evaluated in terms of the criteria (Table 1). In the case of Option 1 (Parallel Models), it was concluded that duplication of functions wastes resources and that special services to particular clients could be provided by a combined modelling system. Although they both use sound methods, neither model could be claimed to be best practice by world or Australian standards. The main advantage of Option 2 (Closely Associated Models) is that the ROM24 team could cover the whole range of road-based traffic, using limited resources to extend the work on detailed traffic modelling as well as freight. The main departure from Option 1 would be to rely on STEM for advanced choice modelling. It is a feasible extension of Option 1, providing greater productivity and bringing the two modelling teams back into collaboration.

Table 1 Summary of option evaluation against best practice criteria

Criteria	Option 1	Option 2	Option 3
Land-use and transport models directly interacting	Little change	Little change	Appropriate software to exchange land-use and transport data
Tour-based modelling	Gradual adoption	Slightly faster adoption	Extended tours with deviations
Simultaneous mode-destination choice; logsum benefits	Depends on STEM achieving destination choice	May achieve destination choice more quickly	Full development of mode-destination choice and logsums
Time of day modelling and peak spreading	Both models have this objective	Both models have this objective	Impact of traffic delays fed back to mode choice and departure timing
Static traffic assignment	Current procedure	Current procedure	Upgraded to provide the base for mesosimulation
Mesoscopic traffic assignment	Being developed by ROM24 team	Accelerate current work	To become the core traffic assignment model
Hybrid mesoscopic and microscopic models	Gradual development	Accelerated development	Embed microsimulation in mesoscopic network simulation
Increased focus on detailed modelling of traffic streams	Current work	Accelerate current work	Allocate more resources to traffic flow and lane modelling

The prognosis for Option 3 (Integrated Models) would be reaching and even surpassing the best practice transport modelling standard already reached in some major cities. Waste of resources through duplication of functions would be eliminated, but the change need not impede specialised and advanced work being done by the ROM24 and STEM teams. Substantial gains could be expected from improved workflow and resource utilisation.

3.7 Selection of Preferred Modelling System Approach

The third option was recommended as the preferred solution for addressing the modelling needs of WA.

The proposed Perth LAnd and Transport INtegrated Urban Model (PLATINUM) is a closely coupled, system of models, exchanging information among them, comprising:

- a five-step, tour-based, multi-modal strategic transport model, supported by a land-use model and outputting to a regional impacts model, for use in long-range planning, scenario analysis and system wide transport policy analysis, also providing road passenger/vehicle origin-destination matrices, by time of day and for planning horizon years to the road transport model;

- a hybrid meso-micro assignment, road transport model of the metropolitan road network including all road-based travel by time of day, and delivering results to a local area impacts model, for use in road project planning and evaluation, traffic management and control, congestion management, local area traffic impacts, event planning and incident management planning, as well as providing information on network travel times, delays and queuing as input to the strategic model;
- an external travel model, interacting with the WA State-wide Transport Model; and
- a freight transport model, comprising an improved Freight Movement Model and a separate model for light goods vehicles.

3.8 Proposing an Implementation Strategy

The implementation strategy contains specification of elements to be developed, linkages and feedbacks to be established, human and financial resourcing to be mobilised, risks to be managed, data to be acquired, maintained and managed, model validation and managing the transition process.

Implementing the preferred approach entails transport modelling for Perth being reorganised into an integrated transport modelling suite, comprising the two main elements of a multi-modal strategic '5-step' model connected to a hybrid meso-micro dynamic assignment model, with both main elements supported by modules for freight transport and external trips (Fig. 2). The strategic model would be used for long-term planning, scenario analysis and transport policy appraisal. The hybrid assignment model would be used for short and medium-term studies, with an emphasis on congestion management, traffic management and road project evaluation.

The development of PLATINUM requires the implementation of the following elements:

- the specification or re-specification of several current model components and alignment of data (consolidate the current 1500 zones STEM and ROM24 networks, re-develop and integrate the land-use model);
- enhancements to the current models (higher granularity in capturing travel behaviour through additional household types, travel purposes, by time of day, using tours instead of trips, and updating parameters of the mode choice model (including travel time variability, crowding and perhaps estimating new models of destination choice); and
- creation of new component models (including departure time, using a hybrid meso-micro dynamic traffic assignment model of all road-based travel by time

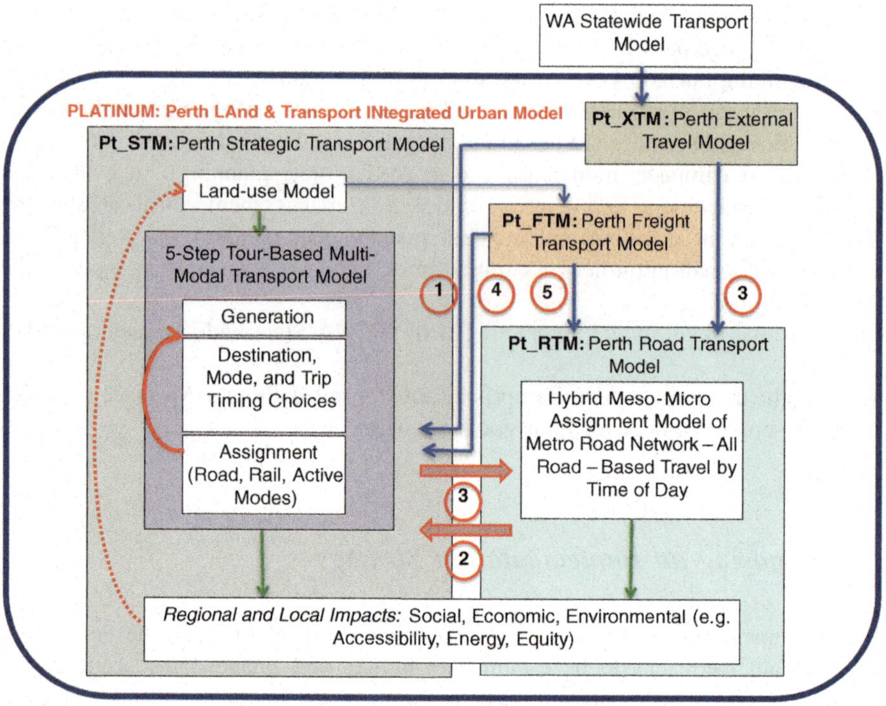

Fig. 2 Structure of the proposed Perth LAnd and Transport INtegrated Urban Model (PLATINUM)

of day, considering quasi-dynamic traffic assignment as a better starting point for meso-simulation, a new model assessing wider benefits of projects, feedback to land use, and authentic integration of all components).

Successful implementation of the strategy will depend to a considerable extent on the functioning of linkages and feedbacks (Fig. 2). Four of these are concerned with land-use and transport interaction. The feedback from regional transport impacts to the land-use model will include the accessibility measures. Between the component transport models, feedback (perhaps iterative) on network travel times, delays and queuing will have substantial impacts on the enhanced mode choice model.

Implementation further requires investment in data acquisition and fusion, both for passengers and freight. Creation of a continuous panel of data would provide measures of change, in addition to updated timely data reflecting changes in travel behaviour, preferences or organisation structures. In addition, public transport smart ticketing information, household travel survey and census data should be aligned with regular traffic and public transport monitoring, supplemented with some automatically collected data using mobile phones, GPS, etc.

The new modelling suite should remain highly responsive to policy changes. The necessary focus of practitioners on calibrating and validating their models against real data should not lead them to overlook any aspect of the capacity to test government policy proposals. This is well understood when the policy is to physically alter the road or public transport system, but there are also other types of policy change. Any potential charge affecting the transport system needs to be modelled before implementation is considered; a model which appears to be reliable when validated against observed flows may well respond poorly to a hypothetical cost change. A weakness of this type can be traced back to the mode choice model, which must be capable of responding fairly accurately to a policy to raise or reduce a charge (either mobility charging or parking). Similarly there must be well-calibrated parameters to ensure that the effect of a policy change to reduce waiting or transit time, or increase comfort and reliability, will produce an appropriate response by the model.

4 Discussion and Conclusion

In common with worldwide trends, key emerging modelling requirements in Perth include the ability to model seamlessly at macro-, meso- and micro-scopic scales, greater levels of disaggregation and segmentation, land-use and transport interactions, behavioural responses, freight movement and time in a dynamic way. These modelling requirements have largely been driven by growing complexity in urban systems throughout the world in relation to changing planning objectives, policies and approaches (Waddell and Ulfarsson 2004). Consistent with worldwide trends, in Perth, objectives have shifted from a rather singular focus on improving road capacity for private vehicles to a much wider array of objectives, developed in response to issues of congestion and environmental consequences, with a much wider and diverse range of multi-modal, demand-side and land-use strategies as alternatives to meet growing transport needs.

While the need for modelling to become less of a 'black box', more open and participatory, has been recognised as a growing pressure (Waddell and Ulfarsson 2004), that modelling be facilitative of participation in the testing and evaluation of alternative policy strategies did not arise as a major requirement in Perth. This is most likely related to the strong role of the state government in planning with local government having little decision-making power and the associated approach of consultation rather than engagement with stakeholders and communities. It may also be a consequence of recognised existing under-application of (modelling in) strategic planning and scenario development processes.

Similarity of modelling requirements across the world has resulted in a wide range of best practice solutions available to draw upon in designing a new modelling system for Perth. Advances in tour-based schemas, time of day modelling, segmentation of demand, dynamic traffic assignment, mesoscopic traffic assignment, 'hybrid' modelling, land market economic theory-based land-use models, freight modelling, quasi-dynamic macro assignment and feedback to land-use choices through accessibility/composite utility values, have all been incorporated into the requirement-responsive design of a modelling system for Perth.

Unique to the development context of Perth, and influencing the development of design options, are:

- the existence of two modelling teams and two different models;
- the modelling teams are relatively well-capacitated and experienced; and
- there is a strong tradition of modelling and use of modelling outputs in transport operations and in evaluation of specific projects and local plans, but not for strategic planning.

Vonk (2006) identified the primary reason for lack of uptake, in particular, of analytical PSS (such as land-use and transport models), is lack of experience, which affects awareness of innovation and judgement of instrument quality. Accordingly, due to the high level of experience of the modelling teams in Perth, the expectation of uptake of the new model design would be positive. However, Vonk (2006) additionally identifies the bottleneck of resistance to change. He ascertains that the more radical the PSS innovation, like many analysing PSS, the higher the resistance. Incremental innovations, as in the case of the more simple information PSS, where planners, managers and specialists essentially continue doing the jobs they are used to, but with greater ease, face less resistance (Vonk 2006). He asserts that only watershed changes in planning context or planning system have the potential to instigate uptake of more radical innovations.

This last point is echoed in Waddell (2011) when he poses the question as to why advanced/innovative activity-based travel and land-use models are placed under such intense critical scrutiny by practitioners when the limitations of four-step travel models in widespread use, particularly in relation to limited degree of validation, are well established. He suggests that users become accustomed to standard practice, finding (incremental) ways around these limitations rather than embracing new innovations, which require a higher burden of proof by practitioners. Waddell (2011) maintains that the challenge in selling a new model is that not only should it perform better empirically but must also be deemed to be worth the effort in terms of overcoming potentially larger costs of data, computation and complexity.

The interpretation for the Perth context is that experience is a two-edged sword. On the one hand, it results in greater awareness and appreciation for the advances being made in practice around the world and their potential to enhance local model development. On the other hand, experience can lock in continued use of existing standard technologies, with ongoing incremental adjustments made to deal with limitations, resulting in a high level of path dependency and resistance to more radical innovations. This is exacerbated in Perth with the existence of two, almost competing, modelling teams and the propensity for modelling output to be used almost religiously, in planning.

For these reasons, and to maximise the opportunity for uptake of best practice advances, the development of model system options and the design and implementation specification of the recommended option for Perth, in addition to incorporating technical advances to meet model requirements, involved a detailed proposal for 'getting from where we are to where we want to be'. Cognisant of recent incremental innovations made by the two modelling teams, and in cases where they are aligned with best practice advances, these were incorporated into the new model design.

In conclusion, the recommendation for future model system design processes is that an explicit step in the assessment process should be devoted to understanding the existing modelling and PSS in use. Neither the model development process of Waddell and Ulfarsson (2004) nor the conceptual frameworks developed to enhance the uptake of PSS (Biermann 2011; Geertman 2006; Geertman and Stillwell 2004; Vonk and Geertman 2008), explicitly incorporate this as a step or a criterion or sub-criterion. Existing modelling considerations could either be undertaken as part of the development context assessment or as a stand-alone step. Designing model system options should explicitly acknowledge current systems in use and solutions should strongly account for the pathway from the existing situation to the new model system. In addition, the two-edged sword of experience should be recognised as both a positive influence in terms of innovation awareness but carefully dealt with in relation to potential negative influences of path-dependent, 'incrementality' at the expense of embracing more radical innovations.

Acknowledgments This research was funded by the Planning and Transport Research Centre (PATREC), a collaborative partnership of the Government of Western Australia (the Department of Transport, the Western Australian Planning Commission and Main Roads WA) and the University of Western Australia, Curtin University and Edith Cowan University. Additional funding was provided by Main Roads Western Australia to undertake an independent review of transport modelling in WA, on which this research was based.

References

Batty, M. (2003). Planning support systems: Techniques that are driving planning. In S. Geertman & J. Stillwell (Eds.), *Planning support systems in practice* (pp. v–viii). Heidelberg: Springer.

Biermann, S. (2011). Planning support systems in a multi-dualistic spatial planning context. *Journal of Urban Technology, 18*(4), 5–37.

De la Barra, T. (1995). *Integrated land use and transportation modeling: Decision chains and hierarchies*. Cambridge: Cambridge University Press.

de Ortúzar, J. D., & Willumsen, L. G. (2011). *Modelling transport* (4th ed.). Chichester: Wiley.

Echenique, M. H., Flowerdew, A. D. J., Hunt, J. D., Mayo, T. R., Skidmore, I. J., & Simmonds, D. C. (1990). The MEPLAN models of Bilbao, Leeds and Dortmund. *Transport Reviews, 10*(4), 309–322.

Geertman, S. (2006). Potentials for planning support: A planning-conceptual approach. *Environment and Planning B: Planning and Design, 33*, 863–880.

Geertman, S., & Stillwell, J. (2004). Planning support systems: An inventory of current practice. *Computers, Environment and Urban Systems, 28*, 291–310.

Geertman, S., & Vonk, G. (2006). Improving the adoption and use of planning support systems in Practice. *Applied Spatial Analysis and Policy, 1*, 153–173.

Government of Western Australia. (2010). *Direction 2031 and beyond. Metropolitan planning beyond the horizon*. Perth: Western Australian Planning Commission.

Government of Western Australia. (2013). *Transport modelling review summary of transport modelling needs*. Unpublished internal report.

Government of Western Australia, Department of Transport. (2013). *MLUFS for MAX*. Internal Report prepared for the Department by Parsons Brinkerhoff Hames Sharley & Economic Research Associates.

Harris, B., & Batty, M. (1993). Locational models, geographic information and planning support systems. *Journal of Planning Education and Research, 12*, 186–198.

Hunt, J. D., Kriger, D. S., & Miller, E. J. (2005). Current operational urban land use-transport modelling frameworks: A review. *Transport Reviews, 25*(3), 329–376.

Iacono, M., Levinson, D., & El-Geneidy, A. (2008). Models of transportation and land use change: A guide to the territory. *Journal of Planning Literature, 22*(4), 323–340.

Lee, D. R. (1973). Requiem for large-scale models. *Journal of the American Institute of Planners, 39*, 163–178.

Lee, D. R. (1994). Retrospective on large-scale urban models. *Journal of the American Planning Association, 60*(1), 35–40.

Miller, E. J. (2003). Microsimulation. In K. G. Goulias (Ed.), *Transportation systems planning methods and applications* (1st ed., Vol. 12, pp. 12–22). Boca Raton, FL: CRC Press.

Putman, S.H. (1983). Integrated Urban Models. London: Pion.

Te Brömmelstroet, M. (2013). Performance of planning support systems. What is it, and how do we report on it? *Computers, Environment and Urban Systems, 41*, 299–308.

Vonk, G.A. (2006). Improving planning support. The use of planning support systems for spatial planning. *Ph.D. thesis published in Netherlands Geographical Studies* 340. Koninklijk Nederlands Aardrijkskundig Genootschap Faculteit Geowetenschappen Universiteit Utrecht: Utrecht.

Vonk, G., Geertman, S., & Schot, P. (2005). Bottlenecks blocking widespread usage of planning support systems. *Environment and Planning A, 17*(5), 909–924.

Vonk, G., Geertman, S., & Schot, P. (2006). Usage of planning support systems: Combining three approaches. In J. Van Leeuwen & H. Timmermans (Eds.), *Innovations in design and decision support systems in architecture and urban planning* (pp. 263–274). Dordrecht, The Netherlands: Springer.

Vonk, G., Geertman, S., & Schot, P. (2007a). New technologies stuck in old hierarchies: An analysis of the diffusion of geo-information technologies in Dutch planning organizations. *Public Administration Review, 67*(4), 745–756.

Vonk, G., Geertman, S., & Schot, P. (2007b). A SWOT-analysis of planning support systems. *Environment and Planning A, 39*(7), 1699–1714.

Vonk, G., & Geertman, S. (2008). Improving the adoption and use of planning support systems in Practice. *Applied Spatial Analysis, 1*, 153–173.

Waddell, P., & Ulfarsson, G. F. (2004). Introduction to urban simulation: Design and development of operational models, Chapter 13. In D. A. Hensher, K. J. Button, K. E. Haynes & P. R. Stopher (Eds.), *Handbook of transport geography and spatial systems*. (Vol. 5, Handbooks in Transport). Pergamon Press.

Waddell, P. (2011). Integrated land use and transportation planning and modelling: Addressing the challenges in research and practice. *Transport Reviews, 31*(2), 209–229.

Index

S. Geertman et al. (eds.), *Planning Support Systems and Smart Cities*,
Lecture Notes in Geoinformation and Cartography,
DOI 10.1007/978-3-319-18368-8

CPSIA information can be obtained at www.ICGtesting.com
Printed in the USA
LVOW05*1830120615

442284LV00004B/6/P

9 783319 183671